Atmospheric Science: A Global Overview

Atmospheric Science: A Global Overview

Edited by Ronin Massey

SYRAWOOD
PUBLISHING HOUSE

New York

Published by Syrawood Publishing House,
750 Third Avenue, 9th Floor,
New York, NY 10017, USA
www.syrawoodpublishinghouse.com

Atmospheric Science: A Global Overview
Edited by Ronin Massey

International Standard Book Number: 978-1-68286-783-9 (Hardback)

Cataloging-in-Publication Data

Atmospheric science : a global overview / edited by Ronin Massey.
    p. cm.
Includes bibliographical references and index.
ISBN 978-1-68286-783-9
1. Atmospheric physics. 2. Atmosphere. 3. Meteorology. I. Massey, Ronin.
QC861.3 .A86 2019
551.5--dc23

TABLE OF CONTENTS

PREFACE

The atmosphere is made up of gaseous layers which surround the Earth. The main layers of the atmosphere are the exosphere, thermosphere, mesosphere, stratosphere and troposphere. The study of the Earth's atmosphere and atmospheric processes, its effect on other systems and their effects on the atmosphere fall under the field of atmospheric sciences. An understanding of the physical and optical properties of the atmosphere is vital to the study of atmospheric science. It can be divided into the fields of meteorology, climatology and aeronomy. The discipline of meteorology is concerned with weather forecasting and focuses on atmospheric physics and chemistry. Climatology studies short-term and long-term atmospheric variations and their changes with time. Aeronomy studies the dissociation and ionization in the upper layers of the atmosphere. This book contains some path-breaking studies in this field. It provides significant information of this discipline to help develop a good understanding of atmospheric science and related disciplines. As this field is emerging at a rapid pace, the contents of this book will help the readers understand the modern concepts and applications of the subject.

This book is a comprehensive compilation of works of different researchers from varied parts of the world. It includes valuable experiences of the researchers with the sole objective of providing the readers (learners) with a proper knowledge of the concerned field. This book will be beneficial in evoking inspiration and enhancing the knowledge of the interested readers.

In the end, I would like to extend my heartiest thanks to the authors who worked with great determination on their chapters. I also appreciate the publisher's support in the course of the book. I would also like to deeply acknowledge my family who stood by me as a source of inspiration during the project.

Editor

Informativeness of wind data in linear Madden–Julian oscillation prediction

Theodore L. Allen,[1*†] Brian E. Mapes[1] and Nicholas Cavanaugh[2]

[1]Department of Meteorology and Physical Oceanography, Rosenstiel School of Marine and Atmospheric Science, University of Miami, FL, USA
[2]Climate and Ecosystem Sciences, Lawrence Berkeley National Laboratory, Berkeley, CA, USA

*Correspondence to:
T. L. Allen, The International Research Institute for Climate and Society, Columbia University, 61 Route 9 W, Palisades, NY 10964–1000, USA. E-mail: tallen@iri.columbia.edu

†Currently at The International Research Institute for Climate and Society, Columbia University.

Abstract

Linear inverse models (LIMs) are used to explore predictability and information content of the Madden–Julian Oscillation (MJO). Hindcast skill for outgoing longwave radiation (OLR) related to the MJO on intraseasonal timescales in the tropics has been examined for a variety of LIMs using OLR and optionally 200 and 850 hPa zonal wind information channels. The dependence of OLR hindcast skill on wind channels was evaluated by randomizing in time, averaging in space, or omitting data entirely. Results show positive prediction skill (relative to climatology) up to 3 weeks and wind information, mostly at the largest scales, adds 1–2 days of skill.

Keywords: linear inverse modeling; Madden–Julian Oscillation; sub-seasonal prediction

1. Introduction

The Madden–Julian Oscillation (MJO) is an intraseasonal zonally propagating atmospheric signal in tropical rainfall and related fields (Madden and Julian, 1971; Madden and Julian, 1972; Zhang, 2005; Wang, 2006; and Lau and Waliser, 2011). Besides impacting weather directly in the tropics it also has impacts in the extra-tropics through teleconnections and can impact short-term climate events (Martin and Schumacher, 2011; Zhang, 2013). The long timescale of the MJO suggests it could be a source of extended-range predictability.

Linear statistical models can have comparable MJO prediction skill to dynamical models (Newman *et al.*, 2009; Xavier *et al.*, 2014; Klingaman and Woolnough, 2014) and offer unique opportunities for decomposition that may reflect on the MJO's incompletely understood dynamics. Cavanaugh *et al.* (2014, hereafter C14) explored the skill of linear inverse models (LIMs) in hindcasting the MJO, and this article extends and complements that work methodologically and scientifically. Klingaman and Woolnough (2014) include LIM results from both C14 and the methods described here, as a baseline for evaluating numerical model hindcasts. Those hindcasts as well as C14 were scored in the time-longitude (latitudinally averaged) space of Wheeler and Hendon (2004)'s Realtime Multivariate MJO index (RMM) encompassing Outgoing Longwave Radiation (OLR) and zonal wind at 850 and 200 hPa (u850 and u200). The relevance of including wind field information in addition to OLR has been questioned,

for both MJO definition and diagnosis (Kiladis *et al.*, 2014) and for aspects of forecasting such as initiation of a new MJO event (Straub, 2013), whose final paragraph notes that "the RMM index is dominated by its circulation components. However, the clouds and rain are of special interest for impacts, so we will score all hindcasts in terms of OLR anomaly (OLR').

The goals of this article are: (1) to illustrate the workings of LIMs in a more intuitive physical channel space (time-longitude sections), rather than C14's truncated space of empirical orthogonal functions (EOFs); (2) to explore how close the resulting large number of channels brings us to the problem of statistical overfitting (the usual justification for such EOF truncation approaches); and (3) to estimate the value of wind information and small-scale information in statistical predictions of intraseasonal cloudiness signals, and consider the implications as a potential partial clue to MJO dynamics.

2. LIM summary and statistical forecasting issues

Linear inverse modeling (Penland and Sardeshmukh, 1995) is a generalization of the simple idea that anomalies in a stationary time series decay with time – exponentially, in the case of a postulated system obeying $dx/dt = -\mathbf{B}\mathbf{X} + \text{noise}$. In a multi-channel LIM (where a 'channel' is meant in the sense of an information stream, i.e. a time series or a column in a dynamical state vector), anomalies can oscillate and propagate among channels as they decay, because the complex exponential function has those behaviors in addition to the simple decay of the real exponential function.

In fitting a LIM from data, one postulates that those data came from a linear stochastically forced system with the form:

$$\frac{dx}{dt} = \mathbf{BX} + \text{noise} \tag{1}$$

where the state vector $\mathbf{X}$ comprises m columns of anomaly values and $\mathbf{B}$ is an $m \times m$ matrix. All linearly predictable dynamical interactions among the system variables are represented in the linear operator $\mathbf{B}$, also known as the deterministic linear feedback matrix or the system sensitivity matrix (Shin et al., 2010). It can be shown for a system of form (Equation (1)) that if the noise term is white (uncorrelated in time, but not necessarily uncorrelated among channels) and Gaussian, then for any specific time lag τ_0, $\mathbf{B}$ is related to the time-lagged covariance matrix $\mathbf{C}(\tau_0)$ by:

$$\mathbf{B} = \frac{1}{\tau_0} \ln \left[\mathbf{C}\left(\tau_0\right) \mathbf{C}(0)^{-1} \right] \tag{2}$$

This result is formally identical to how one would estimate a decay coefficient from lagged autocorrelation in a univariate ODE, but here $\mathbf{C}$ and $\mathbf{B}$ are matrices and the $\ln[\cdot]$ function is the matrix generalization of the ordinary logarithm. The optimum forecast (indicated by the caret) for such a system, optimal in the sense of minimizing squared error, is:

$$\hat{\mathbf{x}}(t + \tau) = \exp(\mathbf{B}\tau) \mathbf{x}(t) = \mathbf{G}_\tau \mathbf{x}(t) \tag{3}$$

where $\mathbf{G}_\tau$ is known as the propagator matrix that evolves initial anomalies, $\mathbf{x}(t)$, forward by any desired lead time (τ).

When working from a finite, real-world data sample, we must view the $\mathbf{B}$ obtained from Equation (2) as an estimate, and view Equation (1) as a postulate of how the real world (which generated the data) acts. One can estimate $\mathbf{B}$ from Equation (2) for various training lags τ_0. The similarity of these various estimates for $\mathbf{B}$ has been viewed as a test (called the 'tau test') of the validity of the postulate that the form (Equation (1)) characterizes the real system adequately (Penland and Sardeshmukh, 1995). Referring again to the simpler univariate case: if the autocorrelation decay rate estimated at different lags is similar, then indeed the decay curve must be close to exponential, which bolsters the case that the data-generating system acts like the simple linear decay equation being postulated (or fitted). We found that Equation (3) gives similar hindcast skill using $\mathbf{B}$ matrices estimated from τ_0 values ranging from 1 to 4 days in Equation (2) (not shown), supporting the LIM approach. Only a little more skill is gained by using lagged regression (LR) rather than LIM (not shown). In LR, one estimates $\mathbf{C}$ separately for each forecast lead time τ, so that Equation (3) simply becomes $\hat{\mathbf{x}}(t+\tau) = \mathbf{C}(\tau)\mathbf{C}(0)^{-1}\mathbf{x}(t)$. LIM results offer similar scientific lessons to LR (not shown), but with more elegance and simplicity, and so will be the main focus of this article.

3. Data and experiments

This study utilizes time-longitude sections of daily data from 1979 to 2011, including interpolated outgoing long wave radiation (OLR) observed from satellites (Liebmann and Smith, 1996) along with zonal wind u at the 850 and 200 hPa levels derived from the NCEP-NCAR Reanalysis project (Kalnay et al., 1996). Each variable was averaged from 15°S to 15°N. A 25-year composite annual cycle (1979–2004) was removed from each variable to produce anomalies, and a 120-day mean prior to each day was subtracted to remove low frequency signals, following Wheeler and Hendon (2004), which means that usable data begins 120 days into the time series. Each channel in the training set thus consists of a time series of more than 7000 daily observations. These high-passed anomalies are denoted with a prime, for example OLR'. These data contain many kinds of variability, but the hindcast skill here mostly bears the hallmarks of the MJO (timescale and eastward propagation), so we have used that moniker in the text and title.

It is helpful to define a LIM baseline or control case: all three variables, in 15 degree longitude bins, using $\tau_0 = 2$ days. Each of the 24 longitude bins thus contains three channels consisting of a daily time series of anomalous MJO index variables (OLR', u850' and u200'). In summary, the baseline LIM has a total of 72 input channels, 3 for each longitude bin. For clean comparisons to this baseline, including wind information denial experiments, we choose to score the hindcasts based on the twenty-four 15-degree OLR bins only. When other channels (u850', u200') are used, their impact is evaluated only in terms of the OLR prediction skill. Likewise, when additional longitude fine structure is included, we score its effect only on 15 degree scale OLR.

In many LIM studies (including C14), principal component series truncation has been used to minimize channel numbers while maximizing the variance represented. However, this encoding of the information channels makes a LIM's workings somewhat opaque. Following the examples of Shin et al. (2010) and Hakim (2013), we leave our channels as spatial boxes, and furthermore they are ordered by adjacent longitudes, so LIM forecasts and their errors can simply be contoured in longitude-time space.

We train LIMs on data from 1979 to 1999 and verify on the independent set 2000–2011, thus eliminating any chance of artificial skill (DelSole and Shukla, 2009). Seasonal masks are also optionally applied to the training period to seek an optimal LIM construction for verification in that specific season.

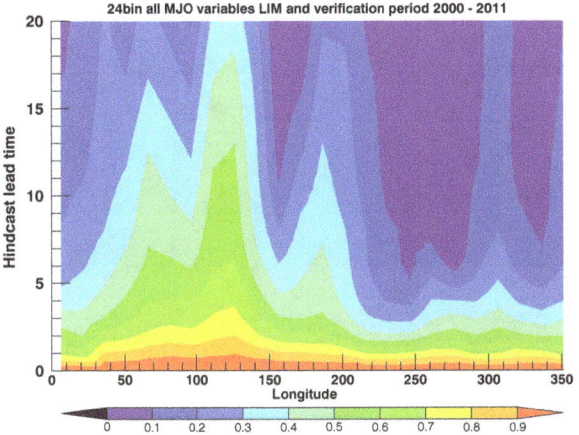

Figure 1. OLR anomaly hindcast correlation skill score for each of the 24 longitude bins from the baseline LIM.

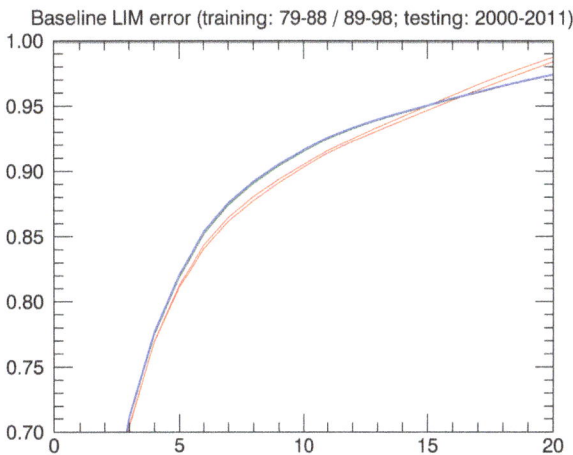

Figure 2. Squared error of the baseline (red) and 144 channel (blue) LIMs trained on two consecutive non-overlapping 10-year epochs between 1979 and 1998. Two curves are shown in each color; their (very close) spacing indicates the level of accuracy for further deductions involving hindcast skill differences in subsequent figures.

4. Results

4.1. Longitude dependence of OLR' predictability

LIM skill can be displayed as a function of longitude (Figure 1). Figure 1 illustrates the correlation coefficient between predicted OLR anomalies from the baseline LIM and observed OLR anomalies for the 2000–2011 period. Regional differences in skill are evident. The highest correlation at all lead times is found in the region of the maritime continent between 110°E and 130°E. Here, the correlation remains above +0.6 for 6 days and remains above +0.5 for 13 days. By contrast, the east Pacific region (longitude 230 in Figure 1) has the lowest one week hindcast correlation skill (r < +0.1). Summarizing Figure 1, three hindcast skill hot-spot peaks are identified within the central Indian Ocean, the maritime continent and central Pacific, with areas of low predictive skill from the east Pacific to the Pacific coast of Central America. These results are consistent with the notion that the MJO is the basis of long-range prediction skill, as presupposed in our title, even though the prediction is really for OLR' including all phenomena.

To have a single scalar skill score, we define the verification error score (to be minimized) in future experiments as a global sum of squared OLR' hindcast errors. The no-skill asymptote of this score is the global climatological variance, which is the skill of a forecast of zero anomaly every day (climatology used as a forecast).

4.2. Impacts of using a large number of channels

Is a 72-channel LIM too large? That is, will statistical overfitting of so many coefficients from finite training data samples lead to poor skill when tested on independent data? The skill of our baseline results, comparable to results from C14's reduced EOF space (Klingaman and Woolnough, 2014), suggests that the answer is no. To further address this question, we push the numbers much further by doubling the number of

longitude bins from 24 to 48, making each longitude bin 7.5 rather than 15 degrees wide. This doubles the number of channels from 72 to 144, quadrupling the number of coefficients in the **G** and **B** matrix estimates. Furthermore, we reduce the training set into two independent and consecutive training periods to estimate the effects of sampling error in our final results graphical space. This experiment reduced the number of data points used per coefficient estimated from 50 to 25.* The change from 15 degree to 7.5 degree longitude resolution has a negligible effect on hindcast error (black curves are only slightly above the red curves in Figure 2), thus providing little evidence of overfitting at 7.5 degree longitude bin resolution. However, overfitting becomes steeply worse once longitude resolution increases to 2.5 degrees (8 data values per coefficient, not shown).

4.3. Estimated value of wind information: illustrating randomization method

Excluding u850 and u200 from the LIM provides physical insight regarding the impact of wind anomalies on the prediction of OLR anomalies associated with the MJO (Figure 3). The skill score omitting or randomizing wind channels during training and hindcasting (blue and green curves) is compared to the baseline LIM (red curves repeated from Figure 2). Figure 3 indicates that the wind information in the baseline LIM does indeed contribute unique and valuable information content to the LIM for OLR' prediction out to 15 days, a result clearly significant with respect to sampling noise (the

---

*Of course, the statistically important information measure is independent degrees of freedom per coefficient estimated, not data points per coefficient. But daily timescale OLR is not excessively autocorrelated, and the rules of thumb for relating degrees of freedom to data points using autocorrelation are imprecise and debatable, so we use data points here for simplicity.

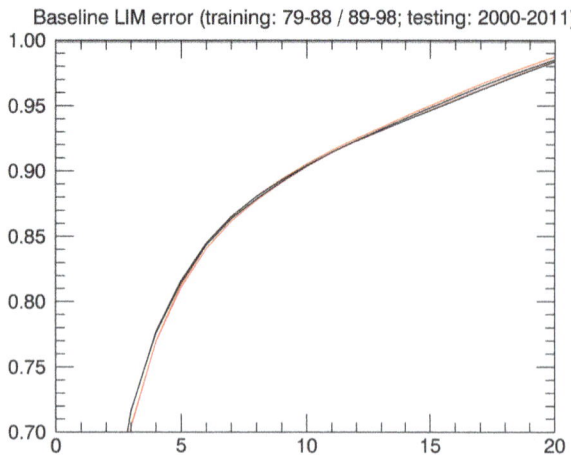

Baseline LIM error (training: 79-88 / 89-98; testing: 2000-2011)

Figure 3. As in Figure 2 but for an OLR only LIM (blue), the baseline LIM with scrambled winds (green), and the baseline LIM (red). Zoomed in insert provided as well.

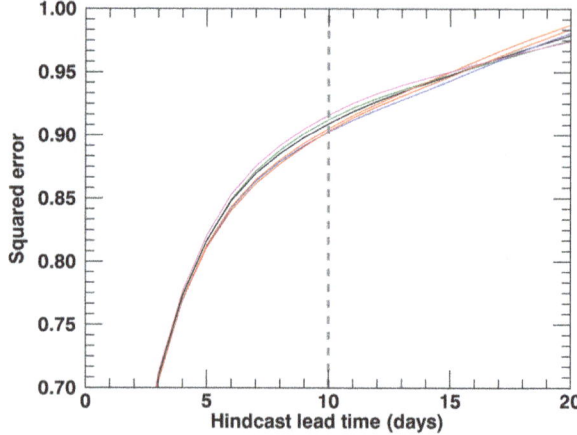

Figure 4. Squared error for five LIMs calculated using 15 degree longitude bins and trained on the following data channels. Listed in order from highest error to lowest error at a 10-day hindcast lead time are (1) OLR only (magenta), (2) OLR and wave numbers 1 and 2 for u850 and u200 (green), (3) OLR and zonal mean wind (black), (4) OLR, zonal mean winds, and wave numbers 1 and 2 (blue), and (5) the baseline LIM with all wind information (double red curves, for two non-overlapping half-length training periods). The vertical dashed grey line provides a visual reference at the 10-day lead time. Zoomed in insert provided as well.

gap between the red curves). Furthermore, we can conclude that omitting vs. randomizing the wind channels has a very similar effect, indicating again that our LIMs are far from the danger of overfitting, with the modest number of channels and large amounts of training and verification data used here.

4.4. Decomposition of the value of wind information

We also want to know what aspects of the wind field contain the important information content for predicting OLR: the zonal mean wind, the first two zonal wavenumbers (indicative of large-scale convergence and divergence) or other aspects? To address this question, we construct a 30-channel LIM built from an anomalous state vector with OLR' for each 15 degree longitude bin (24 channels), the zonal mean from u850' and u200' (two channels), and the first two wavenumbers of u850' and u200' (four channels). We apply the channel randomization technique introduced in the previous section to test the impact of various wind channels on the prediction of OLR'.

To partition the wind channels' information content, Figure 4 shows results from four 30-channel LIMs: (1) OLR data only (randomized zonal mean and the first two wavenumbers of u850 and u200), (2) OLR and the first two wavenumbers (randomized zonal means), (3) OLR and the zonal means (randomized wavenumbers 1 and 2), and (4) OLR with all six wind components. For reference, the 'baseline' LIM with all wind information at 7.5 degree scale is repeated in red, with two lines trained from independent halves of the training epoch as an indicator of the statistical significance of difference due to finite-sample effects.

The OLR' only LIM exhibits the worst skill (greatest error, magenta curve in Figure 4 up to 2 weeks lead time) while the 'baseline' is the best (red curves). The cases in between essentially map the amount of useful information content in various aspects of u850' and u200'. Results may be summarized as follows:

- Value of all wind information ('baseline' vs. OLR only): 1.5–2 days of additional skill between 5 and 10 day hindcast lead time (red curves on Figure 4; as in Figure 3)
- Zonal means plus wavenumbers 1 + 2: about the same as the 'baseline' LIM (blue curve on Figure 4)
- Zonal means alone: about 1/2 of total wind signal (black curve on Figure 4)
- Wavenumbers 1 + 2 alone: about 1/3 of total wind signal (green curve in Figure 4)

The contributions of 'about 1/2' and 'about 1/3' need not sum to unity, because the channels are not orthogonal, merely linearly independent. In particular, wavenumbers 1 + 2 may contain sample-specific noise as well as robustly useful signal, and overfitting of that noise in the training period could yield lower skill (merely 1/3 instead of 1/2 of the total value of all wind information) in the evaluation period.

4.5. Seasonality of predictability

The MJO is seasonally strongest in the northern autumn and winter seasons, with summer intraseasonal variability sometimes given a different name such as MISO or BSISO (Kikuchi *et al.,* 2012; Sharmila *et al.,* 2013). Our findings so far suggest that our dataset is plentiful enough to give robust results even if subdivided. Might OLR' hindcast skill be increased if the training set and verification are confined to certain seasons, rather than pooling all data? To test this notion, the baseline LIM from Figure 2 was subdivided into an all season and boreal winter (DJF) datasets for both training and evaluation. OLR' hindcast skill is much better for DJF training and scoring, compared to the 72 channel all-season LIM (black vs red in Figure 5). If a

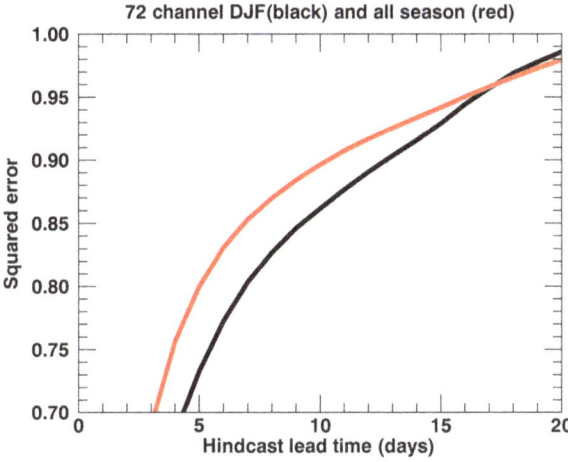

72 channel DJF(black) and all season (red)

Figure 5. As in Figure 2 but for a LIM trained with DJF seasonal (black) and all season (red) LIM.

normalized squared error of 0.85 is used as a no-skill threshold, that is achieved after 7 days for the all season LIM and 9 days for the DJF LIM. Alternatively, if differences are measured vertically on the graph, for a 1-week lead time prediction, the control LIM has a 7% increase in OLR anomaly hindcast error compared to the DJF trained LIM from a 1-week hindcast lead time. All squared errors asymptote toward 1 after a 14-day hindcast lead time. In summary, it is preferable to train the LIM on less data, but on the proper season (DJF), rather than a using a larger set of training data including data from all seasons.

5. Conclusions

Hindcast skill for 15°N–15°S averaged OLR' in time-longitude space has been examined for a variety of LIM models using daily OLR data and optionally 200 and 850 hPa zonal wind from 1979 to 2011. Results show some positive prediction skill (relative to climatology) up to 3 weeks, consistent with Cavanaugh *et al.* (2014) who used a LIM built with a reduced channel space of EOFs from maps (not just latitude belt averages). Klingaman and Woolnough (2014) shows that the present approach, with many more channels but simpler spatial interpretation, is just as skillful, or even more so for the Year of Tropical Convection intercomparison case presented there.

The dependence of OLR' anomaly prediction skill on information in other channels of the dynamical state vector was evaluated. LIM predictive skill is robust to the number of input channels up to 144, giving similar skill whether excess channels are omitted or randomized in their time ordering. Using 15-degree longitude bins for the 3-variable LIM results in a 72 × 72 lagged covariance matrix consisting of 5184 coefficients fitted to two training periods between 1979 and 1999 (3652 days × 72 channels), or about 50 data values per coefficient. Even with double the number of channels (7.5 degree bins; 1/2 as many data-per-coefficient), no

skill loss (evidence of overfitting) was evident. Skill loss and possible overfitting is finally evident with 2.5 degree longitude bin channels or in this case at about 8 data values per coefficient.

Wind data (u850' and u200') adds 1.5–2 days of additional skill. Wavenumbers 1 and 2 contribute less than that in isolation (perhaps because they also contribute 'noise' distractions: sample-dependent patterns that are not repeatable in the verification period). All higher wavenumbers contribute negligibly (the blue line is statistically indistinguishable from the red lines in Figure 4).

In general, winds and OLR are correlated predictors, so their information content is mixed and they are not cleanly separable despite the labels which sound like they are two independent physical quantities. The results here do not necessarily shed light on fundamental MJO dynamics. Combined-variable indices always have debatable relative normalizations for the different variables (Liu *et al.*, 2016). From this study's point of view, predicting anomalous clouds and rain (OLR), the wind information may help the system avoid being misinterpreted by happenstance occurrences of MJO-shaped equatorial cloud patterns that are not actually part of a predictable wave in the real physical memory variables (inertia or water vapor or perhaps SST). Predictability can be limited as much by the strength of distractions and noise as it is by the dynamics of the predictable subsystem, so results about predictability may be results about such noise, not about dynamics. LIM OLR' hindcast prediction error is better during the DJF season and in Indo-Pacific longitudes, both indicative of the region and season where the MJO contributes the most to OLR anomalies and where tropical OLR variance is greatest.

Acknowledgements

The authors gratefully acknowledge financial support from NSF grant 0731520, NASA CYGNSS grant NNX13AQ50G, ONR grant N000141310704, DOE grant DE-SC0006806, NOAA grant NA13OAR4310156, and Government of India EarthMM/ SERP/Univ_Miami_USA/2013/INT-1/002. The authors are also grateful for the two anonymous reviewer's constructive comments.

References

Cavanaugh N, Allen T, Subramanian A, Mapes B, Miller AJ. 2014. The skill of tropical linear inverse models in hindcasting the Madden-Julian Oscillation. *Climate Dynamics* **44**: 897–906, doi: 10.1007/s00382-014-2181-x.

DelSole T, Shukla J. 2009. Artificial skill due to predictor screening. *Journal of Climate* **22**(2): 331–345, doi: 10.1175/2008JCLI2414.1.

Hakim GJ. 2013. The variability and predictability of axisymmetric hurricanes in statistical equilibrium. *Journal of the Atmospheric Sciences* **70**(4): 993–1005, doi: 10.1175/JAS-D-12-0188.1.

Kalnay E, Kanamitsu M, Kistler R. Collins W, Deaven D, Gandin L, Iredell M, Saha S, White G, Woollen J, Zhu Y, Leetmaa A, Reynolds R, Chelliah M, Ebisuzaki W, Higgins W, Janowiak J, Mo KC, Ropelewski C, Wang J, Jenne R, Joseph D. 1996. The NCEP/NCAR 40 year reanalysis project. *The Bulletin of the American Meteorological Society* **77**: 437–471.

Kikuchi K, Wang B, Kajikawa Y. 2012. Bimodal representation of the tropical intraseasonal oscillation. *Climate Dynamics* **38**(9–10): 1989–2000, doi: 10.1007/s00382-011-1159-1.

Kiladis GN, Dias J, Straub KH, Wheeler MC, Tulich SN, Kikuchi K, Weickmann KM, Ventrice MJ. 2014. A comparison of OLR and circulation-based indices for tracking the MJO. *Monthly Weather Review* **142**(5): 1697–1715, doi: 10.1175/MWR-D-13-00301.1.

Klingaman NP, Woolnough SJ. 2014. The role of air-sea coupling in the simulation of the Madden-Julian Oscillation in the Hadley Centre Model: air-sea coupling and the MJO. *Quarterly Journal of the Royal Meteorological Society* **140**(684): 2272–2286, doi: 10.1002/qj.2295.

Lau KH, Waliser DE. 2011. Intraseasonal Variability in the Atmosphere-Ocean Climate System. Praxis, 646 pp.

Liebmann B, Smith CA. 1996. Description of a Complete (Interpolated) Outgoing Longwave Radiation Dataset. *Bulletin of the American Meteorological Society* **77**: 1275–1277.

Liu P, Zhang Q, Zhang C, Zhu Y, Khairoutdinov M, Kim H-M, Schumacher C, Zhang M. 2016. A revised real-time multivariate MJO index. *Monthly Weather Review* **144**: 627–642, doi: 10.1175/MWR-D-15-0237.1.

Madden RA, Julian PR. 1971. Detection of a 40–50 day oscillation in the zonal wind in the tropical Pacific. *Journal of the Atmospheric Sciences* **28**: 702–708, doi: 10.1175/1520-0469(1971)028<0702:DOADOI>2.0.CO;2.

Madden RA, Julian PR. 1972. Description of global-scale circulation cells in the tropics with a 40–50 day period. *Journal of the Atmospheric Sciences* **29**: 1109–1123, doi: 10.1175/1520-0469(1972)029,1109:DOGSCC.2.0.CO;2.

Martin ER, Schumacher C. 2011. Modulation of Caribbean precipitation by the Madden–Julian Oscillation. *Journal of Climate* **24**(3): 813–824, doi: 10.1175/2010JCLI3773.1.

Newman M, Sardeshmukh PD, Penland C. 2009. How important is air–sea coupling in ENSO and MJO evolution? *Journal of Climate* **22**(11): 2958–2977, doi: 10.1175/2008JCLI2659.1.

Penland C, Sardeshmukh P. 1995. The optimal growth of tropical sea surface temperature anomalies. *Journal of Climate* **8**: 1999–2024.

Sharmila S, Pillai PA, Joseph S, Roxy M, Krishna RPM, Chattopadhyay R, Abhilash S, Sahai AK, Goswami BN. 2013. Role of ocean–atmosphere interaction on northward propagation of Indian Summer Monsoon Intra-Seasonal Oscillations (MISO). *Climate Dynamics* **41**(5–6): 1651–1669, doi: 10.1007/s00382-013-1854-1.

Shin S-I, Sardeshmukh PD, Pegion K. 2010. Realism of local and remote feedbacks on tropical sea surface temperatures in climate models. *Journal of Geophysical Research* **115**(D21), doi: 10.1029/2010JD013927.

Straub KH. 2013. MJO Initiation in the Real-Time Multivariate MJO Index. *Journal of Climate* **26**(4): 1130–1151, doi: 10.1175/JCLI-D-12-00074.1.

Wang B. (ed.) 2006. *The Asian Monsoon*. Springer, 787 pp.

Wheeler M, Hendon HH. 2004. An all-season real-time multivariate MJO index: Development of an index for monitoring and prediction. *Monthly Weather Review* **132**: 1917–1932.

Xavier P, Rahmat R, Cheong WK, Wallace E. 2014. Influence of Madden-Julian Oscillation on Southeast Asia rainfall extremes: observations and predictability. *Geophysical Research Letters* **41**(12): 4406–4412, doi: 10.1002/2014GL060241.

Zhang C. 2005. Madden-Julian Oscillation. *Reviews of Geophysics* **43**: RG2003, doi: 10.1029/2004RG000158.

Zhang C. 2013. Madden–Julian Oscillation: bridging weather and climate. *Bulletin of the American Meteorological Society* **94**(12): 1849–1870, doi: 10.1175/BAMS-D-12-00026.1.

2

Anomalous variation in summer tropical cyclone activity by preceding winter Aleutian low oscillation

Jae-Won Choi* and Yumi Cha

Research Planning and Management Division, National Institute of Meteorological Sciences, JeJu, Korea

*Correspondence to:
J.-W. Choi, Research Planning and Management Division, National Institute of Meteorological, 33, Seohobuk-ro, Seogwipo-si, Jeju-do, 63568, Korea.
E-mail: choikiseon@daum.net*

Abstract

This research found a high-positive correlation between Aleutian low oscillation during the winter (November–March) and tropical cyclone (TC) genesis frequency during the following summer (July–September) over a 26-year period (1982–2007). In the years with high Aleutian low oscillation, a number of characteristics were analyzed. In the preceding winter, anomalous pressure patterns, such as south-low and north-high at the low level, formed as the center for the regions near 20°N in the western North Pacific. Sea ice concentration was less than the average around the Sea of Okhotsk and the Bering Sea, which weakened the Aleutian Low in this area. This anomalous pressure pattern continued until the following summer, and it reinforced the anomalous easterlies at the mid-latitudes (20°–40°N) in East Asia and contributed to the high TC passage frequency in the East Asian continent.

Keywords: tropical cyclone; Aleutian low oscillation; sea ice concentration

1. Introduction

Gray (1984) identified six physical parameters that influence tropical cyclone (TC) genesis: (1) low-level relative vorticity, (2) local or planetary vorticity (Coriolis parameter), (3) the inverse of the vertical shear of the horizontal wind between the lower and upper troposphere, (4) ocean thermal energy maintaining temperatures above 26.8 °C at a depth of 60 m, (5) the vertical gradient of the equivalent potential temperature between the surface and 500 mb (hPa), and (6) the middle-troposphere relative humidity. Gray's parameters have been used as common large-scale predictors in statistical models for predicting the seasonal genesis frequency of TCs because of their accuracy in reflecting the seasonal characteristics related to TC activity (Mcdonnell and Holbrook, 2004). In particular, it has been shown that these parameters, when combined, can broadly identify the geographical and seasonal distribution of tropical cyclogenesis in each of the major ocean basins. This combination of parameters is known as the seasonal genesis parameter (Royer et al., 1998).

As seen in the above studies, prediction of TC genesis using the atmosphere and ocean parameters in the main area of TC genesis was performed successfully. However, in addition to TC genesis resulting from the environmental factors in the tropical regions, TCs also often occurred due to interactions among many different teleconnection patterns that existed in areas outside the tropical regions. Therefore, it is important to search for the signals of teleconnection patterns in order to determine a clear relationship with TCs and to influence the genesis of TCs.

The Aleutian low is one of the semi-permanent atmospheric action centers in the Northern Hemisphere, and it plays a vital role in the sea–air interaction in the North Pacific. The variations in its intensity and location play an important role in Northern Hemispheric climate change (Trenberth and Hurrell, 1994). Sun and Wang (2005) showed that the key factor for the Arctic Oscillation (AO), coupled with the Pacific Decadal Oscillation, may lie in the Aleutian low. Park et al. (2012) emphasized the dual contributions from the atmosphere and ocean to the local sea surface temperature (SST) variability in explaining the recent unusual warming in the western North Pacific (WNP). Previous studies have also found that variability of the Aleutian low can influence climate anomalies in remote regions (Zhu and Wang, 2010; Ye and Duan, 2015).

The Arctic condition affects the climate in the high and mid-latitudes of the Northern Hemisphere. Such changes have, in turn, been accompanied by various environmental changes, including Eurasian snow, sea ice, and Northern Hemisphere atmospheric circulation (Budikova, 2009; Li et al., 2013).

Even though the WNP is the ocean basin in which TCs are most active, the seasonal prediction problem in this area is relatively unexplored. The two well-known typhoon centers in this basin – the Regional Specialized Meteorological Center (RSMC) Tokyo and the Joint Typhoon Warning Center – provide only track and intensity forecasts for an individual TC and do not issue seasonal predictions. Thus, the seasonal prediction of TCs is conducted separately by each country in the WNP, and the accurate seasonal prediction of TC genesis can be an important issue in these countries. Therefore, the ultimate aim of this research is to determine whether the teleconnection pattern in the preceding winter is a good predictor for summer TC genesis frequency (TCGF).

2. Data and methods

The information about TC activity was obtained from the best track archives of the RSMC Tokyo Typhoon Center. We used data from the National Centers for Environmental Prediction–National Center for Atmospheric Research (NCEP–NCAR) (Kistler *et al.*, 2001) over a period of 26 years (1982–2007).

This study used the Student's *t*-test to determine significance (Wilks, 1995).

3. Relationship between the winter ALI and TCGF during the following summer

Figure 1 shows the genesis frequency of TCs in the WNP in July, August, and September, and it shows these 3 months collectively as well as for a time series of the Aleutian low index (ALI) in the preceding winter. In each of the 3 months, and in the 3 months collectively, a positive correlation between two variables can be found. The positive correlation has been reinforced by the increase of the month, which indicates a correlation of 0.55 at the 95% confidence level in September. In particular, the TCGF and the 3-month ALI had a correlation coefficient of 0.58 (significant at the 95% confidence level), which is higher than the correlation coefficient in September. This means that TCGF in summer increases if the Aleutian low in the preceding winter is weak and decreases if it is strong.

4. Differences between high ALI years and low ALI years

In order to examine the characteristics of the Aleutian low in the winter that can influence the TCGF in the following summer, the 8 years with the highest ALI (1982, 1985, 1989, 1990, 1991, 1994, 2006, and 2007, hereafter called the high ALI years) and the 8 years with the lowest ALI (1983, 1984, 1986, 1987, 1992, 1998, 2001, and 2003, hereafter called the low ALI years) were defined from the time series, as shown in Figure 1(d). Then, the characteristics between the two phases were compared. Here, we selected a standard deviation of >0.7 and <−0.7 from the normalized ALI time series as high ALI years and low ALI years, respectively. In addition, these 16 years account for approximately 62% of all analysis periods (26 years).

4.1. Tropical cyclone genesis frequency

Figure 2(a) shows the monthly differences in TCGF between the two phases. There was a total of 25 differences between the two phases for the 3 months. These differences mean that approximately three more TCs occurred, on average, during the summers of high ALI years than during the summers of low ALI years. As the months progress, it can be seen that the difference between the two phases also increases: July = 4 TCs,

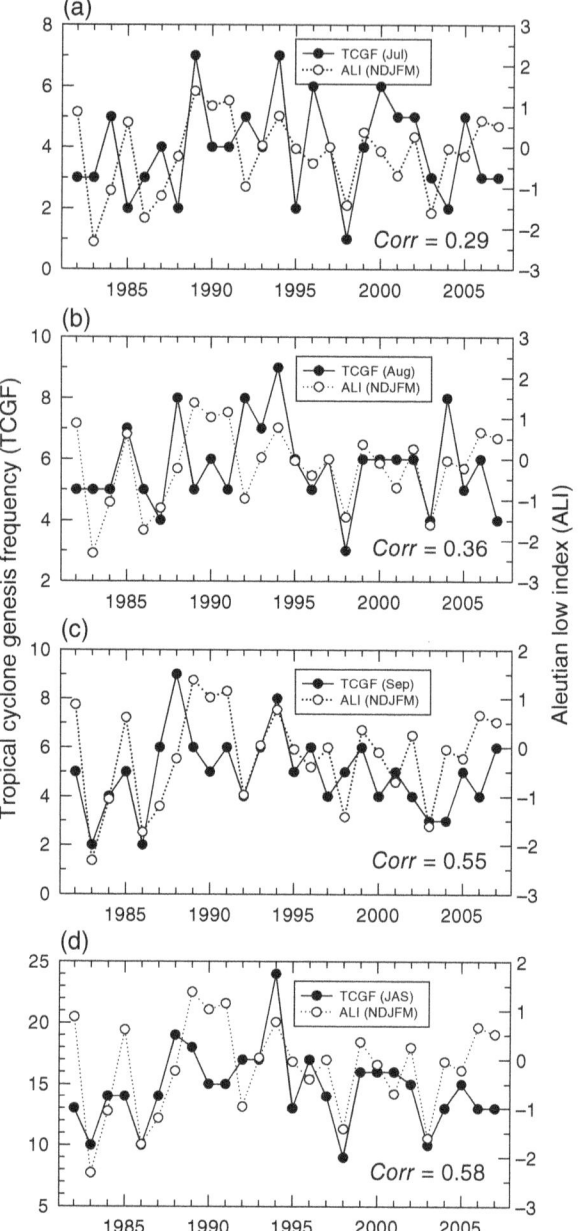

Figure 1. Relationships between Aleutian low index for preceding November–March (NDJFM) and tropical cyclone genesis frequency in the western North Pacific for (a) July, (b) August, (c) September, and (d) all 3 months. The correlation coefficients are significant at the 95% confidence level.

August = 7 TCs, and September = 14 TCs. September, in particular, accounted for more than half of the differences over the 3 months. This had been determined previously because the correlation between the TCGF and the ALI was highest in September.

Figure 2(b) shows the characteristics of the spatial distribution of TCGF in the summer between the two phases. The largest difference between the two phases appears in the east sea (10°–20°N, 130°–155°E) of the Philippines. In this area, TCGF was higher during the high ALI years. The open ocean east of the Philippines is known as the Western Pacific Warm Pool (WPWP), and it has a dominant influence on the climate and the

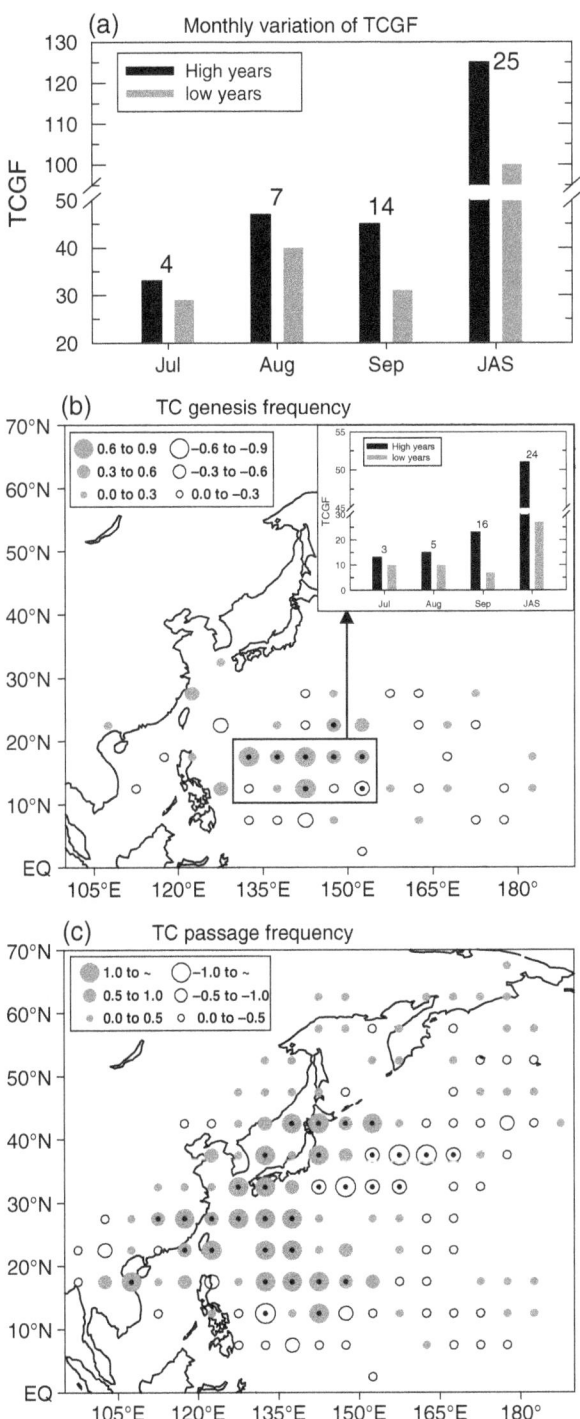

Figure 2. (a) Monthly variation of TCGF in eight highest ALI years and eight lowest ALI years that selected from ALI in Figure 1. The number denotes a difference in TCGF between the two phases. Differences in (b) TC genesis frequency and (c) TC passage frequency between positive ALI years and negative ALI years. Small solid circle indicates that the differences are significant at the 95% confidence level.

interannual (interdecadal) variabilities of TCGF. The SST variability in the WPWP region moves the wave train of atmospheric circulation toward the northeast, and it eventually affects the variation of the Aleutian low (Zhao *et al.,* 2003). Thus, the difference in the TCGF in this region between the two phases is shown

in Figure 2(b) (bar graph). In the 3 months combined, the TCGF is approximately two times higher in the high ALI years than in the low ALI years (high ALI years: 51 TCs, low ALI years: 27 TCs). The largest difference is in September (July: 3 TCs, August: 5 TCs, September: 16 TCs). The difference between the two phases in September accounts for approximately two thirds of the total difference (24 TCs) over the 3 months in this area.

4.2. TC passage frequency (TCPF)

The TC genesis location has an effect on the TC track. Wang and Chan (2002) noted that when TCs occur in the southeast in the subtropical WNP, the TC passage tends to follow a curve along the western periphery of the WNP high. In addition, Ho *et al.* (2005) stressed that when TCs occurred near the Philippines (in other words, when they occurred in the west of the subtropical WNP), they showed a tendency to move to the west or northwest without turning. Therefore, this study analyzed the difference in the average TCPF between the two phases per grid box of $5° \times 5°$ latitude-longitude (Figure 2(c)). The west side, based at about 150°E, showed a higher frequency of TCs in high ALI years, while the east side showed higher frequency for low ALI years. During high ALI years, in particular, it is apparent that the passage moved to Korea and Japan through the East China Sea from the east sea of the Philippines. In the areas of Vietnam and South China, the TCGF is not small. Therefore, we can determine that East Asian countries located on the coast should pay greater attention to the damage caused by TC occurrences during high ALI years.

4.3. Environmental conditions

The characteristics of the environmental conditions that influenced the differences in TCPF and TCGF between the two phases were analyzed with regard to the summer and the preceding winter.

The left panel of Figure 3 shows the differences between the two phases for 850 hPa geopotential heights and horizontal winds during the preceding winter. In the high ALI years, in the north (about 25°N), the anomalous anticyclonic circulations, and in the south, the anomalous cyclonic circulations, are strengthened. This is a positive North Pacific Oscillation (NPO), and it is similar to the pattern analyzed by Walker and Bliss (1932). They discovered that there is a seesaw pattern in the sea level pressure (SLP) between high latitudes from eastern Siberia to western Canada during the winter and subtropical low latitudes below 40°N in the Pacific sector, which is like the North Atlantic Oscillation (NAO). This oscillation between the north and south regions in the North Pacific is the NPO. Wang *et al.* (2007) showed that the summer TCGF in the WNP is higher during a positive NPO phase than during a negative NPO phase in the preceding spring. Therefore, when an anomalous pressure pattern, such as south-low and north-high, strengthens in the North Pacific during the preceding winter and spring, we know that in the following summer, TCGF will be high, while a strong north-low and south-high pattern in the preceding winter and spring would lead to a low TCGF in the following summer. These anomalous pressure patterns were also shown in the 500 hPa geopotential height field (not shown). In particular, the Aleutian low from the region near the Aleutian develops climatologically in the winter and spring. Thus, the anomalous anticyclonic circulations based on this region in high ALI years indicated a weak Aleutian low, while the anomalous cyclonic circulations in low ALI years indicated a strong Aleutian low. When the Aleutian low is weak during the preceding winter and spring, the TCGF increases in the following summer.

The Aleutian low can be associated with the sea ice condition in this region. Therefore, in the area near the Sea of Okhotsk and the Bering Sea, the difference in sea ice concentration between the two phases during the winter was analyzed, as shown in the right panel of Figure 3. In high ALI years, it appears that there was a negative anomaly in most of the areas, with the exception of the northern areas of the Aleutian, while the reverse pattern appears in low ALI years. This means that the TCGF increases in the summer when the sea ice concentration in the area is low in the preceding winter. Fang and Wallace (1994) showed that a higher than average concentration of sea ice in the North Pacific during the winter and spring can strengthen the Aleutian low. Fan (2007) also showed that a greater than average concentration of sea ice in the North Pacific reinforces the Aleutian Low in the North Pacific, which then forms an anomalous pressure pattern, such as south-high and north-low, which has a negative impact on TCGF in the following summer.

The difference that appears in the environmental conditions between these two phases during the winter continues through to the following summer (left panel of Figure 4). Although anomalous cyclonic circulations are located near the Sea of Okhotsk and the Bering Sea (based at 25°N in the south of 45°N) in high ALI years, the anomalous pressure pattern south-low and north-high was still maintained. The opposite pressure pattern appears in low ALI years. These anomalous pressure patterns in the two phases can be clearly seen through outgoing longwave radiation (OLR) analysis (right panel of Figure 4). In high ALI years, the OLR anomaly for each negative and positive is formed as the standard at about 25°N in the region of south and north, respectively. This means that convection is much more active in the subtropical WNP in high ALI years. In contrast, a positive OLR formed from the northeast to southeast, moving from the southeast region of the subtropical WNP to the south region in China, in low ALI years. In conclusion, it is shown that the anomalous south-low and north-high pressure pattern was reinforced, as the preceding winter in high ALI years provided a favorable environment for increased TCGF in the following summer.

We found that the anomalous pressure patterns during the high ALI years and low ALI years (Figure 2(c)) caused the difference in TCPF between the two phases. In the high ALI years, it can be seen that the anomalous easterlies are especially notable at the latitude of 20°–40°N from the winter to the following summer. The steering flows play a role in allowing TCs to move easily toward East Asia. However, when anomalous westerlies are reinforced near 20°N in the low ALI years, the steering flows can interfere with TC movement toward the East Asian region.

The effect of the atmospheric and oceanic environments on TC genesis was also analyzed (Figure 5). These figures show differences in (a) vertical wind shear (200–850 hPa), (b) 850 hPa specific humidity, (c) 850 hPa relative vorticity, (d) vertical meridional circulations (based on the average latitude-pressure cross-section between 100° and 180°E, which was the longitudinal range in which there was frequent TC genesis), and (e) SST between the two phases. The analyses of the vertical wind shear (200–850 hPa) and the 850 hPa specific humidity and relative vorticity show negative anomalies and positive anomalies in the subtropical WNP, respectively (Figures 5(a)–(c)). These results indicate that favorable environments for frequent genesis of TCs are formed in the high ALI years. As for vertical meridional circulations, there are intensified anomalous upward flows between 0° and 20°N (subtropical WNP) when there was frequent genesis of TCs (Figure 5(d)). This shows that there are favorable vertical structures that can increase TCGF in high ALI years. In SST analysis, warm SST anomalies are distinctive over the subtropical WNP, which is a good oceanic environment for TC genesis (Figure 5(e)).

4.4. Evaluation of the prediction performance of the TC statistical model using the preceding winter ALI

Choi *et al.* (2010) developed a multiple linear regression model (MLRM) for the prediction of summer TCGF in the WNP using the three teleconnection patterns. These patterns are representative of the Siberian High Oscillation (SHO) in the East Asian continent, the NPO, and the Antarctic Oscillation (AAO) during the preceding boreal spring (April–May). That is, the predictors in Choi *et al.*'s statistical model are the SHO index, the NPO index, and the AAO index in the preceding boreal spring. This statistical model for the seasonal prediction of TCGF is as follows:

$$TCGF = 1.07^*SHO - 0.71^*AAO + 0.28^*NPO + 10.22$$

In this study, the preceding winter ALI was added to Choi *et al.*'s MLRM as a fourth predictor as follows:

$$TCGF = 0.98^*SHO - 0.62^*AAO + 0.31^*NPO$$
$$+ 0.71^*ALI + 14.43$$

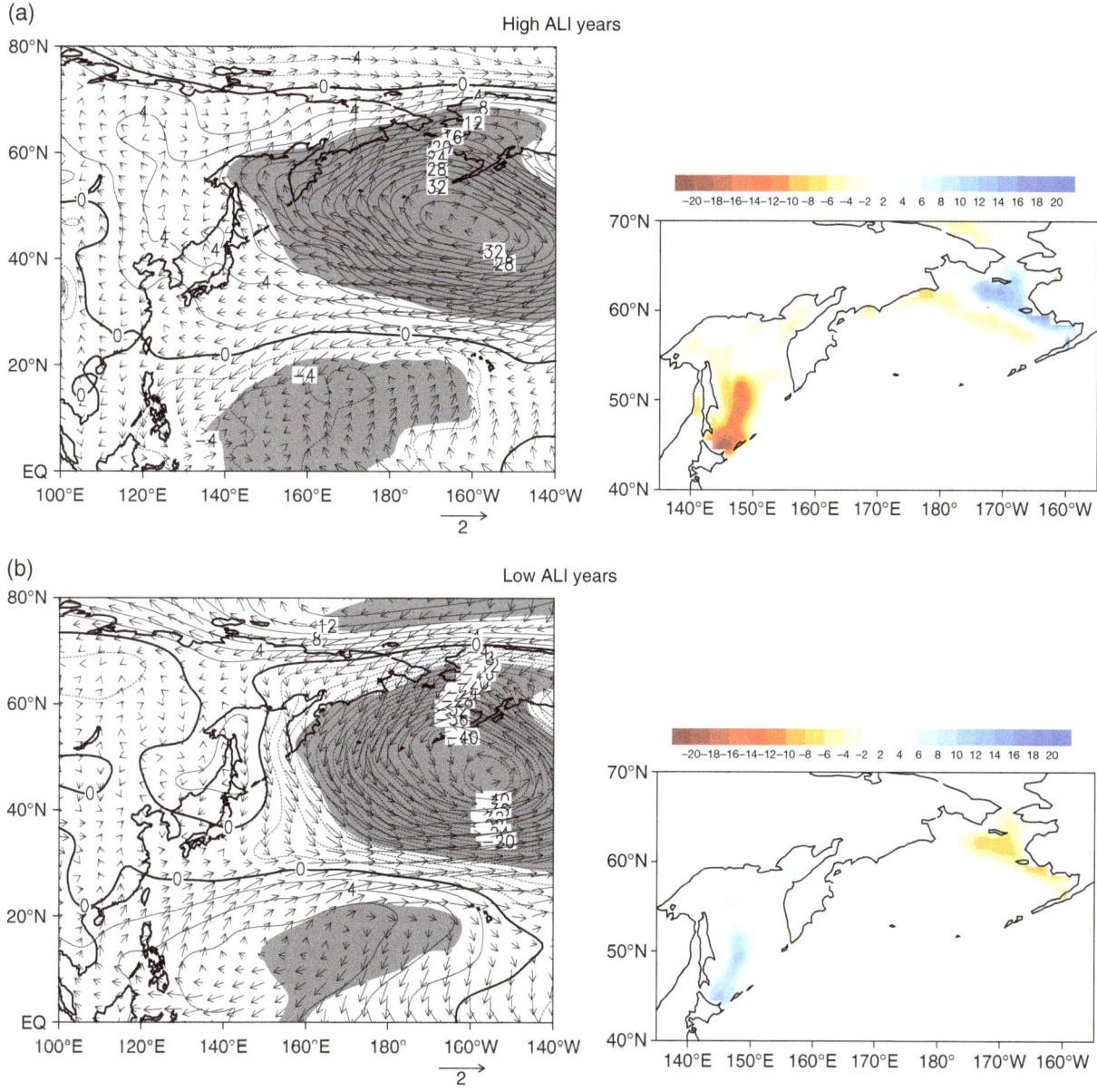

Figure 3. Composites of anomalies of 850 hPa geopotential heights with horizontal wind (left) and sea ice concentration (right) for (a) positive ALI years and (b) negative ALI years for NDJFM. Shaded areas are significant at the 95% confidence level. Contour interval is 2 gpm.

The prediction performance between Choi *et al.*'s MLRM (see Figure 3(d)) and the present study's MLRM (Figure 5(f)) was compared. In Choi *et al.* (2010), the correlation coefficient between the observed TCGF and the hindcasted TCGF by MLRM was 0.73, while in this study, the correlation coefficient was 0.88 (significant at the 99% confidence level). In particular, the prediction performance in extreme years (e.g. 1967, 1994, and 1998) was significantly improved. Therefore, the preceding winter ALI can be used as an effective predictor for TCGF during the following summer.

5. Summary and conclusions

The relationship between the ALI in the winter and TCGF in the following summer was

analyzed based on a period of 26 years (1982–2007). There was a high-positive correlation (corr = 0.58) between the two variables, and the highest correlation among the 3 months July, August, September (JAS) was for September (0.55).

In high ALI years, there was an increase in TCs from the eastern sea areas in the Philippines (WPWP region). These TCs mainly went to Korea and Japan through the East China Sea. Therefore, it was determined that countries located on the East Asian coast must pay more attention to TCs in high ALI years.

The anomalous south-high and north-low pressure pattern that formed in the WNP during the winter continued until the following summer, and the lower than average sea ice in the Sea of Okhotsk and the Bering

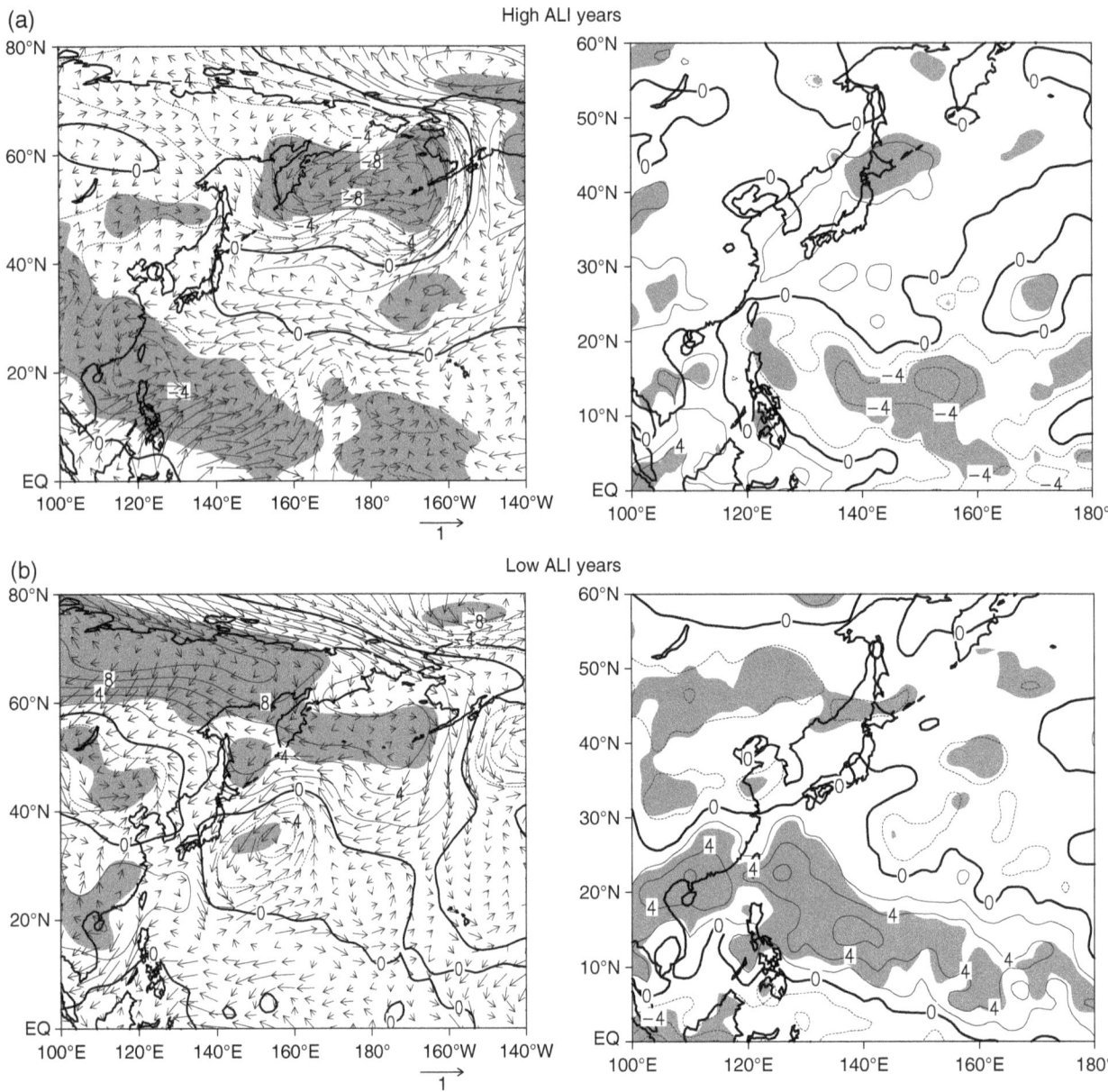

Figure 4. Same as Figure 3, but for summer (July–September). Left and right panels indicate 850 hPa geopotential heights with horizontal wind and OLR, respectively. Shaded areas are significant at the 95% confidence level. Contour interval for OLR is $2\,W\,m^{-2}$.

Sea during the winter reinforced this pressure pattern. Anomalous easterlies moving to the East Asian region as a result of the anomalous south-high and north-low pressure pattern led to more frequent movement of TCs to this area. Therefore, the anomalous pressure pattern formed in the WNP and the sea ice condition in the Sea of Okhotsk and the Bering Sea during the winter are good predictors of TCGF during the following summer.

We also analyzed the atmospheric and oceanic environments that affected TC genesis: the vertical wind shear (200–850 hPa), the 850 hPa specific humidity, the 850 hPa relative vorticity, and the vertical meridional circulations based on the average latitude-pressure cross-section between 100° and 180°E. The analyses of the vertical wind shear (200–850 hPa) and

850 hPa specific humidity and relative vorticity showed negative anomalies and positive anomalies in the subtropical WNP, respectively. With regard to the vertical meridional circulations, there were intensified anomalous upward flows between 0° and 20°N (subtropical WNP). In addition, warm SST anomalies were distinctive over the subtropical WNP. This indicates that there are favorable atmospheric and oceanic conditions that can increase TCGF in high ALI years.

The preceding winter ALI was added to Choi *et al.*'s (2010) MLRM as a predictor, and the prediction performance between the Choi *et al.* MLRM and the present study's MLRM was compared. The correlation between the observed TCGF and hindcasted TCGF was stronger in the present study's MLRM than in the Choi *et al.*

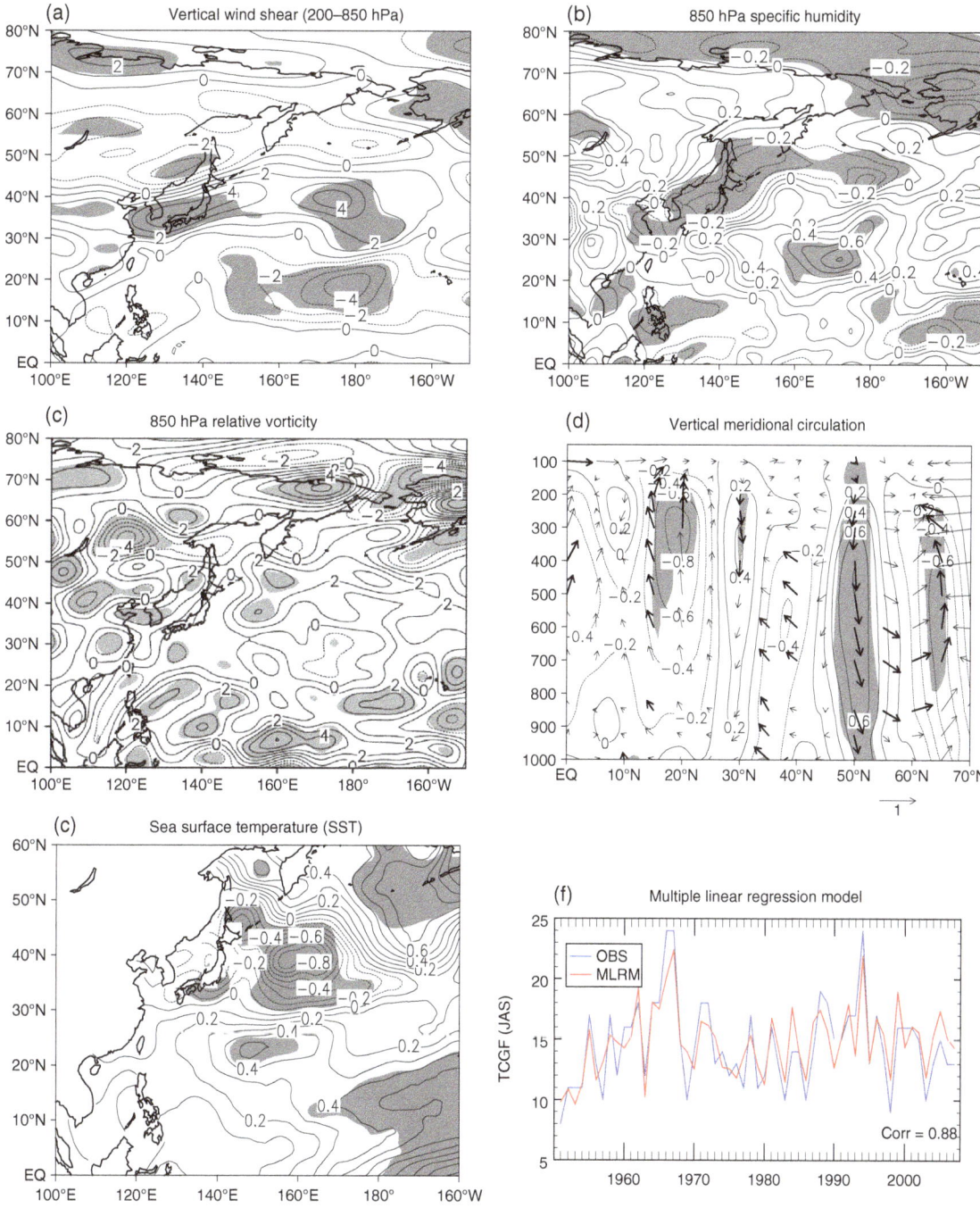

Figure 5. Differences in (a) vertical wind shear (200–850 hPa), (b) 850 hPa specific humidity, (c) 850 hPa relative vorticity (d) latitude-pressure cross-section of vertical velocity (contours) and vertical meridional circulations (vectors) averaged along 100°–180°E, and (e) sea surface temperature between high and low ALI years for JAS. Shaded areas are significant at the 95% confidence level. In (c), the values of vertical velocity are multiplied by −100. Contour intervals are 1 ms⁻¹ for VWS, 0.1 g kg⁻¹ for 850 hPa specific humidity, 10⁻⁶s⁻¹ for 850 hPa relative vorticity, 0.2⁻²hPa s⁻¹ for vertical velocity, and 0.1 °C for SST. (f) Time series of observed TCGF (blue line) and hindcasted TCGF by new multiple linear regression model that ALI predictor was added to model of Choi *et al.* (2010).

MLRM, and, in particular, the prediction performance was significantly improved in extreme years.

References

Budikova D. 2009. Role of Arctic sea ice in global atmospheric circulation: a review. *Global and Planetary Change* **68**: 149–163.

Choi KS, Moon JY, Kim DW, Chu PS. 2010. Seasonal prediction of tropical cyclone genesis frequency over the western North Pacific using teleconnection patterns. *Theoretical and Applied Climatology* **100**: 191–206.

Fan K. 2007. North Pacific sea ice cover, a predictor for the western North Pacific typhoon frequency? *Science in China Series D: Earth Sciences* **8**: 1251–1257.

Fang Z, Wallace JM. 1994. Arctic sea ice variability on a timescale of weeks and its relation to atmospheric forcing. *Journal of Climate* **7**: 1897–1914.

Gray WM. 1984. Atlantic seasonal hurricane frequency. Part II: forecasting its variability. *Monthly Weather Review* **112**: 1669–1683.

Ho CH, Kim JH, Kim HS, Sui CH, Gong DY. 2005. Possible influence of the Antarctic Oscillation on tropical cyclone activity in the western North Pacific. *Journal of Geophysical Research* **110**(D19104). https://doi.org/10.1029/2005JD005766.

Kistler R, Kalnay E, Collins W, Saha S, White G, Woollen J, Chelliah M, Ebisuzaki W, Kanamitsu M, Kousky V, van den Dool H, Jenne R, Fiorino M. 2001. The NCEP-NCAR 50-Year Reanaly-sis: Monthly Means CD-ROM and Documentation. *Bull. Am. Meteorol. Soc.* **82**: 247–268.

Li F, Wang HJ, Liu JP. 2013. The strengthening relationship between Arctic Oscillation and ENSO after the mid-1990s. *International Journal of Climatology* **34**(7): 2515–21. https://doi.org/10.1002/joc.3828.

McDonnell KA, Holbrook NJ. 2004. A poisson regression model of tropical cyclogenesis for the Australian–Southwest Pacific Ocean Region. *Weather and Forecasting* **19**: 440–455.

Park YH, Yoon JH, Youn YH. 2012. Recent warming in the western North Pacific in relation to rapid changes in the atmospheric circulation of the Siberian High and Aleutian Low systems. *Journal of Climate* **25**: 3476–3493.

Royer JF, Chauvin F, Timbal B, Araspin P, Grimal D. 1998. A GCM study of the impact of greenhouse gas increase on the frequency of occurrence of tropical cyclones. *Climatic Change* **38**: 307–343.

Sun JQ, Wang HJ. 2005. Relationship between Arctic oscillation and Pacific decadal oscillation on decadal timescale. *Chinese Science Bulletin* **51**: 75–79.

Trenberth KE, Hurrell JW. 1994. Decadal atmosphere–ocean variations in the Pacific. *Climate Dynamics* **9**: 303–319.

Walker GT, Bliss EW. 1932. World weather V. *Memoirs of the Royal Meteorological Society* **4**: 53–84.

Wang B, Chan JCL. 2002. How strong ENSO events affect tropical storm activity over the western North Pacific. *Journal of Climate* **15**: 1643–1658.

Wang HJ, Sun JQ, Fan K. 2007. Relationship between North Pacific Oscillation and the typhoon and hurricane frequency. *Science in China Series D: Earth Sciences* **50**: 1409–1416.

Wilks DS. 1995. *Statistical Methods in the Atmospheric Sciences*. Academic Press; 467.

Ye SL, Duan WS. 2015. Interannual relationship between the winter Aleutian low and rainfall in the following summer in South China. *Atmospheric and Oceanic Science Letters* **8**: 271–276.

Zhao YP, Wu AM, Chen YL, Hu DX. 2003. The climatic jump of the western Pacific warm pool and its climatic effects. *Journal of Tropical Meteorology* **9**: 9–18.

Zhu Y, Wang H. 2010. The relationship between the Aleutian Low and the Australian summer monsoon at interannual time scales. *Advances in Atmospheric Sciences* **27**: 177–184.

3

Lightning activity in the Mediterranean: Quantification of cyclones contribution and relation to their intensity

Elissavet Galanaki,[1,2]* Emmanouil Flaounas,[1] Vassiliki Kotroni,[1] Konstantinos Lagouvardos[1] and Athanassios Argiriou[2]

[1] National Observatory of Athens, Institute for Environmental Research and Sustainable Development, Greece
[2] Department of Physics, Laboratory of Atmospheric Physics, University of Patras, Greece

*Correspondence to:
E. Galanaki, National Observatory of Athens, Institute for Environmental Research and Sustainable Development, Lofos Koufou, 15236 Athens, Greece.
E-mail: galanaki@noa.gr

Abstract

A 10-year data set of intense Mediterranean cyclones was used for a twofold objective: first to quantify the cyclone's contribution to lightning occurrence in the region and second to investigate potential connection of lightning with cyclones intensity. For this reason, we used cyclone tracks, lightning observations and reanalysis from the European Centre for Medium Range Weather Forecasts, for the 10-year period of 2005–2014. Results showed that intense cyclones provoke <10% of lightning activity over the Mediterranean Sea, however, in certain areas, cyclone contributions might reach 20–30%.

The intense cyclones, which are associated with lightning activity close to their centre, constitute about one third (36%) of the total number of tracked cyclones. Therefore two cyclone groups are identified: those associated with and those without lightning. The first group presents in average 35% more ice and 15% more liquid cloud water content within the upper and lower atmospheric levels, respectively, while is related to approximately three times greater values of convective available potential energy in average. Further analysis shows that the intensities of the cyclones in the two groups present no significant differences, suggesting that deep convection may not be a major mechanism for the occurrence of intense Mediterranean cyclones. Finally, we show that cyclones associated with lightning present the highest lightning activity about 6 h prior to the cyclones maximum intensity.

Keywords: Mediterranean cyclones; deep convection; lightning

1. Introduction

Intense Mediterranean cyclones are linked with environmental risks such as heavy rainfall, windstorms and lightning (Papagiannaki et al., 2013). Especially lightning may inflict fatalities, injuries, fires and economical damage (Mills et al., 2008). Currently, the state-of-the-art lacks of a systematic analysis of the climatology of Mediterranean cyclones associated with lightning.

Lightning observations have been widely used as a proxy for deep convection associated with cold cloud tops and high radar reflectivity. Especially for tropical hurricanes, several studies have shown that strong lightning activity occurred prior to the cyclones maximum intensity. For instance, Price et al. (2009) and Whittaker et al. (2015) showed in a climatological approach that the maximum of lightning impacts occurred approximately 1 day before hurricane winds attain their maximum speed. While numerous studies that investigate the relation between lightning and cyclones are devoted to tropical hurricanes, the relevant literature over the Mediterranean is limited. Recently, Miglietta et al. (2013) studied an ensemble of 14 Medicanes (tropical-like cyclones in the Mediterranean Sea) and their results confirmed that in these systems – as in hurricanes – lightning preceded their

stage of maximum intensity. This result comes also in accordance with several past case studies of cyclones in the Mediterranean, especially when considering rapidly intensifying cyclones and Medicanes (e.g. Lagouvardos and Kotroni, 2007; McTaggart-Cowan et al., 2010). On the other hand, Flaounas et al. (2016) used observation data sets and showed that in two case studies of intense Mediterranean cyclones only one of them was associated with deep convection and lightning, despite that both cyclones shared the same duration, same pressure minima and equally attributed heavy rainfall over the western Mediterranean. As follows, the association of Mediterranean cyclones intensity with deep convection and resulting lightning occurrence still remains an open question.

In this study, we have a twofold objective. First, we aim at quantifying the cyclones contribution to the environmental risk of intense lightning activity in the Mediterranean region and second, we investigate in a climatological framework the dependence of Mediterranean cyclones intensity to the presence of lightning.

2. Data and methods

2.1. Lightning and atmospheric data

In order to acquire realistic results, in this study, we analyse atmospheric reanalysis fields and lightning observations. The atmospheric fields are obtained from the European Centre for Medium Range Weather Forecasting reanalysis of ERA-Interim (ERAI), in a horizontal grid spacing of $0.75° \times 0.75°$ in longitude and latitude (Dee and Uppala, 2009). The lightning observations are provided by the ZEUS long-range lightning detection system, operated by the National Observatory of Athens (Kotroni and Lagouvardos, 2008). ZEUS detects cloud-to-ground lightning strikes with a detection efficiency of the order of 25% and an average location error of the impacts of the order of 7 km (Lagouvardos et al., 2009). It is also noteworthy that ZEUS has a tendency of under-detecting lightning during night-time, without however suffering from missing the detection of thunderstorms within the domain covered by the network that includes the Mediterranean (Lagouvardos et al., 2009).

2.2. Cyclone tracking

To perform cyclone tracking, we used the method developed by Flaounas et al. (2014). This method identifies cyclones as local maxima of relative vorticity at 850 hPa and has been applied to the ERAI reanalysis over the wider Mediterranean region (Figure 1(a)), for the 10-year period of 2005–2014. Only cyclones with a life time of at least 1 day and maximum relative vorticity of $>8 \times 10^{-5}$ s^{-1} have been retained. These criteria have been shown adequate to distinguish intense cyclones from weak vorticity maxima, abrupt wind stirring or heat-lows (Flaounas et al., 2013). In total, we detected 584 intense cyclones (about 60 cyclones per year).

To associate cyclones with lightning, we attributed to each track point (from genesis to lysis) all impacts that took place within 200 km and within 3 h from its location and time. The radius of 200 km first permits to associate cyclones core intensity with the process of deep convection and second permits to associate lightning activity close to the cyclones centres, where rainfall and convection are expected to be maximum (Flaounas et al., 2015b). This process distinguished two cyclone groups: the first group is composed by 211 cases, where all cyclones are related to lightning at least once during their life time, while the second group presents cyclones with no lightning activity at all and comprises 373 cases.

All tracked cyclones occur over the main cyclogenetic regions of the Mediterranean (Figure 1(a) and (b)): over the western Mediterranean, the Ionian and the Aegean Seas (Trigo et al., 1999; Campins et al., 2011). Cyclones in Figure 1(a) and (b) present similar spatial variabilities, however cyclones without lightning tend to present higher rate of occurrence in particular over Northwest Africa, close to Cyprus and the Black Sea. Figure 1(c) shows the seasonal cycle of intense Mediterranean cyclones. In consistency with previous studies (Campins et al., 2011; Flaounas et al., 2013), the intense cyclones frequency of occurrence presents a maximum in autumn and winter (64% of the cases) and a minimum in summer (5% of the cases). It is

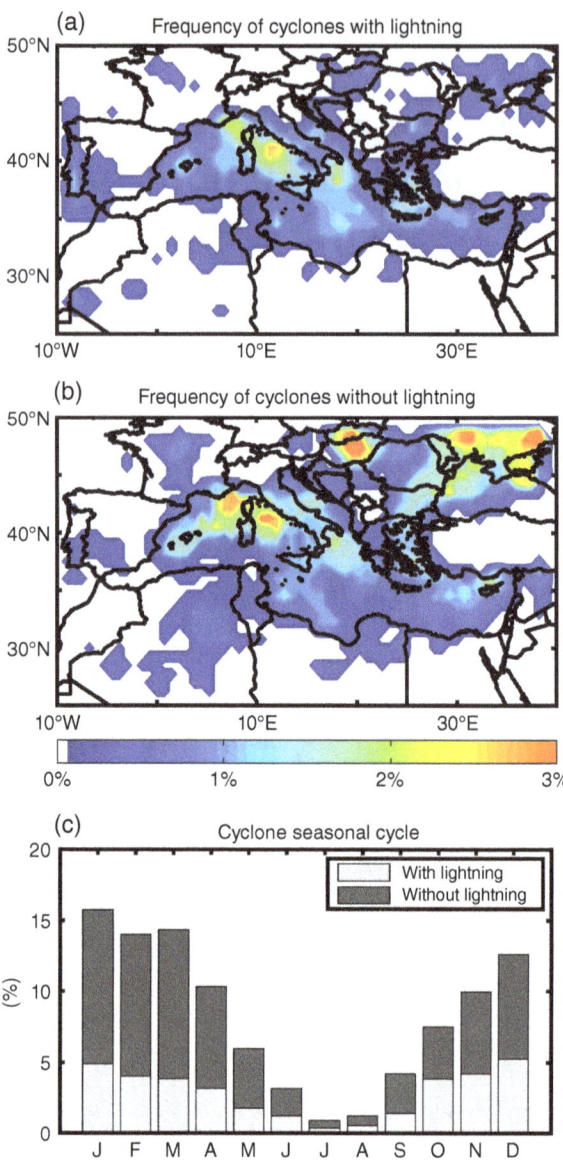

Figure 1. (a) Frequency of intense cyclone centres of occurrence per month in an area of 500×500 km for cyclones associated with lightning. (b) As in (a) but for cyclones that do not present lightning. (c) Seasonal cycle of the intense cyclones associated with and without lightning.

noteworthy that the cyclones associated with lightning are more frequent in autumn and less frequent in winter, while the opposite is true for the cyclones that are not associated with lightning (Figure 1(c)). A plausible explanation lies to the higher Mediterranean Sea surface temperature in autumn, which increases low-level instability and therefore favours convection (Kotroni and Lagouvardos, 2016).

3. Seasonal cyclones contribution to lightning

Figure 2 shows the seasonal contribution of cyclones to the 10-year total of observed lightning over the

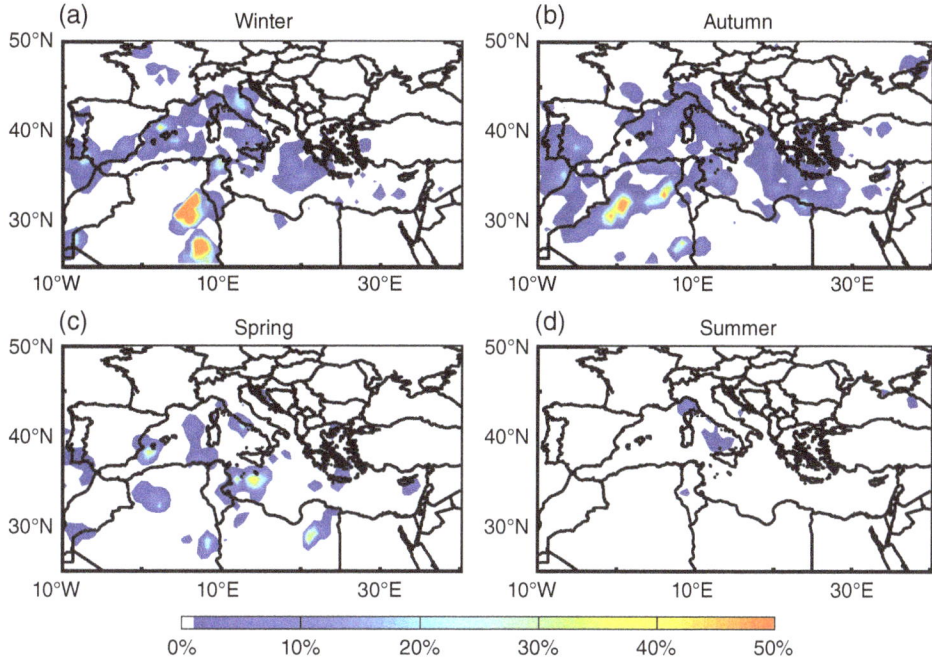

Figure 2. Percentage of the total of lightning related to cyclones for each season.

Mediterranean basin. The overall cyclones contribution is higher in winter and autumn when intense cyclogenesis presents its highest rate of occurrence (Figure 1(c)). Given that the most intense cyclones tend to form over the western and central Mediterranean Sea (Figure 1(a)), it comes as no surprise that these marine areas are mainly affected by cyclone-induced lightning, where intense cyclones provoke no >10% of the total lightning activity in winter and autumn. The higher contributions are observed over northwest Africa (between 0°–10°E and 25°–35°N) where cyclones are related to as much as 60% of the total winter and autumn lightning. Secondary maxima of the order of 10–30% are also observed over several cyclogenesis hot-spots, as for instance in the vicinity of the Gulf of Genoa in winter and autumn, over the Ionian Sea in winter and close to North Africa in spring.

In spring and summer, intense Mediterranean cyclones are at their minimum frequency of occurrence (Figure 1(c)), while lightning activity is more frequent over the land (Altaratz *et al.*, 2003; Tuomi and Mäkelä, 2008; Feudale and Manzato, 2014; Ben Ami *et al.*, 2015; Galanaki *et al.*, 2015). Consequently, during these seasons cyclones contribution to lightning is dramatically reduced, except in spring, in the regions over the North African coast and close to the Atlas Mountain. Indeed, this region is previously shown to consist of a cyclogenesis hot-spot (Sharav cyclones; Alpert and Ziv, 1989; Trigo *et al.*, 1999), where cyclones are responsible for uptaking large loads of dust (Flaounas *et al.*, 2015a). Therefore, our results suggest a possible connection between these systems and dust related lightning activity (Proestakis *et al.*, 2016). In summer, cyclones are at their minimum of occurrence (about 6%; Figure 1(a)) contributing thus

to lightning in specific areas by no >8% of the seasonal total (Figure 2(d)).

Cyclones contribution to lightning activity over the eastern Mediterranean is shown to be rather weak throughout the year (Figure 2), rarely overpassing 50%. This is due to the fact that we track the most intense cyclones in the region, quantifying thus cyclone contributions to lightning only over the favourable areas of intense cyclogenesis, i.e. mainly over the central and western Mediterranean (Figure 1(a)). Considering lower (or no) vorticity thresholds, the overall cyclone contributions to regional lightning activity increases dramatically. However, such results would lack of a robust interpretation of the role of cyclones in forming thunderstorms in the Mediterranean. Indeed, local vorticity maxima – as tracked by our cyclone tracking method – may not always correspond to cyclonic systems (in the sense of meso-scale wind vortices) but rather to local scale wind circulations such as abrupt wind stirring.

4. Intense cyclone occurrence and lightning activity

Strong lightning activity close to the cyclones centre suggests the development of strong convection. This should be reflected in the composition of clouds, in consistency with the theory of non-inductive thunderstorm electrification. Indeed, lightning activity is related with high cloud ice concentrations (Defer *et al.*, 2005; Petersen *et al.*, 2005; Lagouvardos and Kotroni, 2007), while the magnitude of the transferred charge during the electrification process is expected to be related with adequately high cloud liquid water (Avila

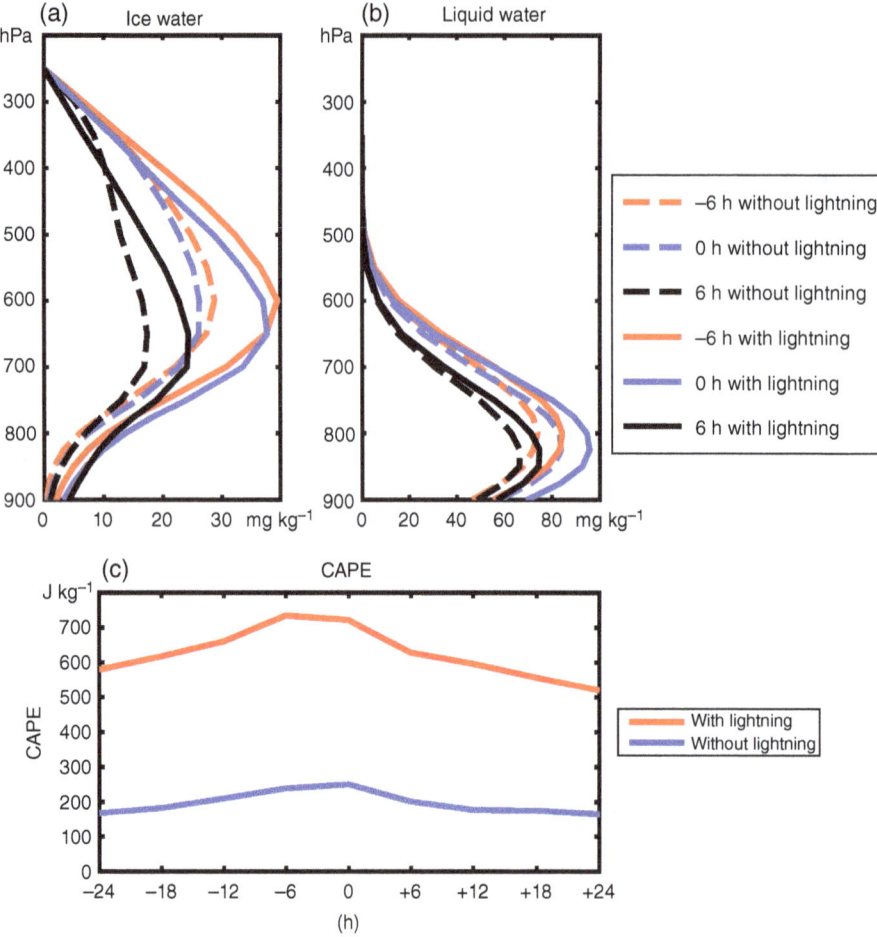

Figure 3. (a) Vertical profile of ice water content averaged within a radius of 200 km around the cyclone centre. (b) As in (a) but for liquid water content. (c) Composite time series of CAPE centred at the cyclones maximum intensity (0 h corresponds to the cyclones maximum relative vorticity).

et al., 1995). In Figure 3(a) and (b), we show the composite vertical profiles of cloud ice and cloud liquid concentrations for the two cyclone groups as derived by the ERAI reanalyses, averaged within 200 km from the cyclones centres. According to our cyclone tracking method, the cyclones mature stage is taking place when relative vorticity at 850 hPa is at its maximum (0 h). Therefore composite vertical profiles in Figure 3 are presented as averages of the three time instances of −6, 0 and +6 h. The cyclones associated with lightning present about 35% more cloud ice (Figure 3(a)) and liquid water (Figure 3(b)) content in the middle atmospheric layers of 500–700 hPa and about 15% more within the lower troposphere (900–700 hPa). It is noteworthy that maximum ice content concentration in Figure 3(a) tends to fall in lower levels from −6 h (∼600 hPa) to +6 h (∼700 hPa) due to gravitational fall of ice within the cyclone clouds (Lagouvardos and Kotroni, 2007). Further analysis on the seasonal dependency of the profiles shown in Figure 3(a) and (b) showed that regardless the season, cyclones with lightning activity always present higher ice concentrations by 46% more in autumn and 33% in summer and higher liquid concentrations by 13% in summer and 23% in winter. It is noteworthy that reanalysis is performed in

a coarse resolution grid where convection is a subgrid scale process. Consequently, uncertainties may arise into the liquid and ice concentration profiles. However, in this study, we focus on the relative difference between the two cyclones groups. These results are consistent with Flaounas *et al.* (2016) where radar observations of ice concentration in a cyclone presenting deep convection were compared to the ones of a cyclone where no deep convection and no lightning activity took place. In order to further examine the atmospheric conditions, which lead to lightning occurrence, we calculated the average values of convective available potential energy (CAPE), which is associated with the intensity of updraughts within convection (Crook, 1996). This calculation was performed within the time frame −24 to +24 h centred at the cyclones maximum intensity (Figure 3(c)). The average CAPE at 0 h for the cyclones producing lightning is of the order of 700 J kg^{-1} in contrast to 200 J kg^{-1} for the cyclones with no lightning. This difference of approximately 500 J kg^{-1} is constantly persistent throughout the cyclones lifetime.

The cloud water content (liquid and ice) in Figure 3(a) tends to decrease in all pressure levels from 0 to +6 h. This suggests that convection tends to develop before cyclones reach their maximum intensity. To address this

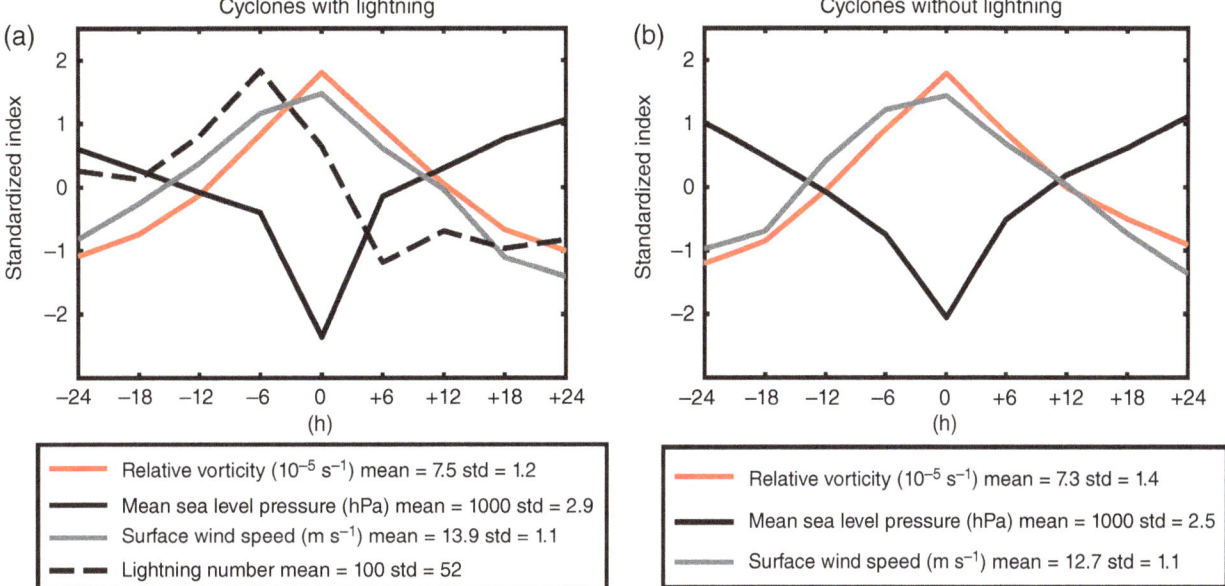

Figure 4. (a) Composite time series of maximum relative vorticity at 850 hPa, minimum MSLP, maximum 10-m wind speed and number of lightning for cyclones presenting lightning within 200 km from their centres centred at cyclones maximum intensity (0 h), normalized by standard deviation (0 h corresponds to the cyclones maximum relative vorticity). (b) As in (a) but for cyclones that do not present lightning. Time series are normalized with respect to their average.

issue, we took as a measure of cyclones intensity the maximum relative vorticity at 850 hPa, the maximum 10-m wind speed and the minimum mean sea level pressure (MSLP) within 200 km from each of the track points of each cyclone from the two cyclone groups. Then, the cyclones intensity composite evolution was calculated by averaging cyclones intensity at each time frame with respect to the cyclones mature stage (from −24 to +24 h). Figure 4 shows the composite time series results (time series are normalized with respect to their time average in order to ease comparison). In addition, Figure 4(a) shows the composite time series of lightning occurrences during the cyclones life time. It seems that there is indeed a lag of 6 h between cyclones maximum intensity (0 h) and cyclones maximum lightning occurrence. After −6 h, the cyclone-induced lightning decreases dramatically, reaching its minimum of activity 12 h later, at +6 h. Further analysis shows that for about 80% of the cyclones presenting lightning, their maximum lightning activity takes place from −24 to 0 h (Figure 5) and for about 60% of the cases lightning activity takes place from −12 to 0 h. At 0 h, when cyclones reach their mature stage, the differences between the intensity of the two cyclone groups are rather weak. In fact, a two sampled *t*-test has been performed and showed that maximum intensity averages between the two cyclone groups are not significantly different, regardless the season of cyclones occurrence and regardless if cyclones intensity is measured in MSLP, 10-m wind speed or relative vorticity. Therefore, our results suggest that convection (as inferred from the presence of lightning activity) may play a secondary role in cyclones reaching high intensities.

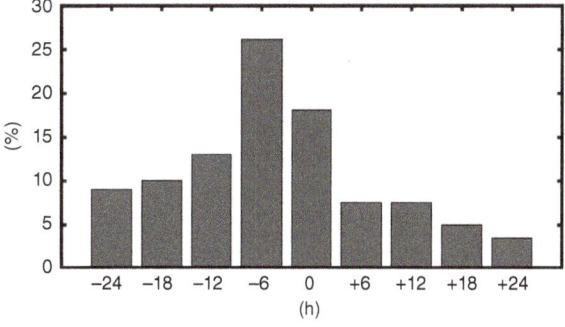

Figure 5. Frequency histogram of the lag between the time of maximum of lightning and cyclones mature stage, i.e. cyclones time of maximum of relative vorticity.

5. Summary and discussion

In this study, we first quantified the contribution of intense cyclones to lightning over the Mediterranean region. Then, we investigated the relation between cyclones that present lightning and cyclones where lightning was absent. For this reason, we used the ERAI reanalyses and cloud-to-ground lightning data from the ZEUS detection system, as well as a cyclone tracking algorithm, applied to the Mediterranean region from 2005 to 2014. Lightning has been attributed to each cyclone if detected within a radius of 200 km from the cyclone centre. To avoid any uncertainties on the number of observed lightning related to cyclones (e.g. underestimations of night-time impacts), results have been presented in a relative way. Our analysis showed that intense cyclones' contribution to lightning is higher in autumn and winter contributing to the total of lightning by 5–30% over the Mediterranean Sea, depending on the area and season. In certain areas, this percentage was significantly high, especially over North Africa where cyclones contribution to lightning may reach up to 60% of the winter total lightning occurrence. The same analysis has been repeated for a 500 km radius, increasing the overall cyclones contribution to lightning activity in the region. However, this did not affect our results on the main areas where cyclones produce lightning, as presented in Figure 2.

In this study, the intense cyclones that were associated with lightning constitute 36% of the total of 584 tracked cyclones. It was also shown that the cyclones associated with lightning tend to have approximately 35% more cloud ice and 15% more liquid cloud water content in the layer 500–700 hPa and are associated with approximately three times greater values of CAPE. Also, the peak of lightning occurrence was found to precede by 6 h in average the time of cyclones maximum intensity. Both cyclone groups presented similar seasonal cycles and areas of occurrence but most importantly they were shown to present similar intensities, suggesting that deep convection may not consist of a necessary condition for the occurrence of intense cyclogenesis in the Mediterranean.

To investigate the role of deep convection in intense cyclones development, we compared the MSLP deepening rates of cyclones presenting strong lightning activity before reaching their mature stage, to the cyclones that either presented their maximum of lightning activity after their mature stage or to the cyclones that did not present any lightning activity at all. However, no significant differences were found. Fink *et al.* (2012) showed that convection contribution to MSLP is mainly related through the column integrated virtual temperature. The vertically integrated temperature may however be connected to cyclones intensification through both advection (baroclinic forcing) and deep convection (diabatic forcing). In several case studies of strong extratropical storms, the authors showed that one of the two mechanisms can largely dominate cyclones intensification. In future studies, we will address the question of the (thermo-)dynamics associated with the two cyclone groups intensities in order to delineate the contribution of deep convection and baroclinicity into the formation of intense cyclones in the region.

Acknowledgements

This article has been funded by the European Union (European Social Fund) under the 'ARISTEIA-II' action of the Operational Programme 'Education and Lifelong Learning' in Greece, Project TALOS-3449. This work also contributes to the HyMeX programme (HYdrological cycle in the Mediterranean Experiment).

References

Alpert P, Ziv B. 1989. The Sharav Cyclone: observations and some theoretical considerations. *Journal of Geophysical Research* **94**: 18495–18514.

Altaratz O, Levin Z, Yair And Y, Ziv B. 2003. Lightning activity over land and sea on the eastern coast of the Mediterranean. *Monthly Weather Review* **131**(9): 2060–2070.

Avila EE, Aguirre Varela GG, Caranti GM. 1995. Temperature dependence of static charging in ice growing by riming. *Journal of Atmospheric Science* **52**: 4515–4522.

Ben Ami Y, Altaratz O, Yair Y, Koren I. 2015. Lightning characteristics over the eastern coast of the Mediterranean during different synoptic systems. *Natural Hazards and Earth System Sciences* **15**(11): 2449–2459.

Campins J, Genovés A, Picornell MA, Jansà A. 2011. Climatology of Mediterranean cyclones using the ERA-40 dataset. *International Journal of Climatology* **31**: 1596–1614.

Crook NA. 1996. Sensitivity of moist convection forced by boundary layer processes to low-level thermodynamic fields. *Monthly Weather Review* **124**: 1767–1785.

Dee DP, Uppala S. 2009. Variational bias correction of satellite radiance data in the ERA-Interim reanalysis. *Quarterly Journal of the Royal Meteorological Society* **135**: 1830–1841.

Defer E, Lagouvardos K, Kotroni V. 2005. Lightning activity in eastern Mediterranean region. *Journal of Geophysical Research, [Atmospheres]* **110**: D24210.

Feudale L, Manzato A. 2014. Cloud-to-ground lightning distribution and its relationship with orography and anthropogenic emissions in the Po Valley. *Journal of Applied Meteorology and Climatology* **53**(12): 2651–2670.

Fink AH, Pohle S, Pinto JG, Knippertz P. 2012. Diagnosing the influence of diabatic processes on the explosive deepening of extratropical cyclones. *Geophysical Research Letters* **39**: L07803.

Flaounas E, Drobinski P, Bastin S. 2013. Dynamical downscaling of IPSL-CM5 CMIP5 historical simulations over the Mediterranean: benefits on the representation of regional surface winds and cyclogenesis. *Climate Dynamics* **40**: 2497–2513.

Flaounas E, Kotroni V, Lagouvardos K, Flaounas I. 2014. CycloTRACK (v1.0) – tracking winter extratropical cyclones based on relative vorticity: sensitivity to data filtering and other relevant parameters. *Geoscientific Model Development* **7**: 1841–1853.

Flaounas E, Kotroni V, Lagouvardos K, Kazadzis S, Gkikas A, Hatzianastassiou N. 2015a. Cyclone contribution to dust transport over the Mediterranean region. *Atmospheric Science Letters* **16**: 473–478.

Flaounas E, Raveh-Rubin S, Wernli H, Drobinski P, Bastin S. 2015b. The dynamical structure of intense Mediterranean cyclones. *Climate Dynamics* **44**: 2411–2427.

Flaounas E, Lagouvardos K, Kotroni V, Claud C, Delanoë J, Flamant C, Madonna E, Wernli H. 2016. Processes leading to heavy precipitation associated with two Mediterranean cyclones observed during the HyMeX SOP1. *Quarterly Journal of the Royal Meteorological Society* **142**: 275–286, doi: 10.1002/qj.2618.

Galanaki E, Kotroni V, Lagouvardos K, Argiriou A. 2015. A ten-year analysis of cloud-to-ground lightning activity over the Eastern Mediterranean region. *Atmospheric Research* **166**: 213–222.

Kotroni V, Lagouvardos K. 2008. Lightning occurrence in relation with elevation, terrain slope and vegetation cover over the Mediterranean. *Journal of Geophysical Research, [Atmospheres]* **113**: D21118.

Kotroni V, Lagouvardos K. 2016. Lightning in the Mediterranean and its relation with sea-surface temperature. *Environmental Research Letters* **11**: 034006.

Lagouvardos K, Kotroni V. 2007. TRMM and lightning observations of a low-pressure system over the Eastern Mediterranean. *Bulletin of the American Meteorological Society* **88**(9): 1363–1367.

Lagouvardos K, Kotroni V, Betz HD, Schmidt K. 2009. A comparison of lightning data provided by ZEUS and LINET networks over Western Europe. *Natural Hazards and Earth System Sciences* **9**(5): 1713–1717.

McTaggart-Cowan R, Galarneau TJ Jr, Bosart LF, Milbrandt JA. 2010. Development and tropical transition of an Alpine Lee Cyclone. Part I: case analysis and evaluation of numerical guidance. *Monthly Weather Review* **138**: 2281–2307.

Miglietta M, Laviola S, Malvaldi A, Conte D, Levizzani V, Price C. 2013. Analysis of tropical-like cyclones over the Mediterranean Sea through a combined modeling and satellite approach. *Geophysical Research Letters* **40**: 2400–2405.

Mills B, Unrau D, Parkinson C, Jones B, Yessis J, Spring K, Pentelow L. 2008. Assessment of lightning related fatality and injury risk in Canada. *Natural Hazards* **47**(2): 157–183.

Papagiannaki K, Lagouvardos K, Kotroni V. 2013. A database of high impact weather related incidents in Greece: a descriptive impact analysis for the period 2001–2011. *Natural Hazards and Earth System Sciences* **13**: 727–736.

Petersen WA, Christian HJ, Rutledge SA. 2005. TRMM observations of the global relationship between ice water content and lightning. *Geophysical Research Letters* **32**: L14819.

Price C, Asfur M, Yair Y. 2009. Maximum hurricane intensity preceded by increase in lightning frequency. *Nature Geoscience* **2**: 329–332.

Proestakis E, Kazadzis S, Lagouvardos K, Kotroni V, Kazantzidis A. 2016. Lightning activity and aerosols in the Mediterranean region. *Atmospheric Research* **170**: 66–75.

Trigo IF, Trevor DD, Grant RB. 1999. Objective climatology of cyclones in the Mediterranean region. *Journal of Climate* **12**: 1685–1696.

Tuomi T, Mäkelä A. 2008. Thunderstorm climate of Finland 1998–2007. *Geophysica* **44**(12): 6780.

Whittaker I, Douma E, Rodger C, Marshall T. 2015. A quantitative examination of lightning as a predictor of peak winds in tropical cyclones. *Journal of Geophysical Research, [Atmospheres]* **120**: 3789–3801.

4

Influences of El Niño on aerosol concentrations over Eastern China

Juan Feng,[1,2] Jianlei Zhu[3,*] and Yan Li[4]

[1] College of Global Change and Earth System Science (GCESS), Beijing Normal University, China
[2] Joint Center for Global Change Studies, Beijing, China
[3] State Key Laboratory of Numerical Modeling for Atmospheric Sciences and Geophysical Fluid Dynamics, Institute of Atmospheric Physics, Chinese Academy of Sciences, Beijing, China
[4] Key Laboratory of Semi-Arid Climate Change of Ministry of Education, College of Atmospheric Sciences, Lanzhou University, China

*Correspondence to:
J. Zhu, State Key Laboratory of Numerical Modeling for Atmospheric Sciences and Geophysical Fluid Dynamics, Institute of Atmospheric Physics, Chinese Academy of Sciences, No. 40, HuaYanLi, ChaoYang District, Beijing 100029, China.
E-mail: zhujl@mail.iap.ac.cn

Abstract

The influence of El Niño on aerosol concentrations is investigated for eastern China over period 1986–2006. Result suggests that the influence of El Niño on aerosol concentration differs between events. The 1987/1988 El Niño event decreased the aerosol concentrations during the mature and decay spring of the event; however, an anomalous northern-increase and southern-decrease dipole structure occurred during the decay summertime over eastern China. In contrast, the event of 1997/1998 was associated with an evident increase in aerosol concentration over its lifespan. These anomalous aerosol concentrations are mainly caused by circulation changes associated with El Niño events.

Keywords: El Niño; aerosol concentrations; eastern China

1. Introduction

Atmospheric aerosols are major air pollutants and have a considerable influence on visibility (Watson, 2002), global climate change (IPCC, 2007), and human health (Dockery *et al.*, 1993; Pope *et al.*, 1995; Schwartz *et al.*, 1996; Bernard *et al.*, 2001). Previous studies have reported that aerosol particles may affect the radiation equilibrium of the earth–atmosphere system by absorbing or scattering solar radiation (Thompson, 1995), and are able to change the physical and microphysical characteristics of clouds, including their optical properties and precipitation rates (e.g. Hansen *et al.*, 1997; Zhang *et al.*, 2007). However, the concentration and distribution of aerosols are also sensitive to meteorological factors, such as temperature (Aw and Kleeman, 2003; Dawson *et al.*, 2007), humidity (Tai *et al.*, 2010), wind (Zhang *et al.*, 2010), and atmospheric boundary layer thickness (Kleeman, 2008) such that large-scale climatic events are able to significant impact aerosol concentrations. A recent study by Zhu *et al.* (2012) reported that the decadal-scale weakening of the East Asian summer monsoon (EASM) resulted in increased aerosol concentrations over northern China, thus the role of circulation in determining background aerosol concentrations cannot be ignored.

The El Niño-Southern Oscillation (ENSO) has a significant impact on seasonal climate around the world. This phenomenon, which develops in the Pacific, generates significant anomalous patterns in regional and global climates (e.g. Harrison and Larkin, 1998; Trenberth and Caron, 2000). Previous studies have shown that, following the decay stage of an El Niño event, the following summer is characterized by positive rainfall anomalies over the Yangtze River valley and Huai-he river valley, and negative anomalies over southern and northern China (e.g. Huang and Wu, 1989; Zhang *et al.*, 1999; Feng and Hu, 2004). This relationship can be demonstrated statistically; however, the influence of El Niño events on the climate over China varies, in that the effects of a strong El Niño event differ to those of moderate events in terms of the patterns of both circulation and rainfall anomalies (Xue and Liu, 2007).

Previously published results indicate that El Niño has a strong influence on the climate over eastern China because China currently experiences a relatively high aerosol loading, particularly in eastern China (Donkelaar *et al.*, 2006; Zhang *et al.*, 2007; Tie and Cao, 2009), and as mentioned above, background circulation plays an important role in determining the distribution of aerosol concentrations. It is therefore important to examine the potential influence of El Niño events on the aerosol concentration over eastern China. Accordingly, the aims of this study are to: explore possible impacts of El Niño on seasonal aerosol concentrations over eastern China, reveal the possible physical processes involved, determine the relative roles of circulation and rainfall anomalies associated with El Niño on aerosol concentrations, and discuss the differences between the influence of strong and moderate El Niño events on aerosol concentrations over eastern China. The remainder of this manuscript is organized as follows: Section 2 describes the model and data sets used in this study; Section 3 discusses the influences of El Niño events on aerosol concentrations over eastern China, as well as the physical processes involved; and a discussion and the conclusions are provided in Section 4.

2. Model and data sets

The global three-dimensional Goddard Earth Observing System chemical-transport model (GEOS-Chem) of tropospheric chemistry (Bey *et al.*, 2001), version 8.02.01, was used to simulate aerosol distribution. The model was driven by assimilated meteorological observations from the GEOS, which was developed by the NASA Global Modeling and Assimilation Office. All the simulations in this study were driven by GEOS-4 meteorological fields on a horizontal resolution of $2° \times 2.5°$ (latitude $\times$ longitude). A detailed description of the treatment of the coupled tropospheric ozone-NO_x-hydrocarbon chemistry, aerosols, and their precursors has been illustrated previously in the literature (e.g. Bey *et al.*, 2001; Park *et al.*, 2003, 2004; Liao *et al.*, 2007). The wet and dry deposition of aerosols in the model follows Liu *et al.* (2001) and Wesely (1989), respectively. Aerosol concentrations of $PM_{2.5}$ were defined as in Malm *et al.* (1994) and were calculated as follows: $\left[PM_{2.5}\right] = 1.37 \times \left[SO_4^{2-}\right] + 1.29 \times \left[NO_3^-\right] + [BC] + [POA] + [SOA]$.

SO_4^{2-}, NO_3^-, BC, POA, and SOA are for the pollutants of sulphate, nitrate, black carbon, primary organic aerosol, and second organic aerosol, respectively. As sea salt and mineral dust aerosol are not considered in this equation, there is an inherent underestimate of $PM_{2.5}$. Global emissions of ozone precursors, aerosol precursors, and aerosol in the GEOS-Chem model are the same as those in Zhu *et al.* (2012).

The timeframe simulated in this study was 1986–2006, which corresponds to the period that the GEOS-4 data sets are available. Here, the global emissions of ozone precursors, aerosol precursors, and aerosols in the GEOS-Chem model generally follow Park *et al.* (2003, 2004, 2006), with the exception of anthropogenic emissions in the southeast Asian domain, which were updated based on the emission scenario described in Streets *et al.* (2003, 2006) and Zhang *et al.* (2007). That is the anthropogenic emissions and biomass-burning emissions are fixed in the simulation, so that any variation of aerosol concentration observed in this study was caused by variations in meteorological conditions, thus reflecting the impact of climatic events on aerosol concentrations. The GEOS-Chem is a widely used and popular atmospheric chemistry model, and many studies have reported that this model can well capture the seasonal and interannual variations of pollutants aerosol over China (e.g. Zhang *et al.*, 2010; Lou *et al.*, 2014; Yang *et al.*, 2014; Feng *et al.*,

2016). The above discussions provide confidence for employing the GEOS-Chem to explore the influences of El Niño events on aerosol concentrations over China.

El Niño events were identified as normalized 3-month running mean Niño3 index exceedances of 0.5 °C that persisted for 8 months. The skin temperature (i.e. it is surface air temperature on land, and sea surface temperature over ocean) was used to calculate the Niño3 index because sea surface temperature is not present in the GEOS meteorological data set. Note that the Niño3 index used in this study is highly correlated with the Niño3 index based on the UK Meteorological Office Hadley Centre's sea ice and SST data sets (Rayner *et al.*, 2003), with a correlation coefficient of 0.99, significant at the 0.01 level. Based on this method, two major El Niño events (i.e. 1987/1988 and 1997/1998) were identified. Note that two equivalent weak EASMs were observed in the summers of 1988 and 1998, with EASM index values of -1.9 and -2.2, respectively (Li and Zeng, 2002; http://ljp.gcess.cn/dct/page/65577), corresponding to the decay phase of the two major El Niño events. As pointed out that the EASM has considerable influences on the aerosol concentrations over eastern China (Zhang *et al.*, 2010; Zhu *et al.*, 2012). However, it is reported the influences of EASM on aerosol concentrations over eastern China depends on its strength. The two El Niño events here associated with two comparable EASM in intensity. Consequently, the influence of the EASM on aerosol concentrations during the summers of 1988 and 1998 would not contaminate the results in this study.

3. Results

Figure 1 shows the anomalies in aerosol column concentrations that were simulated for the two El Niño events as they departed from the climatological mean. Coincident with the 1987/1988 El Niño event, negative aerosol concentration anomalies can be seen over eastern China during the mature phase, with a maximum around the region of the Yangtze and Yellow rivers (Figure 1(a)). A similar pattern of anomalies can be observed for the springtime decay phase; however, the concentrated region around the Yangtze and Yellow rivers is no longer present, and positive anomalies appear in the southern margin areas (Figure 1(b)). A different scenario is depicted for the summertime decay phase. In this case, an apparent dipole structure is visible over eastern China, with positive concentration anomalies over north China, and decreased anomalies over south China (Figure 1(c)). In contrast, simulated increased aerosol concentrations were consistently observed from the mature to the decay phases in summer for the El Niño event of 1997/1998 (Figure 1(d)–(f)). This result is consistent with the surface layer distribution of $PM_{2.5}$ (figures not shown).

These features of the two El Niño events are also apparent in Figure 2, which shows how the anomalies along the zonal mean within eastern China

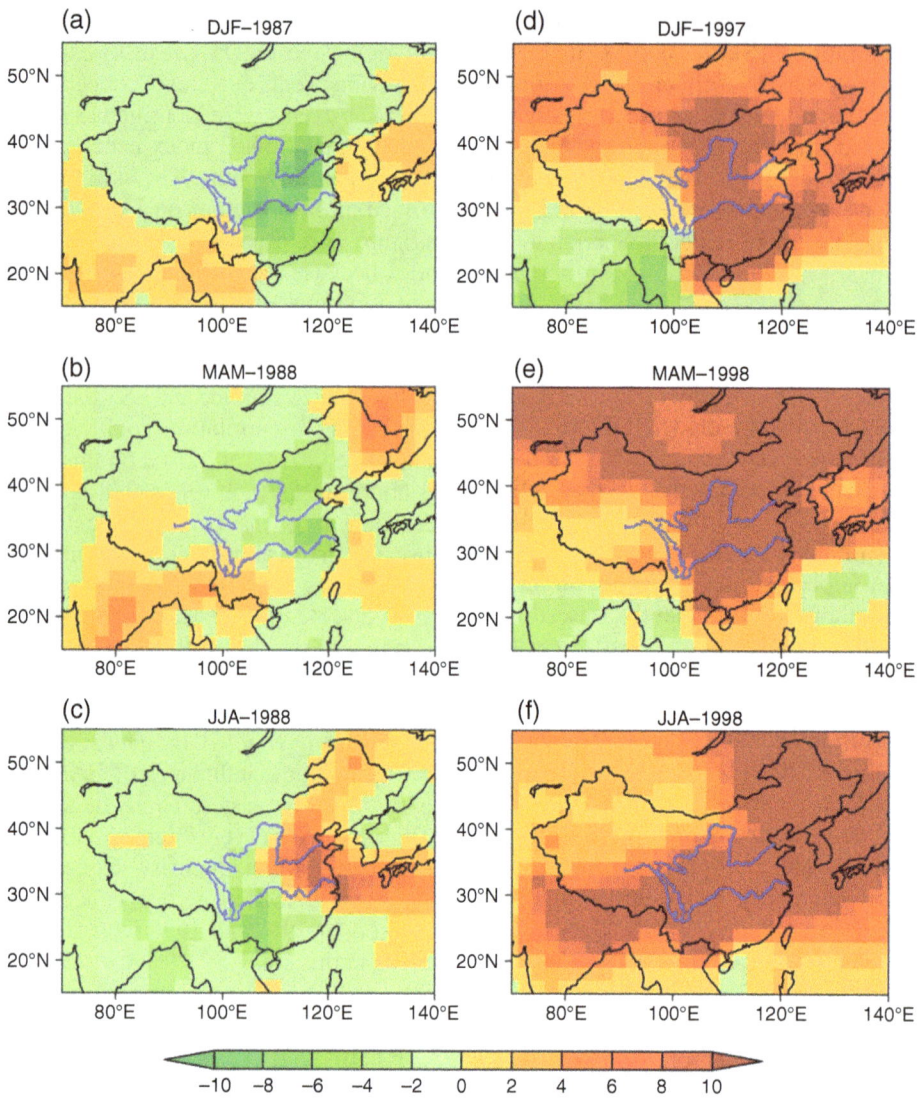

Figure 1. The spatial distribution of simulated column burdens of PM$_{2.5}$ anomalies (mg · m^{-2}) during El Niño events (a–c) 1987/1988 and (d–f) 1997/1998 with respect to the climatological mean averaged from 1986 to 2006.

(105°–120°E) vary with pressure and latitude. From the mature to the decay phase of the 1987/1988 El Niño event, aerosol concentrations over south China (south of 30°N) show a uniform decrease in variation, and increased in the summertime decay phase over north China (north of 30°N). Note that larger values are located mainly in the lower troposphere for both the mature and decay phases (Figure 2(a)–(c)). However, there are clearly homogeneous positive anomalies during the 1997/1998 El Niño event from the mature phase to the decay phase, with a maximum centred around 30°N (Figure 2(d)–(f)). It is worthy of note that the high values in the later El Niño event extend much further than in the 1987/1988 event, and reach the middle levels of the troposphere. Moreover, the anomalies associated with the later El Niño event are much larger than those associated with the 1987/1988 event.

These results indicate that the aerosol concentration anomalies associated with the two El Niño events are different over eastern China. To ascertain

why this phenomenon gives rise to two seemingly disparate effects, the underlying temperature and circulation anomalies associated with the El Niño events must be examined. Based on Figure 3(a) and (d), skin temperature anomalies over the eastern tropical Pacific appear much smaller during the 1987/1988 event during its mature phase and are located further west. As a result of the different warming patterns, different Rossby wave response patterns are created, such that eastern China is influenced by the anomalous cyclonic circulation during the 1987/1988 event, with anomalous northerlies prevailing over most of eastern China (Figure 4(d)). These anomalous northerly winds indicate the climatological northerlies is strengthened (Figure 4(a)), creating comparatively favourable conditions for aerosols transmitted during this period, associating with decreased aerosol concentrations. In contrast, anticyclonic circulation anomalies over the coastal regions occur during the 1997/1998 event, corresponding to anomalous southwesterlies prevailing in eastern China (Figure 4(g)), implying weakened

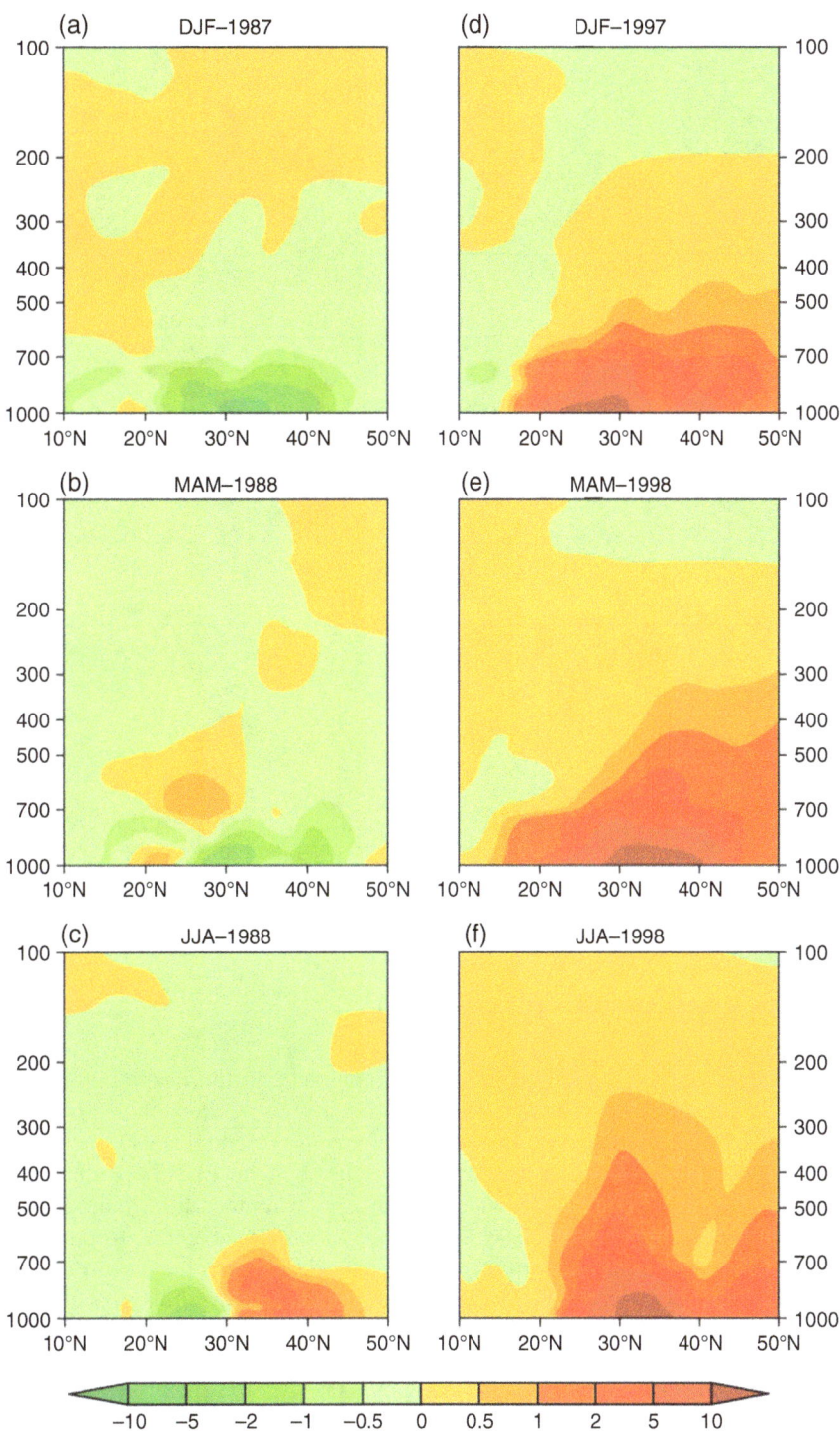

Figure 2. As in Figure 1, but for the pressure-latitude across section of PM$_{2.5}$ concentrations anomalies averaged over longitude range of 105°–120°E ($\mu g \cdot m^{-3}$).

northerlies (Figure 4(a)), which would worsen the meteorological conditions for air quality in the region.

For the springtime decay phase, the 1987/1988 and 1997/1998 events are associated with negative and positive temperature anomalies, respectively, over the eastern tropical Pacific (Figure 3(b) and (e)), and so would induce different circulation anomalies over eastern China because of the different Rossby wave responses, as described by Wang *et al.* (2000) and Feng

and Li (2011). These findings suggest that most of eastern China is controlled by anomalous northerlies in the 1987/1988 event (Figure 4(e)), which intensify northerlies over north China, improving conditions following aerosol emission. However, the anomalous weak northerlies in south China refers to weakened southerlies (Figure 4(b)), creating unfavourable conditions for aerosol dispersion. This result supports the aerosol concentration distributions displayed in

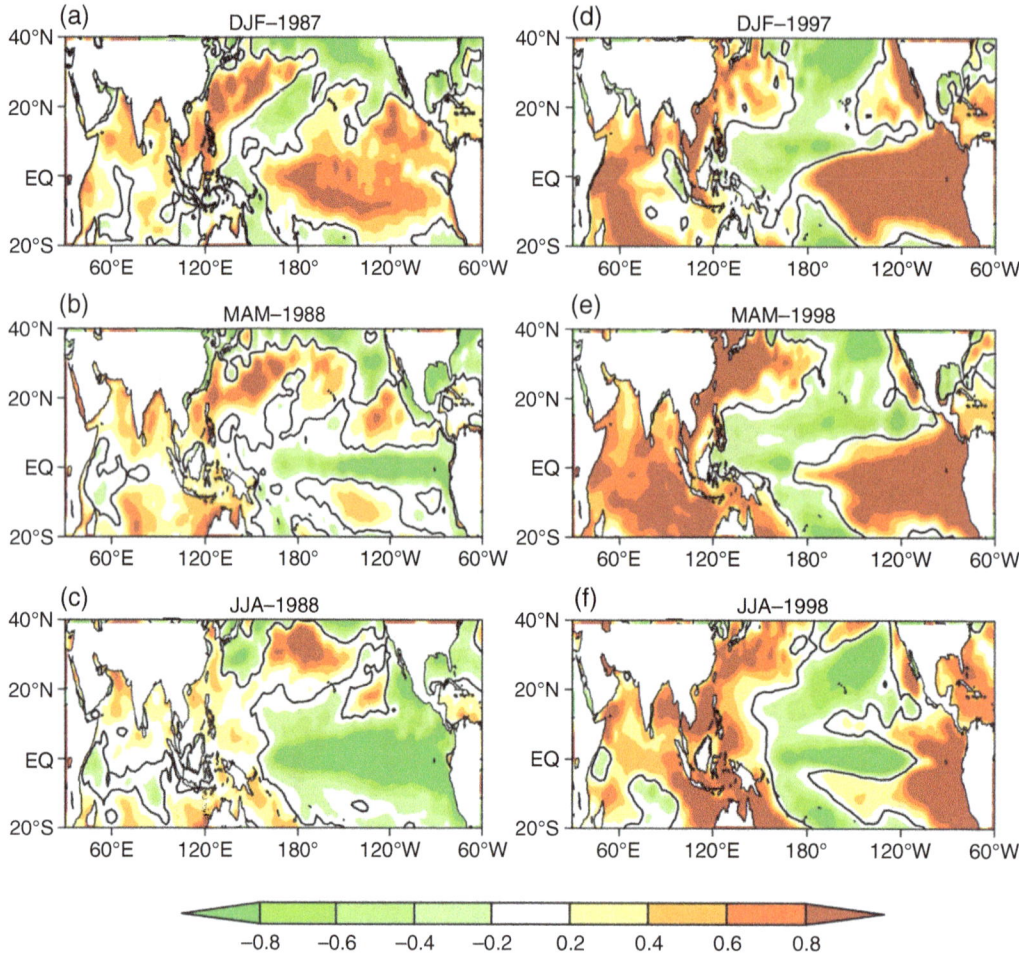

Figure 3. As in Figure 1, but for the skin temperature anomalies (°C).

Figures 1(b) and 2(b), where the focal point in concentration becomes weak, and part of the southwest margin experiences positive aerosol concentration anomalies. Conversely, this region is controlled by anomalous southerlies in the 1997/1998 event (Figure 4(h)), which weaken the northerlies in north China but intensify the southerlies in southern China, the combined effect of which pushes the climatological convergence north of 30°N. This leads to the formation of a strengthened convergence, which can be seen in the distribution of the vertical transport flux (figures not shown), inducing positive aerosol concentration anomalies. This result supports the increased aerosol concentrations north of 30°N in Figure 2(e).

During the summertime decay phase, negative temperature anomalies over the eastern tropical Pacific occur during the 1987/1988 event, while weak positive temperature anomalies arise around the tropical Indo-Pacific Basin (Figure 3(c)). In response to this underlying temperature structure, south China is controlled by an anomalous anticyclonic circulation, with anomalous southerlies (Figure 4(f)), indicating intensified southerlies, consistent with a decrease in aerosol concentration. Elsewhere, an anomalous cyclonic circulation lies to the eastern coastal region of China; while correspondingly, north China is controlled by

anomalous northerlies (Figure 4(f)). This effect parallels to weaken the climatological southerlies (Figure 4(c)), and lead to increased aerosol concentrations. For the El Niño event of 1997/1998, similar warm skin temperature anomalies are present in the tropical eastern Pacific, tropical Indo-Pacific Basin, and northwest Pacific (Figure 3(f)). Under these circumstances, south China is controlled by an intensified westward subtropical high, corresponding to strengthened sinking flows, and north China is controlled by anomalous northerlies, weakening the climatological southerlies in the region (Figure 4(c)). Combined with the anomalous southerlies over south China, an anomalous convergence is formed to the north of 30°N (Figure 4(i)). These factors all contribute to the increased aerosol concentrations in the summertime decay phase of the 1997/1998 event (Figures 1(f) and 2(f)). Besides, the vertical integral of aerosol concentrations transport fluxes at the four boundaries over eastern China are further examined (Figure S1). We found that the decreased aerosol concentrations during mature winter of 1987/1988 mainly attributed to the meridional transport. In addition, the anomalies in the decaying spring and summer largely due to the zonal transport. For the event 1997/1998, the anomalous high aerosol concentrations in both the mature winter and

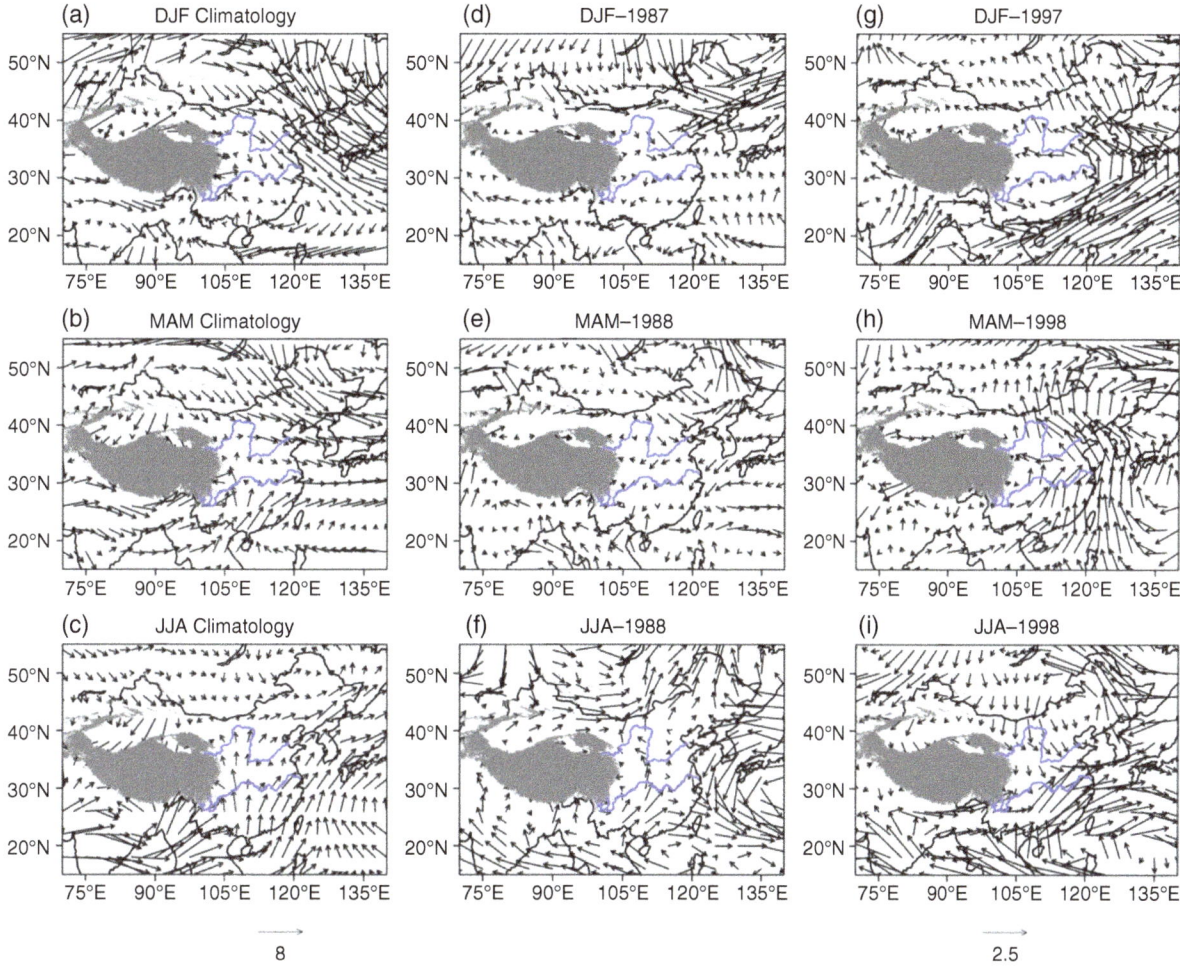

Figure 4. (a–c) The horizontal distribution of climatological mean wind at 850 hPa, (d–f) and (g–i) are the wind anomalies during El Niño events 1987/1988 and 1997/1998 at 850 hPa, respectively (m · s⁻¹).

decay spring and autumn are mainly due to the zonal transport.

Changes in wet deposit associated with El Niño events may also have some effect on aerosol concentration. Figure 5 shows the simulated wet deposit anomalies for the two El Niño events. Positive anomalies are clearly seen over southern China during the winter of 1987, and the spring of 1988, and negative anomalies are found in the decay summer of 1988 over southern China. The positive anomalies of wet deposit are favourable for decreased aerosol concentrations. Thus, the anomalies of wet deposit during mature winter and decay spring is consistent with the anomalies of aerosol concentrations over southern China, however, not for the decaying summer. Meanwhile, negative anomalies are observed in the mature winter, decay spring, and summer during event 1997/1998. This situation agrees with the anomalies of aerosol concentrations during event 1997/1998. This point implies wet deposit plays certain role in influencing the aerosol concentrations. However, the influence of wet deposit varies in events and seasons; this point is consistent with that of Wu (2014), indicating the role of circulation plays a major role in impacting the aerosol concentrations in eastern China.

In addition, we further compared the spatial rainfall anomalies between the *in situ* observations and GEOS-Chem input during the two El Niño events. The rainfall anomalies driving model show similar features to the observations (figures not shown). This result indicates it is not due to the poor quality of rainfall inducing limited role of wet deposit on aerosol concentrations.

4. Discussion and summary

The results from this study indicate that El Niño events (i.e. 1987/1988 and 1997/1998) play an important role in determining aerosol concentrations over eastern China during both the mature and decay phases, and that these influences are distinguishable from each other. For the 1987/1988 El Niño event, an evident decrease in aerosol concentrations was simulated over eastern China for both the wintertime mature phase and the springtime decay phase that followed, whereas an anomalous northern-increasing and southern-decreasing dipole pattern was found to be present during the summertime decay phase. In

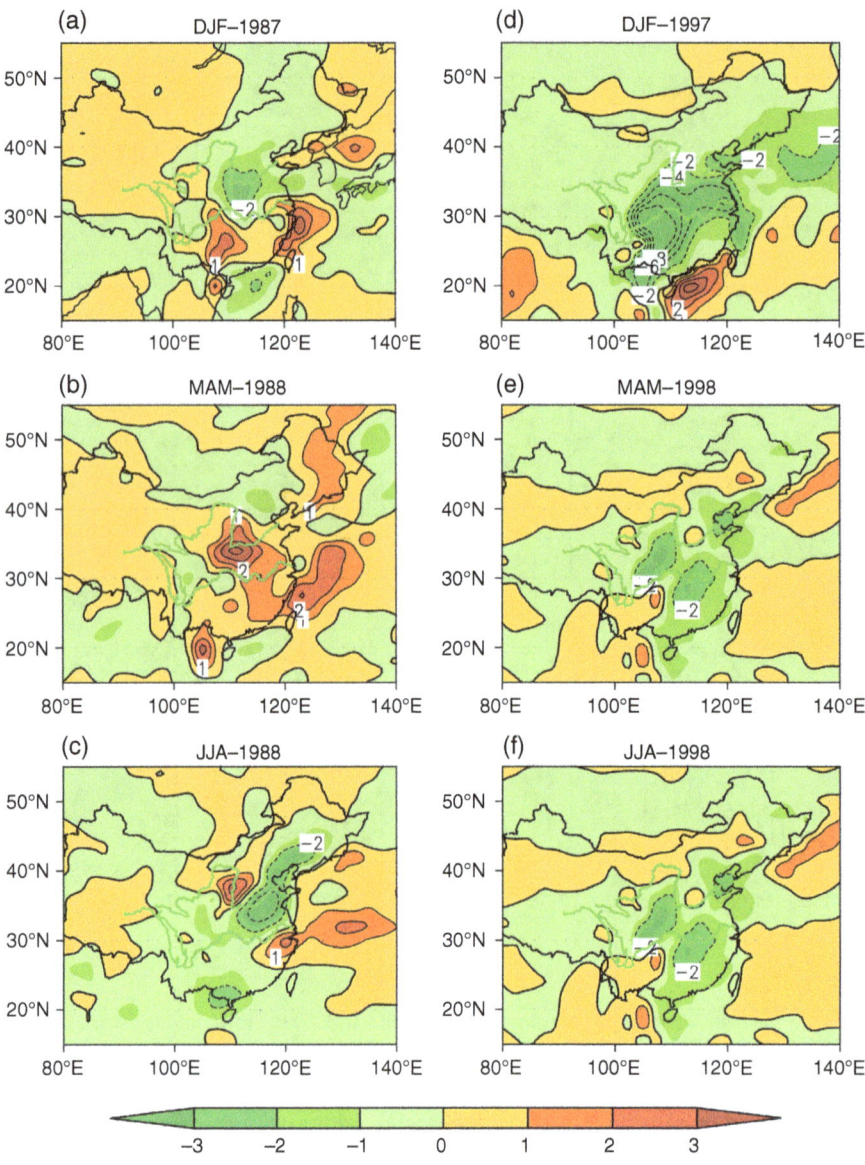

Figure 5. The simulated vertical integral wet deposit flux anomalies during El Niño events (a–c) 1987/1988 and (d–f) 1997/1998 $(kg \cdot s^{-1})$.

contrast, the model simulated increased aerosol concentrations over the 1997/1998 El Niño event cycle, with much larger anomalous values than those associated with the 1987/1988 event. Although it has been reported that each El Niño event may not be identical (Weng *et al.*, 2007), our results show that the impacts of the two El Niño events are nearly opposite. In addition, we have further examined the impacts of El Niño on aerosol concentrations over eastern China by calculating the correlation coefficients between the Niño3 index and areal averaged $PM_{2.5}$ over eastern China. The result shows that the Niño3 index is positively correlated with the aerosol concentrations over north China (30°–40°N, 105°–120°E) during its lifespan, however, only significant during the mature phase over south China (20°–30°N, 105°–120°E). This result agrees well with the event 1997/1998, indicating the strong El Niño event could associate with enhanced aerosol concentrations over north China during its

mature and decay phases, however, not for the event 1987/1988 during its mature phase. Thus, the result here further reveals the complexity of El Niño events, as well as their impact on the regional climate. Moreover, whether this relation is determined by extreme events considering that the extreme event could induce a seeming significant relationship (Feng *et al.*, 2010). Further work based on additional El Niño events will be required if we are to better understand the impact of El Niño events on the aerosol concentration over China. Moreover, we found that the changes in circulation associated with El Niño events play a dominant role in determining the anomalies connected with aerosol concentrations, whereas the role of rainfall appears to be limited.

It is known that some aerosol particles are sensitive to variation in temperature; i.e. low temperatures slow SO_2 oxidation kinetics (Tai *et al.*, 2012), and are not conducive to the volatilization of nitrate and ammonium aerosols leading to their accumulation (Dawson *et al.*, 2007), while decreasing the number of organic carbon (OC) particles (Tai *et al.*, 2012). Furthermore, it has been reported that El Niño events could lead to temperature anomalies over eastern China. As the changes in temperature associated with El Niño events are not uniform over eastern China during an El Niño cycle, and as the chemical processes associated with temperature changes display offset reactions, sensitivity experiments were not performed during this study. However, the findings of Zhu *et al.* (2012) may provide some insight into this issue as they showed that variation in temperature anomalies with magnitudes of approximately $+1\,K$ would induce a reduction of $0.63\,\mu g\,m^{-3}$ (*ca* 1% of the climatological mean) in aerosol concentrations over south China during the summer; i.e. a negligible change when compared with the effect of circulation (up to $10\,\mu g\,m^{-3}$ in Figure 1(f)). This point further supports the conclusion that it is circulation anomalies associated with El Niño events that play the dominant role in influencing the aerosol concentrations.

Finally, it has been reported that America is also significantly influenced by El Niño, and that aerosol concentrations in America peak during the summer (Zhang *et al.*, 2010), corresponding to the decay phase of an ENSO event. Therefore, it is of interest to examine the possible influences of ENSO on aerosol concentrations over this phase as well as during the total event cycle. A comparison of the differences between ENSO-influenced aerosol concentrations in America and China would assist not only the understanding of aerosol concentrations in both countries, but also a better understanding of the dynamics of ENSO in general.

Acknowledgements

We thank two anonymous referees, whose comments improved the article. This work was jointly supported by 'the Fundamental Research Funds for the Central Universities' (2015KJJCB07), National Natural Science Foundation of China (41475076), and 973 Programme (2013CB430200). We thank Harvard Atmospheric Chemistry Modeling Group for providing the GEOS-Chem model.

Supporting information

The following supporting information is available:

Figure S1 The anomalies in vertical integral aerosol concentration fluxes at the (a and b) south, (c and d) north, (e and f) west, and (g and h) east boundaries of the eastern China from 1000 to 100 hPa during El Niño event (a, c, e, and g) 1987/1988 and (b, d, f, and h) 1997/1998.

References

Aw J, Kleeman MJ. 2003. Evaluating the first-order effect of intra-annual temperature variability on urban air pollution. *Journal of Geophysical Research* **108**(D12): 4365, doi: 10.1029/2002jd002688.

Bernard SM, Samet JM, Grambsch A, Ebi KL, Romieu I. 2001. The potential impacts of climate variability and change on air pollution-related health effects in the United States. *Environmental Health Perspectives* **109**: 199–209.

Bey I, Jacob DJ, Yantosca RM, Logan JA, Field BD, Fiore AM, Li QB, Liu HY, Mickey LJ, Schultz MG. 2001. Global modeling of tropospheric chemistry with assimilated meteorology: model description and evaluation. *Journal of Geophysical Research* **106**: 23073–23095.

Dawson JP, Adams PJ, Pandis SN. 2007. Sensitivity of $PM_{2.5}$ to climate in the eastern US: a modeling case study. *Atmospheric Chemistry and Physics* **7**(16): 4295–4309, doi: 10.5194/acp-7-4295-2007.

Dockery DW, Pope CA, Xu XP, Spengler JD, Ware JH, Fay ME, Ferris BG Jr, Speizer FE. 1993. An association between air pollution and mortality in six US cities. *New England Journal of Medicine* **329**: 1753–1759.

Donkelaar VA, Martin R, Park R. 2006. Estimating ground-level $PM_{2.5}$ with aerosol optical depth determined from satellite remote sensing. *Journal of Geophysical Research* **111**: D21201, doi: 10.1029/2005JD006996.

Feng S, Hu Q. 2004. Variations in the teleconnection of ENSO and summer rainfall in northern China: a role of the Indian summer monsoon. *Journal of Climate* **17**: 4871–4881.

Feng J, Li JP. 2011. Influence of El Niño Modoki on spring rainfall over south China. *Journal of Geophysical Research* **116**: D13102, doi: 10.1029/2010JD015160.

Feng J, Li JP, Li Y. 2010. Is there a relationship between the SAM and southwest Western Australian winter rainfall? *Journal of Climate* **23**: 6082–6089.

Feng J, Li JP, Zhu JL, Liao H. 2016. Influences of El Niño Modoki event 1994/1995 on aerosol concentrations over southern China. *Journal of Geophysical Research* **121**: 1637–1651, doi: 10.1002/2015JD023659.

Hansen J, Sato M, Ruedy R. 1997. Radiative forcing and climate response. *Journal of Geophysical Research* **102**(D6): 6831–6864, doi: 10.1029/96jd03436.

Harrison DE, Larkin NK. 1998. Seasonal U.S. temperature and precipitation anomalies with El Niño: historical results and comparison with 1997–98. *Geophysical Research Letters* **25**: 3959–3962.

Huang RH, Wu YF. 1989. The influence of ENSO on the summer climate change in China and its mechanism. *Advances in Atmospheric Sciences* **6**(1): 21–32.

IPCC. 2007. *Climate Change 2007. The Physical Science Basis*. Cambridge University Press: Cambridge, UK.

Kleeman M. 2008. A preliminary assessment of the sensitivity of air quality in California to global change. *Climate Change* **87**: 273–292, doi: 10.1007/s10584-007-9351-3.

Li JP, Zeng QC. 2002. A unified monsoon index. *Geophysical Research Letters* **29**(8): 1274, doi: 10.1029/2001GL013874.

Liao H, Henze DK, Seinfeld JH, Wu SL, Mickey LJ. 2007. Biogenic secondary organic aerosol over the United States: comparison of climatological simulations with observations. *Journal of Geophysical Research* **112**: D06201, doi: 10.1029/2006jd007813.

Liu H, Jacob DJ, Bey I, Yantosca RM. 2001. Constraints from ^{210}Pb and ^{7}Be on wet deposition and transport in a global three-dimensional chemical tracer model driven by assimilated meteorological fields. *Journal of Geophysical Research* **106**: 12109–12128, doi: 10.1029/2000jd900839.

Lou SJ, Liao H, Zhu B. 2014. Impacts of aerosols on surface-layer ozone concentrations in China through heterogeneous reactions and changes in photolysis rates. *Atmospheric Environment* **85**: 123–138.

Malm WC, Sisler JF, Huffman D, Eldred RA, Cahill TA. 1994. Spatial and seasonal trends in particle concentration and optical extinction in the United States. *Journal of Geophysical Research* **99**: 1347–1370.

Park RJ, Jacob DJ, Chin M, Martin RV. 2003. Sources of carbonaceous aerosols over the United States and implications for natural visibility. *Journal of Geophysical Research* **108**: 4355, doi: 10.1029/2002jd003190.

Park RJ, Jacob DJ, Field BD, Yantosca RM, Chin M. 2004. Natural and transboundary pollution influences on sulfate-nitrate-ammonium aerosols in the United States: implications for policy. *Journal of Geophysical Research* **109**: D15204, doi: 10.1029/2003jd004473.

Park RJ, Jacob DJ, Kumar N, Yantosca RM. 2006. Regional visibility statistics in the United States: natural and transboundary pollution influences, and implications for the regional haze rule. *Atmospheric Environment* **40**: 5405–5423.

Pope C, Dockery D, Schwartz J. 1995. Review of epidemiological evidence of health effects of particulate air pollution. *Inhalation Toxicology* **7**: 1–18.

Rayner NA, Parker DE, Horton EB, Folland CK, Alexander LV, Rowell DP. 2003. Global analyses of sea surface temperature, sea ice, and night marine air temperature since the late nineteenth century. *Journal of Geophysical Research* **108**: 4407, doi: 10.1029/2002 JD002670.

Schwartz J, Dockery D, Neas L. 1996. Is daily mortality associated specifically with fine particles? *Journal of the Air & Waste Management Association* **46**: 927.

Streets DG, Bond TC, Carmichael GR, Fernandes SD, Fu Q, He D, Kilmont Z, Nelson SM, Tsai NY, Wang MQ, Woo JH, Yarber KF. 2003. An inventory of gaseous and primary aerosol emissions in Asia in the year 2000. *Journal of Geophysical Research* **108**: 8809, doi: 10.1029/2002jd003093.

Streets DG, Zhang Q, Wang LT, He KB, Hao JM, Wu Y, Tang YH, Carmichael GR. 2006. Revisiting China's CO emissions after the transport and chemical evolution over the Pacific (TRACE-P) mission: synthesis of inventories, atmospheric modeling, and observations. *Journal of Geophysical Research* **111**: D14306.

Tai AP, Mickley LJ, Jacob DJ. 2010. Correlations between fine particulate matter ($PM_{2.5}$) and meteorological variables in the United States: implications for the sensitivity of $PM_{2.5}$ to climate change. *Atmospheric Environment* **44**(32): 3976–3984, doi: 10.1016/j.atmosenv. 2010.06.060.

Tai AP, Mickley LJ, Jacob DJ, Leibensperger EM, Zhang L, Fisher JA, Pye HOT. 2012. Meteorological modes of variability for fine particulate matter ($PM_{2.5}$) air quality in the United States: implications for $PM_{2.5}$ sensitivity to climate change. *Atmospheric Chemistry and Physics* **12**: 3131–3145, doi: 10.5194/acp-12-3131-2012.

Thompson RD. 1995. The impact of atmospheric aerosols on global climate: a review. *Progress in Physical Geography* **19**(3): 336–350, doi: 10.1177/030913339501900303.

Tie X, Cao J. 2009. Aerosol pollution in China: present and future impact on environment. *Particuology* **7**(6): 426–431.

Trenberth KE, Caron JM. 2000. The Southern Oscillation revisited: sea level pressures, surface temperatures, and precipitation. *Journal of Climate* **13**: 4358–4365.

Wang B, Wu RG, Fu XH. 2000. Pacific–East Asian teleconnection: how does ENSO affect East Asian climate? *Journal of Climate* **13**: 1517–1536.

Watson J. 2002. Visibility: science and regulation. *Journal of the Air & Waste Management Association* **52**: 628–713.

Weng HY, Ashok K, Behera SK, Rao SA, Yamagata T. 2007. Impacts of recent El Niño Modoki on dry/wet conditions in the Pacific rim during boreal summer. *Climate Dynamics* **29**: 123–129.

Wesely ML. 1989. Parameterization of surface resistances to gaseous dry deposition in regional-scale numerical models. *Atmospheric Environment* **23**: 1293–1304.

Wu RG. 2014. Seasonal dependence of factors for year-to-year variations of South China aerosol optical depth and Hong Kong air quality. *International Journal of Climatology* **34**(11): 3204–3220.

Xue F, Liu CZ. 2007. The influence of moderate ENSO on summer rainfall in eastern China and its comparison with strong ENSO. *Chinese Science Bulletin* (in Chinese) **53**(5): 791–800.

Yang Y, Liao H, Li JP. 2014. Impacts of the East Asian summer monsoon on interannual variations of summertime surface-layer ozone concentrations over China. *Atmospheric Chemistry and Physics* **14**: 6867–6879.

Zhang RH, Sumi A, Kimoto M. 1999. A diagnostic study of the impact of El Niño on the precipitation in China. *Advances in Atmospheric Sciences* **16**(2): 229–241.

Zhang Q, Streets DG, He RB, Wang YX, Richter A, Burrows JP, Uno I, Jang CJ, Chen D, Yao ZL, Lei Y. 2007. NO_x emission trends for China, 1995–2004: the view from the ground and the view from space. *Journal of Geophysical Research* **112**: D22306, doi: 10.1029/2007jd008684.

Zhang L, Liao H, Li JP. 2010. Impacts of Asian summer monsoon on seasonal and interannual variations of aerosols over eastern China. *Journal of Geophysical Research* **115**: D00K05, doi: 10.1029/2009jd012299.

Zhu JL, Liao H, Li JP. 2012. Increases in aerosol concentrations over eastern China due to the decadal-scale weakening of the East Asian summer monsoon. *Geophysical Research Letters* **39**(9): L09809, doi: 10.1029/2012GL051428.

5

Simulating the 20 May 2013 Moore, Oklahoma tornado with a 100-metre grid-length NWP model

Kirsty E. Hanley,[1]* Andrew I. Barrett[2] and Humphrey W. Lean[1]

[1]MetOffice@Reading, University of Reading, Reading, UK
[2]Department of Meteorology, University of Reading, Reading, UK

Correspondence to:
K. E. Hanley,
MetOffice@Reading, University of Reading, Meteorology Building, Reading RG6 6BB, UK.
E-mail:
kirsty.hanley@metoffice.gov.uk

Abstract

Since 2013, the Met Office have run a 2.2 km horizontal gridlength version of the Unified Model (MetUM) as part of the National Oceanographic and Atmospheric Administration's Hazardous Weather Testbed Spring Forecasting Experiment. In this study, we perform high resolution MetUM simulations of the 20 May 2013 Oklahoma tornado outbreak at horizontal gridlengths between 2.2 km and 100 m. Here we present results showing that at 2.2 km gridlength the MetUM is able to simulate supercell-like storms whereas at O(100 m) gridlength it is able to simulate realistic-looking supercells with tornado-like vortices. This opens up the opportunity for using such simulations to highlight areas of enhanced tornado risk ahead of time.

Keywords: tornado; Unified Model; 100-metre grid length

1. Introduction

Accurate forecasting of severe thunderstorms is crucially important for providing spatially and temporally correct warnings of the convective-scale hazards they can cause, e.g. squall lines and tornadoes. In recent decades, the lead time for tornado warnings has greatly improved and currently averages at about 14 min (Wurman *et al.*, 2012). However, all of the National Oceanographic and Atmospheric Administration's (NOAA's) National Weather Service (NWS) tornado warnings are based upon detection by observers or the presence of a tornado vortex signature in radar data (Brotzge and Donner, 2013) meaning that the threat has to exist before a warning is issued, which limits further improvements in lead time unless an alternative method is found to warn before the threat exists.

Many operational weather centres, including the Met Office, now run order 1 km gridlength models for short-range weather forecasting. Although such models yield qualitatively more realistic precipitation fields than lower resolution simulations with parameterised convection (e.g. Kain *et al.*, 2008; Lean *et al.*, 2008; Weisman *et al.*, 2008; Schwartz *et al.*, 2009; Kendon *et al.*, 2012), these gridlengths are still unable to fully resolve the individual convective elements (e.g. Bryan *et al.*, 2003). As a result, we would not expect kilometre-scale models to be able to resolve tornadoes,

however, they may be able to provide accurate short-term predictions of the storms that produce them.

As model gridlengths are decreased further to order 100 m, we may expect to start resolving features such as tornadoes. Although such gridlength simulations are currently unfeasible to run operationally, they can provide useful insight into tornado dynamics and demonstrate the added benefits of high resolution forecasts. There have been many idealized modelling studies of supercell and tornado dynamics (e.g. Wicker and Wilhelmson, 1995; Markowski *et al.*, 2003; Markowski and Richardson, 2014b; Orf *et al.*, 2014); however, there are few high resolution numerical studies based on real tornadic storms. One such study by Schenkman *et al.* (2014), simulated the 8 May 2003 Oklahoma City supercell using the Advanced Regional Prediction System (ARPS) with four one-way nested grids of 9 km, 1 km, 100 m and 50 m horizontal grid spacing. The 1 km simulation had a 5-min data assimilation cycle performed over a 70-min period, assimilating radar reflectivity and radial velocity from the Weather Surveillance Radar-1988 Doppler (WSR-88D) at Twin Lakes (KTLX) Oklahoma City. The 60-min, 100-m simulation obtained its initial conditions from the 1 km final analysis while the 40-min, 50-m gridlength simulation was nested within the 100 m simulation and obtained its initial conditions from the 100 m forecast at 20 min. Tornado-like vortices were simulated in both the 100 and 50 m simulations: 30 min after the 100-m model was initialised and 10 min after the 50-m run was initialised. The timing, location and intensity of these vortices agreed well with the observed tornado. Previously, Mashiko *et al.* (2009) were able to produce a tornado-like vortex in a 26-min 50-m gridlength simulation of Typhoon Shanshan using the Japan Meteorological Agency Nonhydrostatic Model. The 50-m simulation was nested within a 5 km simulation, the initial conditions for which were provided by an operational regional analysis. Unlike Schenkman *et al.* (2014), no radar data were assimilated; however, the 50-m gridlength simulation was very short and would not aid real-time forecasting and warning of this storm.

In this article, we perform high resolution simulations of the 20 May 2013 tornado outbreak in Moore, Oklahoma using the Met Office Unified Model (MetUM) nested down to 100 m gridlength. Previously, 100 m gridlength versions of the MetUM have been used to study cold pooling in valleys (Vosper *et al.*, 2013), marine stratocumulus (Boutle *et al.*, 2014) and UK convection (Stein *et al.*, 2014; Hanley *et al.*, 2015). The main aim here is to investigate whether an order 100 m gridlength simulation, down-scaled from a free-running 2.2 km gridlength Numerical Weather Prediction (NWP) simulation, can resolve tornado-like vortices and potentially identify enhanced risk regions where tornadoes may occur many hours in advance of what could be obtained if assimilating radar data in the driving model. To our knowledge, this is the first study to simulate a tornado-like vortex over the US Great Plains in a high resolution NWP model at several hours lead-time.

2. Case overview

A 3-day stretch of severe weather across the Great Plains from 18–20 May 2013 produced the most deadly and devastating tornado of the year in the United States on 20 May affecting Moore, Oklahoma. Several supercell thunderstorms developed during early afternoon on 20 May 2013 in central Oklahoma. One of these supercells developed about 50 km southwest of Oklahoma City just after 1900 UTC and rapidly intensified, producing a tornado which touched down at 1956 UTC on the west side of Newcastle, Oklahoma (Atkins *et al.*, 2014). The tornado persisted for about 40 min and produced widespread enhanced-Fujita (EF) scale 3 damage, with localized EF4 and EF5 damage (Figure S1(b), Supporting Information). Several other, less intense, tornadoes were reported during the afternoon. A more detailed overview of the event and the accompanying synoptic conditions was given by Zhang *et al.* (2015).

3. Model description

The experiments are performed using version 8.2 of the MetUM. The MetUM solves non-hydrostatic, deep-atmosphere dynamics using a numerical scheme, which is semi-implicit and semi-Lagrangian (Davies *et al.*, 2005). The model uses Arakawa C-grid staggering in the horizontal and a terrain-following hybrid-height Charney–Phillips vertical grid. The model uses a comprehensive set of parameterisations including surface (Best *et al.*, 2011), mixed-phase cloud microphysics (Wilson and Ballard, 1999) and boundary-layer (Lock *et al.*, 2000). The model also includes a convection scheme (Gregory and Rowntree, 1990), although this is switched off at gridlengths below 4 km. Gridlengths of 2.2 km and finer also use a stability-dependent Smagorinsky-type subgrid turbulence scheme.

During the 2013 NOAA Hazardous Weather Testbed experiment, the Met Office was routinely running both a 4.4 and 2.2 km model (Clark *et al.*, 2014; Kain *et al.*, 2016). The 4.4 km model (US4) was one-way nested within the Met Office global model and covers the Contiguous United States (CONUS). The initial and boundary data for this domain were provided by the 0000 UTC Global analysis and forecast. The 2.2 km model (US2) was one-way nested within the US4 and covers most of the CONUS area. Both domains have 70 vertical levels, with the top at 40 km. The global model uses a hybrid incremental 4D-Var data assimilation system, no further data assimilation was performed on the limited area grids. The setup of the US4 and US2 domains was the same as the Met Office operational European 4 km model and UK 1.5 km model (UKV), respectively.

In this study, a suite of models were one-way nested within the US2 with horizontal gridlengths of 500, 200 and 100 m (Figure S1). The US2 gets its initial conditions and boundary data from the 20 May 2013 US4 model run and is initialised at 0300 UTC. The specification of the 500, 200 and 100 m models hereafter referred to as the 'nested models' is presented in Table 1. All models were integrated forward until 0000 UTC (21 h for the US2, 18 h for the 500 m model, 12 h for the 200 m model and 9 h for the 100 m model). The initialisation times were chosen to be at least 3 h ahead of when convection initiated in reality to allow the storms to develop within each domain.

The nested models configuration is based on the high resolution MetUM simulations performed by Hanley *et al.* (2015) and is very similar to the operational UKV and US2, but with a few differences. Unlike the US2, the

Table 1. Domain details and experimental setup.

Name	Grid length/ km	Domain size	Levels	Start time/ UTC	Time step/s	Convection scheme?
US4	4.4	CONUS	70	0000	100	Yes
US2	2.2	3740 × 2640 km	70	0300	75	No
500 m	0.5	600 × 500 km	140	0600	10	No
200 m	0.2	300 × 300 km	140	1200	6	No
100 m	0.1	150 × 150 km	140	1500	3	No

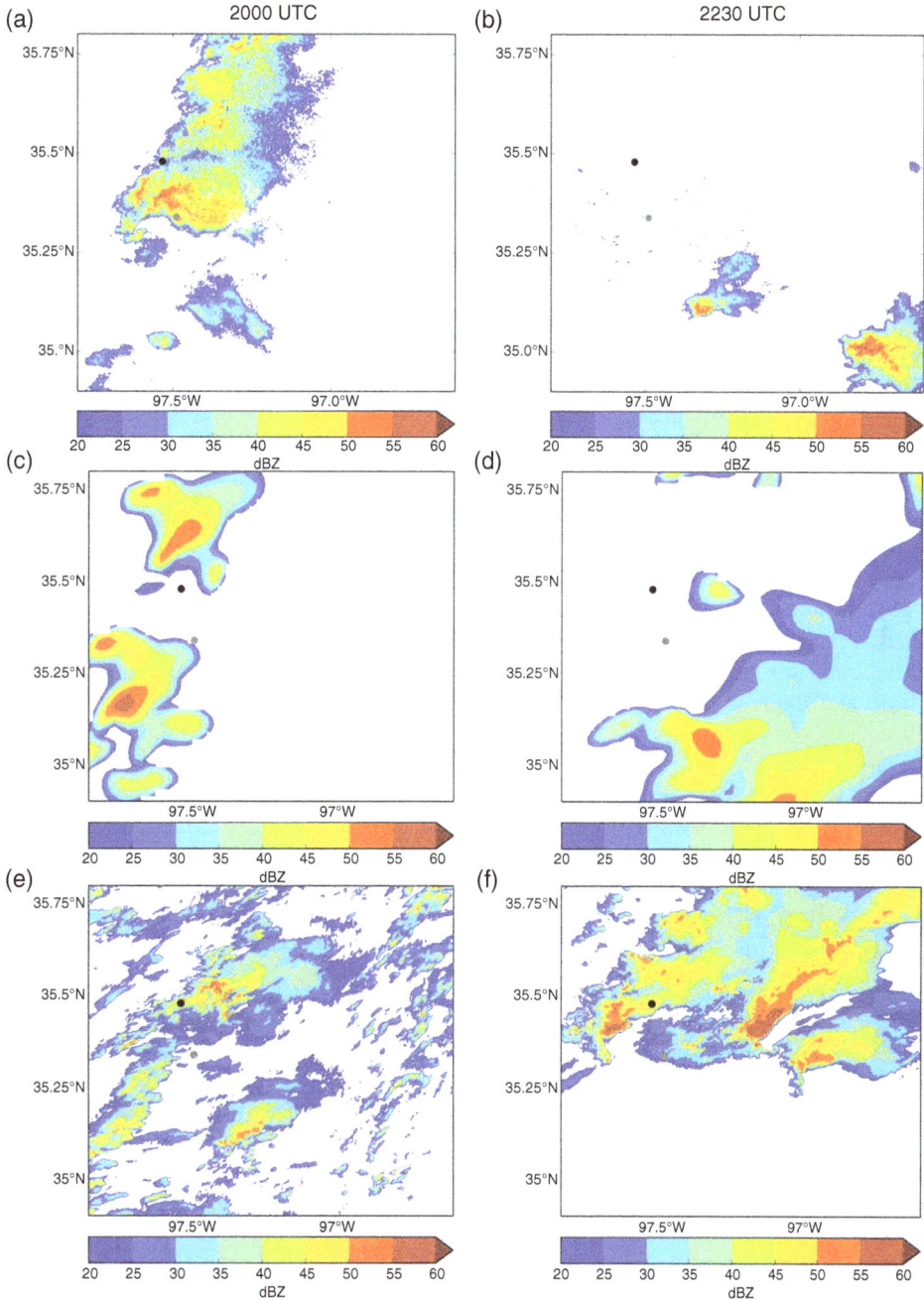

Figure 1. Reflectivity in dBZ at 2000 UTC (left) and 2230 UTC (right) 20 May 2013 from (a) and (b) the WSR-88D Twin Lakes, Oklahoma (KTLX) radar at an elevation angle of 0.5 degrees, (c) and (d) the 2.2 km model and (e) and (f) the 100 m model. The black and grey circles show the locations of Oklahoma City and Moore, respectively.

nested models have 140 vertical levels (corresponding to a spacing of ~75 m at 1 km above ground level compared to 150 m in the US2). Another difference between the models is the critical relative humidity (RH_{crit}) profile used for cloud formation. On the assumption that the subgrid variability of humidity is reduced in smaller grid boxes, the nested models use a larger RH_{crit} than the US2. The final difference of note is that the US2 uses the Smagorinsky subgrid mixing scheme only in the horizontal with vertical mixing done by the boundary layer scheme, whereas the higher resolution models apply the subgrid mixing scheme in both the horizontal and the vertical.

4. Simulation results

In this section, we begin by providing an overview of the 20 May 2013 Oklahoma supercells in the US2 and 100 m simulation. We then provide an analysis of the dynamics of the 200 and 100 m simulations at the tornado-scale, as they both produce tornado-like vortices.

4.1. Storm-scale overview

The MetUM provides surface precipitation rates as a diagnostic. The simulated surface reflectivity, Z, has been derived from the surface rainrate, R, by assuming a Z–R relationship of $Z = 300R^{1.5}$. Figure 1 shows the simulated reflectivity from the US2 and the 100 m model compared with the WSR-88D KTLX radar reflectivity at 2000 and 2230 UTC. On this day, the US2 has convection initiating in the Oklahoma City region at about 1830 UTC (not shown) consistent with radar observations. The 100 m simulation initiates convection earlier but produces wide-spread light rain for several hours before organizing into larger cells by about 2000 UTC (Figure 1(e)). The US2 produces large cells that display some classic supercell features with high reflectivity cores and larger regions of low reflectivity on the forward-flank downdraft (FFD, Figure 1(c)). The size of the cells and the distribution of reflectivity compare reasonably well with the observed radar reflectivity; however, the cells in the US2 tend to be more circular than the observations. The 100 m simulation produces more realistic-looking supercells, at 2000 UTC the location of the Moore supercell is better represented by the 100 m simulation (Figure 1(e)). However, the 100 m simulation is producing too much widespread precipitation at this time and the supercell does not have a hook-echo feature suggesting it isn't tornadic. By 2230 UTC, the precipitation in the vicinity of Oklahoma City has decayed (Figure 1(b)); however, both the US2 and the 100 m simulation are still producing substantial precipitation. At this time, the supercells in the 100 m simulation each have a well-defined hook-echo (Figure 1(f)).

The hook echos in the 100 m simulation develop about 2.5 h later than the Moore tornado. Kilometre-scale ensemble simulations of this case with the WRF-ARW model by Zhang et al. (2015) showed that both the timing and location of the supercells were highly sensitive to small changes in synoptic conditions, so small timing and/or positional errors in the position of the dryline in the driving model could have led to the timing errors seen here in the 100 m simulation.

Downdrafts (rather than updrafts) are considered important in generating near-surface vertical vorticity by transporting vertical vorticity from mid-levels to the surface (e.g. Rotunno and Klemp, 1985; Markowski and Richardson, 2009; Kosiba et al., 2013; Dahl et al., 2014; Markowski and Richardson, 2014b; Naylor and Gilmore, 2014). Due to the deep-layer wind shear and upper-level winds, the bulk of the hydrometeors in a supercell is deposited on the forward flank of the main updraft. Evaporation of rain and melting and sublimation of ice lead to negative buoyancy in this region and the development of a FFD. The FFD in the US2 simulation is coincident with the main area of precipitation (Figure S2). A rear-flank downdraft (RFD) develops when the mesocyclone wraps precipitation around the updraft into drier air at the rear of the storm, leading to latent cooling (e.g. Lemon and Doswell, 1979). The RFD is not well resolved by the US2. Descending air within the RFD can tilt horizontal vorticity into the vertical and advect it towards the ground, leading to vertical vorticity near the ground (e.g. Davies-Jones and Brooks, 1993; Adlerman et al., 1999). Numerical

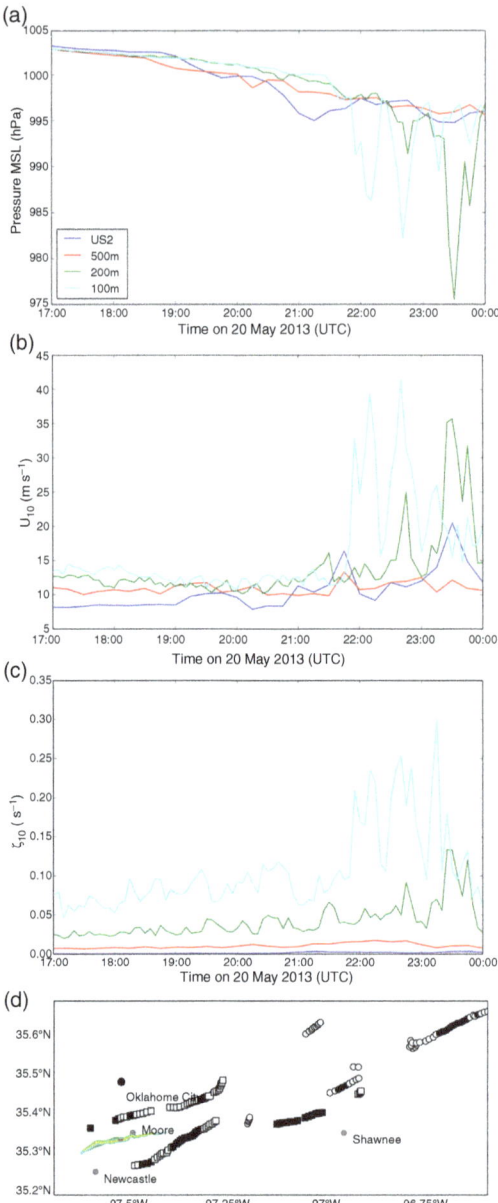

Figure 2. (a) Minimum mean sea level pressure in hPa, (b) maximum 10 m wind speed in m s^{-1} and (c) maximum 10 m vertical vorticity in s^{-1} for the US2 (blue), 500 m (red), 200 m (green) and 100 m (cyan) simulations of 20 May 2013. All data have been sampled over the region of the 100 m domain and is every 5 min. (d) Observed tornado damage path (filled contour) obtained from http://www.srh.noaa.gov/oun/?n=events-20130520 compared with location of maximum 10 m vorticity from the 200 m (circles) and 100 m (squares) simulations. White symbols indicate where the maximum 10 m wind is between 15 and 30 m s^{-1} and black symbols indicate where the maximum 10 m wind exceeds 30 m s^{-1}. Model data are every minute.

simulations of tornadoes by Wicker and Wilhelmson (1995) first indicated that some of the air parcels entering the tornado had indeed passed through the RFD. Most supercells develop near-ground rotation; however, fewer than 20% of supercells produce a tornado (Markowski and Richardson, 2014a). The lack of a RFD and a hook-echo in the US2 simulation may indicate that the gridlength is not sufficient to resolve the mesocyclone and therefore the simulated storm fails to produce a RFD required for tornadogenesis. In the following section, we take a closer look at the 100 m simulation to determine whether the simulated hook-echo features are associated with tornado-like vortices.

4.2. Tornado-scale overview

The 100 and 200 m gridlength simulations both produce tornado-like vortices. The 500 m and US2 simulation both produce supercells, but do not produce tornado-like vortices. Tornado-like vortices were identified from rapid decreases of 10–20 hPa in the minimum mean sea level pressure (mslp) within the domain (Figure 2(a)) for both the 100 and 200 m simulations. The rapid pressure drops are coincident with increases of 10-m wind speed (Figure 2(b)) and 10-m vertical vorticity (Figure 2(c)). These vortices at 2230 UTC (and at other times, not shown) coincide with well-defined hook-echo structures in the surface reflectivity (Figure 1(f)), indicating that both 100 and

200 m simulations are producing tornado-like vortices. Similar rapid drops in pressure are not seen in the US2 and 500 m simulations, because these relatively-coarse gridlengths are not fine enough to resolve tornado-like vortices.

The vortices in both 100 and 200 m simulations have 10-m wind speeds exceeding 30 m s^{-1} that persist for at least 15 min (Figure 2(b)). Both models simulate tornadoes later in the day than the observed EF5 tornado but the 100 m simulation produced a vortex earlier than the 200 m simulation, potentially showing a benefit of better resolving the small-scale features. The 100 m simulation produces a stronger tornado-like vortex than the 200 m simulation. The 100 m simulation has a maximum 10 m wind speed of over 40 m s^{-1} whereas the 200 m simulation has a maximum of just over 35 m s^{-1}. The simulated strength of the tornado-like vortices is significantly weaker than the observed EF5 tornado; this is most likely due to insufficient resolution as the effective resolution of the MetUM is

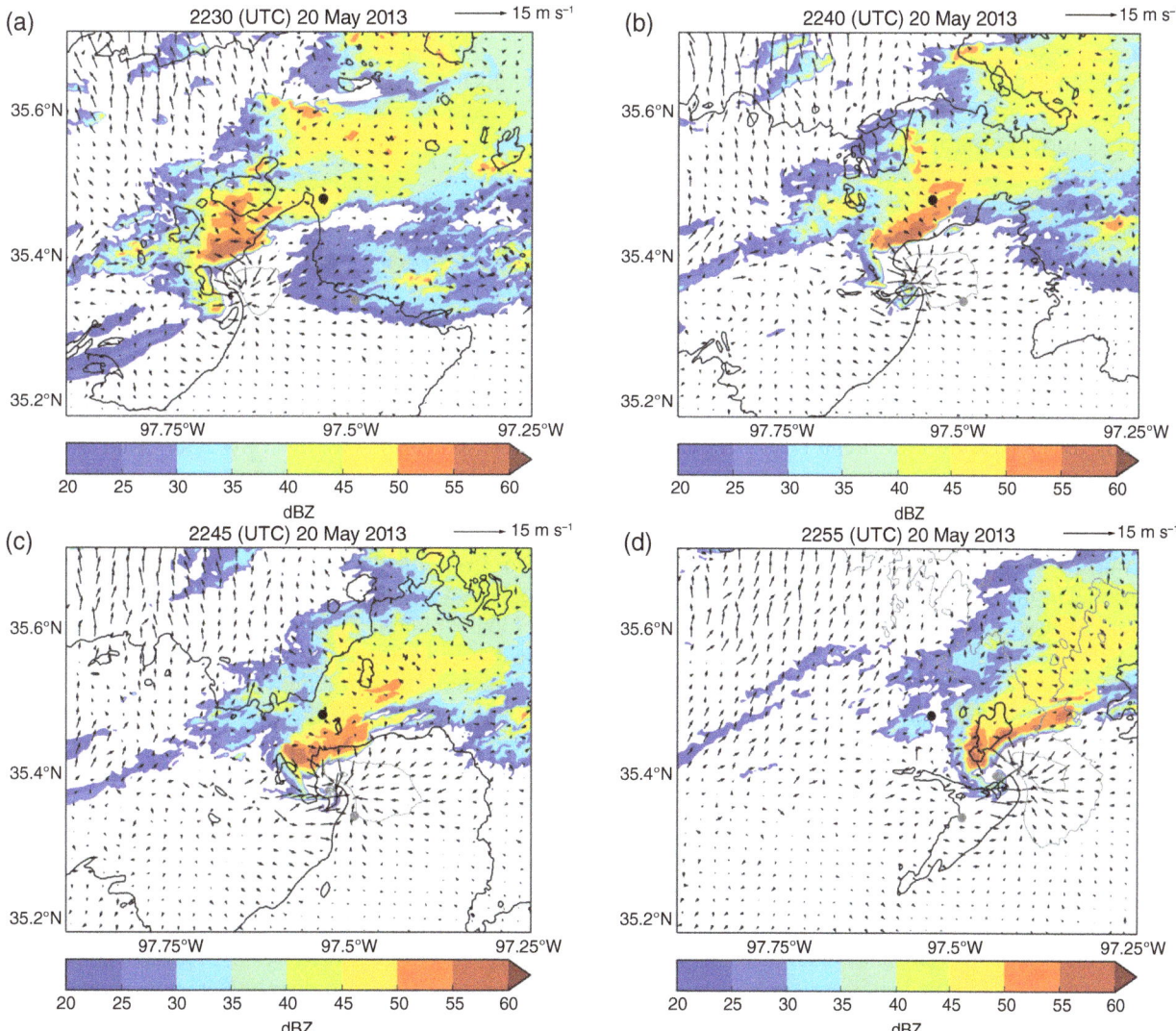

Figure 3. Simulated surface reflectivity in dBZ from the 100 m simulation at (a) 2230, (b) 2240, (c) 2245 and (d) 2255 UTC 20 May 2013. Vectors show the 10 m wind in m s^{-1}, grey contours depict mslp below 1000 hPa and black contours depict mslp above 1000 hPa with a contour interval of 1 hPa. The black and grey circles show the locations of Oklahoma City and Moore respectively.

several gridlengths. The simulated location, direction and length of the tornado paths are very similar to the observed tornado (Figure 2(d)). The points marked show the location of the 10 m vorticity maxima where the 10 m wind speed exceeded $15\,m\,s^{-1}$. The 100 m simulation produces two tornado-like vortices within 10 km of the observed EF5 tornado track. These two tornado-like vortices persist for about 40 min and both produce wind speeds exceeding $30\,m\,s^{-1}$ (reflectivity shown in Figure 3). A third vortex occurs to the east of the observed EF5 tornado (see Figure 1(f)) and persists for about 15 min. The 200 m model also simulates two tornado-like vortices, both of which exceed $30\,m\,s^{-1}$. The first persists for about 10 min and the second persists for over 30 min. Both of these tornado-like vortices occur about 50 km further east, in Lincoln County, where an EF0 tornado was reported on this day. Although the model produces a tornado-like vortex with good agreement with the Moore tornado, Figure 2 shows some differences from the actual event itself with several tornado-like vortices occurring (white and black symbol tracks in Figure 2(d)). This reflects some challenges for tornado-scale forecasting in the future. While the model may produce some tornado-like vortices in the correct location, it may resolve tornado-like vortices that do not occur (or at least were not reported). Thus, providing tornado warnings based on these simulations may increase the warning lead time, however, the false alarm rate could also increase. It may be more valuable to use similar simulations to discriminate between tornadic and non-tornadic storm environments.

Figure 3 focuses on the supercell in the 100 m simulation nearest Moore, which produces an EF1 intensity tornado-like vortex. At 2230 UTC, the hook echo begins to form (Figure 3(a)); at this time a dynamically induced low pressure is present in the inflow region. By 2240 UTC (Figure 3(b)), the hook echo is well established and the dynamic low has deepened. During the next 5 min, the hook echo structure becomes more clearly defined (Figure 3(c)) but also the leading edge of the cold outflow (marked by the black 1000-hPa contour line to the south of the storm) from the RFD starts to separate the storm inflow from the updraft. By 2255 UTC (Figure 3(d)), the hook is quite tightly wrapped and well defined, but the cold outflow from the RFD is running ahead of the hook echo and cutting the storm off from the inflow. Ultimately the tornado-like vortex decays at this time, but the storm forms another tornado-like vortex later on as seen by the break in the tornado path in Figure 2(d) north-east of Moore.

Figure 4(a) shows vertical velocity at 1 km agl from the 100 m simulation at 2230 UTC. There is a strong updraft located near the hook-echo. The updraft is much stronger in the 100 m simulation than the US2 simulation as a result of it being better resolved. There is also a strong RFD in the hook-echo region and a weaker, larger area of downdrafts in the forward-flank region. Figure 4(b) shows a vertical cross-section through the main updraft and RFD from the 100 m simulation at the same time, 2230 UTC. At this time, a funnel cloud has

developed which reaches down to the surface. Looking at the meridional wind component along the same cross-section (Figure 4(c)) shows that the updraft is associated with a wide area of rotation at mid-levels. At lower levels there is rotation coincident with the funnel cloud.

Within the same synoptic environment considered favourable for tornadoes, both tornadic and non-tornadic supercells occur. The reasons why one supercell produces a tornado and another does not are poorly understood. Markowski and Richardson (2009) suggest that tornadogenesis in supercells is a Goldilocks problem whereby the air feeding into the tornado has to be just the right temperature. Downdrafts and their accompanying negative buoyancy are crucial for baroclinic generation of vorticity, but excessive negative buoyancy can prevent near-surface parcels from being dynamically lifted, preventing tornadogenesis. One advantage of having a fairly long simulation is that there are multiple supercells simulated by the 100 m model. In a follow-on study we plan to compare the thermodynamic structure of the tornadic supercells with the non-tornadic supercells. The ability to use such forecasts to discriminate between tornadic and non-tornadic supercells may help reduce the high false alarm rate for both tornado warnings and tornado watches.

5. Discussion

In this study, we have performed high resolution simulations of the 20 May 2013 tornado outbreak. By nesting down to $O(100\,m)$ gridlength, the simulated supercells become more realistic and tornado-like vortices are produced. To our knowledge, this is the first time that tornado-like vortices have been simulated for a real tornado event at several hours lead time and without the use of high resolution data assimilation to force realistic storm development.

These simulations were performed with multiple nesting from the global model down to the 100 m simulation. Data assimilation was only performed on the global domain. Despite the lack of data assimilation, the model was able to simulate a tornado-like vortex within 50 km of the observed Moore tornado, albeit 2 h late. The demonstration here that sufficiently high-resolution simulations can resolve both supercells and tornado-like vortices, without requiring strong constraint from high-resolution radar data assimilation, is encouraging. By improving initial conditions using data assimilation on the limited area domains, there exists a real possibility of using similar high-resolution simulations to provide skilful forecasts that could result in longer-lead-time warnings of severe convective weather, or the ability to discriminate between tornadic and non-tornadic storm environments.

The tornado-like vortices in the 200 and 100 m gridlength simulations were identified from rapid decreases in minimum mslp of the order 10 hPa and were confirmed by the presence of a condensation

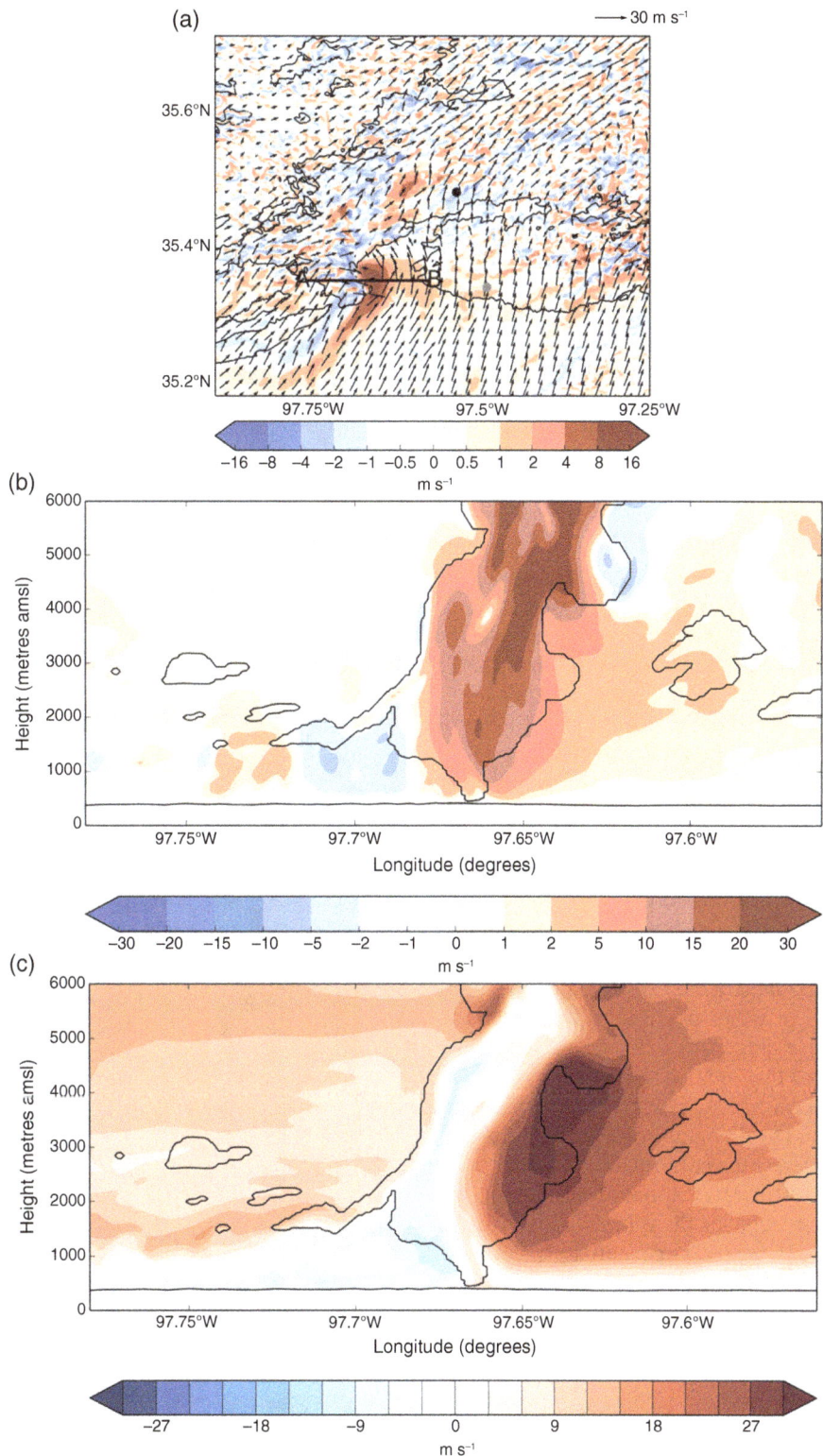

Figure 4. (a) Vertical velocity and wind vectors at 1 km agl in m s^{-1}. (b) Vertical velocity in m s^{-1} and (c) meridional wind component in m s^{-1} along 35.35°N (line A–B in (a)). All from the 100 m simulation at 2230 UTC on 20 May 2013. The black contours in (a) show surface reflectivity of 20 dBZ and the contours in (b) and (c) depict cloud water exceeding 0.001 g kg^{-1}.

funnel reaching down to the surface. Similar experiments were conducted on two further severe weather outbreaks from 2013: 30 and 31 May. On 30 May 2013, there were supercells across Oklahoma but no reported tornadoes; whereas, on 31 May 2013, the

widest tornado in recorded history occurred in central Oklahoma. For both cases, the MetUM did a good job at capturing the supercellular features (not shown). Timeseries analysis similar to Figure 2 showed that for the 31 May 2013, both the 200 and 100 m gridlength simulations produced tornado-like vortices similar in magnitude to those produced in the 20 May 2013 simulations; whereas, on 30 May 2013, no such vortices were simulated. This indicates that the MetUM potentially has some skill in discriminating between tornadic and non-tornadic supercells.

Although the tornado-like vortices simulated by the 200 and 100 m models look realistic, there are issues with the timing and location. This is to be expected because atmospheric predictability at the convective storm scale is limited, which presents some challenges in tornado-scale forecasting in the future. In particular, the model produces several tornado-like vortices in locations where tornadoes were not reported which could lead to increased false alarm rates. One technique for overcoming this uncertainty would be to run an ensemble of 100 m simulations nested within a kilometre-scale ensemble. However, this would be very computationally expensive. A cheaper alternative would be to use the kilometre-scale ensemble to determine the spatial and temporal uncertainty of supercells and then nest one or two 100 m simulations within the ensemble to determine the likelihood of the supercells being tornadic.

NOAA's National Weather Service is moving towards using order 1 km gridlength NWP models in so-called 'warn-on-forecasts' (Stensrud et al., 2013). Although these gridlengths cannot resolve tornadoes, diagnostics such as mid-level updraft helicity and low-level shear are useful for identifying the mid-level rotation associated with supercell storms and previous studies have shown some skill in using updraft helicity to forecast tornado path lengths (Clark et al., 2012, 2013). However, since only a small fraction of supercells actually produce tornadoes, it is not possible to forecast a tornado based on these diagnostics alone. In contrast, we showed in Section 4 that order 100 m NWP models can resolve tornado-like vortices, provided the tornado has a diameter of several gridlengths or greater. This means a relatively small percentage of tornadoes in nature will be resolved in a 100 m NWP model (i.e. most tornadoes have diameters smaller than the effective resolution of the model). However, larger tornadoes are generally more damaging and arguably the most important to forecast (e.g. Brooks, 2004; Agee and Childs, 2014). Further investigation into the difference between the tornadic and non-tornadic supercells at this resolution may provide insight into which supercells in lower resolution simulations are most likely to produce tornadoes. The use of 100 m simulations in this fashion, or through use of real-time (ensemble) simulations, could help improve the prediction of tornado outbreaks by identifying enhanced risk regions where tornadoes may occur and help reduce the high false alarm rate of tornado warnings by discriminating between tornadic and non-tornadic events.

Acknowledgements

The authors would like to thank Mark Weeks for initially setting up the US4 and US2 and Mike Coniglio and Jimmy Correia Jr. for their help in obtaining and visualizing NEXRAD II data. Thanks must also go to Steve Willington for leading the Met Office's involvement in the NOAA Hazardous Weather Testbed.

Supporting information

The following supporting information is available:

Figure S1. (a) Model domains used. The dashed black line shows the outline of the US2, the red line shows the 500 m domain, the blue line shows the 200 m domain and the green line shows the 100 m domain. Coloured contours show the orography of the US2. (b) Orography of the 100 m domain with the observed tornado damage path overlaid (coloured contours) obtained from http://www.srh.noaa.gov/oun/?n=events-20130520.

Figure S2. (a) Updraft vertical velocity in m s^{-1} and wind vectors at 1 km agl and (b) updraft helicity in m^2 s^{-2} between 2 and 5 km from the US2 simulation at 2000 UTC on 20 May 2013. The contours depict surface reflectivity of 20 dBZ (grey) and 50 dBZ (black).

References

Adlerman EJ, Droegemeier KK, Davies-Jones R. 1999. A numerical simulation of cyclic mesocyclogenesis. *Journal of the Atmospheric Sciences* **56**: 2045–2069.

Agee E, Childs S. 2014. Adjustments in tornado counts, F-Scale intensity, and path width for assessing significant tornado destruction. *Journal of Applied Meteorology and Climatology* **53**: 1494–1505.

Atkins NT, Butler KM, Flynn KR, Wakimoto RM. 2014. An integrated damage, visual, and radar analysis of the 2013 Moore Oklahoma EF5 tornado. *Bulletin of the American Meteorological Society* **95**: 1549–1561, doi: 10.1175/BAMS-D-14-00033.1.

Best MJ, Pryor M, Clark DB, Rooney GG, Essery RLH, Menard CB, Edwards JM, Hendry MA, Porson A, Gedney N, Mercado LM, Sitch S, Blyth E, Boucher O, Cox PM, Grimmond CSB, Harding RJ. 2011. The Joint UK Land Environment Simulator (JULES), model description – part 1:energy and water fluxes. *Geoscientific Model Development* **4**: 677–699.

Boutle IA, Eyre JEJ, Lock AP. 2014. Seamless stratocumulus simulation across the turbulent gray zone. *Monthly Weather Review* **142**: 1556–1569, doi: 10.1175/MWR-D-13-00229.1.

Brooks HE. 2004. On the relationship of tornado path length and width to intensity. *Weather and Forecasting* **19**: 310–319.

Brotzge J, Donner W. 2013. The tornado warning process: a review of current research, challenges, and opportunities. *Bulletin of the American Meteorological Society* **94**: 1715–1733.

Bryan GH, Wyngaard JC, Fritsch JM. 2003. Resolution requirements for the simulation of deep convection. *Monthly Weather Review* **131**: 2394–2416.

Clark AJ, Kain JS, Marsh PT, Correia J Jr, Xue M, Kong F. 2012. Forecasting tornado pathlengths using a three-dimensional object identification algorithm applied to convection-allowing forecasts. *Weather and Forecasting* **27**: 1090–1113.

Clark AJ, Gao J, Marsh PT, Smith T, Kain JS, Correia J Jr, Xue M, Kong F. 2013. Forecasting tornado pathlengths using a three-dimensional object identification algorithm applied to convection-allowing forecasts. *Weather and Forecasting* **28**: 387–407.

Clark A, Willington S, Suri D, Kain JS, Coniglio MC, Knopfmeier KH, Weiss SJ, Jirak IL, Lean HW, Roberts R, Weeks M, Dean AR, Melick CJ, Karstens CD, Marsh PT, Correia J Jr. 2014. Comparing the NSSL-WRF Model and Convection-allowing Versions of UKMET's Unified Model during the 2013 and 2014 NOAA/HWT Spring Forecasting Experiments. In 27th Conference on Severe Local Storms. American Meteorological Society: Madison, WI.

Dahl JML, Parker MD, Wicker LJ. 2014. Imported and storm-generated near-ground vertical vorticity in a simulated supercell.

Journal of the Atmospheric Sciences **71**: 3027–3051, doi: 10.1175/JAS-D-13-0123.1.

Davies T, Cullen MJP, Malcolm AJ, Mawson MH, Staniforth A, White AA, Wood N. 2005. A new dynamical core for the Met Office's global and regional modelling of the atmosphere. *Quarterly Journal of the Royal Meteorological Society* **131**: 1759–1782.

Davies-Jones RP, Brooks H. 1993. Mesocyclogenesis from a theoretical perspective. In *The Tornado: Its Structure, Dynamics, Prediction and Hazards.* Church C, Burgess D, Doswell C, Davies-Jones R (eds). American Geophysical Union Press; Washington, DC; 105–114.

Gregory D, Rowntree PR. 1990. A mass flux convection scheme with representation of cloud ensemble characteristics and stability dependent closure. *Monthly Weather Review* **118**: 1483–1506.

Hanley KE, Plant RS, Stein THM, Hogan RJ, Nicol JC, Lean HW, Halliwell C, Clark PA. 2015. Mixing length controls on high resolution simulations of convective storms. *Quarterly Journal of the Royal Meteorological Society* **141**: 272–284, doi: 10.1002/qj.2356.

Kain JS, Weiss SJ, Bright DR, Baldwin ME, Levit JJ, Carbin GW, Schwartz CS, Weisman ML, Droegemeier K, Weber DB, Thomas KW. 2008. Some practical considerations regarding horizontal resolution in the first generation of operational convection-allowing NWP. *Weather and Forecasting* **23**: 931–952.

Kain JS, Clark AJ, Coniglio MC, Knopfmeier KH, Karstens CD, Willington S, Weeks M, Roberts NM, Lean HW, Wilkinson JM, Gilchrist L, Hanley KE, North R, Suri D, Weiss SJ, Jirak IL. 2016. Collaborative efforts between the U.S. and U.K. to advance prediction of high impact weather. *Bulletin of the American Meteorological Society*, in preparation.

Kendon EJ, Roberts NM, Senior CA, Roberts MJ. 2012. Realism of rainfall in a very high resolution regional climate model. *Journal of Climate* **25**: 5791–5806, doi: 10.1175/JCLI-D-11-00562.1.

Kosiba K, Wurman J, Richardson Y, Markowski P, Robinson P, Marquis J. 2013. Genesis of the Goshen County, Wyoming, tornado on 5 June 2009 during VORTEX2. *Monthly Weather Review* **141**: 1157–1181.

Lean HW, Clark PA, Dixon M, Roberts NM, Fitch A, Forbes R, Halliwell C. 2008. Characteristics of high-resolution versions of the Met Office Unified Model for forecasting convection over the United Kingdom. *Monthly Weather Review* **136**: 3408–3424, doi: 10.1175/2008MWR2332.1.

Lemon LR, Doswell CA. 1979. Severe thunderstorm evolution and mesocyclone structure as related to tornadogenesis. *Monthly Weather Review* **107**: 1184–1197.

Lock AP, Brown AR, Bush MR, Martin GM, Smith RNB. 2000. A new boundary layer mixing scheme. Part I: scheme description and single-column model tests. *Monthly Weather Review* **128**: 3187–3199.

Markowski PM, Richardson YP. 2009. Tornadogenesis: our current understanding, forecasting considerations, and questions to guide future research. *Atmospheric Research* **93**: 3–10.

Markowski P, Richardson Y. 2014a. What we know and don't know about tornado formation. *Physics Today* **67**: 26–31, doi: 10.1063/PT.3.2514.

Markowski PM, Richardson YP. 2014b. The influence of environmental low-level shear and cold pools on tornadogenesis: insights from idealized simulations. *Journal of the Atmospheric Sciences* **71**: 243–275, doi: 10.1175/JAS-D-13-0159.1.

Markowski PM, Straka JM, Rasmussen EN. 2003. Tornadogenesis resulting from the transport of circulation by a downdraft: Idealized numerical simulations. *Journal of the Atmospheric Sciences* **60**: 795–823.

Mashiko W, Niino H, Kato T. 2009. Numerical simulation of tornadogenesis in an outer-rainband minisupercell of Typhoon Shanshan on 17 September 2006. *Monthly Weather Review* **137**: 4238–4260, doi: 10.1175/2009MWR2959.1.

Naylor J, Gilmore MS. 2014. Vorticity evolution leading to tornadogenesis and tornadogenesis failure in simulated supercells. *Journal of the Atmospheric Sciences* **71**: 1201–1217.

Orf L, Wilhelmson R, Wicker L. 2014. Numerical simulation of a supercell with an embedded long-track EF5 tornado. Special Symposium on Severe Local Storms: The Current State of the Science and Understanding Impacts. American Meteorological Society, Atlanta, GA

Rotunno R, Klemp J. 1985. On the rotation and propagation of simulated supercell thunderstorms. *Journal of the Atmospheric Sciences* **42**: 271–292.

Schenkman AD, Xue M, Hu M. 2014. Tornadogenesis in a high-resolution simulation of the 8 May 2003 Oklahoma City supercell. *Journal of the Atmospheric Sciences* **71**: 130–154.

Schwartz CS, Kain JS, Weiss SJ, Xue M, Bright DR, Kong F, Thomas KW, Levit JJ, Coniglio MC. 2009. Next-day convection-allowing WRF model guidance: a second look at 2-km versus 4-km grid spacing. *Monthly Weather Review* **137**: 3351–3372.

Stein THM, Hogan RJ, Hanley KE, Clark PA, Halliwell C, Lean HW, Nicol J, Plant RS. 2014. The three-dimensional microphysical structure of convective storms over the southern United Kingdom. *Monthly Weather Review* **142**: 3264–3283, doi: 10.1175/MWR-D-13-00372.1.

Stensrud DJ, Wicker LJ, Xue M, Dawson II DT, Yussouf N, Wheatley DM, Thompson TE, Snook NA, Smith TM, Schenkman AD, Potvin CK, Mansell ER, Lei T, Kuhlman KM, Jung Y, Jones TA, Gao J, Coniglio MC, Brooks HE, Brewster KA. 2013. Progress and challenges with warn-on-forecast. *Atmospheric Research* **123**: 2–16.

Vosper S, Carter E, Lean H, Lock A, Clark P, Webster S. 2013. High resolution modelling of valley cold pools. *Atmospheric Science Letters* **14**: 193–199, doi: 10.1002/asl2.439.

Weisman ML, Davis C, Wang W, Manning KW, Klemp JB. 2008. Experiences with 0-36-h explicit convective forecasts with the WRF-ARW model. *Weather and Forecasting* **23**: 407–437.

Wicker LJ, Wilhelmson RB. 1995. Simulation and analysis of tornado development and decay within a three-dimensional supercell thunderstorm. *Journal of the Atmospheric Sciences* **52**: 2675–2703.

Wilson DR, Ballard SP. 1999. A microphysically based precipitation scheme for the UK Meteorological Office Unified Model. *Quarterly Journal of the Royal Meteorological Society* **125**: 1607–1636.

Wurman J, Dowell D, Richardson Y, Markowski P, Rasmussen E, Burgess D, Wicker L, Bluestein HB. 2012. The second verification of the origins of rotation in tornadoes experiment: VORTEX2. *Bulletin of the American Meteorological Society* **93**: 1147–1170.

Zhang Y, Zhang F, Stensrud DJ, Meng Z. 2015. Practical predictability of the 20 May 2013 tornadic thunderstorm event in Oklahoma: sensitivity to synoptic timing and topographical influence. *Monthly Weather Review* **143**: 2973–2997, doi: 10.1175/MWR-D-14-00394.1.

6

Satellite-based shortwave aerosol radiative forcing of dust storm over the Arabian Sea

Subin Jose, Biswadip Gharai,* P. V. N. Rao and C. B. S. Dutt

Atmospheric and Climate Sciences Group (ACSG)-ECSA, National Remote Sensing Centre (NRSC), ISRO, Hyderabad, India

*Correspondence to:
B. Gharai,
Atmospheric and Climate
Sciences Group (ACSG), Earth
and Climate Science Area
(ECSA), National Remote
Sensing Centre (NRSC), Indian
Space Research Organisation,
Department of Space,
Government of India, Balanagar,
Hyderabad 500 037, Telangana,
India.
E-mail: biswadip_g@nrsc.gov.in

Abstract

Dust storm events over the Arabian Sea (AS) have been detected using Moderate Resolution Imaging Spectroradiometer (MODIS) data. Shortwave Aerosol Radiative Forcing (SWARF) due to dust storm is estimated using synchronous observation of Clouds and Earth's Energy System (CERES) and MODIS aerosol optical depth (AOD). Study established a relationship between them as SWARF = $-39.12 \times$ AOD $- 16.53$ ($0.4 \leq$ AOD ≤ 4.0) with $r^2 = 0.96$. The developed relation can be used for quick, independent estimation of instantaneous SWARF for dust storm over the AS. The relationship can be used to explore the possible effect of dust on climate modulation in this region.

Keywords: AOD; SW flux; SWARF

1. Introduction

Dust aerosols are the second largest natural source of atmospheric aerosols that can produce huge positive or negative feedback on climate (Evan *et al.*, 2009). The principal radiative effect of dust aerosol is to heat the atmosphere by absorption of solar radiation. In addition to the direct effects, there is increasing evidence on the role of dust aerosols in modifying cloud microphysical properties (DeMott *et al.*, 2003; Kaufman *et al.*, 2005). It is estimated that about 1000–3000 Tg year^{-1} of dust emissions are injected into the atmosphere annually, about 30% of which are re-deposited onto the source regions, 20% are transported over regional scales, while the remaining approximately 50% are subject to long-range transport (IPCC, 2007). Hence, International Panel on Climate Change (IPCC) and World Meteorological Organization (WMO) consider it as an essential climate variable and a major component of atmospheric aerosols.

Owing to the limited ground observations in the relevant regions, satellite remote sensing has become an important tool for detecting, monitoring and analyzing dust storms. Significant progress has been made in the development of integrated dust storm monitoring and modeling by making use of advanced numerical models, satellite remote sensing and Geographical Information System (GIS) data (Shao and Dong, 2006). But there exists great uncertainty in quantifying their role in direct and indirect effects on climate. This uncertainty arises because of the challenges faced in determining their physical, optical and chemical properties and the

altitude to which these aerosols are transported (Tegen and Lacis, 1996). In addition, the spatial and temporal variability and the short-term nature of their existence in the atmosphere make accounting for the radiative forcing effect of aerosols uncertain/difficult. Several researchers have used different methods for estimation of direct radiative forcing (DRF) due to aerosols from satellite retrievals (Kaufman *et al.*, 2005; Christopher *et al.*, 2006; Bellouin *et al.*, 2008). Kaufman *et al.* (2005) and Christopher *et al.* (2006) estimated an annual average DRF of -1.4 Wm^{-2}, over clear-sky ocean; the former used satellite retrievals combined with aerosol radiative forcing efficiencies while the latter used a combination of aerosol retrievals and broadband-flux measurements from satellite instruments. Yu *et al.* (2004) used a combination of Georgia Tech/Goddard Global Ozone Chemistry Aerosol Radiation and Transport (GOCART) model simulations and satellite observation to estimate global DRF and found that the weaker dust absorption increases the TOA cooling by 0.4 Wm^{-2}. Bellouin *et al.* (2005) and Chung *et al.* (2005) estimated DRF for all-sky global conditions with values in the range -0.8–0.1 Wm^{-2}. In most of these DRF, thick dust aerosol clouds, biomass burning, etc. are normally not considered, as these thick aerosol clouds are generally removed during the process of stringent cloud screening. The Arabian Sea (AS) has not been extensively studied in terms of heavy dust aerosol loading and its radiative forcing. Hence there is a need to study dust aerosol implication on shortwave (SW) radiation at the top of the atmosphere (TOA) in this region. This becomes even more important in view of the finding by Kaskaoutis *et al.* (2014), that of enhanced dust activity in 2008 over the AS and the occurrence of more dust storms that originated from the region across the border between Iran and Afghanistan.

In the present study, we made an attempt to detect dust storms over the AS and tried to understand its radiative implication in the SW region. Simultaneous measurements of aerosol optical depth (AOD) by moderate resolution imaging spectroradiometer (MODIS) and SW fluxes from clouds and Earth's energy system (CERES), from Aqua satellite during dust storm events have been used to establish a possible relationship between AOD and SWARF. Description of data sets and methodology adopted are discussed below.

2. Dataset and methodology

The AS bounded by $53.75°E$ to $74.75°E$ and $14.25°N$ to $25.25°N$ is considered as the study area as shown in Figure 1. An intense dust storm, which had its origin near the Persian Gulf, passed over the AS on 20 March 2012 (DS-1) as seen visually by Aqua MODIS data (Figure 2(a)). Contemporaneous satellite data have been used for monitoring and estimating its impact on SW radiation at TOA. Two other dust storm events as observed on 13 January 2013 (DS-2) and 13 December 2003 (DS-3) were also investigated to develop a relationship between AOD and SWARF.

2.1. Moderate Resolution Imaging Spectroradiometer

MODIS is one of the key sensors on the Aqua satellite. It provides observation of the Earth in 36 spectral bands ranging from the visible ($0.415\,\mu m$) to infrared ($14.235\,\mu m$) regions of the electromagnetic spectrum with 2 visible bands at a spatial resolution of 250 m, 5 more visible bands at 500 m spatial resolution and the remaining 29 visible and infrared bands at 1000 m resolution (Ackerman *et al.*, 1998). In the present study, we have used MODIS Aqua Level 1B calibrated radiance data at 1 km (MYDO21KM) and Level 2 aerosol data (MYD04_L2) from LAADS website (http://ladsweb.nascom.nasa.gov/data/). MODIS Level 1B calibrated radiance data (reflective and emissive) are used for the detection of dust and cloud. The MODIS aerosol algorithm utilizes its 500 m resolution measured radiances ($0.55–2.1\,\mu m$) to produce AOD, the aerosol effective radius and the fraction of the total optical depth contributed by the sub-micron size mode aerosol (Tanré *et al.*, 1997). The aerosol properties are then reported at 10 km spatial resolution. In the present study, AOD at 550 nm (MYD04_L2) is used to study the effect of dust on TOA CERES SW flux.

2.2. Atmospheric infrared sounder

Atmospheric infrared sounder (AIRS) is a high resolution spectral IR sounder that provides 2378

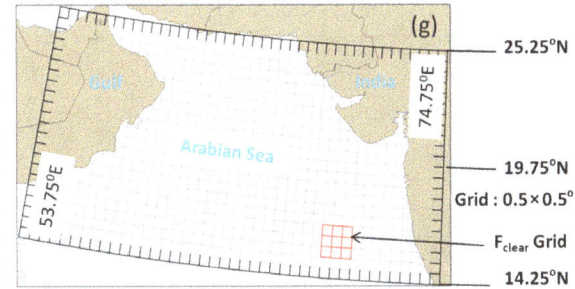

Figure 1. Study area over Arabian Sea superimposed with $0.5 \times 0.5°$ grids.

channels covering IR spectrum from 650 to 2675 cm^{-1} ($3.74–15.4\,\mu m$ wavelengths) at a nominal spectral resolving power ($\lambda/\Delta\,\lambda$) of 1200. The details of this instrument and its performance can be found in Aumann *et al.* (2003). The present study makes use of temperature profiles during the dust storm period using Level 3 daily data with a resolution of $1 \times 1°$, which is available from http://mirador.gsfc.nasa.gov. This retrieved product has been validated using more than 2 years of collocated radiosonde measurements by Divakarla *et al.* (2006).

2.3. Cloud aerosol lidar and infrared pathfinder satellite observation

Cloud aerosol lidar and infrared pathfinder satellite observation (CALIPSO) is a joint US/French mission launched on 28 April 2006. It is a combination of space borne lidar with passive imagery, which gives a 3D perspective of the role of clouds and aerosols in regulating Earth's climate. Cloud Aerosol LIdar with Orthogonal Polarisation (CALIOP), the primary instrument in CALIPSO satellite, is a dual-wavelength polarization-sensitive lidar that provides high resolution vertical profile of aerosols and clouds. It has three receiver channels, one for acquiring backscatter intensity at 1064 nm and other two for acquiring backscatter intensity at 532 nm in parallel and perpendicular polarizations (Winker *et al.*, 2007). The ratio of backscatter intensity in perpendicular to parallel polarization is called particulate depolarization ratio (DPR), which enables us to discriminate between ice clouds and water clouds and identify non-spherical aerosol particles (Huang *et al.*, 2007). In the present study, CALIPSO night time track obtained from www-CALIPSO.larc.nasa.gov/ on 20 March 2012 over the dust storm area is analyzed.

2.4. Clouds and Earth's radiant energy system

CERES on-board Aqua is an instrument that monitors the radiation environment of earth–atmosphere system by providing radiance measurements at the TOA in three different channels: a SW channel ($0.3–5.0\,\mu m$) to measure reflected sunlight, a longwave channel to measure Earth-emitted thermal radiation in the $8–12\,\mu m$ 'window (WN)' region and a total channel

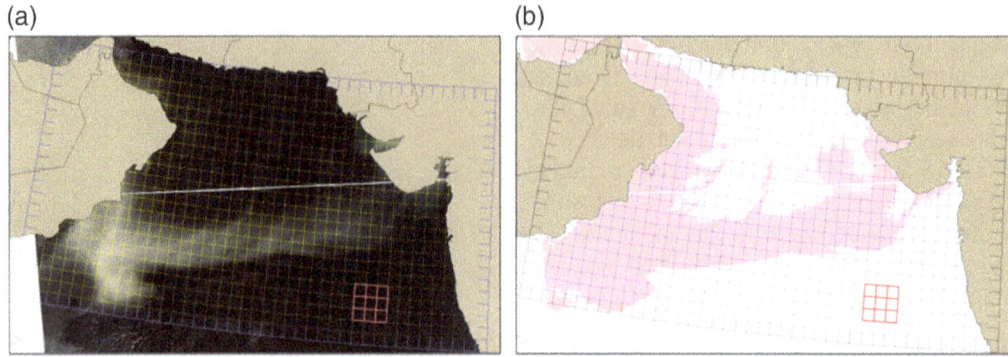

Figure 2. (a) Dust storm as viewed by MODIS Aqua on 20 March 2012. (b) Dust image derived from MODIS Aqua on 20 March, 2012; pink shading indicates dust.

(0.3–200 μm) to measure all wavelengths of radiation, with a spatial resolution of 20 km. The basic instrumentation concept and its performance can be found in Wielicki *et al.* (1996). In the present study, we analyze the role of dust aerosols in modulating the SW radiation at TOA by using the CERES Single Scanner Footprint (SSF) Level 2 instantaneous SW Flux data available at http://ceres.larc.nasa.gov/. The CERES SSF is a unique product for studying the role of clouds, aerosols and radiation in climate. CERES TOA Flux from SSF product combines measurements with scene information from MODIS Aqua (in our study). This data corresponds to the footprints that are co-located in latitude and longitude with reference to the MODIS scene. Each footprint includes radiances (SW, LW and WN) from CERES with temporally and spatially coincident imager-based radiances, cloud properties, aerosols and meteorological information. The measured radiances are inverted to fluxes using angular distribution models (ADM) anisotropic correction factor. The anisotropic correction factor is evaluated at the footprint's viewing zenith angle, relative azimuth angle and solar zenith angle. The ADMs are a function of scene type such as land, ocean, cloud cover, optical depth, etc. In the present study, only the cross-track SSF edition 3A data is used as it provides spatially contiguous data.

3. Results and discussion

The dense dust cloud as seen visually by MODIS Aqua on 20 March 2012 (Figure 2(a)) is detected by taking advantage of the reflecting and thermal bands available in MODIS. MODIS Aqua Level 1B data obtained from the LAADS website and Atmosphere Archive and Distribution System in EOS format are passed through the HDF-EOS Geo Tiff Conversion Tool to reformat and project the data. Projected data are used to detect dust clouds using ERDAS software (v 2014; Intergraph Corporation, Part of Hexagon, USA). The signal attenuation due to absorption by dust in different thermal channels is different. The MODIS channel 31 (11.03 μm) is attenuated more than the MODIS channel 32 (12.02 μm), leading to the 12.02 μm brightness

temperature being stronger than that of the 11.03 μm brightness temperature for a dust cloud. This contrast of signal (brightness temperature difference) is the basis for detecting the dust cloud. Normalized difference dust index, NDDI = (Channel 7 − Channel 3)/(Channel 7 + Channel 3) as given by Qu *et al.* (2006) is used to eliminate the influence of clouds. MODIS channel 26 (1.36–1.39 μm) is used to eliminate the presence of cirrus clouds (Gharai *et al.*, 2013) and the remaining water clouds are eliminated using MODIS channel 20 (3.66–3.84 μm). Dust cloud detected using MODIS data is shown in Figure 2(b). To observe the vertical lifting of dust, cloud and aerosol, profile products obtained from CALIPSO–CALIOP (satellite's lidar image) with a 5 km horizontal resolution is used in the present study. The vertical feature mask obtained from CALIPSO–CALIOP v 3.02 lidar data product available during the study day at 21:13:43–21:27:12 UTC is shown in Figure 3(a). The figure shows that the vertical extent of dust aerosol is about 1.5 km over the dust-detected area. Figure 3(b) shows the vertical profile of aerosol extinction coefficient (km^{-1}) observed at 532 nm represented in black and the DPR in blue color averaged over 20°–21°N latitude of CALIPSO pass. It is clear from the figure that DPR are high (>20%) throughout the profile because the signals originate from highly irregularly shaped particles and can be attributed to dust aerosol (Huang *et al.*, 2007). We also analyzed the temperature profile data obtained from descending pass (equator crossing time is 1.30 am LT) of AIRS on 19 March 2012. It has been observed that at about 3 km (700 haPa) vertical height, the temperature difference over dust cloud band to that of non-dust cloud area is of the order of 5 K (figure not presented). The vertical lift of dust as observed from CALIPSO and AIRS data are, respectively, 1.5 and 3 km as they observed dust cloud at different times. Transport of dust and its trajectories depends on the prevailing meteorology, amount of dust lifting and its mixing with other pollutants along the trajectories. As a result its vertical expansion along the track is highly dynamic in space and time.

A mesh with grid resolution of 0.5 × 0.5° extending from 53.75°E to 74.75°E and 14.25°N to 25.25°N

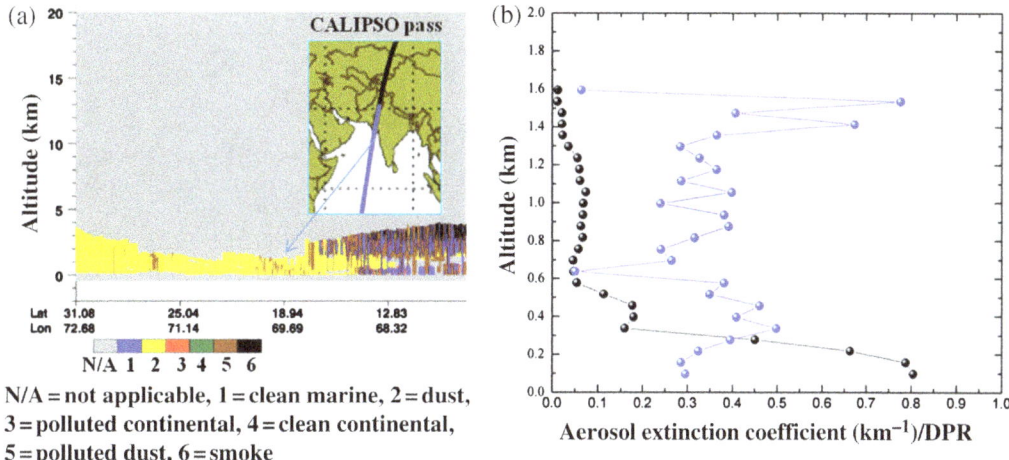

N/A = not applicable, 1 = clean marine, 2 = dust,
3 = polluted continental, 4 = clean continental,
5 = polluted dust, 6 = smoke

Figure 3. (a) Feature mask showing dust aerosol over the AS during 20 March 2012 during 21:13:43–21:27:12 UTC and (b) vertical profile of aerosol extinction coefficient (km^{-1}) at 532 nm in black color and Depolarization Ratio (DPR) in blue color averaged over 20°–21°N latitude over the AS.

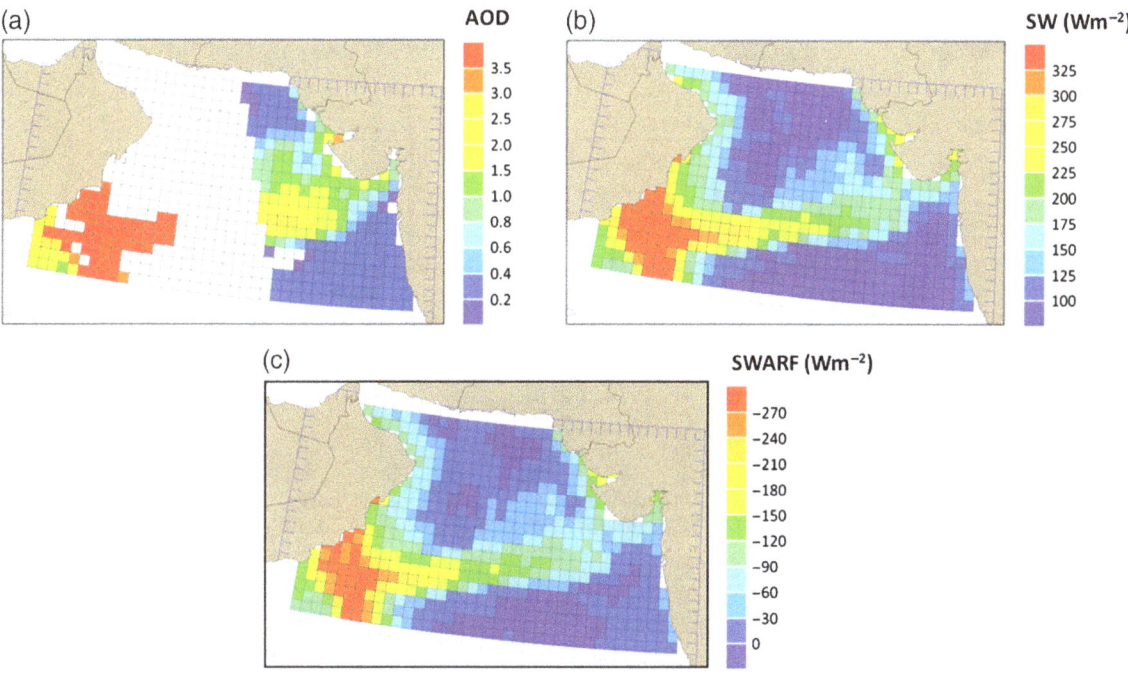

Figure 4. (a) MODIS Aqua Level 2 AOD, (b) CERES SW at TOA and (c) SWARF due to dust cloud. All the data were observed on Aqua platform on 20 March 2012.

has been created using ArcGIS software (v. 9.3; Esri, Redlands, CA, USA) and all the subsequent data analysis are made based on this grid. MODIS Aqua Level 2 AOD data of 20 March 2012 with a foot print of 10 km × 10 km has been averaged within each grid (0.5 × 0.5°) and assigned the average value to the respective grids (Figure 4(a)). Tanré *et al.* (1997) estimated the uncertainties involved in AOD (550 nm) retrievals to be ±0.05 ± 0.05 AOD (550 nm) over ocean. To analyze the impact of the dust cloud on SW radiation at TOA, we have used CERES SSF data available from Aqua platform. CERES SW radiances are converted to SW fluxes using angular distribution models, which take into account the bi-directional characteristics of a reflecting surface (Wielicki *et al.*, 1996). The 20 km

pixel level CERES SW Flux data within each 0.5 × 0.5° grid over the AS have been averaged and assigned the value to the respective grids as shown in Figure 4(b). A land mask is created from the Esri world shape file and is used to restrict our study area within the AS by masking out continental land mass.

In the present study, we have used the CERES Aqua SSF product only where the CERES footprint overlaps the Aqua MODIS imager swath. Each footprint includes radiances from CERES with temporally and spatially coincident imager-based radiances and cloud properties from Aqua MODIS. These radiances depend on, Θ_0 the solar zenith angle; Θ the observer viewing zenith angle and Φ the relative azimuth angle defining the azimuth angle position of the observer

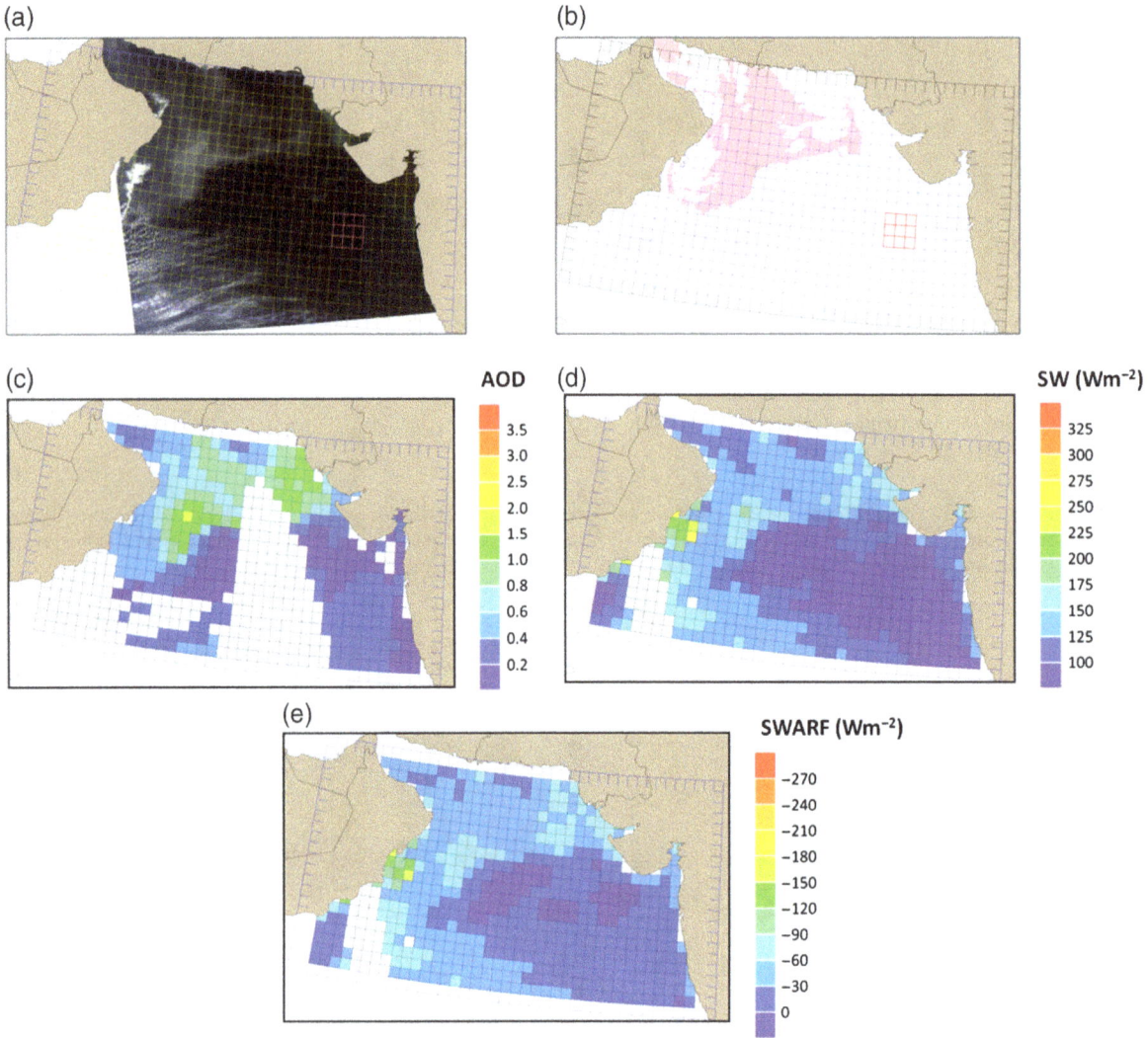

Figure 5. (a) Dust storm as viewed by MODIS Aqua on 13 January 2013. (b) Dust image derived from MODIS Aqua data. (c) MODIS Aqua Level 2 AOD. (d) CERES SW at TOA and (e) SWARF due to dust cloud. All the data were observed on Aqua platform on 13 January 2013.

relative to the solar plane. Since Θ_0, Θ and Φ are identical for the MODIS and CERES observed radiances, we have sought a relationship between AOD and SW fluxes, although they are derived from two independent instruments on-board Aqua.

Christopher and Zhang (2002) defined SWARF at the TOA as the difference between clear (F_{clear}) and aerosol fluxes. In the present study, we used calculated NDDI, and the data from MODIS channels 20 and 26 to remove unwanted cloud patches over the AS. To calculate F_{clear} flux over the AS, we considered the grid cell which are not contaminated by cloud and dust and are at least $1°$ (2 grids) away from the detected dust cloud. We have chosen 3×3, $0.5 \times 0.5°$ grid cell over the AS extending from $15.25°$N to $16.75°$N and $68.75°$E to $70.25°$E as the probable clear flux zone. Although within the defined 3×3 grid area average MODIS aqua AOD is 0.27, we considered this as the background reference aerosol loading and corresponding 3×3 grid cell average SW fluxes calculated as $92.73\,\mathrm{Wm^{-2}}$. In the absence of any dust storm, aerosol over the AS would be of a

mixture of aerosols originating from sea salt, sulfate, black carbon, natural aerosol and other anthropogenic constituents, which make the background aerosol value slightly higher. Kaufman *et al.*, 2001 and Reid *et al.*, 1998 show that Single Scattering Albedo (SSA) of dust (0.97) is higher than smoke (0.86) when observed at 640 nm. The higher SSA value of dust leads to higher TOA SW compared with SW flux when it is due to smoke or other pollutant aerosols (Christopher and Zhang, 2002). Spatial distribution of SWARF due to dust storm (DS-1) is shown in Figure 4(c). Figure shows large negative SWARF over the dust cloud region.

While calculating SWARF due to heavy dust loading ($0.4 \leq \mathrm{AOD} \leq 4.0$), we considered only those grids which are having both AOD and SW flux values and also devoid of clouds within the detected dust zone. To have sufficient data sets, we have investigated two more dust events passed over the AS, respectively, on 13 January 2013 (DS-2) and 13 December 2003 (DS-3). We followed the same methodology to analyze these two additional dust storm events as discussed

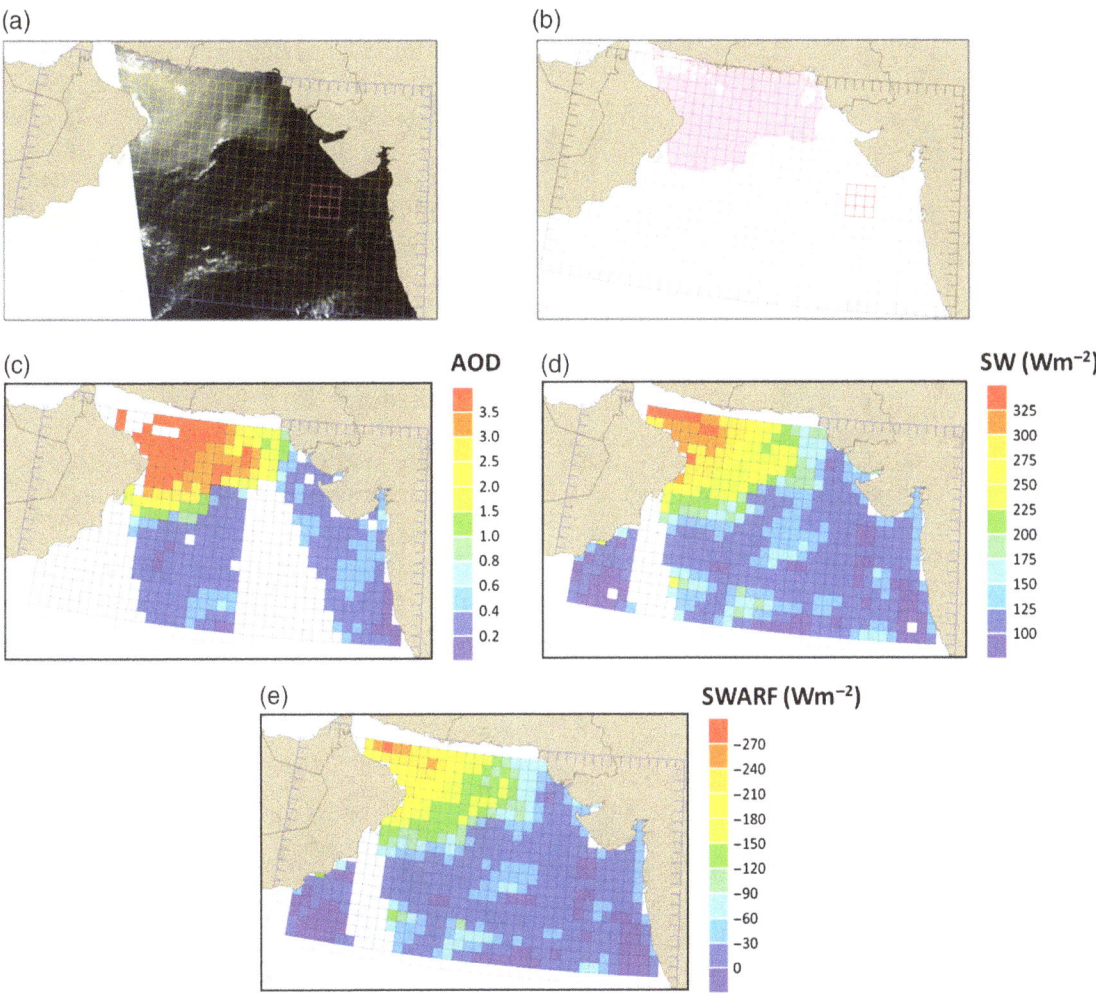

Figure 6. (a) Dust storm as viewed by MODIS Aqua on 13 December, 2003, (b) dust image derived from MODIS Aqua data, (c) MODIS Aqua Level 2 AOD, (d) CERES SW at TOA and (e) SWARF due to dust cloud. All the data were observed on Aqua platform on 13 December 2003.

Table 1. Summary of identified grids considered for the calculation of F_{clear} and SWARF due to dust storm.

SL#	Dust events	Date of occurrence	Latitude (degree)	Longitude (degree)	AOD (550 nm)	SW (Wm⁻²)
1	DS-1	20 March 2012	15.25°N to 16.75°N	68.75°E to 70.25°E	0.27	92.73
2	DS-2	13 January 2013	17.75°N to 19.25°N	68.75°E to 70.25°E	0.20	89.0
3	DS-3	13 December 2003	18.75°N to 20.25°N	68.75°E to 70.25°E	0.36	100.51

DS, dust storm.

above. Dust storms DS-2 and DS-3 as observed by MODIS Aqua are shown in Figures 5(a) and 6(a), respectively and corresponding dust images are shown in Figures 5(b) and 6(b) respectively. The F_{clear} grids are identified for these two dust events based on the same criteria followed for dust storm DS-1. F_{clear} grids (red color in figure) for DS-2 extends from 17.75°N to 19.25°N and 68.75°E to 70.25°E; for DS-3 it extends from 18.75°N to 20.25°N and 68.75°E to 70.25°E. F_{clear} fluxes have been averaged over the identified 3×3 grids, respectively, for dust storm DS-2 and DS-3 are 89.0 and 100.51 Wm⁻². Summary of identified grids considered for the calculation of F_{clear} and SWARF due to dust storm is given in Table 1. The spatial distribution of MODIS Aqua AOD, SW and SWARF for DS-2 and

DS-3 are, respectively shown in Figures 5(c)–(e) and 6(c)–(e). To develop a relationship between MODIS Aqua AOD at 550 nm and SWARF for an intense dust storm, we have restricted AOD values up to 4.0 and all the grid cell which are identified as dust grids and devoid of clouds are only considered. Scatter plot of MODIS aqua AOD at 550 nm versus SWARF for all the three dust storm events are plotted in Figure 7. It has been observed in the figure that aerosol loading is lowest during 13 January 2013 whereas during the dust event on 13 December 2003, aerosols are uniformly distributed across the AOD range and for the dust event on 20 March 2012 aerosols are in the range of moderate to high aerosol. The variations of dust loading during the different dust storm events could be directly

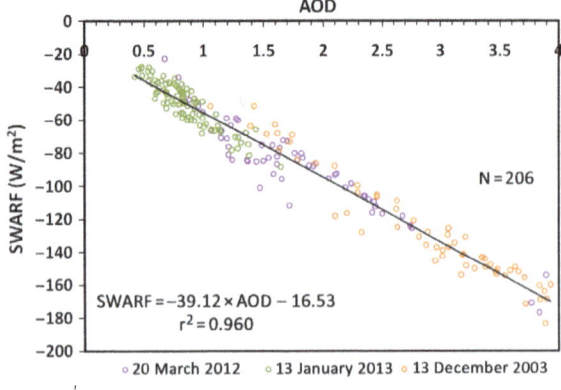

Figure 7. Scatter plot of MODIS Aqua Level 2 AOD and CERES SWARF over the Arabian Sea. Dust cloud from the all three dust events are considered for the plot.

related to the intensity/vertical extent of dust cloud and its spatial distribution. We established an excellent linear relationship between AOD and SWARF as $\text{SWARF} = -39.12 \times \text{AOD} - 16.53$ $(0.4 \leq \text{AOD} \leq 4.0)$ with high correlation coefficient ($r^2 = 0.96$). It may be observed from the figure that overall SWARF efficiency is $-39.12\ \text{Wm}^{-2}$ per unit AOD. This relationship is valid for an intense dust storm over the AS where the AOD lies between 0.4 and 4.0. The above result provides a mean for satellite-based estimation of SWARF due to dust and once dust aerosol loading is known, the radiative forcing can be estimated using the above mentioned relationship. The relationship can also be used to explore the possible effect of dust on climate modulation in the AS region.

Christopher and Zhang (2002) developed a second-order polynomial relationship between MODIS AOD and SWARF over the global oceans, where thick aerosol plumes are not considered in their analysis ($\text{AOD} \leq 0.7$). They reported that SWARF is more sensitive at lower AOD values than that of higher aerosol loading. When aerosol loading is low, the background aerosols are the major contributors to SWARF. Hence at lower aerosol loading, for a small change of background aerosol, the rate of change of SWARF will be significant. On the other hand, when atmosphere is dominated by dust aerosols, which are capable of reflecting solar energy very efficiently, the SWARF will change according to the magnitude of dust loading. Inclusion of lower AOD and exclusion of thick aerosols by Christopher and Zhang (2002), in their study, could be one of the reasons for obtaining a second-order polynomial relationship. Whereas in our study, we found a linear relationship between AOD and SWARF, considering only the intense dust storm events associated with $\text{AOD} \geq 0.4$. Kumar *et al.*, 2015, also found a linear relationship between SWARF and AOD while considering three dust events that occurred during pre-monsoon season of 2010, over central Indo-Gangetic Basin. They used AERONET observations of aerosol data ($0.4 < \text{AOD} < 2$) as input to Santa Barbara DISORT Atmospheric Radiative Transfer (SBDART) model to estimate SWARF during these three dust events.

Owing to the relatively weak dust aerosol signals and large uncertainties associated with bright surface beneath, it is difficult to detect thinner or dispersed dust cloud ($\text{AOD} < 0.4$) with the present dust detection algorithm. Since the detection of areas with dispersed thin dust cloud, which could be due to low intensity dust storms, is difficult, only the intense dust storm events are considered for estimating SWARF. In view of this, the developed relationship in the present study is valid for intense dust storms over the AS.

4. Conclusions

In the present study, we estimated SWARF due to three intense dust storms over the AS and results of the study suggest the following

- Spatial movement of dust storm can be monitored with the procedure developed for dust detection.
- Vertical extent of dust cloud observed by CALIPSO and AIRES over two successive days are, respectively, 1.5 and 3.0 km and the temperature difference between the dust cloud and non-dust areas was around 5 K for the dust storm DS-1.
- An excellent linear relationship between AOD and SWARF has been developed as $\text{SWARF} = -39.12 \times \text{AOD} - 16.53$ $(0.4 \leq \text{AOD} \leq 4.0)$ with $r^2 = 0.96$, which is applicable for intense dust storm over the AS.
- Observed SWARF efficiency is $-39.12\ \text{Wm}^{-2}$ per unit AOD and SWARF reaching $-180\ \text{Wm}^{-2}$ at $\text{AOD} = 4$.
- Quick independent and instantaneous radiative forcing can be estimated using the developed relationship and can also be used to explore the possible role of dust on climate modulation in the AS region.
- In future study, we will attempt to develop a dust detection algorithm for thin dust cloud and thereby address the relationship between AOD and SWARF for low intense dust storm over the AS.

Acknowledgements

The authors are grateful to Director NRSC for encouraging this work. We acknowledge the support of Aerosol Radiative Forcing over India (ARFI) project funded by ISRO-GBP. We also thank all the principal investigators and support staff of MODIS, CERES, AIRES and CALIPSO mission for their support in providing the necessary data for this study. The constructive comments of anonymous reviewers greatly helped to improve the manuscript.

References

Ackerman SA, Strabala KI, Menzel WP, Frey RA, Moeller CC, Gumley LE. 1998. Discriminating clear sky from clouds with MODIS. *Journal of Geophysical Research: Atmospheres (1984–2012)* **103**(D24): 32141–32157.

Aumann HH, Chahine MT, Gautier C, Goldberg MD, Kalnay E, McMillin LM, Susskind J. 2003. AIRS/AMSU/HSB on the aqua mission: design, science objectives, data products, and processing systems. *IEEE Transactions on Geoscience and Remote Sensing* **41**(2): 253–264.

Bellouin N, Boucher O, Haywood J, Reddy MS. 2005. Global estimate of aerosol direct radiative forcing from satellite measurements. *Nature* **438**(7071): 1138–1141.

Bellouin N, Jones A, Haywood J, Christopher SA. 2008. Updated estimate of aerosol direct radiative forcing from satellite observations and comparison against the Hadley Centre climate model. *Journal of Geophysical Research: Atmospheres (1984–2012)* **113**(D10), doi: 10.1029/2007JD009385.

Christopher SA, Zhang J. 2002. Shortwave aerosol radiative forcing from MODIS and CERES observations over the oceans. *Geophysical Research Letters* **29**(18): 1859, doi: 10.1029/2002GL014803.

Christopher SA, Zhang J, Kaufman YJ, Remer LA. 2006. Satellite-based assessment of top of atmosphere anthropogenic aerosol radiative forcing over cloud-free oceans. *Geophysical Research Letters* **33**(15), doi: 10.1029/2005GL025535.

Chung CE, Ramanathan V, Kim D, Podgorny IA. 2005. Global anthropogenic aerosol direct forcing derived from satellite and ground based observations. *Journal of Geophysical Research: Atmospheres (1984–2012)* **110**(D24), doi: 10.1029/2005JD006356.

DeMott PJ, Sassen K, Poellot MR, Baumgardner D, Rogers DC, Brooks SD, Kreidenweis SM. 2003. African dust aerosols as atmospheric ice nuclei. *Geophysical Research Letters* **30**(14), doi: 10.1029/2003GL017410.

Divakarla MG, Barnet CD, Goldberg MD, McMillin LM, Maddy E, Wolf W, Liu X. 2006. Validation of atmospheric infrared sounder temperature and water vapor retrievals with matched radiosonde measurements and forecasts. *Journal of Geophysical Research: Atmospheres (1984–2012)* **111**(D9), doi: 10.1029/2005JD006116.

Evan AT, Vimont DJ, Heidinger AK, Kossin JP, Bennartz R. 2009. The role of aerosols in the evolution of tropical North Atlantic Ocean temperature anomalies. *Science* **324**(5928): 778–781.

Gharai B, Jose S, Mahalakshmi DV. 2013. Monitoring intense dust storms over the Indian region using satellite data–a case study. *International Journal of Remote Sensing* **34**(20): 7038–7048, doi: 10.1080/01431161.2013.813655.

Huang J, Minnis P, Yi Y, Tang Q, Wang X, Hu Y, Liu Z, Ayers K, Trepte C, Winker D. 2007. Summer dust aerosols detected from CALIPSO over the Tibetan Plateau. *Geophysical Research Letters* **34**: L18805, doi: 10.1029/2007GL029938.

IPCC: Climate Change. 2007. *The Physical Science Basis. Contribution of Working Group I to the Fourth Assessment Report of the Intergovernmental Panel on Climate Change*, Solomon S, Qin D, Manning M, Chen Z, Marquis M, Averyt KB, Tignor M, Miller HL (eds). Cambridge University Press: Cambridge.

Kaskaoutis DG, Rashki A, Houssos EE, Goto D, Nastos PT. 2014. Extremely high aerosol loading over Arabian Sea during June 2008: The specific role of the atmospheric dynamics and Sistan dust storm. *Atmospheric Environment* **94**: 374–384, doi: 10.1016/j.atmosenv.2014.05.012.

Kaufman YJ, Tanre´ D, Dubovik O, Karnieli A, Remer LA. 2001. Absorption of sunlight by dust as inferred from satellite and ground based remote sensing. *Geophysical Research Letters* **28**: 1479–1482.

Kaufman YJ, Koren I, Remer LA, Rosenfeld D, Rudich Y. 2005. The effect of smoke, dust, and pollution aerosol on shallow cloud development over the Atlantic Ocean. *Proceedings of the National Academy of Sciences of the United States of America* **102**(32): 11207–11212.

Kumar S, Kumar S, Kaskaoutis DG, Singh RP, Singh RK, Mishra AK, Srivastava MK, Singh AK. 2015. Meteorological, atmospheric and climatic perturbations during major dust storms over Indo-Gangetic Basin. *Aeolian Research* **17**: 15–31, doi: 10.1016/j.aeolia.2015.01.006.

Qu JJ, Hao X, Kafatos M, Wang L. 2006. Asian dust storm monitoring combining Terra and Aqua MODIS SRB measurements. *IEEE Geoscience and Remote Sensing Letters* **3**: 484–486.

Reid JS, Hobbs PV, Ferek RJ, Blake DR, Martins JV, Dunlap MR, Liousse C. 1998. Physical, chemical and optical properties of regional hazes dominated by smoke in Brazil. *Journal of Geophysical Research* **103**: 32059–32080.

Shao Y, Dong CH. 2006. A review on East Asian dust storm climate, modelling and monitoring. *Global and Planetary Change* **52**(1): 1–22.

Tanré D, Kaufman YJ, Herman M, Mattoo S. 1997. Remote sensing of aerosol properties over oceans using the MODIS/EOS spectral radiances. *Journal of Geophysical Research* **102**: 16971–16988.

Tegen I, Lacis AA. 1996. Modeling of particle size distribution and its influence on the radiative properties of mineral dust aerosol. *Journal of Geophysical Research: Atmospheres (1984–2012)* **101**(D14): 19237–19244.

Wielicki BA, Barkstrom BR, Harrison EF, Lee RB III, Smith GL, Cooper JE. 1996. Clouds and the earth's radiant energy system (CERES): an earth observing system experiment. *Bulletin of the American Meteorological Society* **77**: 853–868.

Winker DM, Hunt WH, McGill MJ. 2007. Initial performance assessment of CALIOP. *Geophysical Research Letters* **34**(19), doi: 10.1029/2007GL030135.

Yu H, Dickinson RE, Chin M, Kaufman YJ, Zhou M, Zhou L, Holben BN. 2004. Direct radiative effect of aerosols as determined from a combination of MODIS retrievals and GOCART simulations. *Journal of Geophysical Research: Atmospheres (1984–2012)* **109**(D3), doi: 10.1029/2003JD003914.

7

Characterization of synoptic conditions and cyclones associated with top ranking potential wind loss events over Iberia

Melanie K. Karremann,[1,2]* Margarida L. R. Liberato,[3,4] Paulina Ordóñez[3,5] and Joaquim G. Pinto[1,6]

[1]Institute for Geophysics and Meteorology, University of Cologne, Cologne, Germany
[2]Now at: Institute of Meteorology and Climate Research, Karlsruhe Institute of Technology, Karlsruhe, Germany
[3]Escola de Ciências e Tecnologia, Universidade de Trás-os-Montes e Alto Douro, Vila-Real, Portugal
[4]Instituto Dom Luiz (IDL), Universidade de Lisboa, Lisboa, Portugal
[5]Centro de Ciencias de la Atmósfera, Universidad Nacional Autónoma de México, Mexico City, Mexico
[6]Department of Meteorology, University of Reading, Reading, UK

*Correspondence to:
M. K. Karremann, Institute of
Meteorology and Climate
Research (IMK), Karlsruhe
Institute of Technology,
Hermann-von-Helmholtz-Platz 1,
76344
Eggenstein-Leopoldshafen,
Germany.
E-mail:
melanie.karremann@kit.edu

Abstract

Intense extra-tropical cyclones are often associated with strong winds, heavy precipitation and socio-economic impacts. Over southwestern Europe, such storms occur less often, but still cause high economic losses. We characterize the large-scale atmospheric conditions and cyclone tracks during the top-100 potential losses over Iberia associated with wind events. Based on 65 years of reanalysis data, events are classified into four groups: (1) cyclone tracks crossing over Iberia on the event day ('Iberia'), (2) cyclones crossing further north, typically southwest of the British Isles ('North'), (3) cyclones crossing southwest to northeast near the northwest tip of Iberia ('West'), and (4) so called 'Hybrids', characterized by a strong pressure gradient over Iberia because of the juxtaposition of low and high pressure centres. Generally, 'Iberia' events are the most frequent (31–45% for top-100 *vs* top-20), while 'West' events are rare (10–12%). Seventy percent of the events were primarily associated with a cyclone. Multi-decadal variability in the number of events is identified. While the peak in recent years is quite prominent, other comparably stormy periods occurred in the 1960s and 1980s. This study documents that damaging wind storms over Iberia are not rare events, and their frequency of occurrence undergoes strong multi-decadal variability.

Keywords: windstorms; Iberia; cyclones; impacts; potential wind losses; multi-decadal variability

1. Introduction

Extra-tropical cyclones are determinant for the weather conditions in the mid-latitudes. Embedded in the westerly flow, cyclones typically undergo a strong intensification over the North Atlantic Ocean while travelling towards Europe. Intense cyclones are often associated with strong winds and heavy precipitation (Pfahl, 2014), thus often leading to large socio-economic impacts (Swiss Re, 2008). Over southwestern Europe, such intense cyclones occur less often (Trigo, 2006; Pinto *et al.*, 2009). This can be explained by the cyclone track climatology (Figure 1(a)), which features a reduced number of systems near Iberia. Intense cyclones affecting Iberia typically cause high amounts of precipitation and flooding in that area (Ramos *et al.*, 2014). Still, recent storms like 'Klaus' (cyclone names after Freie Universität Berlin database, www.met.fu-berlin.de/adopt-a-vortex/historie/) (Liberato *et al.*, 2011) were primarily characterized by very strong winds, leading to €541 m insured losses in Spain because of wind gusts (CCS, 2015). In winter 2013/2014, several storms affected the Iberian

Peninsula, including storm 'Dirk' [20131224 (all dates in yyyymmdd)], which caused €30.3 m insured losses in Spain because of both wind gusts and floods (CCS, 2015). While previous studies analysed single storms (e.g. 'Klaus' and 'Xynthia'; Liberato *et al.*, 2011, 2013), a climatological assessment is missing. Based on 65 years of reanalysis data, we characterize the large-scale atmospheric conditions and cyclone tracks associated with the top-100 potential losses over Iberia because of strong wind events (windstorms).

2. Data

The analysis of potential wind loss events over Iberia is performed based on the National Centre for Environmental Prediction/National Centre for Atmospheric Research reanalysis (hereafter NCEP; Kistler *et al.*, 2001). As no gust wind speed is available for this dataset, 6-hourly instantaneous 10 m-wind speeds (hereafter wind) are analysed. NCEP reanalysis is provided on a T62 resolution (Figure 1(b); 1.875°). For each grid point ij, daily maximum wind speeds (largest values for each day between 00, 06, 12 and 18

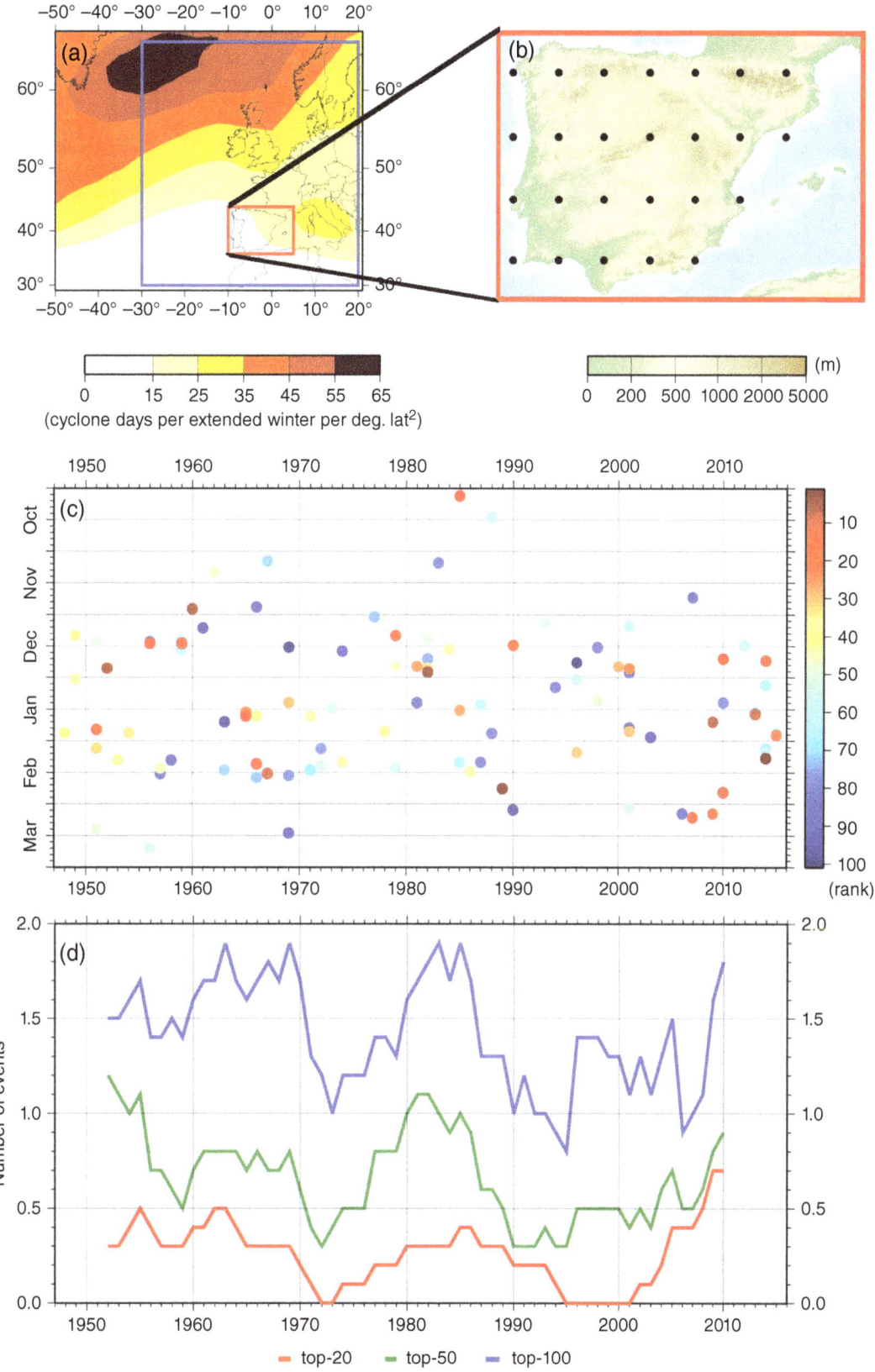

Figure 1. (a) Cyclone track density [cyclone days per extended winter season per deg. lat^2]; red box: geographical location of Iberian Peninsula; blue box: area for identification of cyclone tracks, (b) orography and NCEP grid points of investigated region, (c) time distribution of top-100 potential wind loss events for winters 1949–2015. Year corresponds to January, e.g. 1990: winter 1989/1990. Colours: rank of event (d) 10-year running mean of the number of events for each winter; red: top-20 potential wind losses; green: top-50, blue: top-100 wind loss events; e.g. 1953: running mean of 1949–1958. Average number of events per winter: top-100 potential wind losses: 1.5; top-50: 0.75; top-20: 0.3.

UTC, denoted v_{ij}) for the extended winter (October to March) 1948/1949 to 2014/2015 are selected. Winters are named by the second year, e.g. winter 2014/2015 is named 2015. Based on the climatology, 98th wind percentiles ($v_{98_{ij}}$) are calculated, and are used as a threshold for the loss model. The 6-hourly mean sea level pressure (MSLP) data are used for cyclone tracking and computation of the MSLP gradient over Iberia [$10°W-2.5°E; 35°-45°N$].

Potential loss event rankings may differ between reanalysis datasets (Karremann *et al.*, 2014a). This is also true for cyclone characteristics (Trigo, 2006). A preliminary analysis revealed a similar list of top events for NCEP and ERA-Interim (cf. Table S1, Supporting Information). Our focus on NCEP is motivated by the longer time series, which enables a better representation of long-term variability of events affecting Iberia.

3. Methodology

3.1. Event identification

This study uses a simplified approach of previous empirical models (Klawa and Ulbrich, 2003; Pinto *et al.*, 2012; Karremann *et al.*, 2014a) considering only meteorological parameters to estimate potential losses based on gridded data. The main assumptions are: (1) a critical wind speed needs to be exceeded to cause any loss. In most parts of Europe, this threshold corresponds to $v_{98_{ij}}$ (Klawa and Ulbrich, 2003), (2) a strong non-linearity in the wind – loss relation is assumed as the kinetic energy flux is proportional to the cube of wind speed (Mills, 2005). Following Pinto *et al.* (2012), the potential wind loss (MI) over Iberia (Figure 1(b)) per day is defined as:

$$\text{MI (day)} = \sum_{i=1}^{N} \sum_{j=1}^{M} \left(\frac{v_{ij}}{v_{ij}^{98}} \right)^3 \times I\left(v_{ij}, v_{ij}^{98}\right) \quad (1)$$

$$I\left(v_{ij}, v_{ij}^{98}\right) = \begin{cases} 0 & v_{ij} < v_{ij}^{98} \\ 1 & v_{ij} > v_{ij}^{98} \end{cases}$$

The number of analysed grid points in the longitudinal and latitudinal directions is given by M and N, respectively. Based on all identified MI events, a ranking is established. The top-100 MI events between October 1948 and March 2015 are selected for detailed analysis.

3.2. Cyclone tracking

Cyclone tracks are derived with a cyclone tracking algorithm (Murray and Simmonds, 1991; Pinto *et al.*, 2005). The Laplacian of MSLP is used as proxy for the relative geostrophic vorticity and is used for cyclone identification. Cyclone tracks are compiled by considering the most probable trajectory of the systems between subsequent time frames (estimated from MSLP gradients, past cyclone speed and intensity).

3.3. Assignment of top loss events with cyclone tracks

The top-100 MIs are characterized in terms of large-scale atmospheric conditions and the presence of low pressure centres using NCEP data and weather charts (Berliner Wetterkarte, 2009, 2011, 2014). If cyclone tracks are located on the event day within $30°W-20°E$, $30°N-65°N$ (blue box in Figure1(a)), they are preliminary assigned to the event. These cyclone tracks and associated windstorm footprints are analysed: the cyclone which matches best with the windstorm footprint in terms of timing and overlap with Iberia is subjectively selected as potentially responsible for MIs over Iberia.

If potential loss event days are identified on subsequent dates and both events are associated with the same cyclone, the dates are combined as one event. In this case, MI is recalculated by summing up the maximum exceedance of the 98th percentile within these subsequent days at each grid point ij. Thus, the following MI events (below top-100) are added to the list until the top-100 is complete again. The final list of top-100 MIs is shown in Table S1.

4. Analysis of the top-100 events

The time distribution of top-100 MIs over the 65-year period is displayed in Figure 1(c). About 84% of the events occurred between December and February, and only 2% in October, 6% in November and 8% in March. Colours indicate the ranking of storms. A clear dependency between the ranking of events and the seasonality is not found (Figure 1(c)). On the one hand, the number of events changes strongly from year to year. The largest number of events is identified for 2001 (six events) and 2014 (five events), while for other periods a maximum of two events per winter is found, e.g. in most of the 1970s, 1990s and 2000s. Decadal variability is analysed using a 10-year running average of the number of events per winter for different intensities (Figure 1(d)). Periods with a reduced number of events include 1957–1960, the 1970s and 1994–2004, while periods in the 1960s, 1980s and after 2005 display more events than average. While this result is largely independent from the intensity of events (colours in Figure 1(d)), the peak in recent years is very prominent for the top-20 MIs (red curve). Longer periods with a low number of top-20 MIs are identified in the 1970s, 1990s and early 2000s. Thus, decadal variability in the number of events is identified for all intensities.

For each event, synoptic conditions and associated cyclone tracks are characterized following the methodology described in Section 3. The location of the pressure minima for the top-100 events is depicted in Figure 2(a). The pressure minima of the top-20 MIs (red colours in Figure 2(a)) are mostly identified close to the north of Iberia, roughly between $15°W-5°E$ and $38°-58°N$. For lower ranking events,

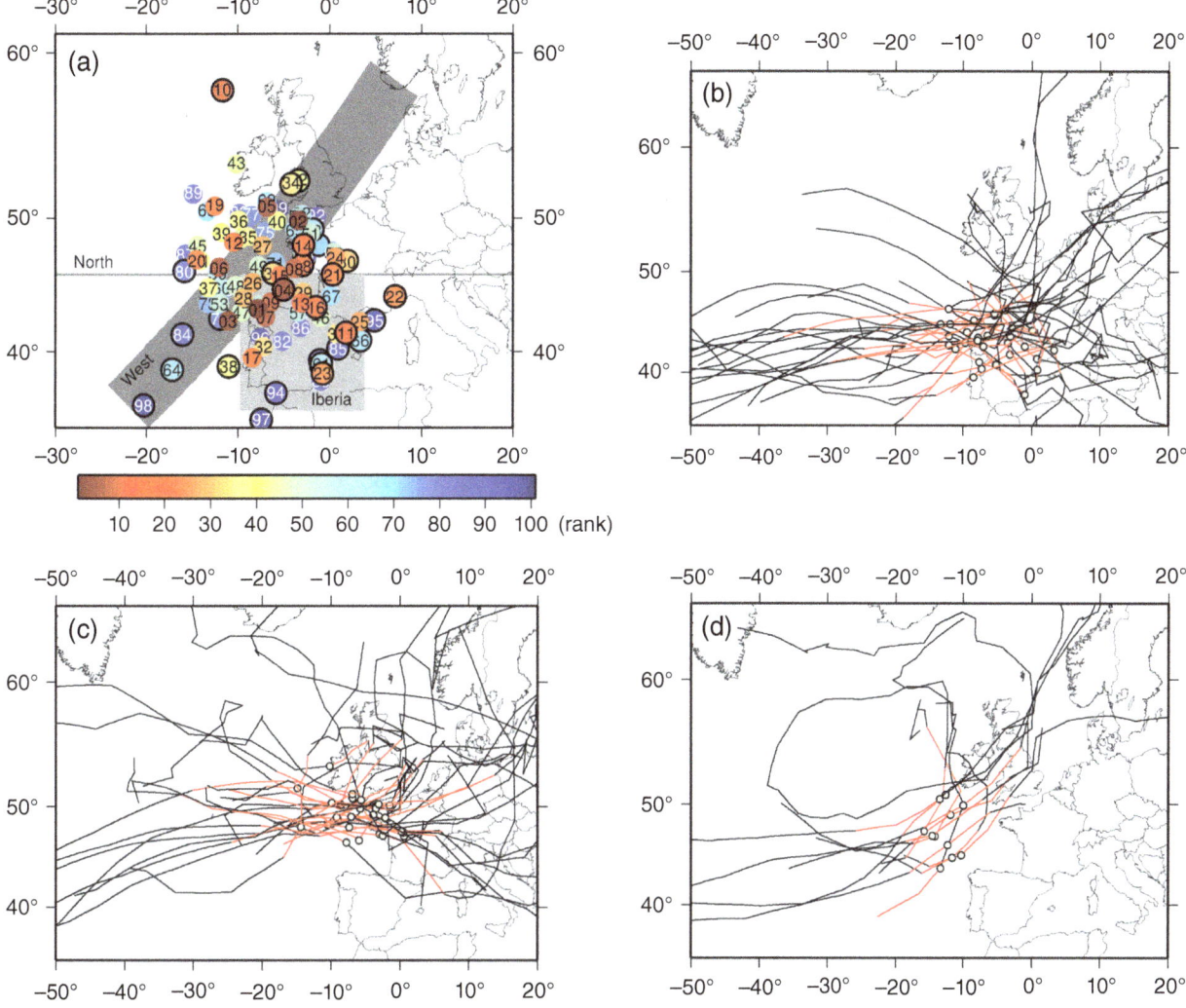

Figure 2. (a) Position of minimum pressure of identified cyclones responsible for potential wind losses over the Iberian Peninsula at the event day. Colours/numbers: rank of event; black circle: cyclones associated with Hybrid type. Identification of regions of the different groups: Iberia (light grey shaded region): cyclones crossing this area during the event day; North: cyclones crossing from west to east in a zonal path within this region during the event day; West (dark grey shaded region): cyclones crossing from southwest to northeast along the dark grey shaded region. (b) Cyclone tracks of group *Iberia* (31 events). (c) Cyclone tracks within *North* (28 events). (d) Cyclone tracks for *West* group (11 events). (b)–(d) red: cyclone track during the event day; circles: corresponding to (a).

the region is wider. It is notable that only four events from the extreme windstorms database [XWS; Roberts *et al.*, 2014; their Table 1 (cf. http://www.european windstorms.org for updates on the XWS database)] are present in our event list, revealing that the top events for Iberia (Table S1) are largely different from those affecting other parts of Europe (cf. also Karremann *et al.*, 2014b).

The identified cyclones were grouped based on their characteristics. Seventy percent of the analysed events are strongly influenced by cyclones:

1. Cyclone tracks crossing over Iberia at the event day (box 'Iberia' in Figure 2(a)) and thus with a direct influence (group '*Iberia*', cf. Figure 2(b)).
2. Cyclone tracks crossing north of Iberia on a zonal track, mostly southwest from the British Isles (region 'North' in Figure 2(a)), and influencing

Iberia primarily because of their extended fronts (group '*North*', Figure 2(c)).
3. Cyclone tracks crossing from southwest to northeast, but west of Iberia (cf. grey area 'West' in Figure 2(a)) and not intersecting the *Iberia* box (group '*West*', Figure 2(d)).

In some cases, the cyclone was not found to be determinant *per se* for the windstorm footprint from the selected event, but rather its co-occurrence with a high pressure centre on the opposite side of Iberia, which led to a strong pressure gradient over the region and consequently strong winds (cf. also Pfahl, 2014). Thus, a final category is defined as follows:

4. Synoptic situation with the juxtaposition of a cyclone and an anticyclone, leading to a pronounced MSLP gradient over Iberia and thus strong winds (group '*Hybrid*', cyclone tracks shown in

Table 1. Number of events [%] per group and intensity.

	Iberia	North	West	Hybrid
Top-20	45	15	10	30
Top-50	36	24	12	28
Top-100	31	28	11	30

Top-20, top-50 and top-100 potential wind loss events between October 1948 and March 2015.

Figure S1). Events are marked with a black circle in Figure 2(a).

The assignment of events into the four groups is shown in Table S1, and composite MSLP fields in Figure S2. For the top-100 MIs, most events are classified as *Iberia* (31%), closely followed by *Hybrid* (30%), *North* (28%) and *West* (11%, Table 1). The relative importance of the *Iberia* group increases with intensity of MI, reaching 45% for top-20 MIs. Conversely, the percentage of *North* storms decreases with intensity, reaching 15% for top-20 MIs. Top ranking events typically affect a larger area (more grid points; Table S1), although the relationship is not strong. While all four groups show a dichotomous pattern with a high and a low pressure system (cf. Pfahl, 2014), the analysis provides evidence that in 70% (all but *Hybrid* group), the cyclones are primarily responsible for the strong winds over Iberia and thus the MI event.

5. Characterization of the four groups

The general characteristics of each group are presented in this section including a representative case study. Like many East Atlantic cyclones, most of the systems considered here are secondary cyclones, which develop on the trailing cold fronts of 'parent' cyclones located further north (Dacre and Gray, 2013).

5.1. Group *Iberia*

The cyclones in this group cross over Iberia (Figure 2(b)), leading to a direct impact. This group includes named storms like 'Klaus' (20090123), 'Xynthia' (20100227) and 'Gong' (20130119). Storm 'Stephanie' (Top#1, 20140209) is selected as representative example. A high pressure at 500 hPa is identified over the subtropical North Atlantic, and a low pressure system north of Scotland (not shown). The corresponding surface low pressure centre ('Ruth', Figure 3(a)) is below 970 hPa, and the Azores high is above 1025 hPa, leading to an intense westerly flow towards West- and Central Europe. 'Stephanie' is a secondary low developing in this strong westerly flow, and is located over the Bay of Biscay on 10 January, 00 UTC (Figure 3(a)). The associated windstorm footprint affected the whole Iberian Peninsula (Figure 3(b)). Hurricane-force winds, snow and rain were reported, causing the strongest damage in western regions where wind uprooted numerous trees, broke windows and blew roofs away while high waves and flooding disrupted roads.

5.2. Group *North*

This group is characterized by cyclones crossing north of Iberia in a zonal track, typically over the British Channel (Figure 2(c)). Such cyclones usually feature extended cold fronts, leading to strong winds further south over Iberia. This group includes named storms 'Martin' (19991227) and 'Anne' (20140104). Storm 'Joachim' (Top#62, 20111216) is selected as representative case. High pressure at 500 hPa is identified over the subtropical North Atlantic, while the mid-level low pressure system is located west of Iceland (not shown). The corresponding surface low pressure centre ('Hergen', Figure 3(c)) is below 970 hPa, and the Azores high is above 1030 hPa. Strong westerly flow dominates at upper levels, in which cyclone 'Joachim' is embedded (not shown). Beside the difference of air masses, a short wave trough influenced its explosive development. The core pressure deepened from 1008 to 980 hPa within 24 h bringing severe wind gusts and heavy rainfall to many parts of northern Iberia, causing numerous incidents such as falling trees, fences and streetlights or landslides.

5.3. Group *West*

This group is characterized by cyclones crossing from the southwest to the northeast, typically close to the northwestward tip of Iberia (Figure 2(d)), and includes storms like the 'Great Storm of 1987' (19871015). Storm 'Qumaira' (Top#59, 20140206) is selected as a representative example. A blocking high pressure system is located over the subtropical North Atlantic and mid-level low pressure centres are located between Greenland and Scotland (not shown). The surface steering low 'Petra' is located over Scotland (Figure 3(e)), while the secondary pressure system 'Qumaira' (980 hPa) moves northeastward towards southern England. 'Qumaira' led to important disruption to all forms of transport in Iberia, together with power cuts and building and tree damage.

5.4. Group *Hybrids*

This group is characterized by a co-occurrence of a high and a low pressure centre on opposite sides of Iberia, leading to a pronounced MSLP gradient and strong winds over the region. A prominent example is 20090305 (Top#21). A low pressure system is located north of Scotland, while a blocking high is found over the North Atlantic (not shown). The juxtaposition of both systems is associated with an intense jet stream from southern Greenland towards Iberia (not shown). At lower levels, the surface high pressure system is above 1040 hPa, while cyclone 'Andreas' located between Scotland and the Faroese has a core pressure of 975 hPa (Figure 3(g)). Further south, over the English Channel, an unnamed low below 985 hPa is found. This example shows that the cyclone over southern England cannot be primarily responsible for the

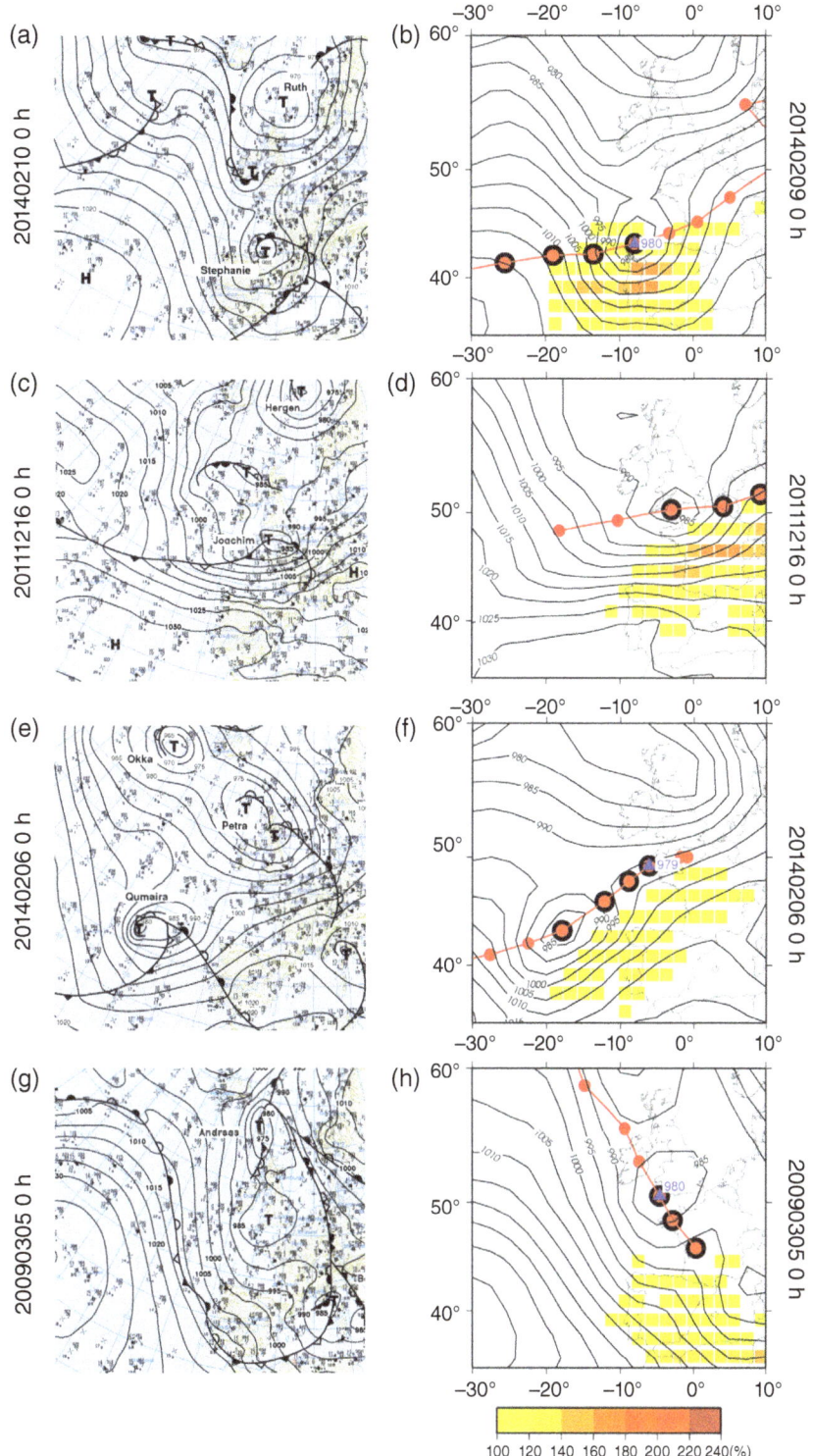

Figure 3. Case study for (a, b) group *Iberia*, (c, d) group *North*, (e, f) group *West*, (g, h) group *Hybrid*. Left panels: surface weather charts adapted with courtesy of Berliner Wetterkarte e.V.; right panels: windstorm footprints and responsible cyclone tracks; black circle: time frames corresponding to event day, blue triangle: pressure minimum at event day [for (d) outside shown area]; colours: exceedance of $v_{98_{ij}}$ in (%); date: yyyymmdd time in UTC.

potential loss event over Iberia; only the combination of the different systems led to a strong pressure gradient and thus strong surface winds. The impacts were severe: many trees were uprooted and a great number of buildings were damaged, also in southern Spain and even Morocco.

6. Comparison of cyclone characteristics

The difference of cyclone characteristics between the groups is now explored. In terms of the minimum core pressure, the lowest minimum mean value for the event day (±1 day) was found for *West* (966 hPa),

Table 2. Mean value and standard deviation for different characteristics for the four groups: pressure minimum at the event day (hPa); Laplace maximum at the event day (hPa deg.lat^{-2}); MSLP gradient at the event day (hPa/100 km); pressure evolution (hPa/24 h) and Laplace evolution (hPa deg.lat^{-2}/24 h) within 24 h.

	Iberia	North	West	Hybrid
Pressure minimum (event day)	983 ± 9.31	976 ± 13.1	966 ± 13.3	983 ± 15.8
Laplace maximum (day)	1.63 ± 0.68	1.51 ± 0.67	1.95 ± 0.95	1.28 ± 0.76
MSLP gradient (day)	1.64 ± 0.46	1.83 ± 0.39	1.81 ± 0.44	1.72 ± 0.44
Pressure evolution (24 h)	13.5 ± 8.46	14.8 ± 8.26	12.7 ± 9.46	13.2 ± 8.93
Laplace evolution (24 h)	0.95 ± 0.60	0.84 ± 0.51	1.13 ± 0.64	0.84 ± 0.64

followed by *North* (976 hPa), *Iberia* and *Hybrid* (both 983 hPa; Table 2). Given the small samples and large spread, the differences between groups are not statistically significant at the 90% significance level. Results are similar for the maximum vorticity, with largest values found for *West* and the lowest for *Hybrid* (1.95 and 1.28 hPa deg.lat^{-2}, respectively). The intensification rates per 24 h, in terms of vorticity, are highest for *West* and *Iberia* [1.13 and 0.95 (hPa deg.lat^{-2}) day^{-1}] and lowest for *North* and *Hybrid* [both 0.84 (hPa deg.lat^{-2}) day^{-1}], while core pressure evolution is highest for *North* (14.8 hPa day^{-1}) and lowest for *West* (12.7 hPa day^{-1}). Results are similar when analysing the 24 h peak intensity change irrespective of the event day (not shown). Unlike other types, the deepening cyclone phase associated with *Hybrid* events does not often coincide with the event day (±1 day). This indicates that cyclones contributing to *Hybrid* events are not necessarily very intense and their development does not always have a clear impact on the events. The mean MSLP gradient around Iberia is lowest for *Iberia* (1.64 hPa/100 km) and largest for *North* and *West* (1.83 and 1.81 hPa/100 km, respectively). The gradient for *Hybrid* is in between (1.72 hPa/100 km). This confirms that despite the weaker cyclones in the *Hybrid* group, the co-occurrence of a high pressure centre on the opposite side of Iberia effectively creates a strong pressure gradient leading to strong winds.

7. Summary and conclusions

The top ranking potential wind loss events affecting Iberia were classified into four groups based on cyclone tracks and large-scale atmospheric conditions: (1) cyclone tracks crossing Iberia on the event day (*Iberia*), (2) cyclones crossing further north, mostly southwest of the British Isles (*North*), (3) tracks crossing southwest to northeast to the northwest of Iberia (*West*), and (4) so called *Hybrids*, days with a large pressure gradient over Iberia because of the juxtaposition of a low and a high pressure centre. Generally, *Iberia* events are the most frequent, ranging from 31% (top-100 MIs) to 45% (top-20 MIs). However, other types like *North* and *Hybrid* are also frequent (28 and 30%, respectively, for top-100 MIs). The number of *North* storms decreases considerably with intensity (15% for top-20 MIs). *West* type storms are rare (10–12%). Seventy percent of the MIs can be primarily

attributed to cyclones. Cyclones associated with *Hybrid* events (30%) are typically weaker than for other cases, but the mean MSLP gradient over Iberia is comparable to other types. Although we have focussed on a single reanalysis dataset, the results would be comparable for others given our focus on large-scale features like MSLP gradients, cyclone tracks and windstorm footprints. Multi-decadal variability of events is identified for all intensities. The peak in recent years is quite prominent in terms of the number of top-20 MIs. Other periods with a large number of storms occurred in the 1960s and 1980s. This study documents that windstorms affecting Iberia may have different characteristics, they are not rare events, and their frequency of occurrence undergoes strong multi-decadal variability.

Acknowledgements

This work was supported through project STORMEx FCOMP-01-0124-FEDER-019524 (PTDC/AAC-CLI/121339/2010) by FEDER funds (COMPETE) and Portuguese National Funds (FCT). We thank the open access fund of the University of Reading for support. We thank NCEP/NCAR for reanalysis, the Berliner Wetterkarte for weather charts, Helen Dacre for comments and Sven Ulbrich for cyclone computations. We thank the three anonymous reviewers for their helpful suggestions.

Supporting information

The following supporting information is available:

Appendix S1. Supplementary material on event selection and assignment to the four groups.

Figure S1. As Figure 2(b) but for group *Hybrid*.

Figure S2. MSLP composites for (a) group *Iberia* (31 dates) (b) group *North* (28 dates) (c) group *West* (11 dates) (d) group *Hybrid* (30 dates). All four MSLP fields (00, 06, 12, 18 UTC) of each event day were included in the composites. For a list of dates see Table S1.

Table S1. List of top-100 potential wind loss events including information on the date, the number of grid points exceeded (GP), the rank (rk) and the corresponding group (G): *Iberia* (I), *North* (N), *West* (W), *Hybrid* (H) for NCEP data. Date format is yyyymmdd. The top-20 events are in bold and italic. Rank of events also featured in the ERA-Interim top-50 (rE) for the period 1979–2014 are added in the last column for comparison. '-' indicates the event was not found within the top-50 for ERA-Interim. 'x' indicates date outside the analysed ERA-Interim period.

References

Berliner Wetterkarte. 2009, 2011, 2014. Published by Society Berliner Wetterkarte e.V., Freie Universität Berlin and German Weather Service. ISSN: 0177-3984. www.berliner-wetterkarte.de.

CCS. 2015. *Estadística – Riesgos extraordinarios. Serie 1971–2014*. Consorcio Compensación de Seguros, Ministerio de Economía y Competitividad: Madrid. 146 pp. NIPO: 720-15-101-5 (in Spanish). www.consorseguros.es.

Dacre HF, Gray SL. 2013. Quantifying the climatological relationship between extratropical cyclone intensity and atmospheric precursors. *Geophysical Research Letters* **40**: 2322–2327, doi: 10.1002/grl.50105.

Karremann MK, Pinto JG, Von Bomhard PJ, Klawa M. 2014a. On the clustering of winter storm loss events over Germany. *Natural Hazards and Earth System Sciences* **14**: 2041–2052, doi: 10.5194/nhess-14-2041-2014.

Karremann MK, Pinto JG, Reyers M, Klawa M. 2014b. Return periods of losses associated with European windstorm series in a changing climate. *Environmental Research Letters* **9**: 124016, doi: 10.1088/1748-9326/9/12/124016.

Kistler R, Kalnay E, Collins W, Saha S, White G, Woollen J, Chelliah M, Ebisuzaki W, Kanamitsu M, Kousky V, van den Dool H, Jenne R, Fiorino M. 2001. The NCEP/NCAR 50-year reanalysis: monthly-means CDROM and documentation. *Bulletin of the American Meteorological Society* **82**: 247–267, doi: 10.1175/1520-0477 (2001)082.

Klawa M, Ulbrich U. 2003. A model for the estimation of storm losses and the identification of severe winter storms in Germany. *Natural Hazards and Earth System Sciences* **3**: 725–732, doi: 10.5194/nhess-13-2239-2013.

Liberato MLR, Pinto JG, Trigo IF, Trigo RM. 2011. Klaus – an exceptional winter storm over Northern Iberia and Southern France. *Weather* **66**: 330–334, doi: 10.1002/wea.755.

Liberato MLR, Pinto JG, Trigo RM, Ludwig P, Ordoñez P, Yuen D, Trigo IF. 2013. Explosive development of winter storm Xynthia over the subtropical North Atlantic Ocean. *Natural Hazards and Earth System Sciences* **13**: 2239–2251, doi: 10.5194/nhess-13-2239-2013.

Mills E. 2005. Insurance in a climate of change. *Science* **309**: 1040–1044, doi: 10.1126/science.1112121.

Murray RJ, Simmonds I. 1991. A numerical scheme for tracking cyclone centres from digital data. Part I: Development and operation of the scheme. *Australian Meteorological Magazine* **39**: 155–166.

Pfahl S. 2014. Characterising the relationship between weather extremes in Europe and synoptic circulation features. *Natural Hazards and Earth System Sciences* **14**: 1461–1475, doi: 10.5194/nhess-14-1461-2014.

Pinto JG, Spangehl T, Ulbrich U, Speth P. 2005. Sensitivities of a cyclone detection and tracking algorithm: individual tracks and climatology. *Meteorologische Zeitschrift* **14**: 823–838, doi: 10.1127/0941-2948/2005/0068.

Pinto JG, Zacharias S, Fink AH, Leckebusch GC, Ulbrich U. 2009. Factors contributing to the development of extreme North Atlantic cyclones and their relationship with the NAO. *Climate Dynamics* **32**: 711–737, doi: 10.1007/s00382-008-0396-4.

Pinto JG, Karremann MK, Born K, Della-Marta PM, Klawa M. 2012. Loss potentials associated with European windstorms under future climate conditions. *Climate Research* **54**: 1–20, doi: 10.3354/cr01111.

Ramos AM, Trigo RM, Liberato MLR. 2014. A ranking of high-resolution daily precipitation extreme events for the Iberian Peninsula. *Atmospheric Science Letters* **15**: 328–334, doi: 10.102/asl2.507.

Roberts JF, Champion AJ, Dawkins LC, Hodges KI, Shaffrey LC, Stephenson DB, Stringer MA, Thornton HE, Youngman BD. 2014. The XWS open access catalogue of extreme windstorms from 1979–2012. *Natural Hazards and Earth System Sciences* **14**: 2487–2501, doi: 10.5194/nhess-14-2487-2014.

Swiss Re. 2008. Natural catastrophes and man-made disasters in 2007: high losses in Europe. In *Sigma, Nr. 1/2008*. Swiss Re Publishing: Zurich. www.swissre.com/sigma/?year=2008#.

Trigo IF. 2006. Climatology and interannual variability of storm-tracks in the Euro-Atlantic sector: a comparison between ERA-40 and NCEP/NCAR reanalyses. *Climate Dynamics* **26**: 127–143, doi: 10.1007/s00382-005-0065-9.

8

Changes in tropical cyclone activity offset the ocean surface warming in Northwest Pacific: 1981–2014

Chiung-Wen June Chang,[1,*] S.-Y. Simon Wang[2] and Huang-Hsiung Hsu[3]

[1] Department of Atmospheric Sciences, Chinese Cultural University, Taipei, Taiwan
[2] Department of Plants, Soils and Climate and Utah Climate Center, Utah State University, Logan, UT, USA
[3] Research Center for Environmental Changes, Academia Sinica, Taipei, Taiwan

*Correspondence to:
C.-W. June Chang, Department
of Atmospheric Sciences, Chinese
Culture University, Taipei 11114,
Taiwan.
E-mail: c.june.chang@gmail.com

Abstract

Tropical cyclones (TCs) leave a cold wake in the sea surface temperature (SST). In the northwest Pacific, TC activity and SST have both increased since the 1980s, but the extent to which ocean surface warming is affected by the changing TC activity is unknown. Analysis of the 1981–2014 period indicates that the intensified effect of TC cold wakes has offset the SST warming trend by 37% during the typhoon season, implying that the observed SST warming might be underestimated. This factor could affect long-term climate simulations that are forced with prescribed SST.

Keywords: northwest pacific; tropical cyclones; cold wakes; warming offset

1. Introduction

Tropical cyclones (TCs) are known to leave behind a cold wake in the sea surface temperature (SST) that extends for hundreds of kilometers (Hart *et al.*, 2007; Dare and McBride, 2011). As an example, Figure 1(a) shows a TC cold wake in which a noticeable cold-water trail of Typhoon Ma-On (Category 4) is observed in the SST with a local maximum cooling of 5.5 °C. The northwest Pacific warm pool is a hotspot for TCs (Figure 1(b); the climatological TC frequency) and it also features the strongest TC-induced cooling than other ocean basins (Sriver and Huber, 2007; Gentemann and Scott, 2014). There, recent warming in the upper ocean has led to increases in average TC intensity (Mei *et al.*, 2015) and frequency of strong typhoons (Wu and Zhao, 2012). The long-term change in TC cold wakes has an important, but perhaps overlooked implication on SST trends – that is, to what extent does the TC-induced cooling and associated change affect the warming trend in SST? The answer to this question cannot be found in the literature and is examined herein.

TC cold wakes are primarily a result of wind-induced upwelling and vertical mixing of cooler subsurface water beneath the TC (Price, 1981; Lin *et al.*, 2009; Mei and Pasquero, 2013). Generally, it takes 1–2 weeks for the TC-induced cooling to rebound and 4–6 weeks to fully regain its climatological values (Hart *et al.*, 2007; Dare and McBride, 2011), depending on the ocean state such as a shallow mixed layer depth (Jacob and Shay, 2003; Vincent *et al.*, 2012) and the translational speed of the TC (Emanuel, 2003). TCs cool the ocean surface through pumping heat into the subsurface (Sriver and Huber, 2007), and the excess heat is either redistributed poleward (Emanuel, 2001) or reabsorbed by the mixed layer and lost to the atmosphere (Jansen *et al.*, 2010).

Projecting future changes in the TC heat pump effect is difficult due to challenges in simulating the TC–ocean interaction by most climate models (Yablonsky *et al.*, 2015).

The pattern of the 1980–2014 trend in the Pacific Ocean SST appears to be asymmetric, consisting of a broad warming area in the west and slight cooling in the east (Guan and Nigam, 2008; Schubert *et al.*, 2009), as is shown in Figure 1(c) for the typhoon season. Regionally, the effect of TC cold wakes can rival the scale of a pre-existing warm SST anomaly. For instance, TC cold wakes in the Caribbean Sea can mitigate coral bleaching by slowing down the build-up of thermal stress associated with seasonal SST warming (Carrigan and Puotinen, 2014). Therefore, as the ocean becomes warmer, the subsequent increases in TC activity and/or intensity could enhance cooling over the ocean surface. If this is the case, then it would produce a negative feedback that offsets the warming trend of SST. The goal of this study is to examine this potential offset and quantify the effect.

2. Data and method

2.1. Data sources

The NOAA 1/4° daily optimum interpolation sea surface temperature (OISST) allows the examination for the long-term change in TC cold wakes starting from 1981. OISST was constructed by combining observations from different platforms including satellites, ships, and buoys with full-year data (Reynolds *et al.*, 2007). To depict TC positions, 6-h interval best track records of TCs were obtained from the Joint Typhoon Warning Center (JTWC). Among other TC track data sets, Wu and Zhao (2012) found that the JTWC data set

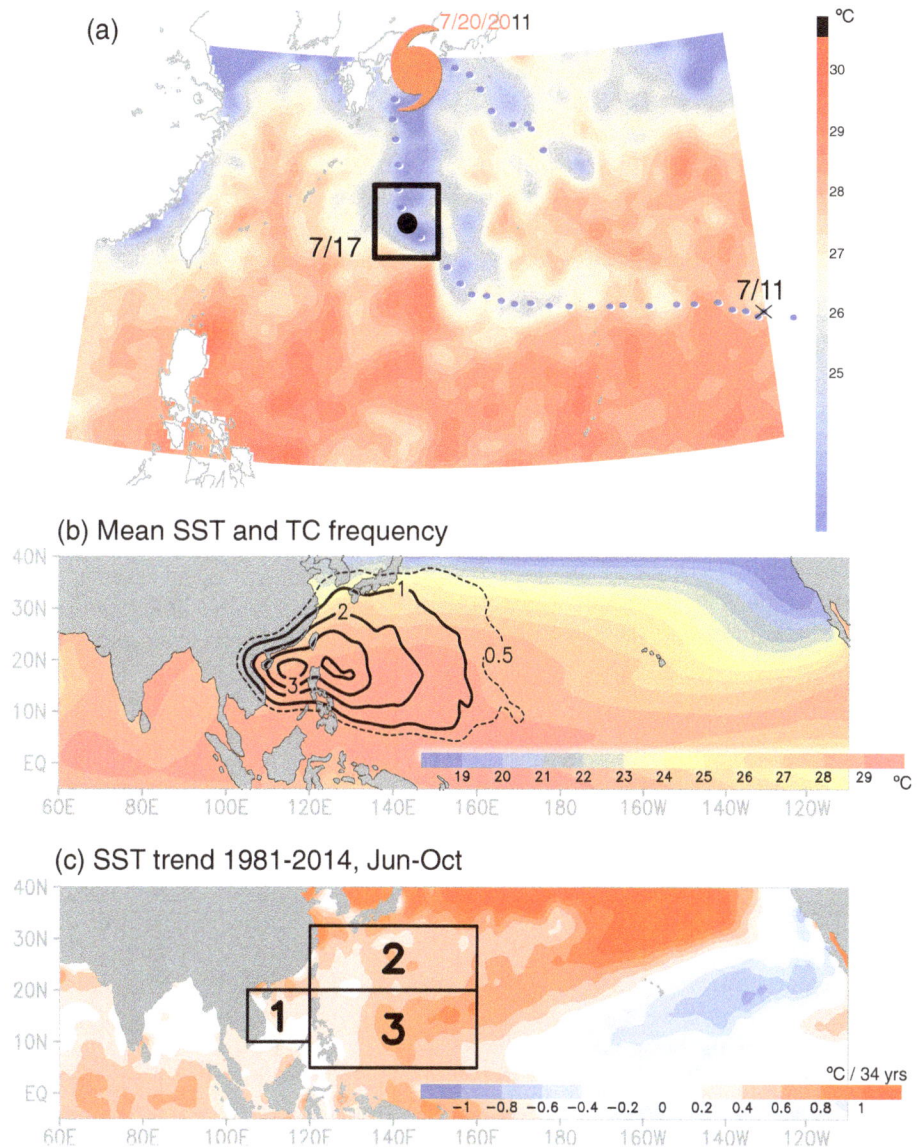

Figure 1. (a) SST on 20 July 2011 for Typhoon Ma On (typhoon symbol) and its track (blue dots); the black box and the center black dot indicates the typhoon's 17 July location surrounded by the 4° × 4° domain used for TC tracking (see text). (b) Seasonal mean SST (shadings) overlaid with the long-term areal frequency of TC best tracks per 2° × 2° grid (contours). (c) Linear trend pattern of SST from 1981 to 2014 during JJASO (total change) overlaid with the three domains used for Figure 3.

produced the TC intensity trends that are in good agreement with those derived dynamically from changes in SST, vertical wind shear, and prevailing tracks. For subsurface ocean temperatures, we utilized the NCEP Global Ocean Data Assimilation System (GODAS) produced by the Geophysical Fluid Dynamics Laboratory's Modular Ocean Model with a horizontal resolution of 1° × 1° enhanced to 1/3° in the north–south direction within 10° of the equator, and 40 vertical levels down to 4000 m (Behringer and Xue, 2004).

2.2. TC cold wake tracking

In order to capture the general size of a TC based upon the grid spacing of OISST data, we tested various settings of longitude × latitude of 2°, 3°, … 6° based upon category 1 TCs, leading to the selection of a 4°

longitude × 4° latitude area. As shown in Figure 1(a), the 4° area effectively covered the cold wake of TC Ma_On; the difference in TC cold wake effect derived between the 3° and 5° areas was negligible. We then centered this 4° area at the daily location of each TC (average location of 00-18Z) and averaged the SST within this area to represent TC cold wakes. Using this method, the data points of TC passages covered 87% of the northwest Pacific Ocean (100°–170°E, 5°–35°N).

The effect of TC cold wakes was computed by averaging SST in each 4° × 4° area for 7 days after the day of TC passage (post-TC days). This method followed the e-folding time of SST recovery of about 10 days estimated by Dare and McBride (2011). To calculate 'TC-free' SST, an average of 7 days from one day before the TC [excluding the day before the TC passage (Huang *et al.*, 2009)] was computed at each corresponding TC location (pre-TC days). However, using 7 days after the TC passage could exaggerate the cooling in the long-term effect of TC cold wakes. Thus, we also computed 30-day average of SST after the TC passage, and 30 days before that for estimating TC-free SST. The analysis was performed for three seasons: June-August (JJA), September-October (SO), and June-October (JJASO), based upon the typhoon season of this region (Chen *et al.*, 2006). Furthermore, since SST in the northwest Pacific peaks around August or September, the calculation of SST cooling (composite) was conducted after removing the seasonal cycle in SST using a second-order polynomial fit on the monthly climatological values.

Note that this composite approach potentially included the 'cyclone-cyclone' interaction as reported by Balaguru *et al.* (2014), in which a TC could reduce intensity of the subsequent TC(s) moving across its cold wake. The same composite methodology was applied to the subsurface temperature data. However, GODAS only provides 5-day mean data and hence we used two pentads (10 days) for post-TC days (including the day of TC) and four pentads (20 days) for pre-TC days to construct the composite.

3. Results

3.1. TC-induced cooling

The basin-integrated pre-TC composite SST (i.e. TC-free) is shown in Figure 2(a) for the JJA, SO, and JJASO seasons. Increasing trends were observed in all three seasons, consistent with the trend pattern in Figure 1(c). The JJA season exhibits the largest SST increase of 0.57 °C from 1981 to 2014, larger than JJASO of 0.38 °C and SO of 0.22 °C. It is important to note that the SST presented here only followed TC tracks and, therefore, were not equivalent to any region-averaged SST values. To examine the effect of TC cold wakes on SST trends, we show in Figure 2(b) the post-TC SST anomalies, computed as the departure from the pre-TC composite to represent the TC cold wake effect. The mean difference in SST is −0.37 °C during JJASO, comparable with the TC-induced cooling found in Dare and McBride (2011) that was estimated within the e-folding time. Sensitivity testing applied to the range of 5–15 days after cyclone passage resulted in a cooling range between −0.48 and −0.32 °C.

Over the 1981–2014 period, the net change in the TC-induced cooling (estimated from the regression coefficient multiplied by 34 years) amounts to −0.23 °C in JJASO, −0.08 °C in JJA, and −0.34 °C in SO. Furthermore, there was a discernable acceleration in the cooling effect after 1995 across all seasons, and by computing a second-order polynomial trend (not shown) it was found that the post-1995 cooling is twice as much as the 1981–2014 one, suggesting concurrent enhancements in both SST warming and TC-induced cooling.

Table 1. SST change estimates from the entire TC tracks.

Season	JJA	SO	JJASO
Seasonal mean warming (°C) − 7 days	0.57	0.22	0.38
Pre-TC (TC-free) (°C) − 7 days	−0.08	−0.34	−0.23
TC cold wake effect (%) − 7 days	12%	60%	37%
Seasonal mean warming (°C) − 30 days	0.56	0.25	0.39
Pre-TC (TC-free) (°C) − 30 days	−0.02	−0.38	−0.22
TC cold wake effect (%) − 30 days	4%	61%	37%

To quantify the effect of long-term change in TC cold wakes, we divided the total change in post-TC SST by total SST (pre-TC + post-TC). For the JJA, SO and JJASO seasons, TC-induced cooling reduced SST trends by 12, 60, and 37%, respectively (Table 1) and these percentages represent the degree of SST reduction from the seasonal-mean warming trends since 1981. One apparent factor linking the rather large reduction in the SO trend is its weaker change in the pre-TC SST (Figure 2(a)). It is plausible that the increased effect of TC cold wakes nearly canceled out the SST warming during SO when TCs are generally stronger and/or last longer (Chen *et al.*, 2006). In terms of interannual variability, years 2013 and 2014 feature three super typhoons in each SO season and their combined cooling effect is visible in Figure 2(b). To test the sensitivity of 2013 and 2014 on the SST trend, we removed these 2 years and the JJASO cooling rate was reduced by 28%, still significant at $p < 0.05$. In Figure 2(c) and (d), we show the similar result derived from 30-day averages of pre-TC SST and post-TC SST anomalies, with the relevant change in SST and percentage of cooling contribution shown in Table 1. This 30-day analysis served as a sensitivity test for the duration of which TC-reduced SST rebounds. In this case, the magnitude of the long-term change in TC cold wakes is similar to that of the 7-day composites, despite a flat trend in JJA's post-TC SST anomalies.

Next, we examined the TC occurrence in terms of accumulated days versus the TC intensity in terms of the Saffir-Simpson hurricane wind scale. The result for the JJASO season is shown in Figure 2(e) and it indicated an overall increase in the TC occurrence (including duration) from 1981 to 2014, amounting to 54 days in all TCs, 72 days in Category 1–5, and 40 days in Category 3–5 over the 34 years, suggesting that the lengthening of cyclone days resulted mainly from stronger TCs. This result echoes previous observation (Emanuel, 2007; Yu *et al.*, 2010; Pun *et al.*, 2013; Mei *et al.*, 2015) that both the TC intensity and duration in the northwest Pacific have increased, especially Category 3–5 TCs (Holland and Bruyère, 2014). Also noteworthy is the marked decadal fluctuation in TC days that is consistent with the documented interdecadal variability of TC activity in this region (Matsuura *et al.*, 2003; Chen *et al.*, 2006). By computing the Pearson correlation (r) between the JJASO TC days (as in Figure 2(e)) and TC-induced cooling (as in Figure 2(b)), we obtained r of −0.31, −0.44 and −0.4 for the total, Category 1–5 and Category 3–5 TCs, respectively. The highest r in

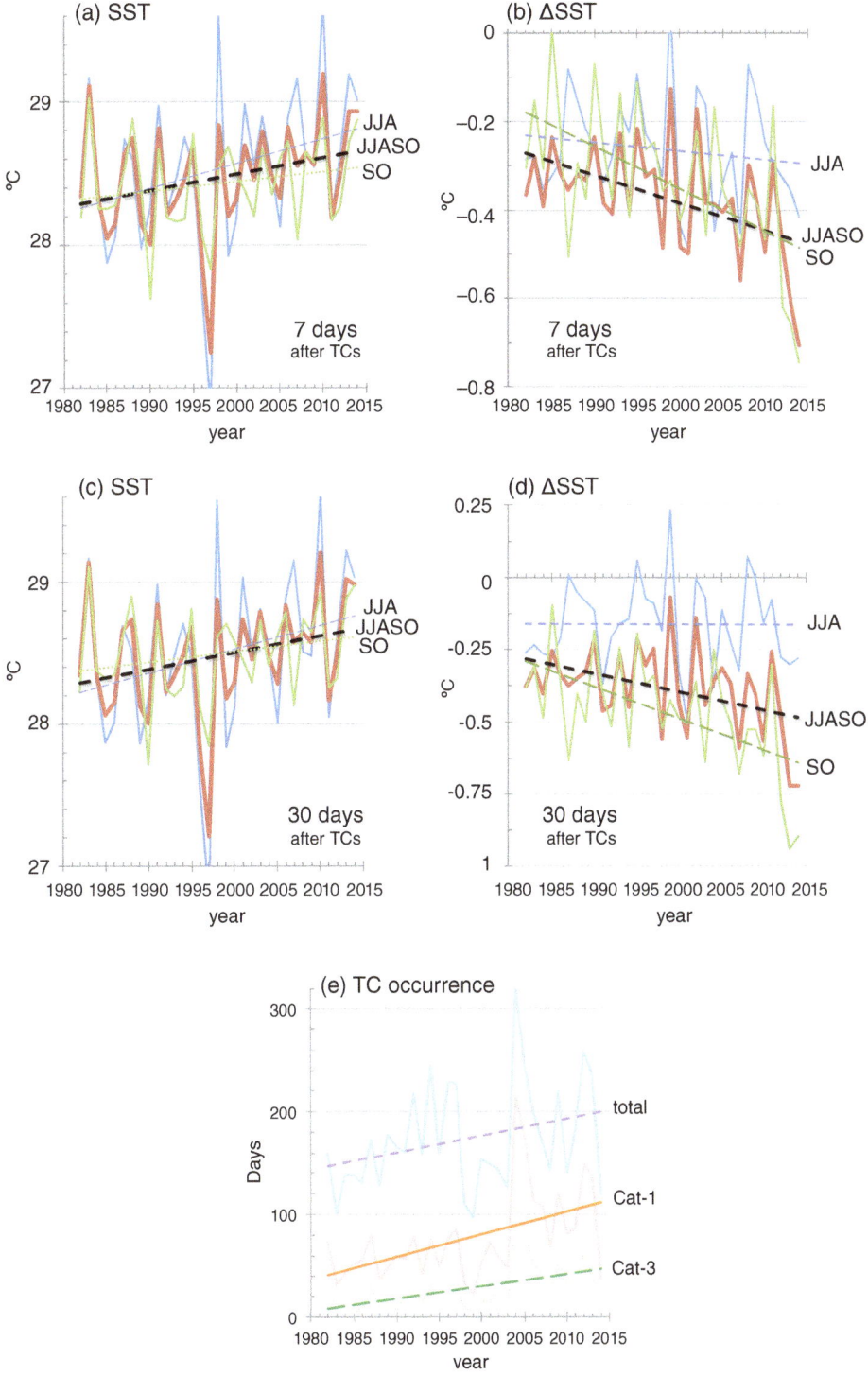

Figure 2. (a) TC-free SST composite during the seasons of JJA (blue), SO (green) and JJASO (red) overlaid with their linear trends in the corresponding color. (b) Post-TC SST anomaly for each season and linear trends estimated from the 7-day composites. (c) and (d) TC-free SST composite during JJA (blue), SO (green) and JJASO (red) overlaid with their linear trends in the corresponding color and Post-TC SST anomaly for each season and linear trends, estimated from the 30-day composites, respectively. (e) Accumulated TC days during JJASO at three intensity scales: total (blue), category 1−5 (pink), and category 3−5 (green), overlaid with linear trends.

Category 1−5 is significant at $p < 0.01$, suggesting that the extent to which TC cold wakes reduce SST in the long term is relevant to both duration and intensity of TCs. This interannual connection lends support to the effect of increased TC-induced cooling on offsetting the local warming trends of SST.

Next, we focused on three regions as indicated in Figure 1(c) as they encompass two areas of relatively weak warming over Northwest Pacific. We then computed the SST trends of each region for the JJASO season. Each region's TC-free SSTs and the associated trend are shown in Figure 3(a); likewise, SST anomalies

Table 2. SST change estimates for the three regions designated in Figure 1(a).

	TC-free (°C)	TC cold wake (°C)	Total (°C)	Contribution (%)
Region 1 – 7 days	0.61	−0.07	0.68	10
Region 2 – 7 days	0.23	−0.20	0.44	46
Region 3 – 7 days	0.35	−0.23	0.58	40
Region 1 – 30 days	0.56	−0.01	0.57	2
Region 2 – 30 days	0.25	−0.20	0.45	45
Region 3 – 30 days	0.39	−0.25	0.63	39

representing TC cold wakes are shown in Figure 3(b). Region 1 (South China Sea) experienced a 'TC-free' warming of 0.61 °C; Region 2 (midlatitude belt) showed a warming of 0.23 °C; and Region 3 (equatorial western Pacific) revealed a warming of 0.35 °C (Table 2). Meanwhile, TC cold wakes resulted in a long-term cooling of −0.07, −0.20, and −0.23 °C in regions 1, 2 and 3, respectively. Combined, TC cold wakes may have offset the post-1981 SST warming by 10% in region 1, 46% in region 2 and 40% in region 3. The mean reduction averaged from these regions is slightly smaller than the basin-scale SST reduction. This is expected since the equatorial western Pacific (Region 3) is characterized with a thick isotherm of 26 °C (D26) and a high tropical cyclone heat potential (TCHP), which are associated with the growth of intense TCs (D'Asaro *et al.*, 2011; Lin *et al.*, 2013). The fact that there was a more than 30% increase in both the D26 and TCHP in region 3 over the past 34 years (not shown) can explain why the equatorial Pacific had the maximum increasing TC-induced cooling among the three regions.

3.2. Implication from subsurface temperature

Next, the change in the basin-scale ocean heat content was analyzed with a focus on the ocean layer down to 200 m, a documented depth limit for the effect of TC cold wakes (Sriver and Huber, 2007). Using GODAS data, we computed subsurface temperatures anomalies (ΔT) between the TC cold wakes and TC-free composites. The results are shown in Figure 4 as vertical cross section of ΔT overlaid with linear trends computed for each depth (contoured). Over the northwest Pacific basin (Figure 4(a)), TCs caused a pronounced vertical redistribution of heat accompanied by noticeable cooling at the upper 25 m and warming within the depth of 30–70 m. The surface cooling has intensified and extended deeper from around 25 m prior to year 2000 to near 50 m after year 2010, indicating an enhanced vertical mixing associated with increased TC-induced cooling. Between the depth of 25 and 100 m, the warming appears to have either weakened (as in Region 2) or deepened (as in Region 3). The change in the TC-induced mixing/cooling is relatively weak in the South China Sea (Figure 4(b)), despite an enhanced warming that took place below 50 m.

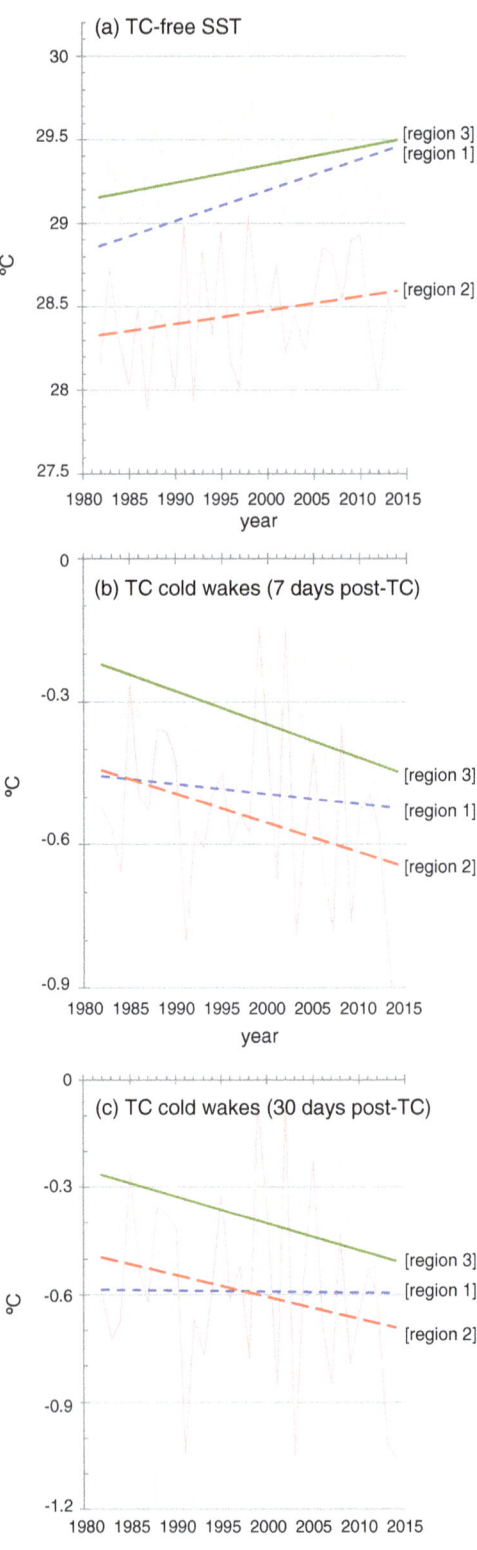

Figure 3. (a) JJASO season TC-free SST and (b) post-TC SST anomalies averaged over the three regions designated in Figure 1c: Region 1 as blue, Region 2 as red, and Region 3 as green overlaid with the respective linear trends, derived from the 7-day composite. (c) post-TC SST anomalies averaged over the three regions designated in Figure 1c: Region 1 as blue, Region 2 as red, and Region 3 as green overlaid with the respective linear trends, derived from the 30-day composite. See Table 2 for their regression slope and percentage contribution.

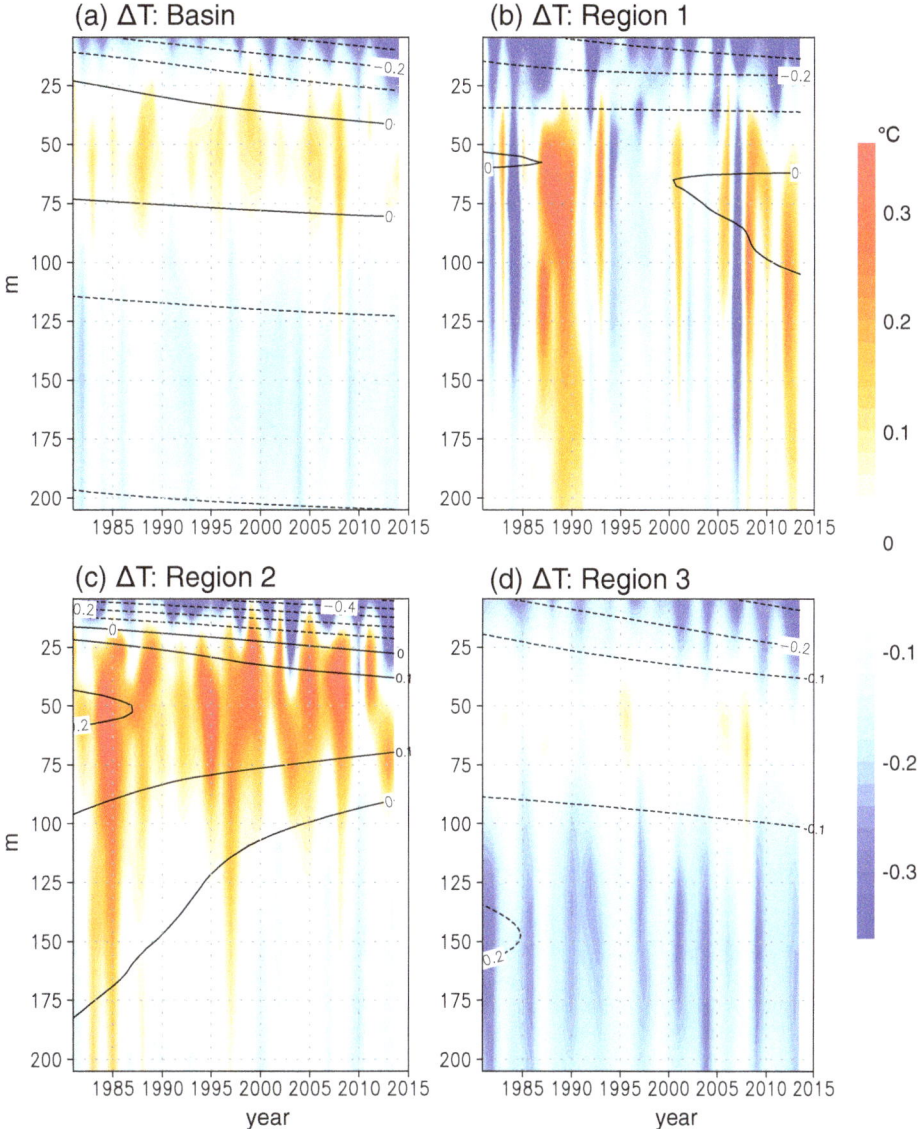

Figure 4. Subsurface temperature difference (ΔT; shadings) between post-TC and TC-free composites in the JJASO season derived for (a) the Northwest Pacific basin, (b) Region 1 (South China Sea), (c) Region 2 (subtropics) and (d) Region 3 (tropics). The linear trends in ΔT were computed at each depth and plotted across the time (from left to right), displayed here as contours with an interval of 0.1 °C (year)$^{-1}$ (i.e. regression slope).

Overall, Figure 4 confirms previous research that TCs produce a heat pump effect in transporting ocean heat downward (Sriver and Huber, 2007), causing heat to be redistributed within the ocean column and remain within the storm region. Over time, there is a discernable change in this 'stored' heat in different regions. It is possible that, according to Emanuel (2001) and subsequent studies, the increase in the stored heat below 50 m in the subtropics (Region 2) is increasingly transported northward leading to the apparent heat reduction (accompanying the increased cooling above 25 m). However, a recent study (Huang *et al.*, 2015) projects that ocean surface warming will reduce the intensity of TC cold wake due to increased stratification that inhibits mixing and prevents deepening of mixed layers. How these two effects counter or balance each other in the future climate poses an intriguing question calling for further research.

4. Summary and discussion

Using the OISST daily data and JTWC TC best tracks, the effect of the long-term changes in TC cold wakes on the ocean surface warming was analyzed for the northwest Pacific over the period of 1981–2014. The results indicated an intensification in TC-induced cooling amounting to −0.23 °C during the JJASO season and this corresponds to the observed increase in stronger TCs. The enhanced cooling arguably offset the seasonal-mean ocean surface warming trend by as much as 37%. The maximum cooling effect of TC cold wakes took place in the SO season when TCs are generally stronger and last longer. In the South China Sea (region 1), where TCs are frequent but not as strong as those in the Philippine Sea (region 3), the long-term change in TC cold wakes on the SST trends was moderate.

The implication of this study is twofold: (a) increased upper ocean warming in the northwest Pacific leads to the general intensification of TCs, as was previously found. Subsequently, (b) more intense TCs can induce stronger cold wakes and, over time, this can offset the warming trend derived from seasonal-mean SST. Together, the long-term effect of intensified TC cold wakes resulted in a 37% reduction of the ocean surface warming (during JJASO) – i.e. an offset that could have been 'added' to the observed warming trend of SST. Another implication is that long-term model simulations forced by time-mean SST as the boundary condition (so-called AMIP style) would not have accounted for the effect of TC-induced cooling, thereby underestimating tropical SST forcing. This bias can then affect the SST variation at both the interannual and interdecadal timescales and associated meridional heat transport. Future research should focus on the interactions between TCs, ocean mixing, and ocean heat uptake and the gained knowledge will improve the projection of SST trends and associated TC activity.

Acknowledgements

C.-W. J. Chang is supported by MOST 104-2111-M-034-002 and H.-H. Hsu is supported by MOST 100-2119-M-001-029-MY5.

References

Balaguru K, Taraphdar S, Leung LR, Foltz GR, Knaff JA. 2014. Cyclone-cyclone interactions through the ocean pathway. *Geophys. Res. Lett.* **41**(19): 6855–6862.

Behringer D, Xue Y. 2004. Evaluation of the global ocean data assimilation system at NCEP: the Pacific Ocean. In *Proceedings of Eighth Symposium on Integrated Observing and Assimilation Systems for Atmosphere, Oceans, and Land Surface*, AMS 84th Annual Meeting, Washington State Convention and Trade Center, Seattle, WA, 11–15 January 2004.

Carrigan AD, Puotinen M. 2014. Tropical cyclone cooling combats region-wide coral bleaching. *Glob. Change Biol.* **20**(5): 1604–1613.

Chen T-C, Wang S-Y, Yen M-C. 2006. Interannual variation of the tropical cyclone activity over the western North Pacific. *J. Clim.* **19**(21): 5709–5720.

Dare RA, McBride JL. 2011. Sea surface temperature response to tropical cyclones. *Mon. Weather Rev.* **139**(12): 3798–3808.

D'Asaro EA, Black PG, Centurioni LR, Harr P, Jayne SR, Lin I-I, Lee CM, Morzel J, Mrvaljevic RK, Niiler PP. 2011. Typhoon-ocean interaction in the western North Pacific: Part 1. *Oceanography* **24**: 24–31.

Emanuel K. 2001. Contribution of tropical cyclones to meridional heat transport by the oceans. *J. Geophys. Res.* **106**(D14): 14771–14781.

Emanuel K. 2003. A similarity hypothesis for air-sea exchange at extreme wind speeds. *J. Atmos. Sci.* **60**(11): 1420–1428.

Emanuel K. 2007. Environmental factors affecting tropical cyclone power dissipation. *J. Clim.* **20**(22): 5497–5509.

Gentemann CL, Scott J. 2014. *Variability in tropical cyclone induced upper ocean cooling*, paper presented at AGU 2014 Ocean Sciences Meeting: Honolulu, HI.

Guan B, Nigam S. 2008. Pacific sea surface temperatures in the twentieth century: an evolution-centric analysis of variability and trend. *J. Clim.* **21**(12): 2790–2809.

Hart RE, Maue RN, Watson MC. 2007. Estimating local memory of tropical cyclones through MPI anomaly evolution. *Mon. Weather Rev.* **135**(12): 3990–4005.

Holland G, Bruyère CL. 2014. Recent intense hurricane response to global climate change. *Clim. Dyn.* **42**(3–4): 617–627.

Huang P, Sanford TB, Imberger J. 2009. Heat and turbulent kinetic energy budgets for surface layer cooling induced by the passage of Hurricane Frances (2004). *J. Geophys. Res. Oceans* **114**: C12023.

Huang P, Lin I-I, Chou C, Huang R-H. 2015. Change in ocean subsurface environment to suppress tropical cyclone intensification under global warming. *Nat. Commun.* **6**: 7188, doi: 10.1038/ncomms8188. 9pp.

Jacob SD, Shay LK. 2003. The role of oceanic mesoscale features on the tropical cyclone-induced mixed layer response: a case study. *J. Phys. Oceanogr.* **33**(4): 649–676.

Jansen MF, Ferrari R, Mooring TA. 2010. Seasonal versus permanent thermocline warming by tropical cyclones. *Geophys. Res. Lett.* **37**(3): L03602.

Lin I, Pun I-F, Wu C-C. 2009. Upper-ocean thermal structure and the western North Pacific category 5 typhoons. Part II: dependence on translation speed. *Mon. Weather Rev.* **137**(11): 3744–3757.

Lin I-I, Goni GJ, Knaff JA, Forbes C, Ali M. 2013. Ocean heat content for tropical cyclone intensity forecasting and its impact on storm surge. *Nat. Hazards* **66**(3): 1481–1500.

Matsuura T, Yumoto M, Iizuka S. 2003. A mechanism of interdecadal variability of tropical cyclone activity over the western North Pacific. *Clim. Dyn.* **21**(2): 105–117.

Mei W, Pasquero C. 2013. Spatial and temporal characterization of sea surface temperature response to tropical cyclones*. *J. Clim.* **26**(11): 3745–3765.

Mei W, Primeau F, McWilliams JC, Pasquero C. 2013. Sea surface height evidence for long-term warming effects of tropical cyclones on the ocean. *Proc. Natl. Acad. Sci. USA* **110**(38): 15207–15210.

Mei W, Xie S-P, Primeau F, McWilliams JC, Pasquero C. 2015. Northwestern Pacific typhoon intensity controlled by changes in ocean temperatures. *Sci. Adv.* **1**(4): e1500014.

Price JF. 1981. Upper ocean response to a hurricane. *J. Phys. Oceanogr.* **11**(2): 153–175.

Pun IF, Lin II, Lo MH. 2013. Recent increase in high tropical cyclone heat potential area in the Western North Pacific Ocean. *Geophys. Res. Lett.* **40**(17): 4680–4684.

Reynolds RW, Smith TM, Liu C, Chelton DB, Casey KS, Schlax MG. 2007. Daily high-resolution-blended analyses for sea surface temperature. *J. Clim.* **20**(22): 5473–5496.

Schubert S et al. 2009. A U.S. CLIVAR project to assess and compare the responses of global climate models to drought-related SST forcing patterns: overview and results. *J. Clim.* **22**(19): 5251–5272.

Sriver RL, Huber M. 2007. Observational evidence for an ocean heat pump induced by tropical cyclones. *Nature* **447**(7144): 577–580.

Vincent EM, Lengaigne M, Madec G, Vialard J, Samson G, Jourdain NC, Menkes CE, Jullien S. 2012. Processes setting the characteristics of sea surface cooling induced by tropical cyclones. *J. Geophys. Res. Oceans* **117**(C2): 05023–05036.

Wu L, Zhao H. 2012. Dynamically derived tropical cyclone intensity changes over the Western North Pacific. *J. Clim.* **25**(1): 89–98.

Yablonsky RM, Ginis I, Thomas B. 2015. Ocean modeling with flexible initialization for improved coupled tropical cyclone-ocean model prediction. *Environ. Model Softw.* **67**: 26–30.

Yu J, Wang Y, Hamilton K. 2010. Response of tropical cyclone potential intensity to a global warming scenario in the IPCC AR4 CGCMs. *J. Clim.* **23**(6): 1354–1373.

9

Hurricane simulation using different representations of atmosphere–ocean interaction: The case of Irene (2011)

P. A. Mooney,[1]* D. O. Gill,[1] F. J. Mulligan[2] and C. L. Bruyère[1]

[1]*Mesoscale and Microscale Meteorology Laboratory, National Center for Atmospheric Research (NCAR), Boulder, CO, USA*
[2]*Department of Experimental Physics, Maynooth University, Kildare, Ireland*

Correspondence to:
P. A. Mooney, Mesoscale and Microscale Meteorology Laboratory, National Center for Atmospheric Research, P.O. Box 3000, Boulder, CO 80307-3000, USA.
E-mail: pmooney@ucar.edu

Abstract

Three approaches to represent sea surface temperatures (SSTs) in atmospheric models have been investigated using the Weather Research and Forecasting model: (1) prescribing SSTs every 6 h from reanalysis, (2) a one-dimensional ocean mixed-layer model and (3) a fully coupled regional ocean model. Hurricane Irene (2011) was chosen as the test case. All three options produced results comparable to observations immediately after storm passage but only options (1) and (3) captured recovery to pre-storm conditions which suggests both are feasible approaches for long-term simulations of tropical cyclones. Option (2) merits further investigation because of its greater computational efficiency and reduced complexity.

Keywords: Weather Research and Forecasting (WRF) model; sea surface temperatures; ocean mixed-layer model; coupled atmosphere–ocean model; WRF–ROMS; hurricanes

1. Introduction

Sea surface temperature plays an important role in the life cycle of a tropical cyclone (TC) and is one of the main factors for cyclogenesis (e.g. Bruyère *et al.*, 2012). As a TC passes over the warm ocean surface, it reduces the temperature of the sea surface leaving a cold wake that continues to cool for up to 2 days afterwards and may extend for hundreds of kilometres adjacent to the storm track (Dare and McBride, 2011). The magnitude of this cooling can be up to 9 °C as shown by Lin *et al.* (2003) for the case of Kai-Tak (2000) in the South China Sea. This cooling depends strongly on the TC intensity, its translational speed and the depth of the ocean mixed layer (Dare and McBride, 2011).

The time required for SSTs to return to their climatological values varies widely between TCs with recovery periods ranging from 1 to 60 days (Hart *et al.*, 2007) with the majority recovering within 30 days after a TC has passed (Dare and McBride, 2011). These lingering cold wakes can impact seasonal TC activity as later storms may interact with them; the probability for cyclones to encounter a cold wake is ~10% on average (Balaguru *et al.*, 2014). This additional mixing may also be important on longer time scales through its impact on the large-scale slowly varying ocean overturning circulation, and may impact the long-term climatology of TCs (Dare and McBride, 2011). Clearly, it is important to include the ocean response to hurricanes in atmospheric models used for long-term studies of TCs. Since most atmospheric models represent the oceanic response solely through changes in SSTs, it is essential to accurately represent them in models.

When the focus of hurricane studies moves from weather forecasting to regional climate studies, computational efficiency becomes increasingly important and the benefits of more realistic representations of physical processes in the model must be critically assessed. This study examines three different methods of representing SSTs in the Weather Research and Forecasting (WRF) model to determine the impact of trading computational efficiency for model configuration, to inform the representation of SSTs in atmospheric models for long-term studies of TCs. Hurricane Irene (2011) was chosen as a case study because most forecast models failed to correctly predict the cyclone intensity due to underestimated storm-induced upper-ocean cooling (Glenn *et al.*, 2016). Simulations are evaluated against the best track, satellite and buoy data.

2. Model domains and details

2.1. The Weather Research and Forecasting (WRF) model

All simulations use the WRF model (Skamarock *et al.*, 2008) over the two domains shown in Figure 1 with two-way nesting, 51 model levels and a model top at 10 hPa. The outer domain in Figure 1 has a grid spacing of 36 km with 340 × 260 (east–west × north–south) grid points, while the inner domain has a 12-km spacing with 802 × 511 grid points. Each simulation covers the 14-day period beginning at 0000 UTC on the 23 August 2011 with initial conditions and 6-h boundary conditions derived from ERA-Interim (Dee *et al.*, 2011).

The WRF parameterization schemes used are the Community Atmosphere Model (Collins *et al.*, 2004)

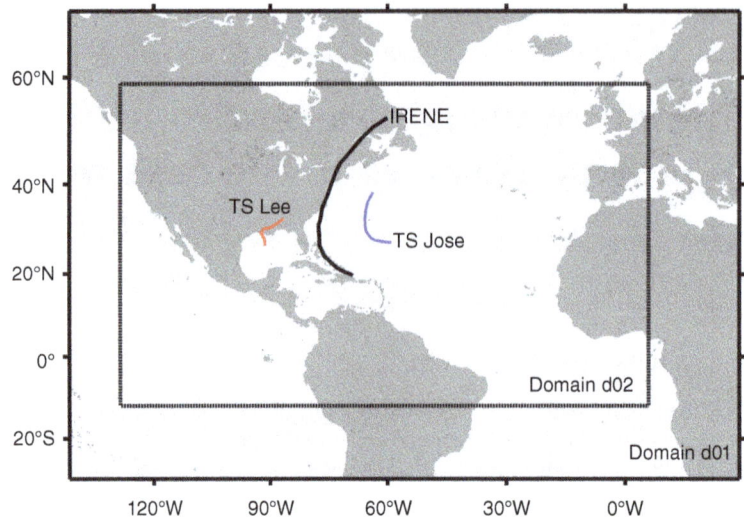

Figure 1. Outer 36-km WRF domain d01 and inner 12-km WRF domain d02. Also shown are the tracks of hurricanes Irene (23 August 2011 to 30 August 2011) and Tropical Storms (TS) Lee (2 September 2011 to 6 September 2011) and Jose (26 August 2011 to 29 August 2011).

longwave and shortwave radiation schemes, the Kain-Fritsch cumulus scheme (Kain and Fritsch, 1990; Kain, 2004), the Yonsei University planetary boundary layer scheme (Hong *et al.*, 2006), the WRF single moment six-class scheme (Hong and Lim, 2006) and the Noah land surface model (Chen and Dudhia, 2001). Analysis of 16 WRF simulations using different combinations of physical parameterization schemes has shown that this combination accurately simulates the track and minimum pressures of Irene (not shown). Three different approaches of updating SSTs are used with this WRF configuration (see Table 1) to simulate the passage of Irene.

2.2. Representation of the ocean surface

2.2.1. ERA-Interim SSTs

In this simulation, WRF uses daily averaged SSTs from ERA-Interim, which are derived from the Operational Sea Surface Temperature and Sea Ice Analysis (OSTIA; Stark *et al.*, 2007). WRF uses the scheme described in Zeng and Beljaars (2005) to modify these daily averages in response to surface winds and changes in the radiative fluxes (e.g. diurnal variations in shortwave radiation).

2.2.2. One-dimensional (1-D) ocean mixed-layer (OML) model

This is a simple 1-D OML model (Davis *et al.*, 2008) based on Pollard *et al.* (1972). Neither horizontal advection nor the pressure gradient are accounted for in this model which simply requires specification of the initial mixed-layer depth (climatological values obtained from www.ifremer.fr/cerweb/deboyer/mld; de Boyer Montégut *et al.*, 2004), the deep layer temperature lapse rate (default 0.14 K/m) and the wind stress at the ocean surface which is provided by the WRF model.

2.2.3. Regional Ocean Modelling System (ROMS)

ROMS (Shchepetkin and McWilliams, 2005) is a three-dimensional (3-D) regional ocean model with terrain following coordinates that solves the free surface, hydrostatic, primitive equations. In this simulation, ROMS is configured using a single domain which covers the same area as the outer WRF domain. The ROMS domain uses a grid spacing of 12 km and 30 stretched vertical levels. Values for the initial conditions and open boundaries are generated from the global HYbrid Coordinate Ocean Model with Naval Research Lab Coupled Ocean Data Assimilation (HYCOM/NCODA; http://tds.hycom.org/thredds/dodsC/GLBa0.08/expt_90 .9; Cummings, 2005).

A 10-day spin-up period is used for ROMS and HYCOM/NCODA tracer and velocity fields are provided using Orlanski-type radiation conditions in conjunction with relaxation. The Flather method was used to obtain boundary values for the free-surface and depth-averaged velocity from HYCOM/NCODA. The Generic Length Scale vertical mixing scheme (Warner *et al.*, 2005) was used to calculate the vertical turbulent mixing and specify the quadratic drag formulation for the bottom friction. Other parameters for ROMS are shown in Table 2.

The coupled WRF–ROMS modelling system used in this study is the Coupled Ocean–atmosphere–Wave–Sediment Transport (Warner *et al.*, 2010) modelling system. These models exchange data once per hour: WRF receives SSTs from ROMS every hour while providing ROMS with wind stress, sea level pressure and surface heat fluxes.

2.3. Data sets

ERA-Interim was obtained on a global 0.75° grid from the European Centre for Medium Range Weather Forecasting data server: http://data.ecmwf.int/data.

Table 1. Summary of the ocean surface representation in each of the WRF simulations.

WRF–ERAI	WRF with updated daily averaged SSTs from ERA-Interim; a diurnal cycle is imposed on the input SST data.
WRF–OML	WRF run with a simple 1-D ocean mixed layer model
WRF–ROMS	WRF fully coupled to a 3-D hydrostatic, primitive equation regional ocean model system (ROMS).

Table 2. Model parameters used in ROMS.

L	1015	Number of I-direction interior rho-points
M	775	Number of J-direction interior rho-points
h_{max}	5000 m	Maximum depth of computational domain
h_{min}	50 m	Minimum depth of computational domain
θ_s	5	Sigma coordinate stretching factor
θ_b	0.4	Sigma coordinate bottom stretching factor
dt (baroclinic)	30 s	Baroclinic time step
dt (barotropic)	1 s	Barotropic time step
Outflow	10 days	
Inflow	0.5 days	

The OSTIA is provided by GHRSST, UKMO and MyOcean and obtained from http://podaac.jpl.nasa.gov/dataset/UKMO-L4HRfnd-GLOB-OSTIA. This is a Level 4 SST analysis produced daily on a global 0.054° grid by the UK Met Office using optimal interpolation. Best track data for observed hurricanes (Figure 1) were obtained from the International Best Track Archive for Climate Stewardship (Knapp *et al.*, 2010). Buoy data (locations shown in Figure 2) were obtained from the US National Data Buoy Center: http://www.ndbc.noaa.gov/to_station.shtml.

3. Results

Figure 2 shows the difference between pre-storm SSTs (23 August 2011) and post-storm SSTs (27 August 2011; 1 September 2011; 5 September 2011) for the OSTIA satellite data (row one) and the three simulations. This sequence of SST differences shows the cooling generated by Irene and the recovery to pre-storm conditions. Figure 2(a) shows that Irene's passage caused SST cooling of approximately 2–3 °C by the 27 August 2011, while the observed recovery to pre-storm conditions is shown in Figure 2(b) and (c). A wide wake with greater cooling on the right side of the track is clearly evident.

Corresponding results from WRF–ERAI (Figure 2(d)–(f)) are in excellent agreement with the OSTIA results. Small differences arise from the imposed diurnal cycle (described above) and the interpolation of OSTIA's SSTs (12 km) to the coarser ERA-Interim grid (~80 km). Figure 2(e) shows that WRF–ERAI SSTs have begun recovery to pre-storm conditions five days after the passage of Irene. Four days later, the SSTs have almost returned to pre-storm conditions (Figure 2(f)).

The behaviour of SSTs in the WRF–OML simulation is shown in Figure 2(g)–(i). Irene's passage generates a slightly colder wake than observations. Of interest to this study is the absence of a recovery to pre-storm conditions. The cold wake in the WRF–OML simulation shows almost no deterioration even after nine days. Recovery to pre-storm conditions is primarily driven by surface fluxes and horizontal advection (Vincent *et al.*, 2012) which is not represented in the OML model. The WRF–OML has two additional cold wakes associated with tropical storms Jose and Lee shown in Figure 1. Lee is also present in OSTIA (Figure 2(c)), WRF–ERAI (Figure 2(f)) and WRF–ROMS (Figure 2(k)–(l)).

The characteristics of Irene's cold wake in the WRF–ROMS simulation are similar to those observed in the satellite SSTs – wide with greater cooling on the right side of the track. Similar to the WRF–ERAI simulation and satellite SSTs, the cold wake in WRF–ROMS shows some recovery after five days with SSTs partially returning to pre-storm conditions after nine days. However, the WRF–ROMS simulation generates a colder wake than both WRF–ERAI and OSTIA.

This is also evident in Figure 3 which shows the mean SSTs over a square area, whose width is four times the radius of maximum winds and centred on the location of maximum cyclone intensity for each simulation (see Figure 2). The evolution of the SSTs in Figure 3 shows the rapid cooling as Irene passes and the subsequent recovery or the absence of a recovery in the case of WRF–OML. SSTs in WRF–ROMS and WRF–OML show very similar initial cooling, and both show greater cooling than either WRF–ERAI or OSTIA.

While the lower SSTs in WRF–ROMS and WRF–OML may be caused by a slower than observed translational speed, it must be noted that satellite SST measurements are based on a very thin layer of the ocean surface while simulated SSTs can represent a layer several centimetres deep (Costa *et al.*, 2012). For this reason, the simulated SSTs are also compared to measurements from the three buoys (m44014, m44065 and m44013) shown in Figure 2.

A comparison of simulated and observed SSTs, 10-m wind speeds and mean sea level pressure (MSLP) at each buoy is shown in Figure 4. At buoy m44014, the WRF–ERAI simulation shows greater cooling than the other simulations (Figure 4(a)) but it still fails to reach the cold SSTs observed at the buoy. Both WRF–ERAI and WRF–ROMS SSTs show a slow recovery to pre-storm conditions which agrees with the observed rate of warming at the buoy. WRF–ROMS and WRF–OML temperatures show little initial cooling (approximately 1 °C) with no return to pre-storm conditions in the case of WRF–OML. At buoy m44065, SSTs cool by approximately 4 °C as the hurricane passes and continue to cool at a much slower rate for the next day

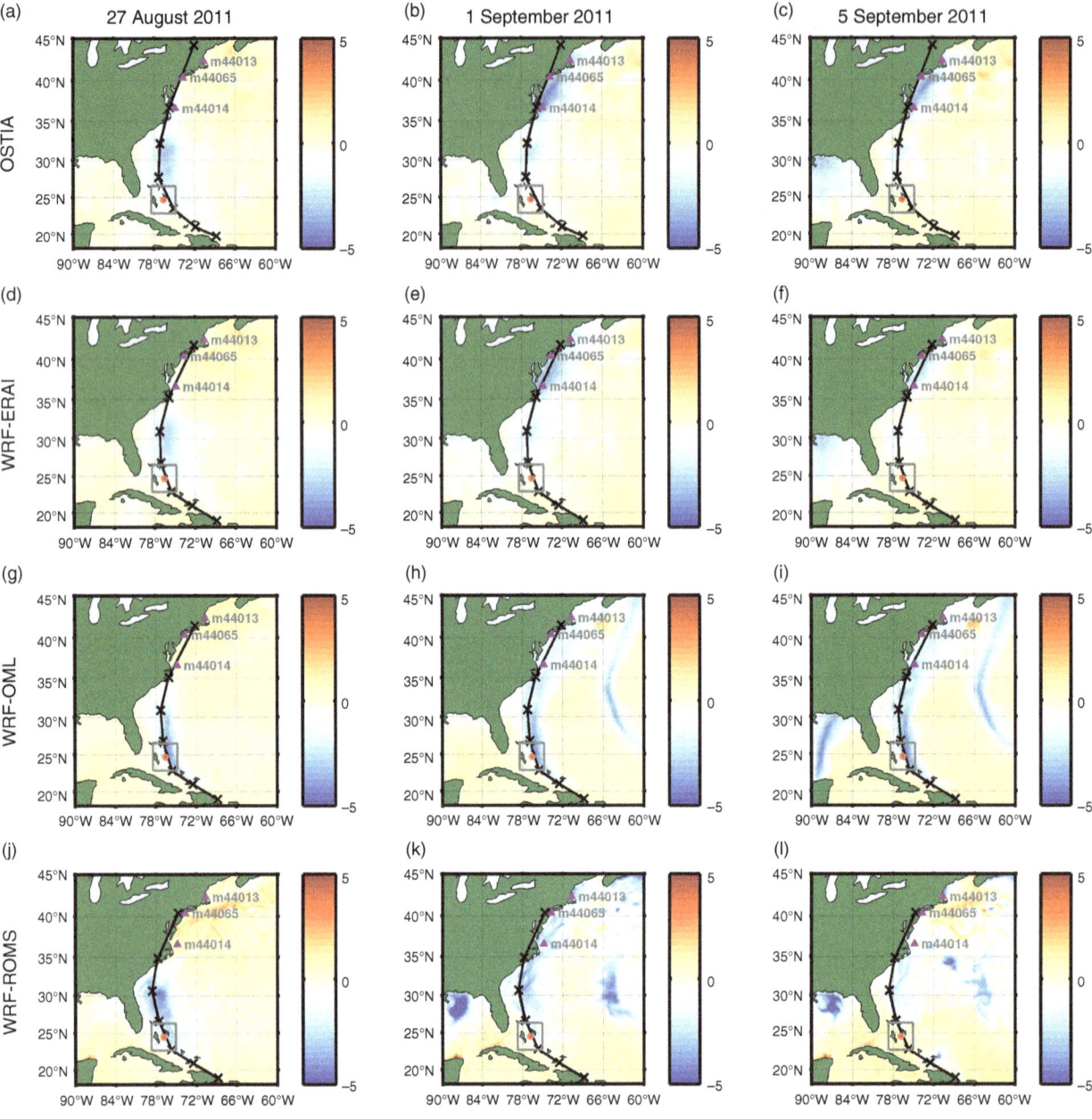

Figure 2. Post-storm cooling in OSTIA sea surface temperatures (a) difference between the 27 August 2011 and the 23 August 2011, (b) difference between the 1 September 2011 and the 23 August 2011, (c) difference between the 4 September 2011 and the 23 August 2011. (d)–(f) Same as (a)–(c) except for WRF–ERAI. (g)–(i) Same as (a)–(c) except for WRF–OML. (j)–(l) Same as (a)–(c) except for WRF–ROMS. SSTs in the square box centred on the red dots are averaged at each time step and the results are displayed in Figure 3.

or so (see Figure 4(b)). This is followed by rapid warming which quickly reduces to a rate of approximately 0.25 °C/day. A week after Irene passed, SSTs at buoy m44065 are still 2 °C colder than pre-storm temperatures. WRF–ERAI SSTs display most of this observed behaviour with the exception of the rapid warming immediately following the passage of Irene. Although WRF–ROMS does simulate the correct temperature three days after the hurricane passed, its SSTs do not show the observed 4 °C drop in temperature immediately following Irene's passage nor the rapid warming subsequent to it. WRF–OML shows cooling that is intermediate between WRF–ERAI and WRF–ROMS

but again shows no evidence of recovery to pre-storm conditions.

Although the SSTs at buoy m44013 experience greater cooling (almost 8 °C) than buoy m44065, they behave in a similar way – rapid cooling, followed by a brief period of rapid warming and then a slow oscillatory recovery. This behaviour is well captured by WRF–ROMS that outperforms the other simulations at this buoy, and shows almost 6 °C of cooling. WRF–ERAI SSTs show a similar rate of recovery in the week following Irene's passage, however, it shows only 2 °C of cooling as the hurricane passes and no rapid recovery. The WRF–OML shows very

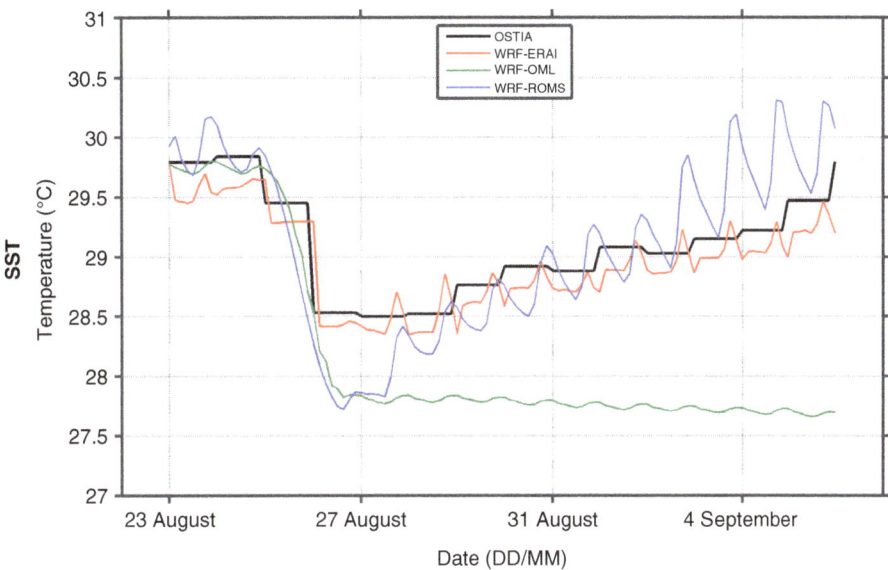

Figure 3. Sea surface temperatures from each simulation and OSTIA averaged over a square box whose width is four times the radius of maximum winds at maximum storm intensity (∼4 × 90 km) centred on the location where the hurricane reaches maximum intensity (see red dot near 25°N on Figures 2(a), (d), (g) and (j)).

little cooling (approximately 1 °C) as the hurricane passes.

Figure 4(d)–(f) and Figure 4(g)–(i) show the measured and simulated 10-m wind speeds and MSLP at each buoy. The MSLPs from WRF–ERAI and WRF–OML are in good agreement with those measured at the buoy, except the simulated minimum occurs slightly later than the observed minimum. This is due to the slower than observed translational speed of the simulated cyclones. Similarly, WRF–ROMS shows a delay in the timing of the minimum pressure, which is also 10–15 hPa weaker than the minimum measured at the buoys.

Table 3 shows the root mean squared error (RMSE) of each simulated track (determined from the location of minimum pressure) from the best-observed track. Somewhat surprisingly WRF–ROMS has the greatest departure from the best-observed track. This is partially due to the slower translational speed of the WRF–ROMS simulation.

RMSEs for minimum sea level pressures are also included in Table 3. Clearly the hurricane intensity is well captured by all three simulations, which is in good agreement with the behaviour in pressure noted earlier at the buoys.

4. Discussion and Conclusions

The processes that cause cooling of the upper ocean mixed layer can be divided into those that are responsible for cooling the ocean surface within hours of the TC arrival and those that continue the cooling for up to two days after the TC has passed (Hart *et al.*, 2007). The dominant mechanisms for cooling the ocean surface are transient upwelling and wind-driven oceanic turbulence that causes vertical mixing and entrainment

of colder water from below the thermocline into the overlying mixed layer (Vincent *et al.*, 2012). Other processes include enhanced surface sensible and latent heat fluxes from the ocean to the atmosphere driven primarily by the winds near the radius of maximum wind (Price, 1981), horizontal transport of warm water away from the storm centre (Leipper, 1967), precipitation falling into the ocean surface and radiative losses (Brand, 1971). Those processes which cool the sea surface temperature in the vicinity of the storm have the greatest impact as they reduce the amount of heat and moisture available to the tropical storm which in turn limits its intensity (Yablonsky *et al.*, 2015). In addition to influencing the individual TC, the cold wake may also impact other storms that interact with it at a later stage (Balaguru *et al.*, 2014).

In WRF, the coupled ocean model (WRF–ROMS) is the most physically realistic simulation of the air–sea interaction and it produces SSTs similar to observed values, but it is computationally expensive. For example, WRF–ROMS used 1.5 times as many processors and 60% more computational time than WRF–ERAI.

The WFR–ERAI is a good, less-expensive alternative to WRF–ROMS and as a result it continues to be in widespread use by the regional climate modelling community. However, it cannot represent the feedback between the atmosphere and ocean. This could have implications for long-term climate simulations of TCs, where TCs in regional models do not coincide with TC tracks in SSTs from the parent model. This can impact TC genesis, track and intensity in regional climate simulations.

WRF with the one-dimensional ocean mixed-layer model (WRF–OML) is capable of simulating the cold wake in the SSTs caused by the passage of hurricane Irene which makes it useful for short-term studies.

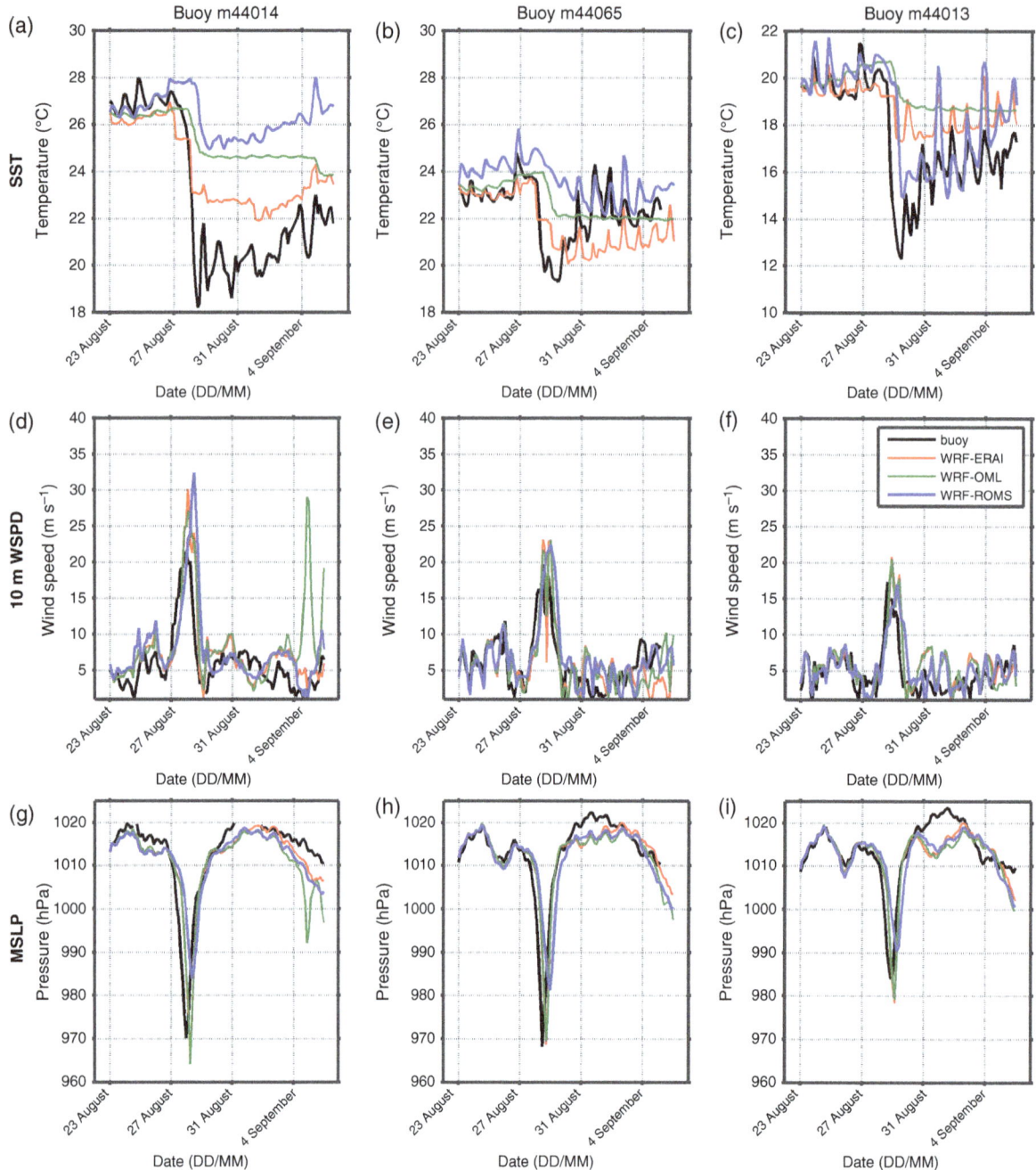

Figure 4. (a)–(c) Simulated and measured sea surface temperatures at buoys m44014, m44065 and m44013 covering the period 23 August 2011 to 6 September 2011. (d)–(f) Same as (a)–(c) except for 10-m wind speeds. (g)–(i) Same as (a)–(c) except for mean sea level pressure.

Recovery to climatological values is primarily driven by surface fluxes and horizontal advection (Vincent *et al.*, 2012). As the OML model does not include horizontal advection, it is foreseeable that its SSTs fail to recover to pre-storm conditions. While this suggests that the OML model is unsuitable for longer term studies, its computational efficiency and physically based representation of oceanic cooling make it an attractive option for future work. One approach which would retain computational efficiency will focus on representing the recovery of SSTs by adding a relaxation term based on empirical data.

Table 3. Root mean squared errors (RMSEs) for the tracks and intensity of hurricane Irene simulated by the three simulations listed in Table 1.

	WRF–ERAI	**WRF–OML**	**WRF–ROMS**
Tracks (km)	56	58	113
Intensity (hPa)	9	9	8

Acknowledgements

The authors are grateful to Dr. John Warner (USGS, Woods Hole) for providing access to the COAWST modelling system. NCAR is sponsored by the National Science Foundation

(NSF). This work was partially supported by NSF EASM Grant AGS-1048829 and the Research Partnership to Secure Energy for America. The authors acknowledge high-performance computing support from Yellowstone (ark:/85065/d7wd3xhc) provided by NCAR's Computational and Information Systems Laboratory, sponsored by the NSF. Part of the data analysis in this work was undertaken with climate data operators (CDO).

References

Balaguru K, Taraphdar S, Leung LR, Foltz GR, Knaff JA. 2014. Cyclone-cyclone interactions through the ocean pathway. *Geophysical Research Letters* **41**: 6855–6862, doi: 10.1002/2014GL061489.

de Boyer Montégut C, Madec G, Fischer AS, Lazar A, Iudicone D. 2004. Mixed layer depth over the global ocean: an examination of profile data and a profile-based climatology. *Journal of Geophysical Research* **109**: C12003, doi: 10.1029/2004JC002378.

Brand S. 1971. The effects on a tropical cyclone of cooler surface waters due to upwelling and mixing produced by a prior tropical cyclone. *Journal of Applied Meteorology* **10**(5): 865–874.

Bruyère CL, Holland GJ, Towler E. 2012. Investigating the use of a genesis potential index for Tropical Cyclones in the North Atlantic Basin. *Journal of Climate* **25**: 8611–8626, doi: 10.1175/JCLI-D-11-00619.1.

Chen F, Dudhia J. 2001. Coupling an advanced land surface-hydrology model with the Penn State-NCAR MM5 modeling system Part I: model implementation and sensitivity. *Monthly Weather Review* **129**: 569–585.

Collins WD, Rasch PJ, Boville BA, Hack JJ, McCaa JR, Williamson DL, Kiehl JT, Briegleb B, Bitz C, Lin S-J, Zhang M, Dai Y. 2004. Description of the NCAR Community Atmosphere Model (CAM3.0). NCAR Technical Note, TN-464+STR, NCAR, Boulder, CO, USA, 102–143.

Costa P, Gómez B, Venâncio A, Pérez E, Pérez-Munuzuri V. 2012. Using the Regional Ocean Modeling System (ROMS) to improve the sea surface temperature predictions of the MERCATOR Ocean System. In *Advances in Spanish Physical Oceanography*, Espino M, Font J, Pelegrí JL, Sánchez-Arcilla A (eds). Scientia Marina 76S1: Barcelona, Spain; 165–175. ISSN: 0214-8358; doi: 10.3989/scimar.03614.19E.

Cummings JA. 2005. Operational multivariate ocean data assimilation. *Quarterly Journal of the Royal Meteorological Society* **131**: 3583–3604, doi: 10.1256/qj.05.105.

Dare RA, McBride JL. 2011. Sea surface temperature response to tropical cyclones. *Monthly Weather Review* **139**(12): 3798–3808.

Davis CAW, Wang W, Chen SS, Chen Y, Corbosiero K, DeMaria M, Dudhia J, Holland G, Klemp J, Michalakes J, Reeves H, Rotunno R, Snyder C, Xiao Q. 2008. Prediction of landfalling hurricanes with the advanced hurricane WRF model. *Monthly Weather Review* **136**: 1990–2005, doi: 10.1175/2007MWR2085.1.

Dee DP, Uppala SM, Simmons AJ, Berrisford P, Poli P, Kobayashi S, Andrae U, Balmaseda MA, Balsamo G, Bauer P, Bechtold P, Beljaars ACM, van de Berg L, Bidlot J, Bormann N, Dlesoc C, Dragani R, Fuentes M, Geer AJ, Haimberger L, Healy SB, Hersbach H, Hólm EV, Isaken L, Källberg P, Kohler M, Matricardi M, McNally AP, Monge-Sanz BM, Morcrette J-J, Park BK, Peubey C, de Rosnay P, Tavoloato C, Thépaut J-N, Vitart F. 2011. The ERA-Interim reanalysis: configuration and performance of the data assimilation system. *Quarterly Journal of the Royal Meteorological Society* **137**: 553–597, doi: 10.1002/qj.828.

Glenn SM, Miles TN, Seroka GN, Xu Y, Forney RK, Yu F, Roarty H, Schofield O, Kohut J. 2016. Stratified coastal ocean interactions with tropical cyclones. *Nature Communications* **7**: 10887, doi: 10.1038/ncomms10887.

Hart RE, Maue RN, Watson MC. 2007. Estimating local memory of tropical cyclones through MPI anomaly evolution. *Monthly Weather Review* **135**(12): 3990–4005.

Hong S-Y, Lim J-OJ. 2006. The WRF single moment six class scheme (WSM6). *Journal of the Korean Meteorological Society* **42**: 129–151.

Hong S-Y, Noh Y, Dudhia J. 2006. A new vertical diffusion package with an explicit treatment of entrainment processes. *Monthly Weather Review* **134**: 2318–2341.

Kain JS. 2004. The Kain-Fritsch convective parameterization: an update. *Journal of Applied Meteorology* **43**: 170–181.

Kain JS, Fritsch JM. 1990. A one-dimensional entraining/detraining plume model and its application in convective parameterization. *Journal of the Atmospheric Sciences* **47**: 2784–2802.

Knapp KR, Kruk MC, Levinson DH, Diamond HJ, Neumann CJ. 2010. The International Best Track Archive for Climate Stewardship (IBTrACS): unifying tropical cyclone data. *Bulletin of the American Meteorological Society* **91**: 363–376.

Leipper DF. 1967. Observed ocean conditions and Hurricane Hilda. *Journal of the Atmospheric Sciences* **24**: 182–196.

Lin I, Liu WT, Wu C-C, Wong GTF, Hu C, Chen Z, Liang W-D, Yang Y, Liu KK. 2003. New evidence for enhanced ocean primary production triggered by tropical cyclone. *Geophysical Research Letters* **30**: 1718, doi: 10.1029/2003GL017141.

Pollard RT, Rhines PB, Thompson RORY. 1972. The deepening of the wind-mixed layer. *Geophysical Fluid Dynamics* **4**(1): 381–404, doi: 10.1080/03091927208236105.

Price JF. 1981. Upper ocean response to a hurricane. *Journal of Physical Oceanography* **11**(2): 153–175.

Shchepetkin AF, McWilliams JC. 2005. The regional ocean modeling system: a split-explicit free-surface, topography-following coordinates ocean model. *Ocean Model* **9**: 347–404, doi: 10.1016/jocemod.2004.08.002.

Skamarock WC, Klemp JB, Dudhia J, Gill DO, Barker DM, Duda MG, Huang X-Y, Wang W, Powers JG. 2008. A description of the advanced research WRF Version 3. NCAR Technical Note, NCAR, Boulder, CO, USA.

Stark JD, Donlon CJ, Martin MJ, McCulloch ME. 2007. OSTIA: an operational, high resolution, real time, global sea surface temperature analysis system, Oceans 07 IEEE Aberdeen, conference proceedings. In *Marine Challenges: Coastline to Deep Sea*. IEEE: Aberdeen, Scotland.

Vincent EM, Lengaigue M, Madec G, Vialard J, Samson G, Jourdain NC, Menkes CE, Julien S. 2012. Processes setting the characteristics of sea surface cooling induced by tropical cyclones. *Journal of Geophysical Research* **117**: C02020, doi: 10.1029/2011JC007396.

Warner JC, Sherwood CR, Arango HG, Signell RP. 2005. Performance of four turbulence closure methods implemented using a generic length scale method. *Ocean Modelling* **8**: 81–113.

Warner JC, Armstrong B, He R, Zambon JB. 2010. Development of a Coupled Ocean–Atmosphere–Wave–Sediment Transport (COAWST) modeling system. *Ocean Modelling* **35**(3): 230–244.

Yablonsky RM, Ginis I, Thomas B, Tallapragada V, Sheinin D, Bernardet L. 2015. Description and analysis of the ocean component of NOAA's Operational Hurricane Weather Research and Forecasting Model (HWRF). *Journal of Atmospheric and Oceanic Technology* **32**(1): 144–163, doi: 10.1175/JTECH-D-14-00063.1.

Zeng X, Beljaars A. 2005. A prognostic scheme of sea surface skin temperature for modeling and data assimilation. *Geophysical Research Letters* **32**: L14605, doi: 10.1029/2005GL023030.

10

Thundersnow in Brazil: A case study of 22 July 2013

Giovanni Dolif Neto,[1,2]* Patrick S. Market,[3] Alexandre Bernardes Pezza,[4,5] Carlos Augusto Morales Rodriguez,[6] Leonardo Calvetti,[7] Pedro Leite da Silva Dias[8] and Gustavo Carlos Juan Escobar[1]

[1]Centre for Weather Forecasting and Climate Studies, National Space Research Institute (CPTEC-INPE), Cachoeira Paulista, SP, Brazil
[2]Centre for Natural Disaster Monitoring and Early Warning – CEMADEN, São José dos Campos, SP, Brazil
[3]Department of Soil and Atmospheric Sciences, University of Missouri, Columbia, MO, USA
[4]School of Earth Sciences, The University of Melbourne, Australia
[5]Environmental Science Department, Greater Wellington Regional Council (GWRC), Wellington, New Zealand
[6]Atmospheric Sciences Department, Institute of Astronomy, Geophysics and Atmospheric Sciences, University of São Paulo (IAG-USP), São Paulo, SP, Brazil
[7]Meteorological Service of Parana State (SIMEPAR), Curitiba, PR, Brazil
[8]National Laboratory for Scientific Computing (LNCC), Institute of Astronomy, Geophysics and Atmospheric Sciences, University of São Paulo (IAG-USP), Petrópolis, RJ, Brazil

*Correspondence to:
G. Dolif Neto, Centro Nacional de Monitoramento e Alertas de Desastres Naturais – CEMADEN, Rodovia Presidente Dutra Km 39, 12630-000 Cachoeira Paulista, SP, Brazil.
E-mail:
giovanni.dolif@cemaden.gov.br

Abstract

On 23 July 2013, a snowstorm hit southern Brazil causing material damage and two deaths. Radar reflectivity and lightning data revealed a rare thundersnow occurrence. This study revealed that a Rossby wave propagation followed a typical pattern of cold incursions in South America, but some fundamental differences can be pointed out: (1) further northward Rossby wave amplification; (2) strong upward vertical motion within a deep nearly saturated layer and (3) a conditional symmetric instability layer, in response to strong vertical shear, beneath a layer of weak conditional instability, and above a significant near-surface vertical depth where temperatures hover around $0\,°C$.

Keywords: thundersnow; cold surge; southern Brazil; lightning; radar; symmetric instability

1. Introduction

Cold Surges (CSs) can have a devastating impact on the economy in many countries of South America, with Brazil's multimillion dollar Arabic coffee market being traditionally (in the past) one of the most severely affected (DaMatta and Ramalho, 2006; Camargo, 2010). The most documented impact of CSs in South America is indeed economic damage to the agriculture sector (Marengo *et al.*, 1997). Livestock loss, transportation disruptions and increased energy demands are commonly associated with CSs.

In these early studies, frost has been the most relevant motivating factor, as frost has had a sustained impact on the Brazilian agriculture over the years. On the other hand, snow events have been very rare and largely unreported in Brazil, although they have been known for causing significant damage in the past (Pezza and Ambrizzi, 2005b). The reporting of thundersnow, however, has only one known published study, Dolif Neto *et al.* (2009).

The July 2013 CS event broke a large number of records including temperature, first time snow occurrence (Figures S1 and S2) and thundersnow. The city of São Paulo had the coldest maximum temperature on record with only $8.7\,°C$, and Curitiba measured snow for the first time since 1975. In Guarapuava city, media news reported damage to buildings and traffic

accidents resulting in two deaths (G1. News Portal, 2014). Guarapuava is a small city (180 000 inhabitants) located in the central portion of the Brazilian Parana State at 1120 m above sea level (25.3 S and 51.4 W – denoted by black filled star in Figure 2).

Previous studies of CSs revealed that the primary mechanism behind CSs is a surface anticyclone/cyclone couplet over the affected region, which results from a significant meridional amplification of a Rossby wave (Pezza and Ambrizzi, 2005a; Müller and Berri 2007; Müller and Berri, 2012; Sprenger *et al.*, 2013). Cold air reaches low latitudes, channeled also by the north–south orientation of the Andes (Garreaud, 2000; Vera and Vigliarolo, 2000; Pezza and Ambrizzi, 2005a; Escobar, 2007). As a result of the interaction with the Andes, the baroclinic Rossby wave is amplified, acquiring a north/south elongated shape (Gan and Rao, 1994; Seluchi *et al.*, 1998), which is also observed in parts of North America (Schultz *et al.*, 1998) and Asia (Lau and Chang, 1987) where the topography is sufficiently pronounced.

In Brazil, Dolif Neto *et al.* (2009) studied a rare event comparing low-latitude thundersnow events in Brazil and in the United States. Although their results revealed the importance of the synoptic-scale driver, the smaller-scale influences such as orographic lifting and increased instability were also highlighted. This late 2005 winter episode in southern Brazil produced a significant frequency of cloud-to-ground flashes, without appreciable snow accumulation. Some broad features were highlighted as being consistent with the literature for the Northern Hemisphere: (1) significant and well-defined synoptic-scale weather systems at low latitudes, (2) a strong baroclinic zone with a well-defined ($>60\,\text{m s}^{-1}$) jet structure aloft, (3) cold air with appreciable depth and areal extent reaching regions very far to the north and (4) a moist neutral to conditionally unstable layer above the frontal zone. A recent study (Bech *et al.*, 2013) of a low-latitude thundersnow event in Spain also revealed important synoptic and mesoscale features, namely upper level cold trough and gravity wave dynamics.

In this study, the broad features of the 22–23 July 2013 thundersnow episode are investigated in comparison to previous studies. In Section 2, data and methodology are detailed. Results are addressed in Section 3. Discussion and concluding remarks will be offered in Section 4.

2. Data and methodology

Radar and lightning data were employed in order to reveal evidences of thundersnow occurrence in Southern Brazil on 22–23 July 2013. Radar reflectivity (Z) measurements from Meteorological Service of Parana State (SIMEPAR) S-band weather Doppler radar have been used to diagnose the main precipitation characteristics. The radar is located at Teixeixa Soares and makes continuous volume scans with 14 elevations and 250-m gate resolution every 10 min. Lightning discharges are from RINDAT (Beneti *et al.*, 2000) and STARNET (Morales *et al.*, 2011). Those two networks measure mainly cloud-to-ground lightning strikes by measuring both the magnetic and electric field emitted by the atmospherics discharge in the low frequency (LF) and very low frequency (VLF) spectrum, respectively. The lightning strikes detected within a radius of 50 km centered in Guarapuava city were considered. Supporting Information also presents lightning and radar data for Candói city, neighboring Guarapuava city (Figure S3).

For synoptic and frontal zone cross-section analyses, output from the US National Weather Service's Global Forecasting System – GFS (n.d.) was employed. This approach helps promote a more uniform, larger-scale comparison to other meteorological episodes with similar characteristics. Horizontal model resolution is $0.5° \times 0.5°$ and there are 64 unequally spaced vertical sigma levels. The horizontal grid spacing works out to be ~50 km at the latitude of Santa Catarina, suitable for resolving waves at the low meso-α scale. For a surface pressure of 1000 hPa, 15 levels are below 800 hPa, and 24 levels are above 100 hPa with the top level at 3 hPa.

3. Results

Thundersnow occurrence was evidenced through ground measurements at Guarapuava city, in the central

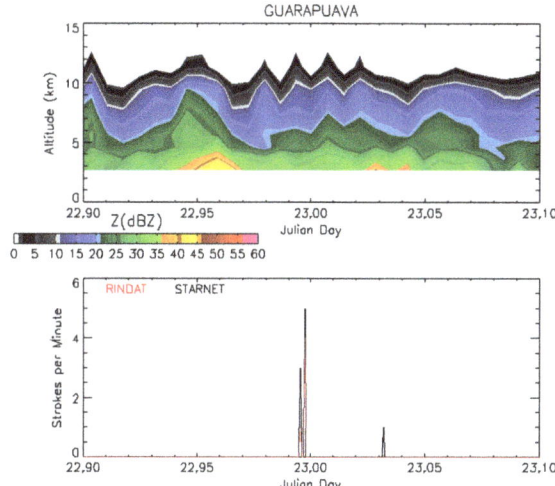

Figure 1. Vertical radar reflectivity profiles over Guarapuava city in Brazilian Parana State from 22 July 2013 2136 UTC until 23 July 2013 0224 UTC. Vertical black and red bars indicate the number of cloud-to-ground strokes detected by RINDAT and STARNET, respectively, over that area.

portion of Parana State. The time evolution of the reflectivity and cloud-to-ground strikes between 22 July 2013 2136 UTC and 23 July 2013 0224 UTC is shown in Figure 1. The cooccurrence of high (Z ~ 40 dBZ) radar reflectivities and cloud-to-ground strikes, defining the thundersnow, is evident twice: on Julian day 22.99 (22 July 2013 2355 UTC) and 23 July 2013 0045 UTC. The same evidence of thundersnow was found for Candói city, neighboring Guarapuava city (Figure S3).

Figure 2 shows surface atmospheric pressure fields associated with the CS that resulted in the thundersnow occurrence. The region marked as a star denotes the approximate location of Guarapuava, within the high terrain, steep plateau which climatologically receives cold outbreaks from Argentina. The four maps (day − 3 to day 1) show a very strong blocking anticyclone in the south-eastern Pacific Ocean. This high pressure is very vigorous (stronger than 1040 hPa), and projects a ridge well into the Atlantic Ocean on day −2. As a result of this intrusion, the Atlantic sector receives a continuous influx of very cold air coupled with cyclonic circulation over the Southern Atlantic Ocean getting more organized on day −2, increasing the pressure gradient with the high pressure over the South American continent. On day 0, the affected area is under the domain of anticyclonic conditions, with strong continental advection of cold air forced by the projection of the Pacific blocking high inland. This evolution of the synoptic pressure fields suggests a large equatorward amplification of a Rossby wave.

Besides this, the evolution of the 1000–500-hPa thickness from day −3 to day 0 is shown in Figure 3, where the jet stream is shown in shading (wind velocity over $40\,\text{m s}^{-1}$). The black thick line represents the 5400-gpm layer, which has been traditionally associated with the occurrence of snow in southern Brazil based on the experience of authors. It becomes obvious

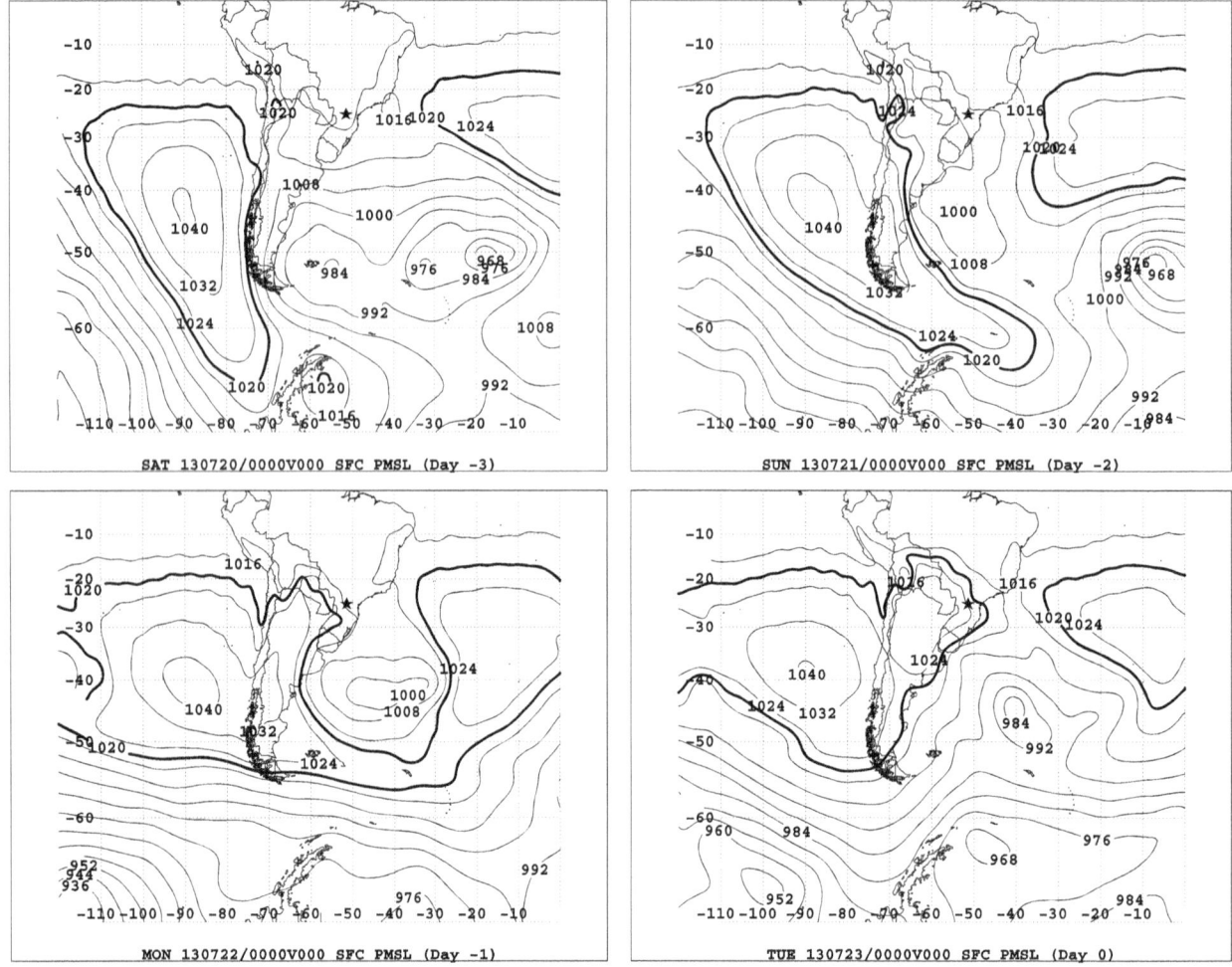

Figure 2. Horizontal distribution of atmospheric pressure at mean sea level from 'day −3', 20 July 2013, to 'day 0', 23 July 2013. The black filled star denotes the approximate location of Guarapuava. The 1020 hPa isobar is highlighted by the bold line.

that the event was characterized by the overlap of very low thickness as well as a very strong upper level jet from day −1. The position of Guarapuava, marked on the map (black filled star), approximately coincides with the equatorward side of the jet entrance, where the cold front is positioned and which dynamically would offer the ideal conditions for deep convection (Shapiro and Keyser, 1990; Moore and Vanknowe, 1992). This configuration suggests that the environment was highly conducive for the outburst of convective systems embedded in a very cold layer of mid-level air.

An environment conducive to convection, besides the presence of a mesoscale cyclonic curvature of 1020 hPa isobar northeast of Guarapuava (Figure 1), reinforces the importance of mesoscale processes. The mesoscale environment is then analyzed. A cross section of equivalent potential temperature (θ_e), omega (ω), moist potential vorticity (MPV), and relative humidity (Figure 4) reveals the well-developed baroclinic structure of a frontal zone. We note a deep nearly saturated layer (with respect to water) over the area of interest, with upward vertical motions in excess of −8 μb s^{-1} and regions of elevated instability with appreciable vertical shear. The cross section also shows a shallow moist

neutral region atop a region of conditional symmetric instability (CSI), diagnosed with negative MPV (Moore and Lambert, 1993; Schultz and Schumacher, 1999). This CSI layer exists in response to large changes in vertical winds and thus in the vertical gradient of geostrophic pseudo-angular momentum. However, this layer also exists above a significant near-surface vertical depth where temperatures hover around 0 °C. Further aloft, there is a small, weak area of conditional instability *very* high in the profile. The GFS skew-T profile for that location (25.3°S, 51.4°W; Figure 5) corroborates the shallow moist nearly neutral region beginning at 500 hPa, which is a quite high lifting parcel level (LPL) compared with what has typically been observed at higher latitudes (Market *et al.*, 2006). Yet, the temperature of the LPL is −10 °C, which suggests the presence of sufficient super cooled liquid water to supply any weak updraft that might form. Also, no convective available potential energy (CAPE) was found for a parcel lifted from that level. These model solutions suggest a region of significant frontogenesis (not shown) collocated with an area of weak CSI beneath a layer of weak CI.

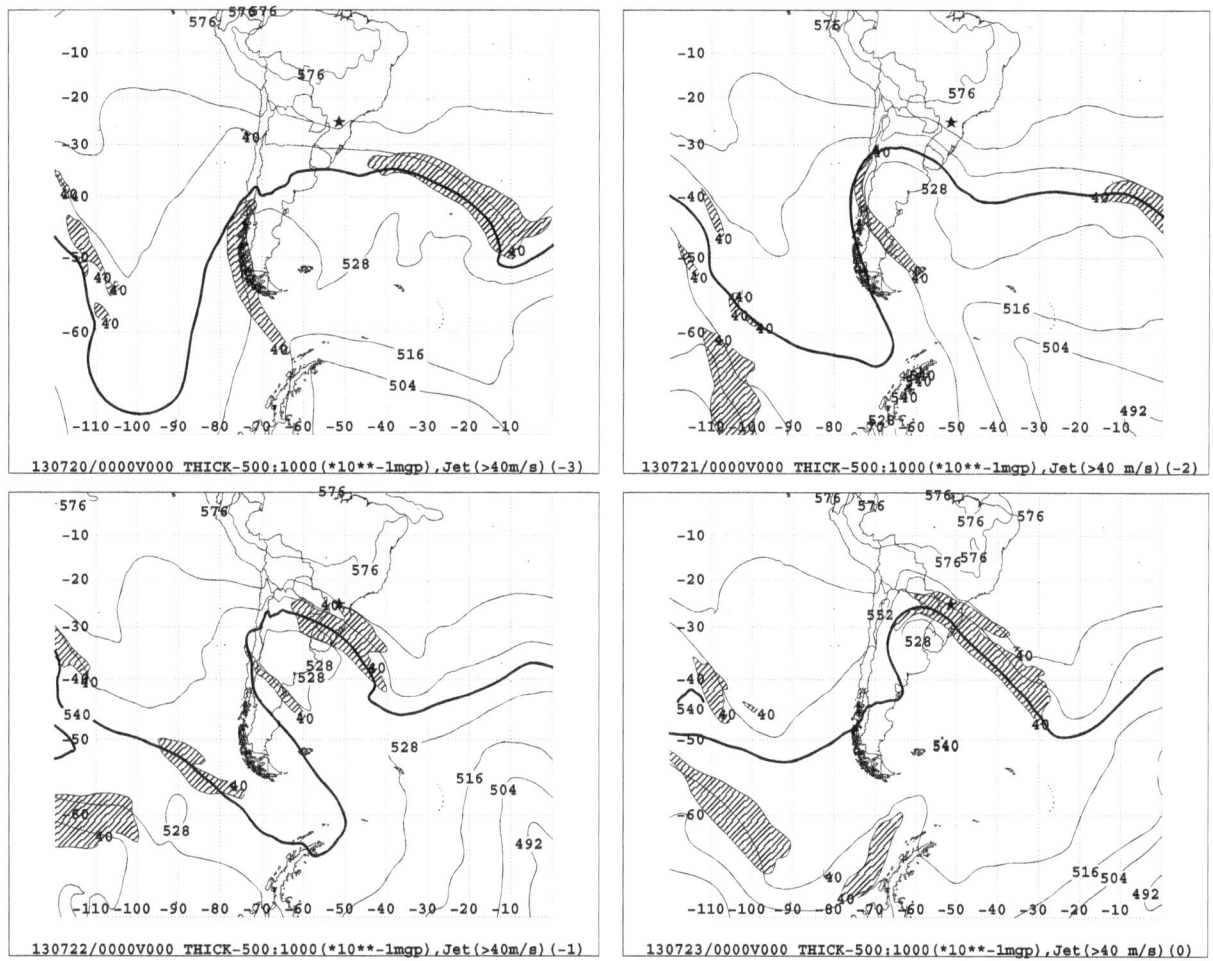

Figure 3. Horizontal distribution of thickness between pressure levels of 1000 and 500 hPa, and jet stream at 250 hPa ($>40\,m\,s^{-1}$ – shaded), from 'day −3', 20 July 2013, to 'day 0', 23 July 2013. The 5400 mgp thickness isoline is highlighted by the bold line. The black filled star denotes Guarapuava city in Brazilian Parana State.

4. Discussion and conclusions

On the 22 and 23 July 2013, a CS resulted in a very rare snowfall accompanied by thundersnow in southern Brazil. In Curitiba, one of the most important southern metropolises of Brazil, the phenomenon had not been observed since 1975. In Guarapuava, state of Parana, the accumulated snow caused damage to roofs and disrupted roads, resulting in two deaths in a traffic accident.

Radar precipitation analysis revealed radar reflectivity (Z) values between 10 and 30 dBZ over Guarapuava in most of the period when snowfall was observed. STARNET and RINDAT detected 9 and 5 cloud-to-ground strokes, respectively, close to 23 July 2013 0000 UTC as shown in Figure 1. Reflectivity (Z) values around 40 dBZ were found prior to the first strokes and in the exact moment of the last discharges.

The most important feature on the synoptic scale was the large amplification of the Rossby wave from the Pacific Ocean which is typical of South American CSs (e.g. Pezza and Ambrizzi, 2005a). In association with the particular case here discussed, however, the inclination of the mid-tropospheric level trough component of the wave was such (to the west) as to favor

the displacement of the largest anomalies within the southern part of Brazil. The stationarity of such pattern was then a pivotal component in allowing the cold advection to last sufficiently long so as to trigger the event, while the upper level jet gave the right conditions for the convective nature of the air mass, with fairly cold air in the mid-troposphere.

Dolif Neto *et al.* (2009) found a similar synoptic setup associated with the 2005 thundersnow case in Santa Catarina state, without significant accumulation in that case. Our results suggest that the greater wave amplification in 2013, in synergy with a more northward displacement of the upper level jet was responsible for the more unusual conditions. In both cases, the slow propagation of the upper level trough was crucial to provide the necessary persistence for the cold air to unfold.

This situation is one of the classic modes of cold air incursion leading to frosts in South America (Escobar *et al.*, 2004; Pezza and Ambrizzi, 2005a; Escobar, 2007; Pezza *et al.*, 2010), but is unusual in a sense that convective snow would primarily be associated with cyclonic conditions (Market et al., 2002). However, Figure 2 also shows an evident shortwave trough projecting around

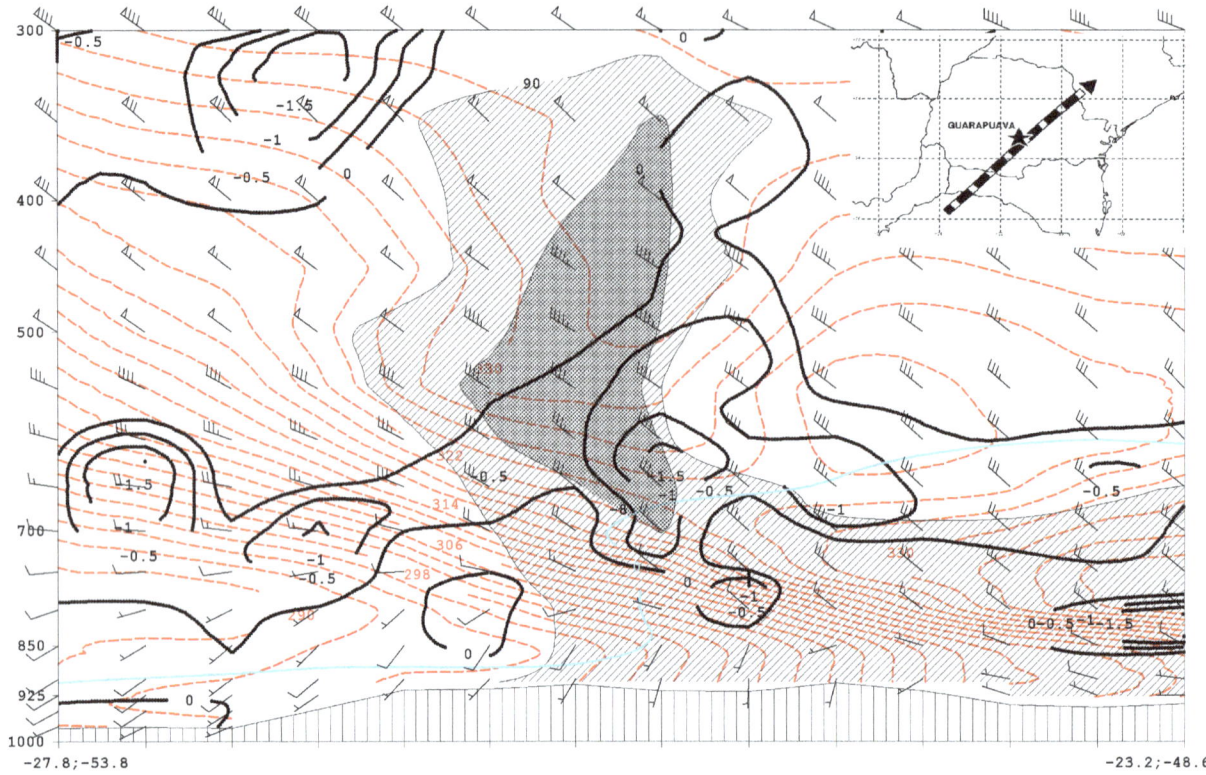

Figure 4. Cross section from the GFS initial condition fields for South America valid at 23 July 2013 0000 UTC over Guarapuava along from 27.8°S, 53.8°W to 23.2°S, 48.6°W). In upper corner the black filled star denotes Guarapuava city, and the arrow indicates the horizontal location of the cross section from Foz do Iguaçu to São Paulo. Winds are plotted in station model format (in m s^{-1}), with relative humidity greater than 90% (with respect to slanted lines fill), equivalent potential temperature (dashed red lines, every 2 K), pressure vertical velocity (dotted fill, $< -8\,\mu$b s^{-1}; zig–zag lines fill, $> 4\,\mu$b s^{-1}), freezing isotherm (cyan line), moist potential vorticity (black bold lines; negative values, every 0.5×10^{-6} m K s^{-3} Pa^{-1}).

the location of Guarapuava (day zero), suggesting the presence of a mesoscale upper level disturbance within the anticyclonic region.

Mesoscale analysis using a cross section over Guara-puava revealed a well-developed baroclinic structure of a frontal zone. The cross section also shows a shallow moist nearly neutral region atop a region of CSI. The CSI exists in response to large vertical shear what is caused by the structure of the upper level jet stream. This highlights the importance of the synoptic-scale Rossby wave amplification and interaction to the sur-rounding atmosphere in order to produce such a jet stream.

Equatorward Rossby wave breaking (RWB) over South America is typically associated with intense CSs causing frosts in southern and southeastern Brazil (Sprenger *et al.*, 2013). Comparing our studied case to other CSs in South America or other thunder-snow events in Brazil and in the United States (Dolif Neto *et al.*, 2009) or in Spain (Bech *et al.*, 2013), the convective structures and synoptic pattern observed are more intense and more amplified. The marked Rossby wave amplification, resulted from RWB, was crucial to enable snowfall reaching subtropical latitudes in Parana State. Furthermore, the vigorous convective regime favored the occurrence of thundersnow, with a significant snow accumulation. In comparison to classical thundersnow cases in the United States

(Market *et al.*, 2002), the present event occurred much further from the cyclone's center (at least 500 km further), what highlights the importance of mesoscale processes.

In summary, the CS accompanied by thundersnow and heavy snow appears to be a combination of: (1) fur-ther northward Rossby wave amplification; (2) strong upward vertical motion within a deep nearly saturated layer and (3) a CSI layer, in response to strong vertical shear, beneath a layer of weak conditional instability, and above a significant near-surface vertical depth where temperatures hover around 0 °C.

Finally, we conclude that a typical atmospheric pat-tern of CSs in South America presented some particular features that permitted the rare occurrence of a thunder-snow event and heavy snow in southern Brazil on 22 and 23 July 2013. Indeed, further studies will be nec-essary to better establish the underlying microphysical properties and atmospheric electricity that were also pivotal within this development. Future improvements in the Brazilian network of *in situ* remote sensing would be a desirable outcome in order to study and predict future thundersnow cases.

Acknowledgements

The authors would like to thank the two anonymous reviewers whose efforts produced an improved manuscript.

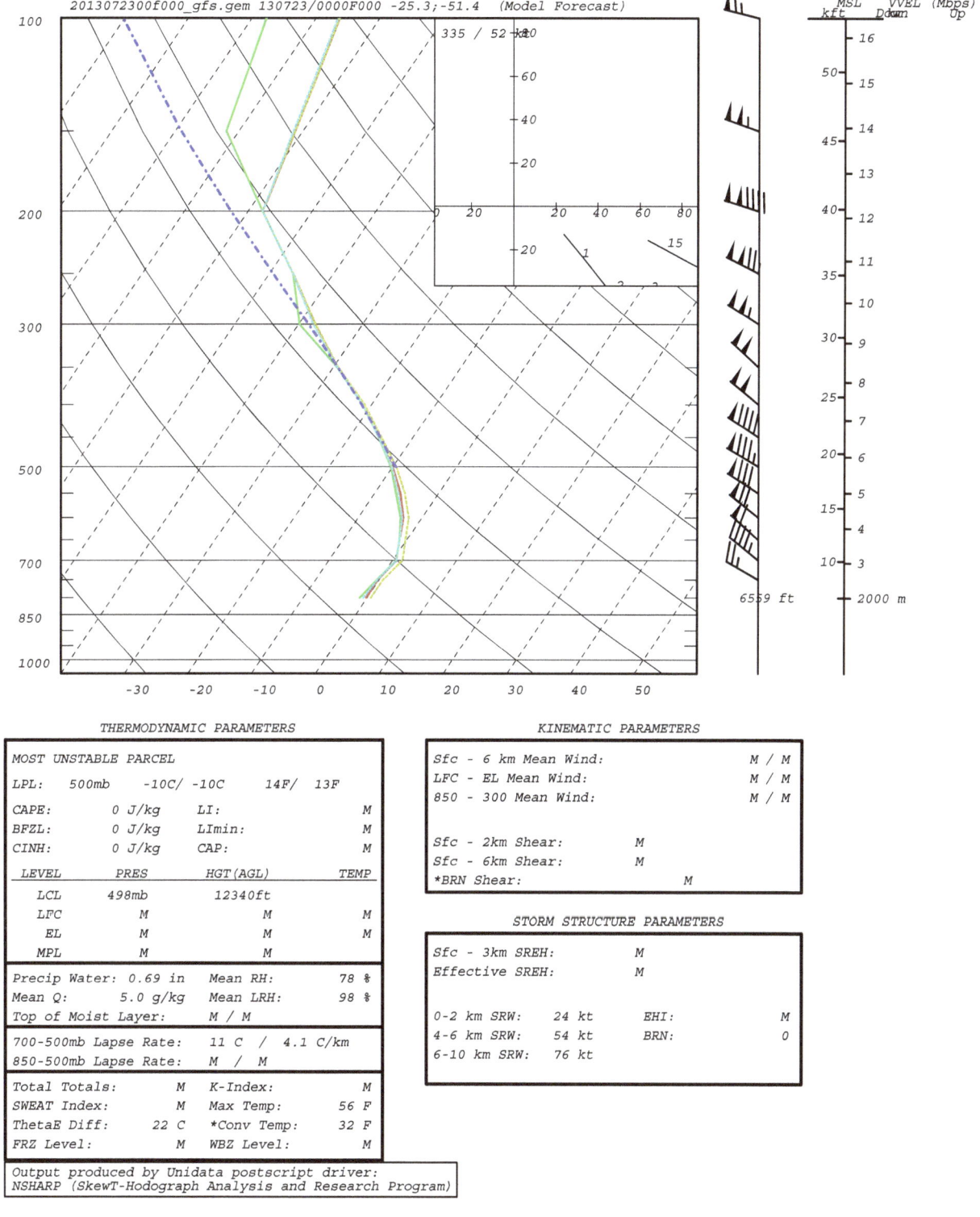

Figure 5. Skew-T log p profile from the GFS initial condition fields for South America, at 25.3°S and 51.4°W valid at 23 July 2013 0000 UTC.

Supporting information

The following supporting information is available:

Figure S1. *Diário Catarinense* newspaper front page from 24 July 2013 showing an unforgettable (in portuguese: 'Inesquecível') view of the *Serra do Tabuleiro* mountains covered by a deep layer of snow. The picture was taken from Florianópolis city, which is at sea level.

Figure S2. Remote sensing image from TERRA/Modis on 23 July 2013 1320 UTC, and Aerial photograph (upper left corner – Picture taken by Eliana Panty Schwarz) over the same area and within the same day.

Figure S3. Vertical radar reflectivity profiles over Candói city in Brazilian Parana State, neighboring Guarapuava city, from 22 July 2013 2136 UTC until 23 July 2013 0224 UTC. Vertical black bars indicate the number of cloud to ground strokes detected by STARNET within a radius of 10 km.

References

Bech J, Pineda N, Rigo T, Aran M. 2013. Remote sensing analysis of a Mediterranean thundersnow and low-altitude heavy snowfall event. *Atmospheric Research* **123**: 305–322, doi: 10.1016/j.atmosres.2012.06.021.

Beneti CAA, Leite EA, Garcia SAM, Assunção LAR, Cazetta Filho A, Reis RJ. 2000. RIDAT – Rede integrada de detecção de descargas atmosféricas no Brasil: situação atual, aplicações e perspectivas.(in Portuguese, abstract in English). In XI Congresso Brasileiro de Meteorologia, Rio de Janeiro, Brazil.

Camargo MBP. 2010. The impact of climatic variability and climate change on Arabic coffee crop in Brazil. *Bragantia* **69**(1): 239–247, doi: 10.1590/S0006-87052010000100030.

DaMatta FM, Ramalho JDC. 2006. Impacts of drought and temperature stress on coffee physiology and production: a review. *Brazilian Journal of Plant Physiology* **18**: 1, doi: 10.1590/S1677-0420 2006000100006.

Dolif Neto G, Market PS, Becker AE, Pettegrew B, Melick C, Schultz C, Barbieri CE. 2009. A comparison of two cases of low-latitude thundersnow. *Atmosfera* **22**(3): 315–330.

Escobar GCJ. 2007. Synoptic classification associated with cold waves in são Paulo city (in Portuguese, abstract in English). *Revista Brasileira de Meteorologia* **22**(2): 241–254.

Escobar GCJ, Compagnucci RH, Bischoff SA. 2004. Sequence Patterns of 1000 hPa and 500 hPa geopotential height fields associated with cold surges in Buenos Aires. *Atmosfera* **12**(3): 69–89.

G1. *News Portal.* 2014. Globo Comunicações e Participações S.A. http://g1.globo.com/pr/parana/noticia/2013/07/neve-causa-estragos-e-acidentes-em-guarapuava-no-interior-do-parana.html (accessed: 15 October 2014).

Gan MA, Rao VB. 1994. The influence of the Andes Cordillera on transient disturbances. *Monthly Weather Review* **122**: 1141–1157, doi: 10.1175%2F1520-0493%281994%29122%3C1141%3ATIOTAC%3E2.0.CO%3B2.

Garreaud RD. 2000. Cold air incursions over subtropical South America: mean structure and dynamics. *Monthly Weather Review* **128**: 2544–2559, doi: 10.1175/1520-0493(2000)128<2544:CAIOSS>2.0.CO;2.

GFS. n.d. Global forecast system. http://www.emc.ncep.noaa.gov/GFS/doc.php (accessed: 19 November 2013).

Lau K-M, Chang C-P. 1987. Planetary scale aspects of winter monsoon and teleconnections. In *Monsoon Meteorology*, Chang C-P, Krishnamurti TN (eds). Oxford University Press: Oxford; 161–202.

Marengo AJO, Cornejo A, Satyamurty P, Nobre CA, Sea W. 1997. Cold surges in tropical and extratropical South America: the strong event in june 1994. *Monthly Weather Review* **125**: 2759–2786, doi: 10.1175/1520-0493(1997)125<2759:CSITAE>2.0.CO;2.

Market PS, Halcomb CE, Ebert RL. 2002. A climatology of thundersnow events over the contiguous united states. *Weather and Forecasting* **17**: 1290–1295, doi: 10.1175/1520-0434(2002)017<1290:ACOTEO>2.0.CO;2.

Market PS, Oravetz AM, Gaede D, Bookbinder E, Lupo AR, Melick CJ, Smith LL, Thomas R, Redburn R, Pettegrew BP, Becker AE. 2006. Proximity soundings of thundersnow in the central United States. *Journal of Geophysical Research* **111**, doi: 10.1029/2006JD007061.

Moore JT, Lambert TE. 1993. The use of equivalent potential vorticity to diagnose regions of conditional symmetric instability. *Weather and Forecasting* **8**: 301–308.

Moore JT, Vanknowe GE. 1992. The effect of jet-streak curvature on kinematic fields. *Monthly Weather Review* **120**(11): 2429–2441.

Morales CA, Neves JR, Anselmo E. 2011. Sferics timing and ranging network – STARNET: evaluation over South America. In Proceedings of the 14th International Conference on Atmospheric Electricity - ICAE, Rio de Janeiro, Brazil.

Müller GV, Berri GJ. 2007. Atmospheric circulation associated with persistent generalized frosts in central-southern South America. *Monthly Weather Review* **135**: 1268–1289, doi: 10.1175/MWR3344.1.

Müller GV, Berri GJ. 2012. Atmospheric circulation associated with extreme generalized frosts persistence in central-southern South America. *Climate Dynamics* **38**(5–6): 837–857, doi: 10.1007/s00382-011-1113-2.

Pezza AB, Ambrizzi T. 2005a. Dynamical conditions and synoptic tracks associated with different types of cold surge over tropical South America. *International Journal of Climatology* **25**(2), 215–241.

Pezza AB, Ambrizzi T. 2005b. Cold waves in South America and freezing temperatures in São Paulo: historical background (1888–2003) and case studies of cyclone and anticyclone tracks. *Revista Brasileira de Meteorologia* **20**(1), 141–158.

Pezza AB, Simmonds I, Coelho CA. 2010. The unusual Buenos Aires snowfall of July 2007. *Atmospheric Science Letters* **11**(4): 249–254.

Schultz DM, Schumacher PN. 1999. The use and misuse of conditional symmetric instability. *Monthly Weather Review* **127**: 2709–2732; Corrigendum, 128, 1573..

Schultz DM, Keyser D, Bosart LF. 1998. The effect of large-scale flow on low-level frontal structure and evolution in midlatitude cyclones. *Monthly Weather Review* **126**: 1767–1791, doi: 10.1175/1520-0493 (1998)126<1767:TEOLSF>2.0.CO;2.

Seluchi ME, Serafini VY, Le Treut H. 1998. The impact of the Andes on transient atmospheric systems: a comparison between observations and GCM results. *Monthly Weather Review* **126**: 895–912, doi: 10.1175%2F1520-0493%281998%29126%3C0895%3ATIOTAO%3E2.0.CO%3B2.

Shapiro MA, Keyser DA. 1990. *Fronts, jet streams, and the tropopause.* US Department of Commerce, National Oceanic and Atmospheric Administration, Environmental Research Laboratories, Wave Propagation Laboratory, Boulder, CO.

Sprenger M, Martius O, Arnold J. 2013. Cold surge episodes over southeastern Brazil – a potential vorticity perspective. *International Journal of Climatology* **33**: 2758–2767, doi: 10.1002/joc.3618.

Vera CS, Vigliarolo PK. 2000. A diagnostic study of cold-air outbreaks over South America. *Monthly Weather Review* **128**: 3–24, doi: 10.1175/1520-0493(2000)128<0003:ADSOCA>2.0.CO;2.

11

A dynamical link between deep Atlantic extratropical cyclones and intense Mediterranean cyclones

Shira Raveh-Rubin[1]*[iD] and Emmanouil Flaounas[2]

[1]Institute for Atmospheric and Climate Science, ETH Zurich, Switzerland
[2]National Observatory of Athens, Greece

*Correspondence to:
S. Raveh-Rubin, Institute for
Atmospheric and Climate
Science, ETH Zurich, CHN
M12.3, Universitätstrasse 16,
8092 Zurich, Switzerland.
E-mail: shira.raveh@env.ethz.ch

Abstract

Breaking of atmospheric Rossby waves has been previously shown to lead to intense Mediterranean cyclones, one of the most prominent environmental risks in the region. Wave breaking may be enhanced by warm conveyor belts (WCBs) associated with extratropical cyclones developing over the Atlantic Ocean. More precisely, WCBs supply the upper troposphere with air masses of low potential vorticity that, in turn, amplify ridges and thus favor Rossby wave breaking. This study identifies the mechanism that connects Atlantic cyclones and intense mature Mediterranean cyclones through ridge amplification by WCBs, and validates its climatological relevance. Using European Centre for Medium-Range Weather Forecasts (ECMWF) ERA-Interim reanalyses and a feature-based approach, we analyze the 200 most intense Mediterranean cyclones for the years 1989–2008 and show that their majority (181 cases) is indeed associated with this mechanism upstream. Results show that multiple Atlantic cyclones are associated with each case of intense Mediterranean cyclone downstream. Moreover, the associated Atlantic cyclones are particularly deeply intensifying compared with climatology.

Keywords: trajectories; cyclogenesis; ridge amplification; warm conveyor belt

1. Introduction

Potential vorticity (PV) streamers correspond to high-PV stratospheric air that intrudes into lower atmospheric levels in a filamentary structure during Rossby wave breaking (Appenzeller and Davies, 1992, Wernli and Sprenger, 2007). PV streamers are widely recognized as upper-tropospheric precursors for Mediterranean cyclogenesis, and they play a primary role in the development and intensification of these cyclones (Tafferner, 1990, Fita *et al.*, 2006, Funatsu *et al.*, 2007, Homar *et al.*, 2007, Tous and Romero, 2013). Flaounas *et al.* (2015) recently analyzed the baroclinic life cycle of the 200 most intense Mediterranean cyclones in a 20-year climatology and confirmed that PV streamers systematically precede cyclogenesis. In fact, their results showed that the streamers correspond to the downstream, southward, deflection of the polar jet, similarly to the anticyclonic wave breaking (e.g. Thorncroft *et al.*, 1993). This type of upper-level wave breaking is also of key relevance for heavy precipitation events in the region (Massacand *et al.*, 1998, Martius *et al.*, 2006, 2007, Raveh-Rubin and Wernli, 2015).

The synoptic scale conditions related to the 200 most intense Mediterranean cyclones (as identified in Flaounas *et al.*, 2015) are shown in Figure 1(a) as composite averages of pressure at sea level and of PV and wind on the 330-K isentropic surface. Composites are centered at the time when the Mediterranean cyclones reach their maximum intensity. In addition, Figure 1(b)

shows the composite fields as monthly anomalies (deviations from the 20-year monthly climatology). A coherent structure of anticyclonic wave breaking is shown in Figure 1(a), characterized by a ridge over the eastern North Atlantic and a trough over Western Europe. The ridge is associated with negative PV anomalies, while the streamer is associated with positive PV anomalies (Figure 1(b)). The latter are centered over the Mediterranean, to the northwest of the cyclone centers [negative sea-level pressure (SLP) anomalies over the central Mediterranean Sea in Figure 1(b)], consistent with the downstream deflection of the polar jet (Figure 1(a), at the eastern flank of the ridge).

Despite the importance of PV streamers for the genesis of intense Mediterranean cyclones, our knowledge lacks a systematic analysis on the atmospheric mechanism that contributes to the wave breaking, and in turn, the intrusion of PV streamers over the Mediterranean. In a numerical case study of a Mediterranean cyclone and PV streamer which caused Alpine flooding, Massacand *et al.* (2001) showed that an upstream Atlantic cyclone and its associated diabatic heating contributed to the upper-tropospheric Rossby wave breaking. This was achieved by the cross-isentropic transport of low-PV air to the upper troposphere, thereby enhancing the upper-level ridge and forming the elongated PV streamer downstream. In a case study in September 2008, Rossby wave breaking was provoked by the extratropical transition of hurricane Hanna over the North Atlantic (Grams *et al.*, 2011). The wave breaking was similar to the one shown in Figure 1 and was

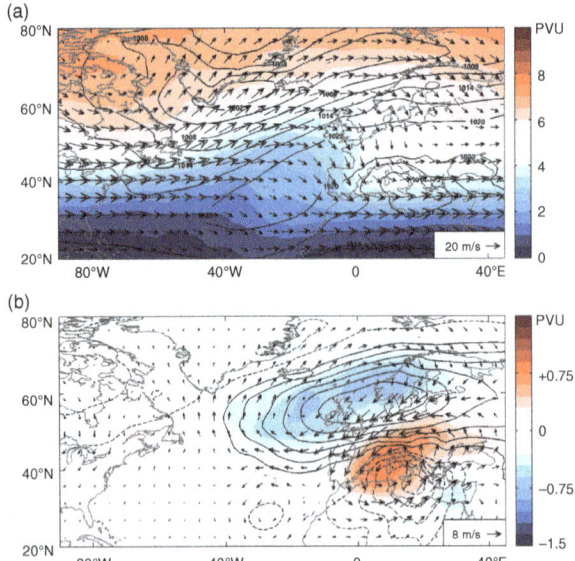

Figure 1. (a) ERA-Interim composite fields of sea-level pressure (black contours), PV (shaded), and wind (arrows) on the 330-K isentropic surface, centered at the time of maximum intensity of the 200 Mediterranean cyclones. (b) as (a) but for monthly anomalies (contours every 1 hPa, dashed for negative and solid for positive values).

enhanced by an upstream ridge, which was amplified due to rising moist air masses (see their Figure 4), similar to the case in Massacand *et al.* (2001). Such transport of low-PV air has been achieved through warm conveyor belts (WCBs; strongly ascending airstreams along the warm sector of extratropical cyclones), which undergo significant latent heating (Browning, 1990, Wernli and Davies, 1997). The wave breaking formed a PV streamer over the Mediterranean basin which, in turn, instigated Mediterranean cyclogenesis south of the Alps. A similar scenario of intense Mediterranean cyclogenesis has been shown by Chaboureau *et al.* (2012), Pantillon *et al.* 2015) and noted as well in other regions (Pomroy and Thorpe, 2000, Grams and Archambault, 2016).

The case studies demonstrate the concept of downstream development of baroclinic waves (Simmons and Hoskins, 1979, Chang, 1993, Orlanski and Chang, 1993, Orlanski and Sheldon, 1993, Wernli *et al.*, 1999). These idealized studies, which employ dry primitive equations, explain the development of a downstream cyclone by the export of energy from the primary system via the ageostrophic flux divergence. Recently, in an idealized moist baroclinic channel setup, Schemm *et al.* (2013) have suggested that moist diabatic processes related to WCBs enhance the downstream development. Madonna *et al.* (2014b) quantified climatologically the co-occurrence of Rossby-wave breaking and WCB outflows and found that 60% of the WCBs occur together with a PV streamer, but less than 15% of PV streamers are accompanied by WCBs. Once occurring together, the most prevalent situation (35% of the cases) is for WCBs to be followed by PV streamers downstream. A common feature to the studies

incorporating moist dynamics and the case studies described above is a primary cyclone with associated WCB outflow which enhances the upper-tropospheric ridge downstream, causing eventually Rossby-wave breaking and downstream cyclogenesis.

The motivation of this study is thus to investigate whether the aforementioned mechanism dynamically relating two cyclones via wave breaking due to ridge amplification by the WCB of the first cyclone is systematically present whenever intense cyclogenesis takes place in the Mediterranean basin.

2. Methodology

2.1. Data and approach

Our methodology is designed to trace the dynamical link between North Atlantic cyclones and those developing downstream in the Mediterranean. Therefore, an objective feature-based approach has been developed and applied to the 200 most intense Mediterranean cyclones for the years 1989–2008, taken from Flaounas *et al.* (2015). The set of intense cyclones was determined based on their 850-hPa relative vorticity, and their occurrence frequency peaks in winter, while still significant in autumn and spring (Flaounas *et al.*, 2015). These cyclones have been identified in a regional climate simulation forced at its boundaries and nudged within its domain by the ERA-Interim reanalysis of the European Centre for Medium-range Weather Forecasts (Dee *et al.*, 2011). Consistently, in this study, the dynamical conditions that lead to these Mediterranean cyclones are analyzed with 6-hour ERA-Interim atmospheric fields, interpolated to a regular 1° × 1° horizontal grid.

Our methodology follows a three-component feature-based approach, applied to each of the 200 cases. Given the time and location of the mature stage of each cyclone (i.e. the time when cyclones reach their maximum intensity), we first track the associated ridge that formed upstream of the wave breaking, i.e. the PV streamer typically northwest of the center of the Mediterranean cyclone (Figure 1(b)). Then we identify the WCB that is potentially associated with the amplification of the tracked ridge (if there is any), and finally, we detect the cyclones that go along with these WCBs. In more detail, the analysis is based on the following three atmospheric features:

1. *Tracked ridges*: At the time when Mediterranean cyclones reach their maximum intensity (time 0 h hereafter), we locate the associated ridge that is located at a cyclone northwest side. The ridge is defined by enclosed contours of negative PV anomalies of less than −2 PVU (PV units) on the 330-K surface. The ridge is then tracked backwards in time up to 1 week prior to time 0 h, or until 120°W. Tracking is achieved first by detecting closed contours of −2 PVU anomalies on 330 K every 6 hours until −168 h, followed by connecting the PV anomalies

that overlap at consecutive time steps. If more than one PV anomaly overlaps at consecutive time steps, we retain the one with the largest overlapping area. Tracking the ridges provides a dataset of the location and extent of the negative PV anomalies that eventually favor the wave breaking.

2. *Warm conveyor belts:* WCBs are identified objectively in a Lagrangian framework as airstreams ascending at least 600 hPa in 48 h in the vicinity of a cyclone (Madonna *et al.*, 2014a). This dataset includes all identified WCB trajectories in the ERA-Interim data period, their three-dimensional position, potential temperature, and ice water content traced along the trajectories from the starting time of the vertical ascent, to the end of the ascent after 48 h and beyond. For the purpose of this study, both the WCB ascent and outflow phases are important to identify. The ascent phase of a WCB is defined as the time of the first appearance of ice water content along the WCB trajectory. This methodology to define WCBs ascent has been previously applied in Flaounas *et al.* (2016). The outflow portion of the WCB trajectory is of particular interest, as it is the outflow of the WCBs which might amplify the PV ridges and thus influence wave breaking (e.g., Massacand *et al.*, 2001, Madonna *et al.*, 2014b). Performing several sensitivity tests, a suitable approach to identify WCB outflow is to consider the section of a WCB trajectory from 36 to 72 h (i.e., last 12 h of the ascent and the subsequent 24 h) with potential temperature higher than 310 K. The choice of the potential temperature threshold is consistent with the fixed isentropic level used to track the negative PV anomalies (see above) and ensures that only outflows reaching the highest levels are considered.

3. *Cyclone tracks:* cyclone tracks are extracted in ERA-Interim by applying the method of Wernli and Schwierz (2006). In this method, cyclones are identified as SLP local minima within enclosed SLP contours from a standard $0.75° \times 0.75°$ grid in longitude and latitude.

Our methodological approach has two steps and is applied separately for each of the 200 intense Mediterranean cyclone cases, providing a meaningful connection among the three datasets: In a first step, at every 6-h time step that precedes a Mediterranean cyclone's maximum intensity (time 0 h), WCB outflow positions are matched with the tracked negative PV anomalies. This is done by counting how many WCB trajectory outflows are located in the area of the tracked ridge. These trajectories are 'tagged', to allow the identification of the North Atlantic cyclones they are associated with in the second step. The association is valid if the WCB ascent is closer than 10° from a cyclone center.

2.2. Two illustrative examples

Figure 2 shows two examples of intense Mediterranean cyclones which are consistent with the chain of events described in the introduction. In our first example, a Mediterranean cyclone reaches its mature stage over northwest Africa at 0000 UTC 10 November 2001 (0 h, left panels). The cyclone is clearly formed alongside an elongated stratospheric PV streamer, located northwest of the cyclone, having a southwest-northeast direction and corresponding to the eastern side of an eastwards tilted ridge (outlined by blue contour at 0 h). This configuration is similar to the composite presented in Figure 1(a), as well as to the anticyclonic wave breaking scenario LC1 in Thorncroft *et al.* (1993). The evolution of the wave breaking is shown in the left panels of Figure 2, from −60 to 0 h. Strong negative PV anomalies (blue contour) tend to propagate and progressively extend toward the east from the central North Atlantic (blue contour at −60 h) until forming the ridge over the eastern North Atlantic (blue contour at 0 h). In parallel, the outflows of WCBs reach the flanks of the ridge (magenta dots) potentially contributing to its amplification and eventual wave breaking due to advection of low PV air masses from the lower troposphere. Several North Atlantic cyclones have been found to be associated with this series of WCBs (shown in green line).

In the second example, a deep cyclone occurred over Italy at 0600 UTC 13 December 1990 (0 h, right panels in Figure 2), also associated with a trough and a PV streamer. The trough itself was stationary over the region until being amplified and deformed into a narrow PV streamer. The ridge amplification took place from −60 to −12 h, again supported by WCB outflow of air with low PV. However, the clear ridge and PV streamer shape broke down by the time the Mediterranean cyclone attained maximum intensity. Here, two North Atlantic cyclones were found to produce the WCBs leading to the ridge-streamer pattern, with tracks that, unlike the first example, are confined to the western North Atlantic.

These two case studies exemplify a considerable case-to-case variability concerning the Mediterranean cyclone itself, the location and shape of the upper-tropospheric wave breaking, the WCB outflow locations and the associated Atlantic cyclone tracks, as well as the dynamical evolution and the mutual interactions of these atmospheric features.

3. The climatological connection between North Atlantic and intense Mediterranean cyclones

Performing our methodology for all 200 intense Mediterranean cyclones, we found that a vast majority, 181 cases (90.5% of Mediterranean cyclones) were dynamically associated with upstream North Atlantic cyclones. For the rest of the cases, the mechanism was not confirmed. In fact, an average of 4.3 North Atlantic cyclones have been identified as being related to the Mediterranean cyclones with a standard deviation of 2.1 cyclones. This indicates that a series of cyclones and their associated WCBs, rather than a single primary

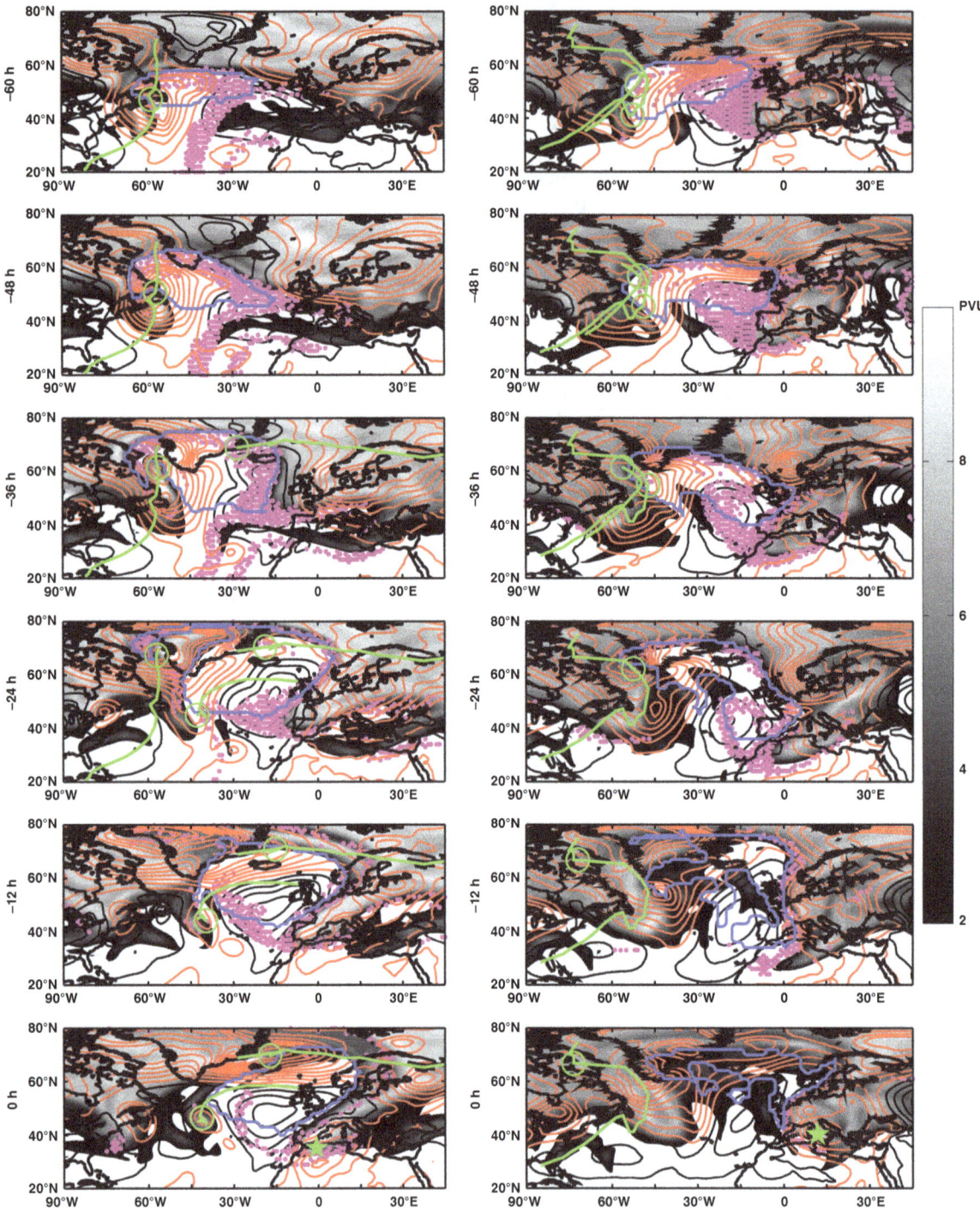

Figure 2. Sea-level pressure (contours with 5-hPa intervals, red contours depict values of less than 1015 hPa), PV on the 330-K isentropic surface (PVU, shaded), WCB outflow positions (magenta dots), and tracked negative PV anomaly on 330-K surface (−2 PVU in blue contour) for different times relative to 0000 UTC 10 November 2001 (left panels) and 0600 UTC 13 December 1990 (right panels). The associated North Atlantic cyclones tracks are plotted in green (only the associated cyclone tracks are plotted), and the position of the cyclone at the corresponding time is marked by a green circle. The downstream Mediterranean cyclone is marked with a green star at time 0 h.

cyclone, build up the North Atlantic ridge required for the downstream formation of the PV streamer instigating the Mediterranean cyclone, during the 168-h period prior to maximum intensity. However, not all the identified cyclones contribute equally in terms of associated WCB trajectories that reach the ridge. To distinguish those cyclones that contribute significantly

to the WCB outflows, we filter out cyclones with less than 25% of the total associated WCBs count for that event. This reduces the number of associated cyclones by almost a factor of two.

All associated extratropical cyclone tracks are presented in Figure 3(a). Despite the large variability of their tracks, there is a high track density between 40 and 60°N, where cyclones also tend to attain their deepest central SLP. This is consistent with the climatology of winter cyclones in the region (Wernli and Schwierz, 2006) and thus no different areas of occurrence are favored for Atlantic cyclones associated with Mediterranean cyclogenesis. It is also evident that five cyclones underwent extratropical transition, including the event at 0000 UTC 10 November 2001 (Figure 2, left column). From a composite perspective, Figure 3(b) shows the spatial variability of the WCBs ascent and outflow positions. Consistent with the cyclones' location at their time of maximum intensity (Figure 3(a)), the WCBs tend to start their ascent over the western North Atlantic and to reach the upper troposphere further northeast near 50°N/40°W where they amplify the PV ridges. Outflow locations are indeed collocated with the average ridge location in the composite PV at 330 K (Figure 1(b)). WCB densities are located consistently with their climatological distribution (Figure 4 in Madonna *et al.*, 2014a).

The mechanism identified and explained in this study suggests a series of processes that stem from North Atlantic cyclones and result in intense Mediterranean cyclogenesis, with a certain degree of diabatic contribution to the wave breaking via WCBs. The preferred location of the associated group of Atlantic cyclones, and their key contribution for the downstream development, suggests that relatively strong extratropical cyclones are more likely to play such a role. This hypothesis is addressed by comparing the intensification rates of the associated North Atlantic cyclones to a climatological reference, formed by 20 random samples of North Atlantic cyclones. The random samples were chosen by maintaining the same seasonality of the 200 Mediterranean cyclones and the same number of associated Atlantic cyclones. In addition, the random cyclones were bounded to attain their minimum SLP within the area 70°–5°W, and 30°–70°N, where the associated Atlantic cyclones attain their deepest SLP. Figure 3(c) shows the probability density function of the cyclones intensification in Bergeron units for the randomly selected cyclones (solid line), compared with the one of the North Atlantic cyclones that are dynamically related to the 200 most intense Mediterranean cyclones (dashed line). This metric refers to the cyclones maximum deepening rate within 24 h, taking into consideration their latitudes (*Sanders and Gyakum*, 1980). There is a fair shift of the associated cyclones distribution to faster deepening cyclones, with respect to the random cyclones dataset. Our results are in line with Binder *et al.* (2016), who analyzed the cyclones deepening rate to show a climatological correlation between cyclone intensification and WCB air mass (see their Figure 1). Indeed, it is the strongly deepening North Atlantic cyclones that we found to contribute significantly to downstream Rossby wave breaking. However, a large variability of the cyclone intensification-WCB relationship is governed by dynamical interactions between the low-level diabatic PV production in the WCB, the cyclone, and the upper-tropospheric jet. Therefore, it

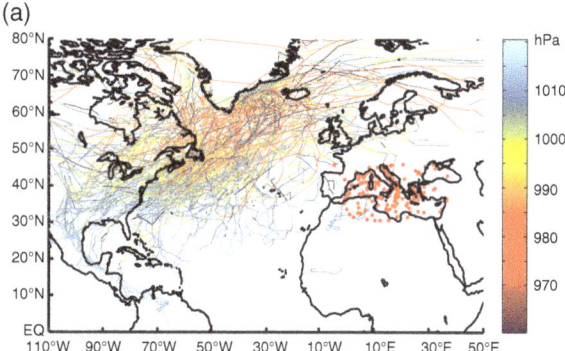

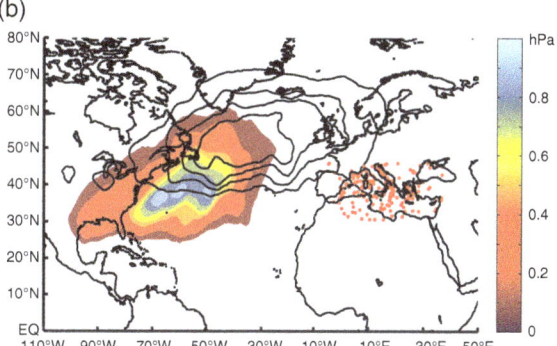

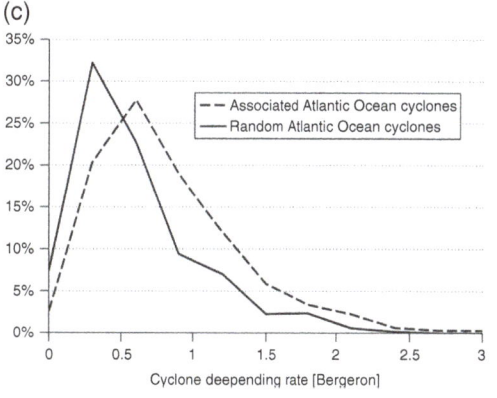

Figure 3. (a) All North Atlantic cyclone tracks associated with the 200 Mediterranean cyclones. Tracks are colored according to central sea-level pressure (in hPa). Red dots depict the locations of the associated Mediterranean cyclone's maximum intensity. (b) Colors show the distribution of WCB ascent (counts per 1°×1° grid element and scaled by the maximum count 3604, such that the maximum value is 1. Black contours show the distribution of WCB outflows, normalized by the maximum count of 184 (contours for scaled values of 0.2, 0.4, 0.6, and 0.8). Red dots depict the locations of the associated Mediterranean cyclones at their time of maximum intensity. (c) Probability density distributions of the deepening rate, expressed in Bergeron units of associated (dashed line) and climatological (solid line) sets of North Atlantic cyclones. See text for details on the sample selection.

would be misleading to draw a direct conclusion that it is only the strongest Atlantic cyclones that provoke Mediterranean cyclogenesis. Finally, we found that the WCBs of the North Atlantic cyclones ascend, on average, 4 days before the Mediterranean cyclones' mature stage (with a standard deviation of 2 days). This time lag is fairly consistent with the time lags of about 5 days, reported both the idealized simulations by Schemm *et al.* (2013) and the real case study analyzed by Grams *et al.* (2011).

4. Summary and discussion

The key role of WCB air as a dynamical link between intense North Atlantic and Mediterranean cyclones is demonstrated in this study. The vast majority (90.5%) of intense Mediterranean cyclones have been found to be associated with upstream North Atlantic cyclones, which are accompanied by ascending WCBs transporting low-PV air to the upper troposphere. This amplifies the upper-level ridges, which in turn contribute to the downstream Rossby wave breaking and cyclogenesis in the Mediterranean. In fact, it is the strongly deepening Atlantic cyclones, located between Newfoundland and Iceland that are particularly favorable for initiating the mechanism of downstream genesis of intense Mediterranean cyclones, presumably because these cyclones develop substantial WCB airstream flow compared with all North Atlantic cyclones.

The examination of the upstream development of the 200 intense Mediterranean cyclones is challenging due to the large spatiotemporal variability among the cases. For instance, Figure 2 shows two examples where Rossby wave breaking leads to downstream Mediterranean cyclogenesis. However, the ridges evolve, amplify, and propagate differently. A sophisticated feature-based approach, as outlined in this study, has been successful in highlighting a coherent mechanism that is common to more than 90% of all cases, and to provide a quantification of its occurrence. The high relevance of the mechanism to the climatological set is robust with respect to the Mediterranean cyclone identification and tracking technique. We have carried out the methodological procedure for a set of the 200 most intense Mediterranean cyclones in the ERA-Interim dataset, based on the identification and tracking technique of Wernli and Schwierz (2006). This set of 200 Mediterranean cyclones has only 39 overlapping events with the set analyzed in this study, yet, the mechanism is relevant for a comparable percentage of events (88%). Nevertheless, our methodology failed to associate North Atlantic with Mediterranean cyclones in 19 cases out of 200. This is partly due to uncertainties in our feature tracking procedure or due to the absence of WCBs in cases when Rossby wave breaking occurs without a strong diabatic contribution. While WCBs and cyclone tracks are defined by specific physical criteria, the definition of ridges and PV streamers may present considerable uncertainties, especially when tracking these features in time. In our current study, the areas of interest are relatively close, taking into account the characteristic lengths of the tracked features and hence our methodology provided meaningful results. It would be of great interest to identify the relation of cyclone tracks to wave breaking in the whole Northern Hemisphere. In this context, it is for future work to investigate the statistical relationship between cyclone intensity, latent heating within WCBs and downstream ridge amplification, wave breaking, and secondary cyclogenesis. Understanding the chain of atmospheric processes that lead to the formation of Mediterranean cyclones may significantly contribute to the improvement of medium-range (of the order of 3–5 days) efficient forecasting of extreme weather in the Mediterranean.

Acknowledgements

We are deeply grateful to Heini Wernli (ETH) for the fruitful discussions and feedback on the manuscript, as well as for providing the cyclone tracks and supporting the research visit of EF to ETH for carrying out this work. We thank Hanin Binder and Michael Sprenger (ETH) for providing the ERA-Interim WCB climatology. MeteoSwiss and ECMWF are acknowledged for providing access to the ERA-Interim data. S. Raveh-Rubin is funded by Swiss National Science Foundation, Marie Heim-Vögtlin Programme (PMPDP2_158347/1). We thank the two anonymous reviewers who helped to improve the clarity of the manuscript.

References

Appenzeller C, Davies HC. 1992. Structure of stratospheric intrusions into the troposphere. *Nature* **358**: 570–572, doi: 10.1038/358570a0.

Binder H, Boettcher M, Joos H, Wernli H. 2016. The role of warm conveyor belts for the intensification of extratropical cyclones in Northern Hemisphere winter. *Journal of the Atmospheric Sciences* **73**: 3997–4020, doi: 10.1175/JAS-D-15-0302.1.

Browning KA. 1990. Organization of clouds and precipitation in extratropical cyclones. In *Extratropical Cyclones: The Erik Palmen Memorial Volume*, Newton CW, Holopainen EO (eds). American Meteor Society; 129–153.

Chaboureau J-P, Pantillon F, Lambert D, Richard E, Claud C. 2012. Tropical transition of a Mediterranean storm by jet crossing. *Quarterly Journal of the Royal Meteorological Society* **138**: 596–611, doi: 10.1002/qj.960.

Chang EKM. 1993. Downstream development of baroclinic waves as inferred from regression analysis. *Journal of the Atmospheric Sciences* **50**: 2038–2053, doi: 10.1175/1520-0469(1993) 050<2038:DDOBWA>2.0.CO;2.

Dee DP, Uppala SM, Simmons AJ, Berrisford P, Poli P, Kobayashi S, Andrae U, Balmaseda MA, Balsamo G, Bauer P, Bechtold P. 2011. The ERA-Interim reanalysis: configuration and performance of the data assimilation system. *Quarterly Journal of the Royal Meteorological Society* **137**: 553–597, doi: 10.1002/qj.828.

Fita L, Romero R, Ramis C. 2006. Intercomparison of intense cyclogenesis events over the Mediterranean basin based on baroclinic and diabatic influences. *Advances in Geosciences* **7**: 333–342, doi: 10.5194/adgeo-7-333-2006.

Flaounas E, Raveh-Rubin S, Wernli H, Drobinski P, Bastin S. 2015. The dynamical structure of intense Mediterranean cyclones. *Climate Dynamics* **44**: 2411–2427, doi: 10.1007/s00382-014-2330-2.

Flaounas E, Lagouvardos K, Kotroni V, Claud C, Delanoë J, Flamant C, Madonna E, Wernli H. 2016. Processes leading to heavy precipitation associated with two Mediterranean cyclones observed during the HyMeX SOP1. *Quarterly Journal of the Royal Meteorological Society* **142**: 275–286, doi: 10.1002/qj.2618.

Funatsu BM, Claud C, Chaboureau J-P. 2007. Potential of Advanced Microwave Sounding Unit to identify precipitating systems and associated upper-level features in the Mediterranean region: case studies. *Journal of Geophysical Research – Atmospheres* **112**: D17113, doi: 10.1029/2006JD008297.

Grams C, Archambault H. 2016. The key role of diabatic outflow in amplifying the midlatitude flow: a representative case study of weather systems surrounding western North Pacific extratropical transition. *Monthly Weather Review* **144**: 3847–3869, doi: 10.1175/MWR-D-15-0419.1.

Grams CM, Wernli H, Boettcher M, Campa J, Corsmeier U, Jones SC, Keller JH, Lenz C-J, Wiegand L. 2011. The key role of diabatic processes in modifying the upper-tropospheric wave guide: a North Atlantic case-study. *Quarterly Journal of the Royal Meteorological Society* **137**: 2174–2193, doi: 10.1002/qj.891.

Homar V, Jansà A, Campins J, Genovés A, Ramis C. 2007. Towards a systematic climatology of sensitivities of Mediterranean high impact weather: a contribution based on intense cyclones. *Natural Hazards and Earth System Sciences* **7**: 445–454, doi: 10.5194/nhess-7-445-2007.

Madonna E, Wernli H, Joos H, Martius O. 2014a. Warm conveyor belts in the ERA-Interim dataset (1979-2010). Part I: climatology and potential vorticity evolution. *Journal of Climate* **27**: 3–26, doi: 10.1175/JCLI-D-12-00720.1.

Madonna E, Limbach S, Aebi C, Joos H, Wernli H, Martius O. 2014b. On the co-occurrence of warm conveyor belt outflows and PV streamers. *Journal of the Atmospheric Sciences* **71**: 3668–3673, doi: 10.1175/JAS-D-14-0119.1.

Martius O, Zenklusen E, Schwierz C, Davies HC. 2006. Episodes of Alpine heavy precipitation with an overlying elongated stratospheric intrusion: a climatology. *International Journal of Climatology* **26**: 1149–1164, doi: 10.1002/joc.1295.

Martius O, Schwierz C, Davies HC. 2007. Breaking waves at the tropopause in the wintertime northern hemisphere: climatological analyses of the orientation and the theoretical LC1/2 classification. *Journal of the Atmospheric Sciences* **64**: 2576–2592, doi: 10.1175/JAS3977.1.

Massacand AC, Wernli H, Davies HC. 1998. Heavy precipitation on the Alpine southside: an upper-level precursor. *Geophysical Research Letters* **25**: 1435–1438, doi: 10.1029/98GL50869.

Massacand AC, Wernli H, Davies HC. 2001. Influence of upstream diabatic heating upon an Alpine event of heavy precipitation. *Monthly Weather Review* **129**: 2822–2828, doi: 10.1175/1520-0493(2001)129<2822:IOUDHU>2.0.CO;2.

Orlanski I, Chang EK. 1993. Ageostrophic geopotential fluxes in downstream and upstream development of baroclinic waves. *Journal of the Atmospheric Sciences* **50**: 212–225, doi: 10.1175/1520-0469(1993)050<0212:AGFIDA>2.0.CO;2.

Orlanski I, Sheldon J. 1993. A case of downstream baroclinic development over western north America. *Monthly Weather Review* **121**: 2929–2950, doi: 10.1175/1520-0493(1993)121<2929:ACODBD>2.0.CO;2.

Pantillon F, Chaboureau J-P, Richard E. 2015. Remote impact of North Atlantic hurricanes on the Mediterranean during episodes of intense rainfall in autumn 2012. *Quarterly Journal of the Royal Meteorological Society* **141**: 967–978, doi: 10.1002/qj.2419.

Pomroy HR, Thorpe AJ. 2000. The evolution and dynamical role of reduced upper-tropospheric potential vorticity in Intensive Observing Period One of FASTEX. *Monthly Weather Review* **128**: 1817–1834, doi: 10.1175/1520-0493(2000)128<1817:TEADRO>2.0.CO;2.

Raveh-Rubin S, Wernli H. 2015. Large-scale wind and precipitation extremes in the Mediterranean: a climatological analysis for 1979–2012. *Quarterly Journal of the Royal Meteorological Society* **141**: 2404–2417, doi: 10.1002/qj.2531.

Sanders F, Gyakum JR. 1980. Synoptic-dynamic climatology of the "bomb". *Monthly Weather Review* **108**: 1589–1606, doi: 10.1175/1520-0493(1980)108,1589:SDCOT.2.0.CO;2.

Schemm S, Wernli H, Papritz L. 2013. Warm conveyor belts in idealized moist baroclinic wave simulations. *Journal of the Atmospheric Sciences* **70**: 627–652, doi: 10.1175/JAS-D-12-0147.1.

Simmons AJ, Hoskins BJ. 1979. The downstream and upstream development of unstable baroclinic waves. *Journal of the Atmospheric Sciences* **36**: 1239–1254, doi: 10.1175/1520-0469(1979)036<1239:TDAUDO>2.0.CO;2.

Tafferner A. 1990. Lee cyclogenesis resulting from the combined outbreak of cold air and potential vorticity against the Alps. *Meteorology and Atmospheric Physics* **43**: 31–47, doi: 10.1007/BF01028107.

Thorncroft CD, Hoskins BJ, McIntyre ME. 1993. Two paradigms of baroclinic-wave life-cycle behavior. *Quarterly Journal of the Royal Meteorological Society* **119**: 17–55, doi: 10.1002/qj.49711950903.

Tous M, Romero R. 2013. Meteorological environments associated with medicane development. *International Journal of Climatology* **33**: 1–14, doi: 10.1002/joc.3428.

Wernli H, Davies HC. 1997. A Lagrangian-based analysis of extratropical cyclones I: the method and some applications. *Quarterly Journal of the Royal Meteorological Society* **123**: 467–489, doi: 10.1002/qj.49712353811.

Wernli H, Schwierz C. 2006. Surface cyclones in the ERA-40 Dataset (1958–2001). Part I: novel identification method and global climatology. *Journal of the Atmospheric Sciences* **63**: 2486–2507, doi: 10.1175/JAS3766.1.

Wernli H, Sprenger M. 2007. Identification and ERA-15 climatology of potential vorticity streamers and cutoffs near the extratropical tropopause. *Journal of the Atmospheric Sciences* **64**: 1569–1586, doi: 10.1175/JAS3912.1.

Wernli H, Shapiro MA, Schmidli J. 1999. Upstream development in idealized baroclinic wave experiments. *Tellus A* **51**: 574–587, doi: 10.1034/j.1600-0870.1999.00003.x.

12

Distinct linkage between winter Tibetan Plateau snow depth and early summer Philippine Sea anomalous anticyclone

Hong-Chang Ren,[1] Weijing Li,[1,2]* Hong-Li Ren[2,3] and Jinqing Zuo[1,2]

[1] *Collaborative Innovation Center on Forecast and Evaluation of Meteorological Disasters, Nanjing University of Information Science & Technology, China*
[2] *Laboratory for Climate Studies, National Climate Center, China Meteorological Administration, Beijing, China*
[3] *Joint Center for Global Change Studies (JCGCS), Beijing, China*

*Correspondence to:
W. Li, Laboratory for Climate Studies, National Climate Center, China Meteorological Administration, No. 46 Zhongguancun Nandajie, Beijing 100081, China.
E-mail: liwj@cma.gov.cn*

Abstract

This article demonstrates that the above-normal Tibetan Plateau snow depth (TPSD) in winter appears to be followed by an intensified Philippine Sea anticyclone in June (PSAC-J), and vice versa. This linkage is clearly independent of the relationship between the El Niño-Southern Oscillation (ENSO) and the PSAC-J. Moreover, winter TPSD anomalies are typically associated with a PSAC-J pattern that shifts northwards compared with that associated with ENSO. A better understanding of the combined effects of the winter TPSD and ENSO on the PSAC-J could improve our ability to predict both the East Asia summer monsoon and variations in the Meiyu–Changma–Baiu rainbelt.

Keywords: Tibetan Plateau snow depth; ENSO; Philippine Sea anomalous anticyclone

1. Introduction

The anomalous low-level Philippine Sea anticyclone (PSAC), which develops during the boreal winter (typically following the mature phase of El Niño events) and persists through the following spring and early summer, is one of the key factors impacting East Asian and western Pacific climate variability. Previous studies suggest that PSAC is an important bridge that links El Niño/Southern Oscillation (ENSO) events to East Asian climate variability (Chang *et al.*, 2000; Wang *et al.*, 2000; Chou *et al.*, 2003; Lau and Nath, 2006). The PSAC (cyclone) is initiated and maintained by a positive feedback resulting from the thermodynamic coupling of atmospheric Rossby waves and the oceanic mixed layer in the presence of strong remote forcing from the equatorial eastern–central Pacific associated with El Niño (La Niña) events (Wang *et al.*, 2000). Tropical Indian Ocean (TIO) sea-surface temperature (SST) anomalies, which usually follow the mature phase of ENSO, also make a contribution to the persistence of the PSAC (Watanabe and Jin, 2002; Yoo *et al.*, 2006; Yang *et al.*, 2007; Schott *et al.*, 2009; Xie *et al.*, 2009, 2010; Ding *et al.*, 2010; Chowdary *et al.*, 2011). The capacitor effect of the Indian Ocean allows the PSAC to survive into the early summer following the El Niño winter peak, and thus allows El Niño to affect the climate in East Asia. The PSAC is accompanied by surface southwesterly anomalies on its northwestern flank, which favour heavy precipitation over the area between the Yangtze River valley and southern Japan.

However, the influences of ENSO events and TIO SST anomalies on the East Asian climate are not always apparent, and the aforementioned relationship cannot explain all of the variance of the PSAC. During the period 1960–2012, only half of the 12 strongest anomalous PSAC years correspond with a decaying El Niño (6 years: 1983, 1988, 1995, 1998, 2003, 2010) or TIO warming (6 years: 1969, 1983, 1988, 1998, 2003, 2010). This implies that other external forcing factors may also exert an important effect on the variability of PSAC. Among them, Tibetan Plateau (TP) snow depth anomalies are believed to play an important role in the formation and variability of the East Asian summer monsoon (EASM) and western Pacific atmospheric circulation (Tao and Ding, 1981; Hsu and Liu, 2003; Wu and Qian, 2003; Zhang *et al.*, 2004; Zhao *et al.*, 2007). An increase in winter TP snow depth tends to precede a weakened EASM, which can generate more precipitation between the Yangtze River valley and southern Japan (Zhang and Tao, 2001; Wu and Qian, 2003; Ding *et al.*, 2009). Recent research has proposed that decadal changes in the TP snow depth have shifted the pattern of summer precipitation over East Asia (Si and Ding, 2013). Additionally, TP snow cover is negatively correlated with typhoon formation and landfall numbers over the western North Pacific (Xie *et al.*, 2005). In this study, we focus on the distinct impact of winter TP snow depth anomalies on the early summer PSAC by comparing its impact with those of ENSO on the PSAC.

2. Data

Monthly mean surface snow depth observations used in this study were provided by the National

Table 1. Correlations (R) of the monthly PSAC index from May to August with the previous winter (DJF) TPSD index, DJF Niño-3.4 (5°S–5°N, 170°–120°W) SST index, and spring (MAM) TIO (20°S–20°N, 40°–100°E) SST index for the periods of 1960–2012.

Index	PSAC			
	May	June	July	August
DJF TPSD	0.16	0.39***	0.02	0.17
DJF Niño-3.4	0.54***	0.30**	0.50***	0.29**
MAM TIO	0.55***	0.27**	0.34**	0.41***

Note: **95% confidence level; ***99% confidence level.

Meteorological Information Center (NMIC) of the China Meteorological Administration (CMA). To ensure the representativeness of the data records and for quality control purposes, this study analysed data recorded at 25 stations since 1960 (Figure S1, Supporting Information). The TP snow depth index (the TPSD index) is the mean snow depth recorded at the 25 stations over the TP.

In addition, we used monthly atmospheric reanalysis data from the European Center for Medium-Range Weather Forecasts (ECMWF), including the 40-year ECMWF Reanalysis dataset (ERA40) (Uppala *et al.*, 2005) and the ERA-Interim dataset (Dee *et al.*, 2011). Both ERA40 and ERA-Interim data have a horizontal resolution of 2.5° latitude by 2.5° longitude. ERA40 data were analysed for the period 1960–1978, and ERA-Interim data were analysed for the period 1979–2012. SST data were obtained from the Hadley Centre HadISST dataset with a horizontal resolution of 1° × 1° (Rayner *et al.*, 2003).

3. Results

To examine the relationship between the winter (December–January–February; DJF) TP snow depth and the PSAC, we calculated the correlations between the winter TPSD and monthly PSAC indices from the following May to August over the period 1960–2012. Here, the PSAC index is defined after Wang and Zhang (2002) as sea-level pressure anomalies over the Philippine Sea (10–20°N, 120–150°E).

Table 1 shows that the winter TPSD index is closely correlated with the PSAC index in June. The correlation coefficient was 0.39 for the entire 53-year period (1960–2012), and is significant at the 99% confidence level (Student's *t*-test). The time series of winter TPSD and June PSAC indices (Figure 1(a)) shows that the anomalous winter TPSD tends to be in phase with the June PSAC, and this is confirmed by the 21-year sliding correlation analysis between the two indices (Figure 1(b)). These positive correlations have increased markedly since the mid-1980s.

The above results indicate that the June PSAC is closely linked to TP snow depth anomalies from the previous winter. As demonstrated by Wang *et al.* (2000) and Lau and Nath (2006), summer PSAC

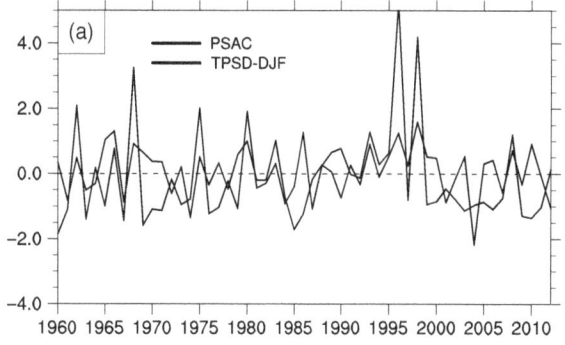

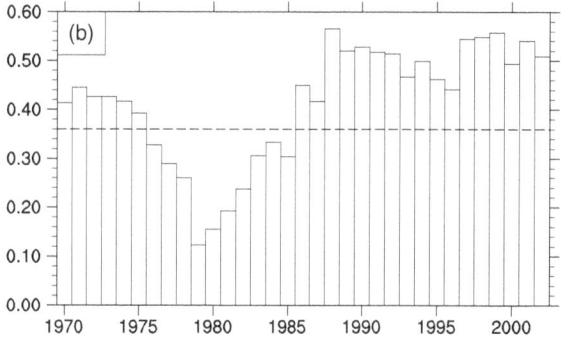

Figure 1. (a) Time series of the DJF TPSD and June PSAC indices. (b) Sliding correlations between the DJF TPSD and June PSAC indices with a 21-year moving window for the period 1960–2012. Dashed line denotes the significance at the 90% confidence level based on Student's *t*-test.

variability is also related to ENSO. However, we found that the June PSAC index has a moderate correlation coefficient with the preceding winter Niño3.4 SST index (0.30; Table 1), and is lower than that with the winter TPSD index (0.39). We further examined the correlation of the June PSAC indices with the spring (March–April–May; MAM) TIO SST index, defined as regional-mean SST anomalies over 20°S–20°N and 40°E–100°E (Xie *et al.*, 2009). The correlation coefficient for PSAC and preceding spring TIO SST index is 0.27 (Table 1), which is much lower than that associated with the winter TP snow depth.

ENSO may also contribute to changing the winter TP snow depth through a stationary wave teleconnection (Shaman and Tziperman, 2005). Thus, a question that arises is: can the close relationship between winter TPSD and June PSAC be attributed to the effect of ENSO on the TP snow depth anomalies? To answer this question, we examined the correlation coefficients between the June PSAC and winter TPSD indices after linearly removing winter Niño3.4 and spring TIO SST signals (Figure 2). Interestingly, the correlation between winter-TPSD and June PSAC indices changed little after removing the winter Niño3.4 SST signal, or after removing the linear effect of spring TIO SST anomalies (Figure 2). These results indicate that TPSD results in the most significant correlation coefficient of the three PSAC-related factors, and the high correlation remains almost constant after removing the linear effects of the prior ENSO cycle and spring TIO SST anomalies. In addition, the results of the partial

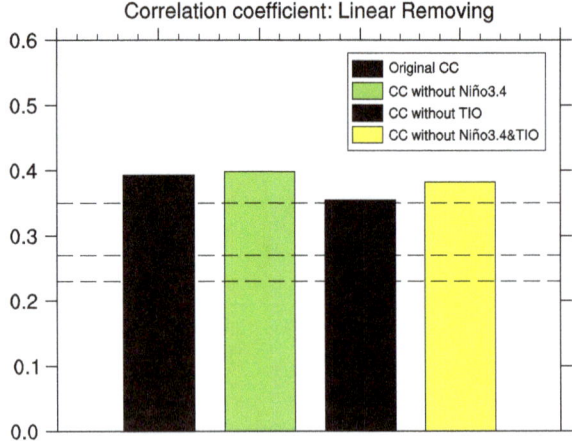

Figure 2. Correlation coefficients between the original winter TPSD and June PSAC indices (red), and the indices being linearly removed the signals of winter Niño3.4 (green), spring TIO SST (blue) and both the winter Niño3.4 and spring TIO SST (yellow). Dashed lines indicate the 90, 95, and 99% confidence levels from the bottom, respectively.

correlation analysis were inconsistent with the above results (Figure S2). Therefore, we conclude that winter TP snow depth anomalies have a significant relationship with the June PSAC, and this relationship appears to be independent of the effects of ENSO and TIO SST anomalies.

Another question arises regarding what are the differences between the impacts of winter TP snow depth anomalies and ENSO events on the June PSAC? Figure 3(a) shows wind anomalies at 850 hPa in June regressed against the TPSD index from the previous winter. This demonstrates that above normal winter TP snow depth corresponds to an anomalous anticyclone over the Philippine Sea and adjacent regions in June, which is consistent with the results from the correlation analysis. In addition, we investigated the composites of the June wind anomalies at 850 hPa for deep (above 0.5 standard deviations) and shallow (under −0.5 standard deviations) snow depth (Figure 3(b) and (c)). Deeper winter TP snow depth tends to be followed by a stronger anticyclone in the northern Philippine Sea in June (Figure 3(b)). In contrast, shallower snow depth appears to correspond to an anomalous cyclonic wind pattern (Figure 3(c)), but this is much weaker compared with the deep snow case.

The regressed pattern of 850-hPa wind anomalies in June onto the previous winter Niño3.4 index (Figure 3(d)) also shows an anomalous anticyclone over the western Pacific. A comparison between Figure 3(a) and (d) indicates that the central position of the anomalous PSAC associated with winter TPSD differs to that associated with the winter Niño3.4. The former is mainly centred over the northern Philippine Sea at about 20°N, and is accompanied by stronger southwesterly anomalies over southeast China, whereas the latter is confined to about 15°N and is concurrent with a weaker southerly anomaly over coastal southern China. Composites were also compiled

for 850-hPa wind anomalies in June for El Niño (Figure 3(e)) and La Niña (Figure 3(f)) events. The comparison between Figure 3(b) and (e) is in excellent agreement with the above result, and its sign reverses in Figure 3(c) and (f).

To further examine different features of the PSAC associated with previous winter TP snow depth anomalies and ENSO events, composite patterns of 850-hPa wind anomalies in June are shown with respect to the different phases of the winter TPSD index and ENSO (Figure 4). The years used for the composites in Figures 3 and 4 are shown in Table S1. For neutral ENSO events, a significant anomalous cyclone appears when snow is shallow (Figure 4(b)), and an anticyclone appears when snow is deep (Figure 4(c)). This result confirms the crucial role of winter TP snow depth anomalies in inducing the June PSAC. When La Niña events are accompanied by a normal TP snow depth anomaly in winter, there is an anomalous cyclone over the Philippine Sea and adjacent regions in the following June (Figure 4(g)). Therefore, it appears that either TP snow depth or ENSO become the dominant factor while the other is in a neutral state. We note that there is no significant anticyclone over the Philippine Sea and adjacent regions in June during El Niño events accompanied by a normal winter TP snow depth (Figure 4(d)). This is mainly because these cases are mostly El Niño-Modoki events in the preceding winter, the anticyclone associated with El Niño-Modoki events is centred over southern Japan during the decaying summer period (Yuan et al., 2012).

It is interesting that when the winter TP snow depth anomaly and ENSO have an opposing influence on the June PSAC, the dominant control appears to be the one in its positive phase (El Niño or deep snow). When El Niño events accompanies shallow TP snow in winter, it tends to lead to a positive PSAC in the following June (Figure 4(e)), and La Niña events accompanied by deep winter TP snow also results in a positive PSAC in the following June (Figure 4(i)). In addition, comparison between Figure 4(e) and (i) indicates that the PSAC pattern is noticeably different in the two scenarios. The PSAC pattern in Figure 4(e) is similar to that associated with ENSO (Figure 3(d)), whereas the PSAC pattern in Figure 4(i) is almost the same as that associated with the winter TP snow depth anomaly (Figure 3(a)). This indicates that El Niño or deep TP snow depth dominates the June PSAC, and overcomes the contrary influence of a shallow TP snow depth or La Niña. When an El Niño (La Niña) event accompanies deep (shallow) TP snow depth in winter, there tends to be an anomalous anticyclone (cyclone) in the Philippine Sea (Figure 4(f) and (h)).

Figure 3(a) and (d) demonstrates that the June PSAC pattern associated with the winter TPSD index differs to that associated with the winter Niño3.4 index in terms of the position of the PSAC. The former is located to the north and accompanied by stronger southwesterly surface wind anomalies on its northwestern flank, favouring more water vapour transport to the

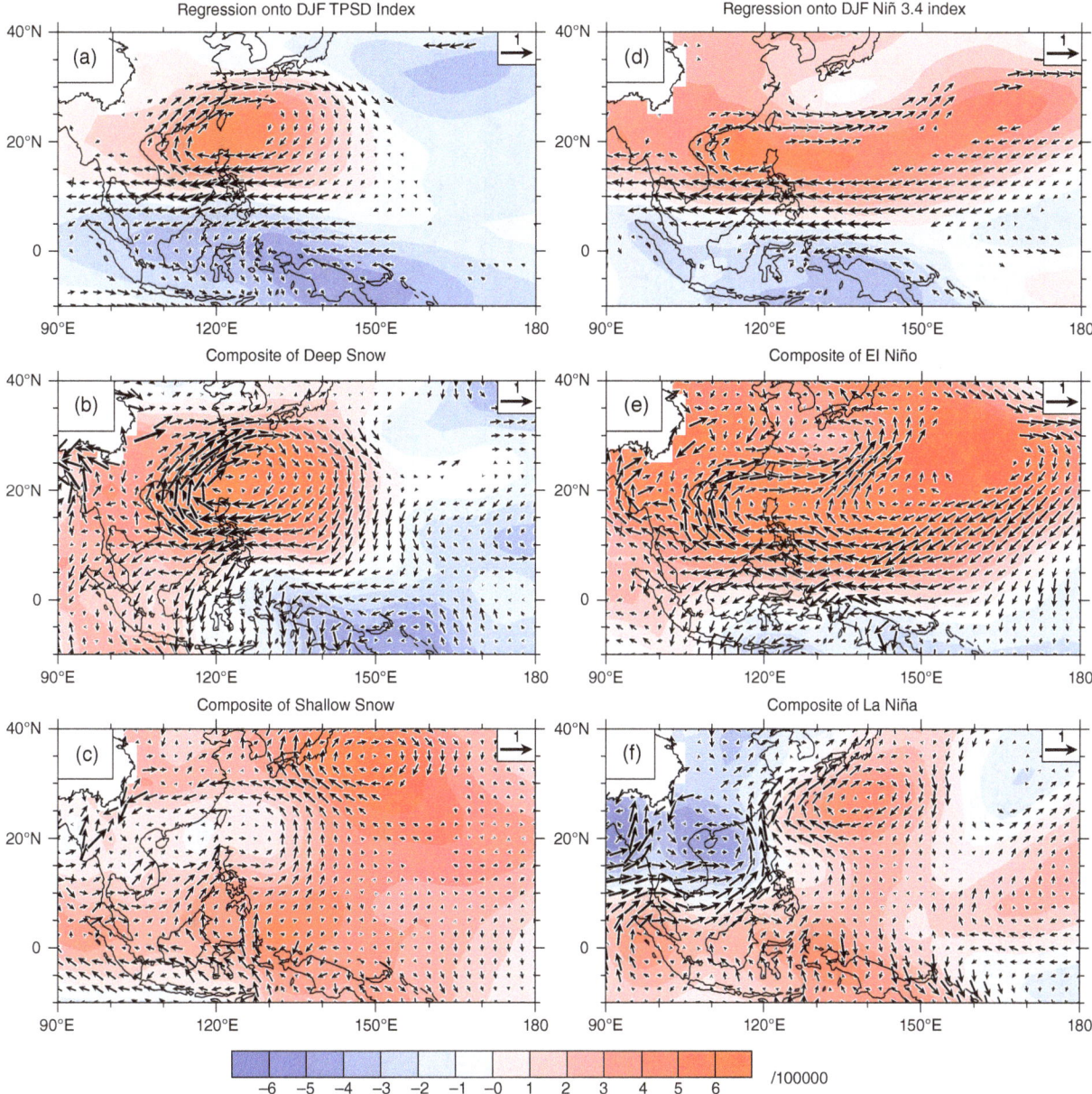

Figure 3. Regressions of June wind and stream function anomalies at 850 hPa onto the (a) DJF TPSD index and (b) Niño3.4 index. The composite anomalies of wind and stream function anomalies at 850 hPa with respect to (c) deep snow and (e) shallow snow conditions, and (d) El Niño, (f) La Niña events. Vectors (unit: m s^{-1}) are significant at the 90% confidence level.

Yangtze River valley and southern Japan. The north-ward location of the PSAC is shown in Figure 4(c), (f), and (i) while the TP snow depth is above normal in the preceding winter. The pattern in Figure 3(c) more closely resembles that in Figure 4(h) and (g), rather than Figure 4(b). Therefore, the composite result of shallow snow depth alone may mainly reveal the signals accompanied by La Niña events. These asymmetric relationships are important for our understanding of East Asia climate variability through the early summer.

From the above results, we conclude that winter TP snow depth anomalies have a significant relationship with PSAC in June. Such a relationship is independent of the ENSO impact, and the corresponding spatial patterns are distinct from those of ENSO in terms of the central position of the PSAC pattern.

4. Discussion and conclusions

This study has shown that a significant relationship exists between winter TP snow depth and the subsequent early summer PSAC, as well as the mutual cooperation consequence of both TP snow depth and the ENSO. This relationship is independent of ENSO. When either winter TP snow depths are greater or El Niño occurs, an anomalous anticyclone develops over the Philippine Sea in the following June, regardless of whether other factors are in a neutral or negative phase (shallow TP snow depth or La Niña). The anomalous PSAC associated with the TP snow depth anomaly tends to be located to the north and is accompanied by stronger southwesterly wind anomalies on its

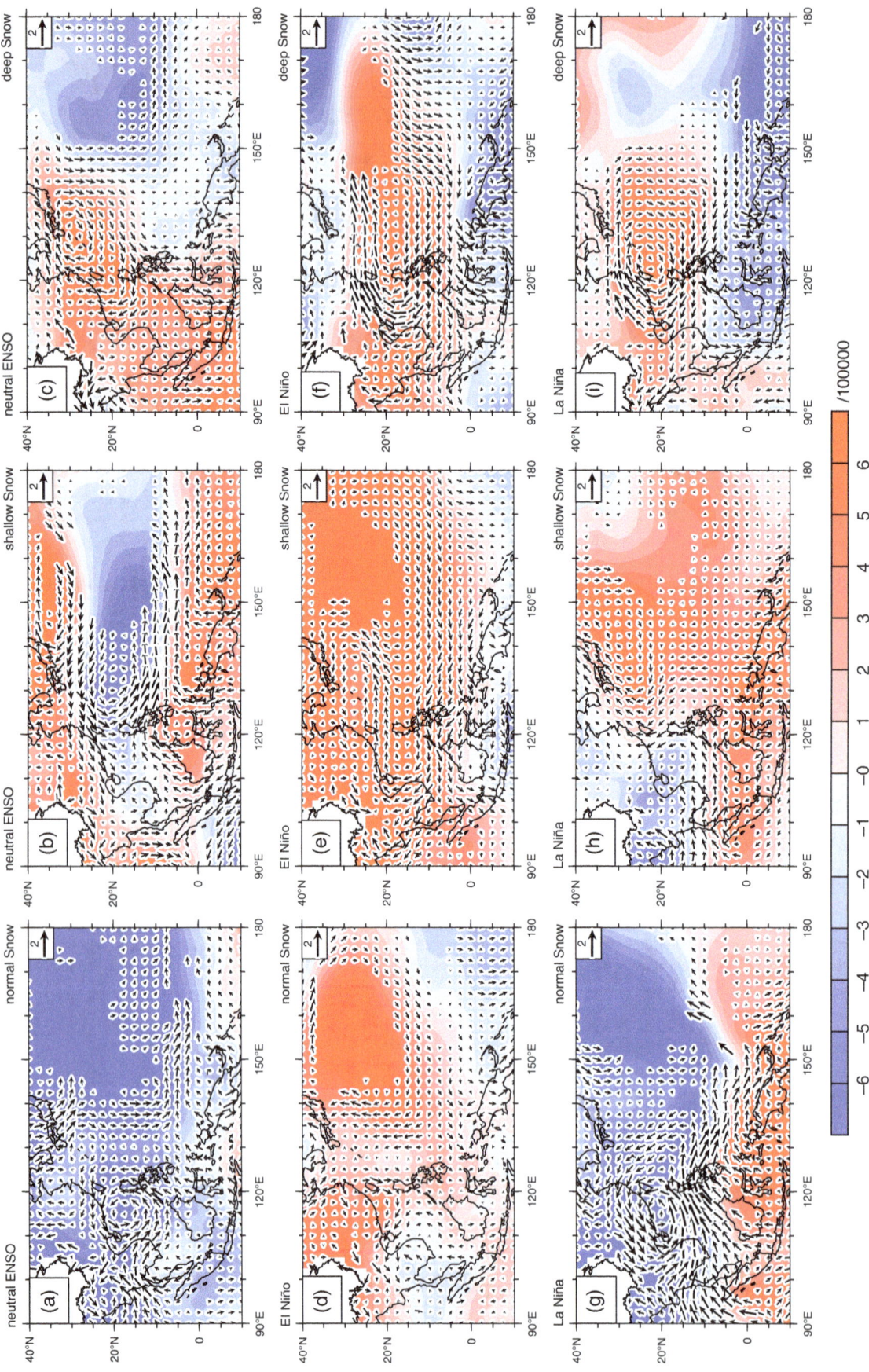

Figure 4. Composites anomalies of June 850-hPa wind (vectors, unit: m s^{-1}) and stream function (shading, levels 10^5 m^2 s^{-1}) with respect to the phase of winter ENSO and TP snow depth anomalies. Only the vectors with significant at 90% confidence level are shown.

northwestern flank. These wind anomalies play an important role in East Asian climate variability.

Thermodynamic processes over the TP may explain the distinct relationship identified in this article. Variation in the snow depth can modulate surface albedo as well as hydrological processes on the TP (e.g. Yamazaki 1989; Yasunari *et al.*, 1991; Chou *et al.*, 2003). Above-normal snow depth tends to cause a weak land–sea thermal contrast between the TP heat source and the heat sink over the adjacent ocean, resulting in a weak EASM (Zhang and Tao, 2001). It may also lead to a southerly shift in the position of the western Pacific subtropical high (WPSH), and a negative SST anomaly in the northwestern Pacific (Chen *et al.*, 2000). The above atmospheric circulation anomaly pattern favours high pressure and an anticyclonic wind field over the Philippine Sea and adjacent regions. The dominant influence of the winter TP snow depth anomaly on East Asian climate variability can be explained by the aforementioned mechanisms. The relationship between TPSD and PSAC can be reproduced by the recent version of the Community Atmospheric Model (CAM; version 5.3) coupled to the Community Land Model (CLM; version 4.0), developed by the National Center for Atmospheric Research (NCAR; Figure S3). However, the physical processes associated with the influence of TP thermal forcing on the multiscale climate variability of East Asia remain to be identified. As more accurate observational data are obtained and model simulations improve, more in-depth studies will be required to advance our understanding of the influence of TP thermal forcing on East Asian–western Pacific climate variability.

Furthermore, it is interesting to note that the distinct relationship between winter TPSD and June PSAC is particularly significant in June. Many studies have reported that in spring and summer, a huge heat source exists over the TP, and the thermal forcing of the TP reaches maximum in June (Ye and Gao, 1979; Weng, 1986; Wu and Zhang, 1998). In the mean time, snow on the TP generally melts in spring and early summer (Figure S4), and it should be noted that snow depth decreases significantly in June. Previous studies have demonstrated that the impact of snow on the atmospheric circulation is most significant during the snowmelt period and this is due to both hydrological and albedo effects (Xu and Dirmeyer, 2011). Therefore, it may reach a conclusion that the impact of TP snow on the atmospheric circulation is strongest in June. In addition, the South Asian high moves abruptly northwestwards to the TP from May to June (Krishnamurti, 1985; He *et al.*, 1987; Yanai *et al.*, 1992), which acts as a role of bridge that facilitates the significant influence of TP thermal anomalies on the atmospheric circulation at lower latitudes, such as WPSH (Mason and Anderson, 1963; Krishnamurti *et al.*, 1973). Therefore, the correlation coefficient between TPSD and PSAC increases rapidly from May (0.16) to June (0.39), but becomes insignificant after June.

Acknowledgements

This research was jointly supported by the National Basic Research Program of China (grant no. 2013CB430203), the China Meteorological Special Programs (grants nos. GYHY201506013 and GYHY201306033), the National Science Foundation of China (grant no. 41205058), and the Research Innovation Project for College Graduates of Jiangsu Province (KYLX_0841). The authors declare no conflict of interest regarding this manuscript.

Supporting information

The following supporting information is available:

Appendix S1. Supporting information for the numerical experiment.

Figure S1. Distribution of surface observation stations (red dots) over the Tibetan Plateau.

Figure S2. Correlation coefficients between the original winter TPSD and June PSAC indices (red), and those that partially exclude the effect of winter Niño3.4 (green), spring TIO SST (blue), and both the winter Niño3.4 and spring TIO SST (yellow). Dashed lines indicate the 90% (bottom), 95% (middle), and 99% (top) confidence levels.

Figure S3. Difference in geopotential height (shading; gpm) and horizontal wind (vector; $m\,s^{-1}$) between the (a) deep and (b) shallow snow experiments and the control run in June. Dots indicate the difference of geopotential height is significant at the 90% confidence level (Student's t-test).

Figure S4. Climatological annual cycle of the Tibetan Plateau snow depth (a) and monthly snow depth variation (b), which is defined as the difference in snow depth between the current and previous months ($cm\,day^{-1}$).

Table S1. Years used for composite figures.

References

Chang C-P, Zhang Y, Li T. 2000. Interannual and interdecadal variations of the East Asian summer monsoon and tropical Pacific SSTs. Part I: roles of the subtropical ridge. *Journal of Climate* **13**: 4310–4325, doi: 10.1175/1520-0442(2000)013<4310:IAIVOT>2.0.CO;2.

Chen Q, Gao B, Zhang Q. 2000. Studies on relation of snow cover over the Tibetan Plateau in winter to the winter–summer monsoon change. *Scientia Atmospherica Sinica* **24**: 477–492.

Chou C, Tu J-Y, Yu J-Y. 2003. Interannual variability of the western North Pacific summer monsoon: differences between ENSO and non-ENSO years. *Journal of Climate* **16**: 2275–2287, doi: 10.1175/2761.1.

Chowdary JS, Xie S-P, Luo J-J, Hanfner J, Behera S, Masumoto Y, Yamagata T. 2011. Predictability of Northwest Pacific climate during summer and the role of the tropical Indian Ocean. *Climate Dynamics* **36**: 607–621, doi: 10.1007/s00382-009-0686-5.

Dee DP, Uppala SM, Simmons AJ, Berrisford P, Poli P, Kobayashi S, Andrae U, Balmaseda MA, Balsamo G, Bauer P, Bechtold P, Beljaars ACM, van de Berg L, Bidlot J, Bormann N, Delsol C, Dragani R, Fuentes M, Geer AJ, Haimberger L, Healy SB, Hersbach H, Hólm EV, Isaksen L, Kållberg P, Köhler M, Matricardi M, McNally AP, Monge-Sanz BM, Morcrette J-J, Park BP, Peubey B-K, Peubey C, de Rosnay P, Tavolato C, Thépaut J-N, Vitart F. 2011. The ERA-Interim reanalysis: configuration and performance of the data assimilation system. *Quarterly Journal of the Royal Meteorological Society* **137**: 553–597, doi: 10.1002/qj.828.

Ding Y, Sun Y, Wang Z, Zhu Y, Song Y. 2009. Inter-decadal variation of the summer precipitation in China and its association with decreasing Asian summer monsoon. Part II: possible causes. *International Journal of Climatology* **29**: 1926–1944, doi: 10.1002/ joc.1759.

Ding R, Ha K-J, Li J. 2010. Interdecadal shift in the relationship between the East Asian summer monsoon and the tropical Indian Ocean. *Climate Dynamics* **34**: 1059–1071, doi: 10.1007/s00382-009-0555-2.

He H, McGinnis JW, Song Z, Yanai M. 1987. Onset of the Asian summer monsoon in 1979 and the effect of the Tibetan Plateau. *Monthly Weather Review* **115**: 1966–1995.

Hsu H-H, Liu X. 2003. Relationship between the Tibetan Plateau heating and East Asian summer monsoon rainfall. *Geophysical Research Letters* **30**: 2066, doi: 10.1029/2003GL017909, 20.

Krishnamurti TN. 1985. Summer monsoon experiment: a review. *Monthly Weather Review* **113**: 1590–1626.

Krishnamurti TN, Daggupaty SM, Fein J, Kanamitsu M, Lee JD. 1973. Tibetan high and upper tropospheric tropical circulations during northern summer. *Bulletin of the American Meteorological Society* **54**: 1234–1249.

Lau N-C, Nath MJ. 2006. ENSO modulation of the interannual and intraseasonal variability of the East Asian monsoon: a model study. *Journal of Climate* **19**: 4508–4530, doi: 10.1175/JCLI3878.1.

Mason RB, Anderson CE. 1963. The development and decay of the 100 mb summertime anticyclone over southern Asia. *Monthly Weather Review* **91**: 3–12.

Rayner NA, Parker DE, Horton EB, Folland CK, Alexander LV, Rowell DP, Kent EC, Kaplan A. 2003. Global analyses of sea surface temperature, sea ice, and night marine air temperature since the late nineteenth century. *Journal of Geophysical Research: Atmospheres (1984–2012)* **108**: 4407.

Schott FA, Xie S-P, McCreary JP. 2009. Indian Ocean circulation and climate variability. *Reviews of Geophysics* **47**: RG1002, doi: 10.1029/ 2007RG000245.

Shaman J, Tziperman E. 2005. The effect of ENSO on Tibetan Plateau snow depth: a stationary wave teleconnection mechanism and implications for the South Asian monsoons. *Journal of Climate* **18**: 2067–2079, doi: 10.1175/JCLI3391.1.

Si D, Ding Y. 2013. Decadal change in the correlation pattern between the Tibetan Plateau winter snow and the East Asian summer precipitation during 1979–2011. *Journal of Climate* **26**: 7622–7634, doi: 10.1175/JCLI-D-12-00587.1.

Tao S-Y, Ding Y. 1981. Observational evidence of the influence of the Qinghai-Xizang (Tibet) Plateau on the occurrence of heavy rain and severe convective storms in China. *Bulletin of the American Meteorological Society* **62**: 23–30, doi: 10.1175/1520-0477(1981)062<0023:OEOTIO>2.0.CO;2.

Uppala SM, Kållberg PW, Simmons AJ, Andrae U, Da Costa Bechtold V, Fiorino M, Gibson JK, Haseler J, Hernandez A, Kelly GA, Li X, Onogi K, Saarinen S, Sokka N, Allan RP, Andersson E, Arpe K, Balmaseda MA, Beljaars ACM, Van De Berg L, Bidlot J, Bormann N, Caires S, Chevallier F, Dethof A, Dragosavac M, Fisher M, Fuentes M, Hagemann S, Hólm EH, Hoskins BJ, Isaksen L, Janssen PAEM, Jenne R, McNally AP, Mahfouf J-F, Morcrette J-J, Rayner NA, Saunders RW, Simon P, Sterl A, Trenberth KE, Untch A, Vasiljevic D, Viterbo P, Woollen J. 2005. The ERA-40 re-analysis. *Quarterly Journal of the Royal Meteorological Society* **131**: 2961–3012, doi: 10.1256/qj.04.176.

Wang B, Zhang Q. 2002. Pacific-East Asian teleconnection. Part II: how the Philippine Sea anomalous anticyclone is established during El Niño development. *Journal of climate* **15**: 3252–3265.

Wang B, Wu R, Fu X. 2000. Pacific–East Asia teleconnection: how does ENSO affect East Asian climate. *Journal of Climate* **13**: 1517–1536, doi: 10.1175/1520-0442(2000)013<1517:PEATHD>2.0.CO;2.

Watanabe M, Jin F-F. 2002. Role of Indian Ocean warming in the development of Philippine Sea anticyclone during ENSO.

Geophysical Research Letters **29**: 116-1–116-4, doi: 10.1029/2001 GL014318,2002.

Weng D. 1986 An analysis of the characteristic features of the surface heat source and heat balance over the Qinghai-Xizang Plateau from May to August, 1979, Proceedings of the International Symposium on the Qinghai-Xizang Plateau and Mountain Meteorology. Science Press, Beijing, 201–216.

Wu TW, Qian ZA. 2003. The relation between the Tibetan winter snow and the Asian summer monsoon and rainfall: an observational investigation. *Journal of Climate* **16**: 2038–2051, doi: 10.1175/1520-0442 (2003)016<2038:TRBTTW>2.0.CO;2.

Wu G, Zhang Y. 1998. Tibetan Plateau forcing and the timing of the monsoon onset over South Asia and the South China Sea. *Monthly Weather Review* **126**: 913–927.

Xie L, Yan T, Pietrafesa LJ, Karl T, Xu X. 2005. Relationship between western North Pacific typhoon activity and Tibetan Plateau winter and spring snow cover. *Geophysical Research Letters* **32**: L16703, doi: 10.1029/2005GL023237.

Xie S-P, Hu K, Hafner J, Tokinaga H, Du Y, Huang G, Sampe T. 2009. Indian Ocean capacitor effect on Indo-western Pacific climate during the summer following El Niño. *Journal of Climate* **22**: 730–747, doi: 10.1175/2008JCLI2544.1.

Xie S-P, Du Y, Huang G, Zheng X-T, Tokinaga H, Hu K, Liu Q. 2010. Decadal Shift in El Niño Influences on Indo-Western Pacific and East Asian Climate in the 1970s. *Journal of Climate* **23**: 3352–3368, doi: 10.1175/2010JCLI3429.1.

Xu L, Dirmeyer P. 2011. Snow-atmosphere coupling strength in a global atmospheric model. *Geophysical Research Letters* **38**: 13.

Yamazaki K. 1989. A study of the impact of soil moisture and surface albedo changes on global climate using the MRI-GCM-I. *Journal of the Meteorological Society of Japan* **67**: 123–146.

Yanai M, Li C, Song Z. 1992. Seasonal heating of the Tibetan Plateau and its effects on the evolution of the Asian summer monsoon. *Journal of the Meteorological Society of Japan* **70**: 319–351.

Yang J, Liu Q, Xie S-P, Liu Z, Wu L. 2007. Impact of the Indian Ocean SST basin mode on the Asian summer monsoon. *Geophysical Research Letters* **34**: L02708, doi: 10.1029/2006GL028571.

Yasunari T, Kitoh A, Tokioka T. 1991. Local and remote responses to excessive snow mass over Eurasia appearing in the northern spring and summer climate: a study with the MRI GCM. *Journal of the Meteorological Society of Japan* **69**: 473–487.

Ye T, Gao Y. 1979. *The meteorology of the Qinghai-Xizang (Tibet) plateau*. Science Press: Beijing; 278 pp.

Yoo S-H, Yang S, Ho C-H. 2006. Variability of the Indian Ocean sea surface temperature and its impacts on Asian-Australian monsoon climate. *Journal of Geophysical Research, [Atmospheres]* **111**: D03108, doi: 10.1029/2005JD006001.

Yuan Y, Song Y, Zhang Z. 2012. Different evolutions of the Philippine Sea anticyclone between the Eastern and Central Pacific El Niño: possible effects of Indian Ocean SST. *Journal of Climate* **25**: 7867–7883, doi: 10.1175/JCLI-D-12-00004.1.

Zhang S, Tao S. 2001. The influences of snow cover over the Tibetan Plateau on Asian summer monsoon. *Chinese Journal of Atmospheric Sciences* **25**: 372–390.

Zhang Y, Li T, Wang B. 2004. Decadal change of the spring snow depth over the Tibetan Plateau: the associated circulation and influence on the East Asian summer monsoon. *Journal of Climate* **17**: 2780–2793, doi: 10.1175/1520-0442(2004)017<2780:DCOTSS> 2.0.CO;2.

Zhao P, Zhou Z, Liu J. 2007. Variability of Tibetan spring snow and its associations with the hemispheric extratropical circulation and East Asian summer monsoon rainfall: an observational investigation. *Journal of Climate* **20**: 3942–3955, doi: 10.1175/JCLI4205.1.

13

The warm-core structure of Super Typhoon Rammasun derived by FY-3C microwave temperature sounder measurements

Xiang Wang,[1,2,*] Yifang Ren[3] and Xun Li[4]

[1]Center of Data Assimilation for Research and Application, Nanjing University of Information Science & Technology, China
[2]Key Laboratory of South China Sea Meteorological Disaster Prevention and Mitigation of Hainan Province, Haikou, China
[3]Jiangsu Meteorological Service, China Meteorological Administration, Nanjing, China
[4]Hainan Meteorological Service, China Meteorological Administration, Haikou, China

*Correspondence to:
 X. Wang, Center of Data Assimilation for Research and Application, Nanjing University of Information Science & Technology, No. 219, Ningliu Road, Pukou District, Nanjing 210044, Jiangsu, China.
E-mail: wangxiang@nuist.edu.cn

Abstract

The microwave temperature sounder (MWTS) on board the third FengYun satellite (FY-3C) is well suited for the detection of tropical cyclones because the observed radiance is insignificantly affected by non-precipitating clouds. In this study, a stepwise multiple linear regression technique is proposed to retrieve the atmospheric temperature profiles based on MWTS measurements. Under clear-sky conditions, the root-mean-square error of retrieved temperature is no >1.4 K, which is low enough to monitor the thermal structure within typhoons. An application of this method to the Super Typhoon Rammasun resolves the warm-core eye and temperature gradients across the eye-wall very well.

Keywords: microwave temperature sounder; warm core; FengYun satellite

1. Introduction

The warm-core thermal structure is one of the most notable characteristics of typhoons. Typhoons are cyclonic circulations, and the establishment of the high-level warm-core structure is the main marker of typhoon generation. Wang et al. (2010) made some efforts to study the formation of typhoon warm cores and noted that the warm core is a result of huge release of latent heat from the warm and moist inflow during the upward movement of air in the storm's core. Palmen and Newton (1969) found that the precipitation from typhoons is correlated with the area of the warm core: the larger the area of the warm core, the greater the precipitation. In general, the maximum of temperature anomaly for most typhoons appears in the upper troposphere (approximately 200–400 hPa). The warm-core thermal structure forms at the development stage of typhoon, and it reaches its maximum strength at the mature stage. Once the warm-core structure weakens and disappears, the typhoon will dissipate.

It is well known that typhoons develop over the open ocean; due to the lack of traditional observations, there has been a major increase in the use of satellite observations, especially microwave measurements, to detect typhoons. As early as 1978, the Nimbus-6 scanning microwave spectrometer measurements were utilized to study Typhoon June, and a warm-core structure was first imaged (Rosenkranz et al., 1978). During the same year, through microwave sensors, Kidder et al. (1978) verified that the warm anomaly was correlated to the

typhoon's minimum sea level pressure and maximum wind speed. Later on, microwave radiance data from a microwave sounding unit was used to estimate the intensity of a large sample of typhoons (Velden and Smith, 1983; Velden, 1989; Velden et al., 1991). Kidder et al. (2000) and Zhu et al. (2002) used brightness temperature data from an advanced microwave sounding unit to retrieve atmospheric temperature profiles and found that the root-mean-square error of the data was <2 K. With the launch of the Suomi NPP satellite, the sounding channels of the advanced technology microwave sounder (ATMS) onboard the Suomi NPP have also been utilized to detect the warm-core structure of typhoons (Zhu and Weng, 2013).

A new generation of polar-orbiting satellite series, the FY-3 fleet, will consist of seven satellites; each one will be launched approximately every 2 years (Zhang et al., 2009). The first operational satellite, FY-3C, was successfully launched into morning-configured orbit on 23 September 2013. MWTS on board FY-3C are designed with many more channels, finer spatial resolution, and better sensor precision than previous instruments (Dong et al., 2009; Wang and Li, 2014), which makes MWTS measurements more capable to detect the thermal features of typhoons.

2. MWTS instrument and typhoon case description

MWTS is a total power cross-track radiometer, and it has a maximum scan angle of 49.5°. Along each

Table I. Channel characteristics of MWTS.

Channel	Central frequency (GHz)	Bandwidth (MHz)	NEDT (K)	Peak weighting (hPa)
1	50.300	180	1.20	Surface
2	51.760	400	0.75	Surface
3	52.800	400	0.75	950
4	53.596	400	0.75	700
5	54.400	400	0.75	400
6	54.940	400	0.75	270
7	55.500	330	0.75	180
8	57.290 (f_0)	330	0.75	90
9	$f_0 \pm 0.217$	78	1.20	50
10	$f_0 \pm 0.322 \pm 0.048$	36	1.20	20
11	$f_0 \pm 0.322 \pm 0.022$	16	1.70	12
12	$f_0 \pm 0.322 \pm 0.010$	8	2.40	5
13	$f_0 \pm 0.322 \pm 0.005$	3	3.60	2

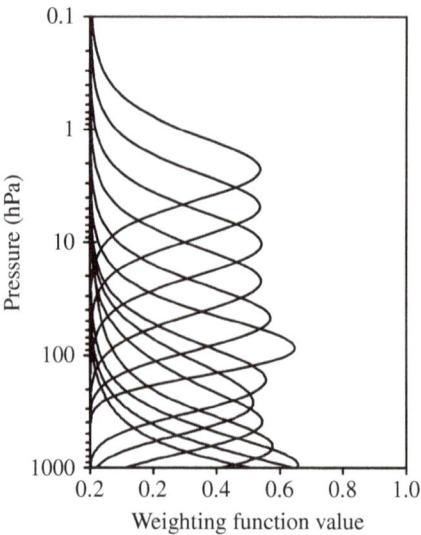

Figure 1. Vertical distribution of FY-3C MWTS channels 1–13 weighting functions calculated using US standard atmosphere.

scan line, there are 90 fields of view (FOVs), and the scan stepping angle and beam width of each FOV is 2.2°; therefore, the nominal spatial resolution at nadir is approximately 33 km, which can depict the thermal structure within typhoon better than previous instruments (Weng *et al.*, 2012). Compared to the MWTS onboard the FY-3A/B, the MWTS on board the FY-3C has a total of 13 channels in the oxygen band, whose central frequencies are located from 50 to 60 GHz. The purpose of MWTS sounding channels is to record temperature information from the low troposphere to the upper stratosphere because the absorption and emission of microwave radiation by atmospheric oxygen enables the MWTS to passively sound temperature as a function of altitude. Some selected channel characteristics, such as channel central frequency, 3-dB bandwidth, sensor measurement precision (noise-equivalent delta temperature), and the height of peak weighting functions, are listed in Table 1. The weighting functions for the 13 MWTS channels (1–13 from bottom to top), which are calculated using a standard US atmospheric profile, such as the Community Radiative Transfer Model (CRTM) input, are given in Figure 1. The CRTM was developed by the US Joint Centre for Satellite Data Assimilation (Han *et al.*, 2007; Weng, 2007). Measurements from MWTS channels 3 to 10 will be used to retrieve the atmospheric temperature profile in this article.

Super Typhoon Rammasun formed as a tropical depression over the northwest Pacific Ocean on the morning of 11 July. It intensified gradually and moved westwards steadily in the following few days. Rammasun developed into a severe typhoon on 15 and 16 July, moving across the central part of the Philippines and entering the South China Sea. After weakening over land, Rammasun reorganized over the South China Sea and intensified into a super typhoon on 18 July, reaching its peak intensity with an estimated sustained wind of 52 m s^{-1} and minimum sea level pressure of 935 hPa. Figure 2 shows the spatial distributions of brightness temperatures observed from MWTS channel 1 within Super Typhoon Rammasun at 1353 UTC on 18

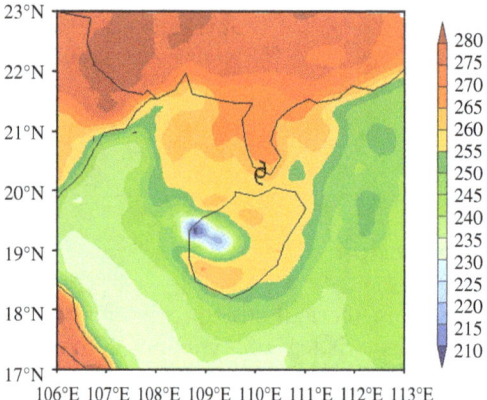

Figure 2. Distributions of observed brightness temperature from FY-3C MWTS channel 1 around Typhoon Rammasun at 1353 UTC on 18 July 2014. The centre of typhoon is indicated by a typhoon symbol in white.

July 2014. Basically, the typhoon structure is captured; namely, the spiral cloud presence renders the brightness temperature much colder than the warm core due to smaller thermal emission from clouds.

3. MWTS-derived warm-core structure

As it is well known that the radiance in microwave bands is linearly proportional to the entire layer of atmospheric temperature, and the weighting functions of all MWTS sounding channels are essentially steady, the atmospheric temperature at a given pressure may be expressed as a linear combination of brightness temperatures measured at different sounding channels (Janssen, 1993). Following the similar algorithm proposed in Zhu *et al.* (2002), Equation ((1) is utilized to retrieve the atmospheric temperature profiles; that is

$$T(p) = \beta_0(p, \theta) + \sum_{i=1} \beta_i(p, \theta) T_b(v_i, \theta) \quad (1)$$

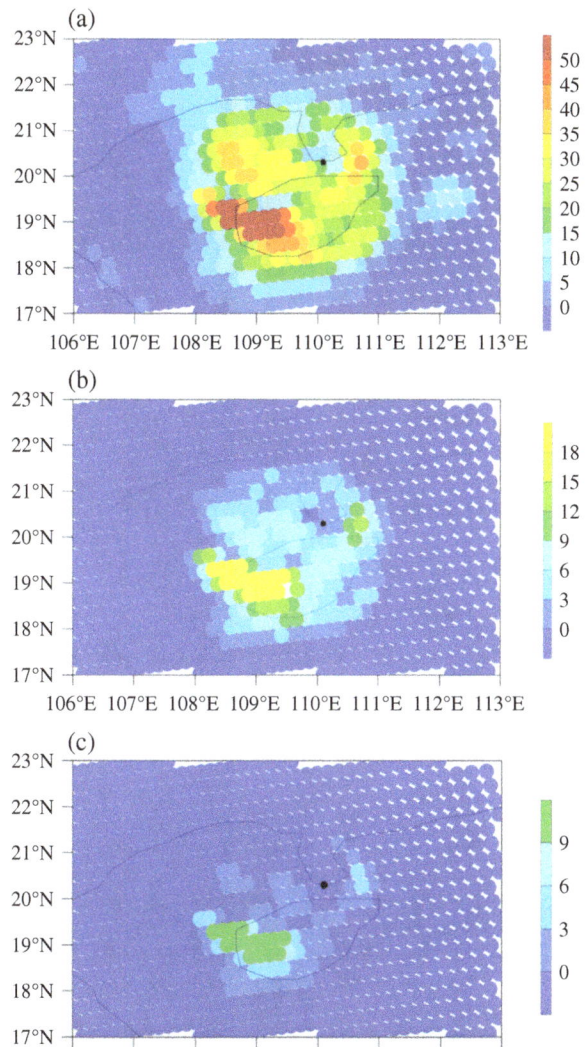

Figure 3. CESI derived from FY-3C (a) MWHS channel 7 and MWTS channel 3, (b) MWHS channel 6 and MWTS channel 5, and (c) MWHS channel 5 and MWTS channel 6 for Typhoon Rammasun at 1353 UTC on 18 July 2014. The centre of typhoon is indicated by a typhoon symbol in black.

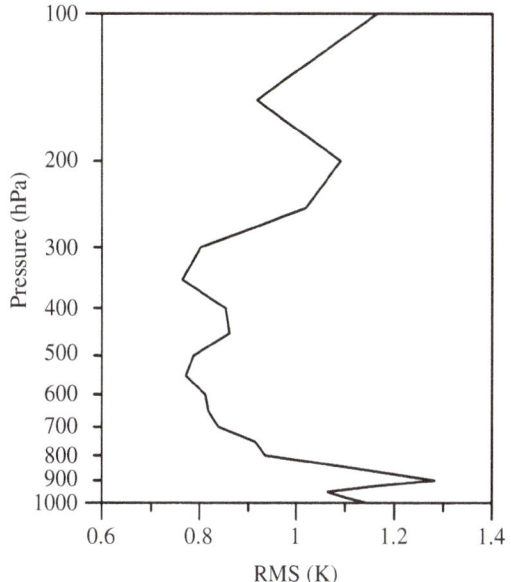

Figure 4. Vertical distribution of the errors of the MWTS retrieved temperatures.

where p is the given pressure, v_i is the central frequency for the chosen channel, θ represents the scan angle, and T_b is the MWTS brightness temperature.

Practically, it may be not economical to utilize all sounding channel measurements because some of them may be correlated, providing redundant information. Therefore, the coefficients β_0 and β_i (responding to the selected channels) are determined using a stepwise linear regression method at a 1% significance level. In fact, however, I tried stepwise regression and found that all channels are needed. The data collected from the collocated MWTS observations and Global Forecast System reanalysis field over ocean and land between 30°S and 30°N latitudes during 11–17 July 2014 are separately used to obtain the regression coefficients. Under a clear-sky condition, brightness temperatures from channels 3 to 10 are used to retrieve the temperatures at 21 pressure levels from 100 to 1000 hPa. Because the brightness temperatures of the surface and lower

troposphere could be contaminated by large cloud and rain droplets, channels 3 and 4 are not used for the retrieval under precipitation conditions. In this article, the cloud emission and scattering index (CESI) method is first used to identify the cloud regions. The CESI can be defined (Han *et al.*, 2015):

$$\text{CESI} = T^{\text{reg}}_{\text{MWHS}} - T^{\text{obs}}_{\text{MWHS}} \qquad (2)$$

where $T^{\text{obs}}_{\text{MWHS}}$ is the microwave humidity sounder (MWHS) measurements and $T^{\text{reg}}_{\text{MWHS}}$ is calculated from the MWTS measurements using linear regression. Figure 3 shows the spatial distributions of CESI within Super Typhoon Rammasun at 1353 UTC on 18 July 2014 derived from three selected paired channels: MWHS channel 5 (centred at $118.75 \pm 0.8\,\text{GHz}$) and MWTS channel 6, MWHS channel 6 (centred at $118.75 \pm 1.1\,\text{GHz}$) and MWTS channel 5, and MWHS channel 7 (centred at $118.75 \pm 2.5\,\text{GHz}$) and MWTS channel 3. As shown in the figure, Super Typhoon Rammasun's eye, eye-wall, and rain-bands are clearly captured in the CESI distributions. In this article, the observation will be treated as cloudy as long as one of the CESI for the three pairs is greater than zero.

The performance of the temperature retrieval from the MWTS brightness temperature is verified from a vertical profile of the root-mean-square error, which is computed using all of the collocated data set. As shown in Figure 4, the root-mean-square error is within 1.4 K, which is good enough to monitor the warm-core structure within typhoons. In addition, the author has verified against an independent subset of the same data used for training and found the similar results.

The vertical cross-sections of temperature anomaly of Typhoon Rammasun at the mature stage (1353 UTC on 18 July 2014) along 110.3°E are shown in Figure 5, respectively. The temperature anomaly is defined as a

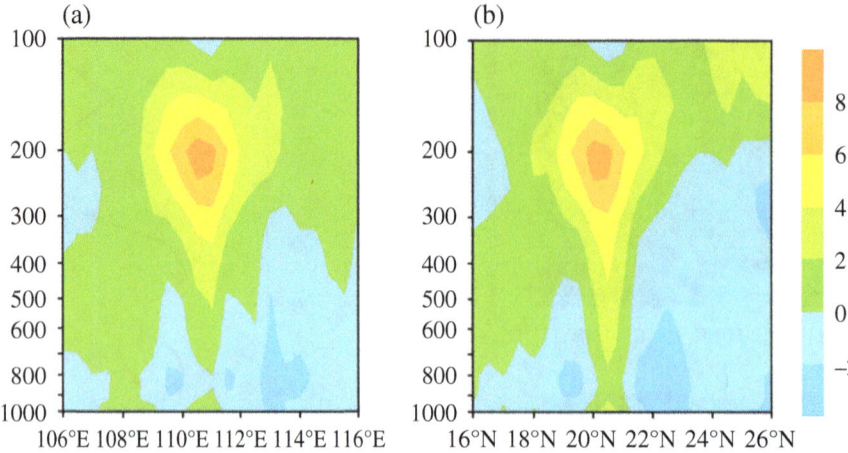

Figure 5. Vertical cross-sections of temperature anomalies for tropical cyclone Rammasun retrieved from FY-3C MWTS along (a) 20.3°N and (b) 110.3°E at 1353 UTC on 18 July 2014.

deviation from the unperturbed environment. A warm core can be identified throughout the troposphere with a maximum anomaly of 8–10 °C near 200 hPa and extending to the sea surface, which is similar to other observations. From the anomaly field, the radius of the Typhoon eye at the sea surface is observed to be approximately 100 km, and the eye tilts outward with height.

4. Summary

The three-dimensional warm-core structure is one of the most important parameters in monitoring typhoon intensity, studying typhoon inner core dynamics, and constructing the initial vortex for a typhoon simulation. In this study, we investigated the derivation of this structure from FY-3C MWTS measurements. To retrieve atmospheric temperature profiles from MWTS measurements, a stepwise regression algorithm was developed by using the collocated MWTS data and Global Forecast System (GFS) reanalysis field for each scan angle. The root-mean-square errors are <1.4 K at all pressure levels from 100 to 1000 hPa. An application to the Super Typhoon Rammasun shows that the warm-core eye and temperature gradients across the eye-wall can be captured very well. There is no doubt that the MWTS is extremely promising for improving our knowledge of typhoons and hurricanes.

Acknowledgements

Sincere thanks are given to the National Satellite Meteorological Centre of the Chinese Meteorology Administration for providing the FY-3C MWTS data. The authors would also like to thank Mr Han for providing data.

This work was jointly supported by the Key Laboratory of South China Sea Meteorological Disaster Prevention and Mitigation of Hainan Province grant number SCSF201402, The Startup Foundation for Introducing Talent of NUSIT S8113066001, and the National Natural Foundation of China grant 41365005.

References

Dong C, Yang J, Zhang W, Yang Z, Lu N, Shi J, Zhang P, Liu Y, Cai B. 2009. An overview of a new Chinese weather satellite FY-3A. *Bulletin of the American Meteorological Society* **90**: 1531–1544.

Han Y, Weng F, Liu Q, VanDelst P. 2007. A fast radiative transfer model for SSMIS upper atmosphere sounding channel. *Journal of Geophysical Research* **112**(11): D11121, doi: 10.1029/2006JD008208.

Han Y, Zou X, Weng F. 2015. Cloud and precipitation features of Super Typhoon Neoguri revealed from dual oxygen absorption band sounding instruments on board FengYun-3C satellite. *Geophysical Research Letters* **42**: 916–924, doi: 10.1002/2014GL062753.

Janssen MA. 1993. *Atmospheric Remote Sensing by Microwave Radiometry. Wiley Series in remote sensing.* John Wiley & Sons: New York, NY; 572.

Kidder SQ, Gray WM, VonderHaar TH. 1978. Estimating tropical cyclone central pressure and outer winds from satellite microwave data. *Monthly Weather Review* **106**: 1458–1464.

Kidder SQ, Goldberg MD, Zehr RM, DeMaria M, Purdom JFW, Velden CS, Grody NC, Kusselson SJ. 2000. Satellite analysis of tropical cyclones using the advanced microwave sounding unit (AMSU). *Bulletin of the American Meteorological Society* **81**: 1241–1260.

Palmen E, Newton CW. 1969. *Atmospheric Circulation Systems*. Elsevier: New York, NY; 603 pp.

Rosenkranz PW, Staelin DH, Grody NC. 1978. Typhoon June (1975) viewed by a scanning microwave spectrometer. *Journal of Geophysical Research* **83**: 1857–1868.

Velden CS. 1989. Observational analyses of North Atlantic tropical cyclones from NOAA polar-orbiting satellite microwave data. *Journal of Applied Meteorology* **28**: 59–70.

Velden CS, Smith WL. 1983. Monitoring tropical cyclone evolution with NOAA satellite microwave observations. *Journal of Climate and Applied Meteorology* **22**: 714–724.

Velden CS, Goodman BM, Merill RT. 1991. Western North Pacific tropical cyclone intensity estimation from NOAA polar-orbiting satellite microwave data. *Monthly Weather Review* **119**: 159–168.

Wang X, Li X. 2014. Preliminary investigation of FengYun-3C microwave temperature sounder (MWTS) measurements. *Remote Sensing Letters* **5**(12): 1002–1011, doi: 10.1080/2150704X.2014.988305.

Wang Z, Montgomery MT, Dunkerton TJ. 2010. Genesis of pre-Hurricane Felix (2007). Part II: warm core formation, precipitation evolution, and predictability. *Journal of the Atmospheric Sciences* **67**: 1730–1744.

Weng F. 2007. Advances in radiative transfer modeling in support of satellite data assimilation. *Journal of the Atmospheric Sciences* **64**: 3799–3807, doi: 10.1175/2007JAS2112.1.

Weng F, Zou X, Wang X, Yang S, Goldberg MD. 2012. Introduction to Suomi national polar-orbiting partnership advanced technology microwave sounder for numerical weather predict ion and tropical cyclone applications. *Journal of Geophysical Research* **117**: D19112, doi: 10.1029/2012JD018144.

Zhang P, Yang J, Dong C, Lu N, Yang Z, Shi J. 2009. General introduction on payloads, ground segment and data application of Fengyun 3A. *Frontiers of Earth Science in China* **3**: 367–373.

Zhu T, Weng F. 2013. Hurricane Sandy warm-core structure observed from advanced technology microwave sounder. *Geophysical Research Letters* **40**: 3325–3330, doi: 10.1002/grl.50626.

Zhu T, Zhang D-L, Weng F. 2002. Impact of the advanced microwave sounding unit measurements on hurricane prediction. *Monthly Weather Review* **130**: 2416–2432.

14

New Saharan wind observations reveal substantial biases in analysed dust-generating winds

Alexander J. Roberts,[1]*[ID] John H. Marsham,[1] Peter Knippertz,[2] Douglas J. Parker,[1] Mark Bart,[3] Luis Garcia-Carreras,[1] Matthew Hobby,[1] James B. McQuaid,[1] Philip D. Rosenberg[1] and Daniel Walker[4]

[1] Institute for Climate and Atmospheric Science, School of Earth and Environment, University of Leeds, UK
[2] Institute of Meteorology and Climate Research, Karlsruhe Institute of Technology, Germany
[3] Air Quality Ltd., Auckland, New Zealand
[4] National Centre for Atmospheric Science, University of Leeds, Leeds, UK

*Correspondence to:
A. J. Roberts, School of Earth and Environment, University of Leeds, LS2 9JT, UK.
E-mail: a.j.roberts1@leeds.ac.uk

Abstract

For the remote Sahara, the Earth's largest dust source, there has always been a near-absence of data for evaluating models. Here, new observations from the Fennec project are used along with Sahelian data from the African Monsoon Multidisciplinary Analysis (AMMA) to give an unprecedented evaluation of dust-generating winds in the European Centre for Medium-Range Weather Forecasts ERA-Interim reanalysis (ERA-I). Consistent with past studies, near-surface, high-speed winds are lacking in ERA-I and the diurnal variability is under-represented. During the summer monsoon season, correlations of ERA-I with observed wind-speed are low (~0.35 in Sahel and 0.25–0.4 in the Sahara). Fennec data show for the first time that: (1) correlations are reduced even in the Sahara, not directly influenced by the monsoon, (2) the systematic underestimation of observed winds by ERA-I in the summertime Sahel extends into the central Sahara: potentially explaining the failure of global models to capture the observed global dust maximum that occurs over the summertime Sahara (such as CMIP5), and demonstrates that modelled winds must be improved if they are to capture this key feature of the climatology.

Keywords: fennec; AMMA; dust; reanalysis; monsoon; Sahara;

1. Introduction

The Sahara/Sahel is the world's largest dust source but wind remains a key uncertainty in modelling emission (Knippertz and Todd, 2012). The relative importance of dust raising phenomena and their climatologies are poorly understood (Knippertz and Stuut, 2014). Important processes include the breakdown of the nocturnal low-level jet (NLLJ) and haboobs (Heinold et al., 2013). The NLLJ is often underestimated in models (Fiedler et al., 2013; Largeron et al., 2015) and haboobs are missed by models with parametrized convection (Marsham et al., 2011).

Where observations are sparse, the observational constraint on reanalyses can be insufficient, leading to significant errors (Agustí-Panareda et al., 2010; Garcia-Carreras et al., 2013; Roberts et al., 2015). This becomes something of circular problem, in that for regions with very few observations model developers, and researchers commonly use reanalyses as de-facto observations. This can be particularly important when we consider the known limitations of models to capture key dust uplift mechanisms.

Large dust biases exist in climate models, such as their inability to capture the observed Saharan summertime maximum (Heinold et al., 2016; Todd and Cavazos-Guerra, 2016). The severe shortage of data from the region has meant systematic analysis of modelled/analysed dust-generating winds has been very limited. Observations are mostly restricted to the inhabited margins and have systematic sampling biases, especially of the diurnal cycle (Cowie et al., 2014).

Here, we compare low-level winds from the ERA-I reanalysis (Persson and Grazzini, 2007; Dee et al., 2011) with new observations from the Sahara from the Fennec field campaign, and also Sahelian observations from the African Monsoon Multidisciplinary Analysis (AMMA) campaign. The value of ERA-I comparisons lies in its widespread use both operationally and for research. It also shares many features (and therefore process errors) with coarse resolution weather and climate models.

2. Methods

Observations of wind-speed are from stations in the Sahara and Sahel (Figure 1(a)). The Fennec campaign was an international consortium project aimed at improving the understanding of meteorology and climate in the Sahara, specifically with a focus on the lifting of desert dust and with observations from both land-based and aircraft platforms in 2011 and 2012 (Washington et al., 2012). Fennec AWSs were distributed in 2011 and operated into 2013 (Hobby et al.,

2013). The AMMA field campaign was an international project with the goal of improving the understanding of the West African Monsoon. Both the meteorological dynamics of the system, and the socio-economic impacts were investigated (Lebel *et al.*, 2010). AMMA provides data from the Atmospheric Radiation Measurement (ARM) Mobile Facilities (AMFs) at Niamey, Niger (2006) and three AMMA-CATCH (Couplage de l'Atmosphère Tropicale et du Cycle Hydrologique) stations: Agoufou (2005–2011), Bamba (2005–2010), and Kobou (2008–2010), all in Mali (Lebel *et al.*, 2010). The AMMA-CATCH observations are of particular use due to the long-term nature of the observations.

To resolve the diurnal cycle, 6-hourly ERA-I reanalyses are augmented with forecasts (giving 3-hourly data). Due to gaps in data reanalysis data has been subsampled to match available observations. Fennec observations are at 2 m, and AMMA data at 3 m, ERA-I 10 m winds from the closest grid-box are adjusted to observation heights using the wind profile power law $u = u_r(z/z_r)^\alpha$. u_r is the observed wind-speed at the reference height (z_r), z is the height to be adjusted to, and α is a stability coefficient (nominally 0.143; Touma, 1977). A range of α values is used to represent uncertainty in the conversion of (higher stability at night; α between 0.1 and 0.4, and lower stability in the day; α between 0.1 and 0.2). This uncertainty encompasses that from using a logarithmic wind profile approach with a range of friction velocities and surface roughness lengths. The choice to use the wind profile power law was a pragmatic one, and was driven by the ability to make some meaningful assumptions about the low-level stability whilst also having no wind profile data and virtually no information about surface characteristics.

This study assumes some parity when comparing time-averaged (1-h) winds at a single point with an instantaneous ERA-I grid-box value (1 h at $1-10\,\mathrm{m\,s^{-1}}$ is 3.6–36 km, comparable with a grid-box). This remains an imperfect comparison but this cannot explain the large differences between ERA-I and observations shown here.

Since dust uplift is a cubic function we study both u and wind-speed cubed (u^3) (Marsham *et al.*, 2013b). For the sake of simplicity, however, we do not apply thresholds. Thresholds vary significantly between regions and seasonally (Cowie *et al.*, 2014) and are unknown for these sites. What thresholds to use with ERA-I is also unclear since ERA-I misses rare high wind-speed events. Introducing a threshold would increase the effect of extreme events, increasing disagreements between ERA-I and observations (Cowie *et al.*, 2015).

3. Results

AMMA and Fennec data are analysed separately since their seasonal cycles are generally different (Appendix S1, Supporting information) with the AMMA stations

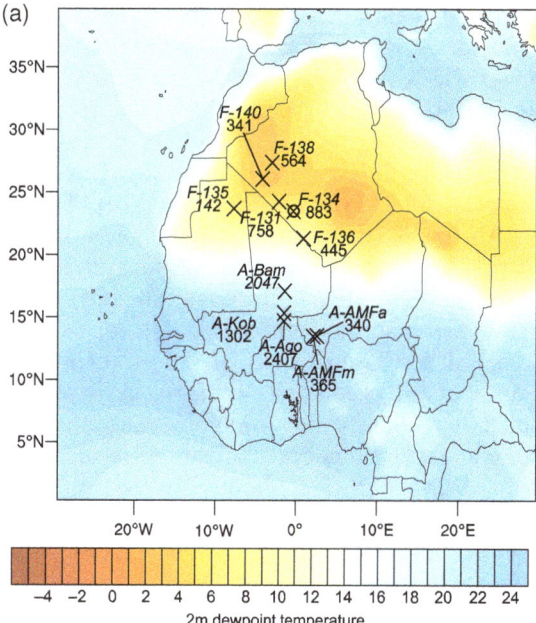

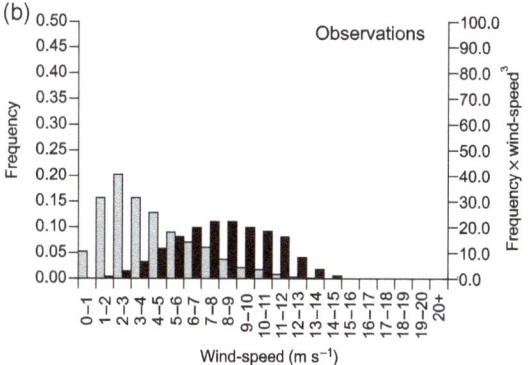

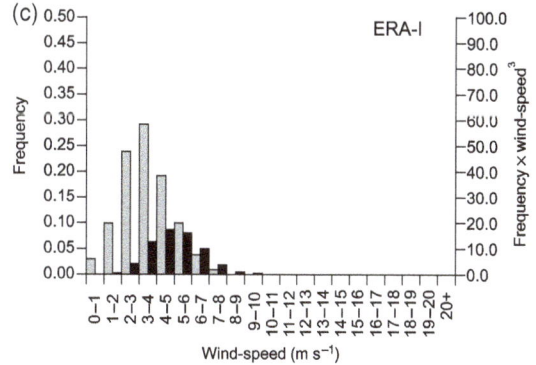

Figure 1. (a) Position of stations, including station identifier and number of days of observations. F, Fennec; A, AMMA. Colour-scale indicates ERA-I 1979–2014 mean August 2 m dewpoint temperature indicating maximum extent of the monsoon. (b) and (c) wind-speed distributions (grey) and distributions multiplied by mean wind-speed cubed (black) for F-134, circled on (a).

being more directly influenced by the West African Monsoon (WAM).

3.1. Wind-speed and wind-power distribution

Figure 1(b) (and Figure 1(c)) show distributions of all observed (ERA-I) u data from the Fennec AWS 134

(F-134). Also shown is the distribution multiplied by the bin mean u^3 giving the wind power. As expected from Cowie *et al.* (2015) and Largeron *et al.* (2015) there is a missing tail of high u events. The area of the wind-power bars (black) should be closely linked with dust uplift (Marsham *et al.*, 2013b). This is much smaller in the ERA-I panel (Figure 1(c)). The peak ERA-I power occurs at $4-5\,\mathrm{ms^{-1}}$ (range of $3-8\,\mathrm{ms^{-1}}$ across all station grid boxes) compared with $7-8\,\mathrm{ms^{-1}}$ in observations ($5-10\,\mathrm{ms^{-1}}$ across all stations). In many dust schemes tuning helps to compensate for such issues to match observed dust loads. However, the misrepresentation of rare events and the diurnal cycle remains a problem that tuning cannot overcome. The effect of the scaling uncertainty on the distributions is small, shifting the distribution right slightly for weaker stability and left for stronger stability (not shown). The area of the black bars in the ERA-I plots is always far less than that from observed stations and the ERA-I distributions always are always lacking the high wind-speed tail.

3.1.1. Scatter and best-fit statistics

Figures 2(a) and (b) show scatter graphs for January at a 3-hourly sampling frequency. Figures 2(c) and (d) show the same for daily means. Seasonal variation of best-fit statistics is shown in Figures 2(e) and (f). Rare high wind events are missed in ERA-I, there are no equivalent events missing in observations. This is characteristic of all other months and consistent with ERA-I capturing synoptic-scale features, but missing unresolved uplift processes.

Correlation coefficients for 3-hourly data are lower than those for daily data in both groups (Figures 2(e) and (f)), showing the importance of sub-daily processes on correlation (e.g. NLLJs and haboobs). Correlations for the Fennec (AMMA) group vary seasonally from 0.24 to 0.71 (0.29–0.64) for 3-hourly u, and from 0.22 to 0.68 (0.15–0.65) for 3-hourly u^3 (not shown). Therefore, across both groups ERA-I explains 6–50% of the variance in 3-hourly u and 2–46% of the variance in 3-hourly u^3. A double dip in correlation is present in the AMMA group (Figure 2(e)), temporally coinciding with the passage of the monsoon front. The weakest correlation for the Fennec stations is in August (Figure 2(f)), when the monsoon front is at its most northerly position. Features associated with the edge of the monsoon flow such as strong gradients in humidity and the effects of moist convection are poorly represented in reanalyses (Roberts *et al.*, 2015). The misrepresentation of moist convection during the wet monsoon affects the entire WAM circulation (Garcia-Carreras *et al.*, 2013; Marsham *et al.*, 2013b) leading to lower correlations for Fennec stations beyond the edge of the monsoon flow. Consistent with the low correlations during the monsoon, best-fit line gradients are highest and closest to 1 in winter and lowest (around 0.6) in summer (Figures 2(e) and (f)). Y-intercepts are greater indicating that on average, during the monsoon season ERA-I underestimates (overestimates) light

(strong) winds. However, it is important to note that rare, strong wind-speeds that dominate dust uplift are absent in ERA-I but present in observations.

3.2. Diurnal behaviour

Figure 3 shows that both groups across all months have a too weak diurnal cycle even accounting for stability uncertainty from scaling 10 m winds (dashed red lines). The maximum values (generally during the day), are greatly underestimated in ERA-I. This is particularly pronounced for the Saharan/Fennec stations, consistent with the missing rare winds shown in Figure 1, that contribute disproportionately to u^3.

Night-time ERA-I winds are generally stronger than observed at the Sahel/AMMA stations, consistent with Largeron *et al.* (2015). Possibly caused by artificially enhanced turbulent diffusion in models (Sandu *et al.* 2013) that drives mixing of NLLJ momentum to the surface. At the Fennec stations night-time ERA-I winds are often weaker than observed. This is possibly due to observed winds containing the effects of intermittent shear-driven mixing of momentum to the surface (Schepanski *et al.*, 2015). The shorter inertial oscillation period over the Sahara compared to the Sahel gives greater shear-induced mixing and therefore higher mean surface winds (Heinold *et al.*, 2015). However, differences could also be the result of large-scale pressure gradient errors, which are poorly constrained in reanalyses across the whole region (Roberts and Knippertz, 2014).

The timing of strongest winds in ERA-I is generally correct at the Sahel/AMMA stations but is often too late at the Sahara/Fennec stations. This suggests that modelled NLLJ breakdown takes longer than reality, consistent with inaccurate boundary layer growth (Garcia-Carreras *et al.*, 2015), or enhanced turbulent diffusion artificially raising the height of the NLLJ. The observed 0900 UTC peak in winds (breakdown of the NLLJ) grows in strength in the Sahara/Fennec from March to May and remains strong till October. There is no corresponding increase in u^3 values in ERA-I, again indicating that the winds associated with the NLLJ breakdown are not properly represented in ERA-I in the remote Sahara.

Observed AMMA evening u^3 increases in May, June and July (from approximately 10 to $50\,\mathrm{m^3\,s^{-3}}$), this is not present in ERA-I and much less pronounced for the Fennec stations. This suggests that high wind-speeds are missing due to poor representation of convectively generated cold pools (Marsham *et al.*, 2011).

3.3. Seasonal behaviour

Figures 4(a) and (b) show the seasonal development of u^3 for AMMA and Fennec data. Figure 4(c) shows the same for F-136 at Bordj-Badji Mokhtar (BBM). F-136, unlike the other Saharan/Fennec stations, is at the northern limit of haboob uplift in convection-permitting models (Figure 11 in Pantillon *et al.*, 2015) and has been

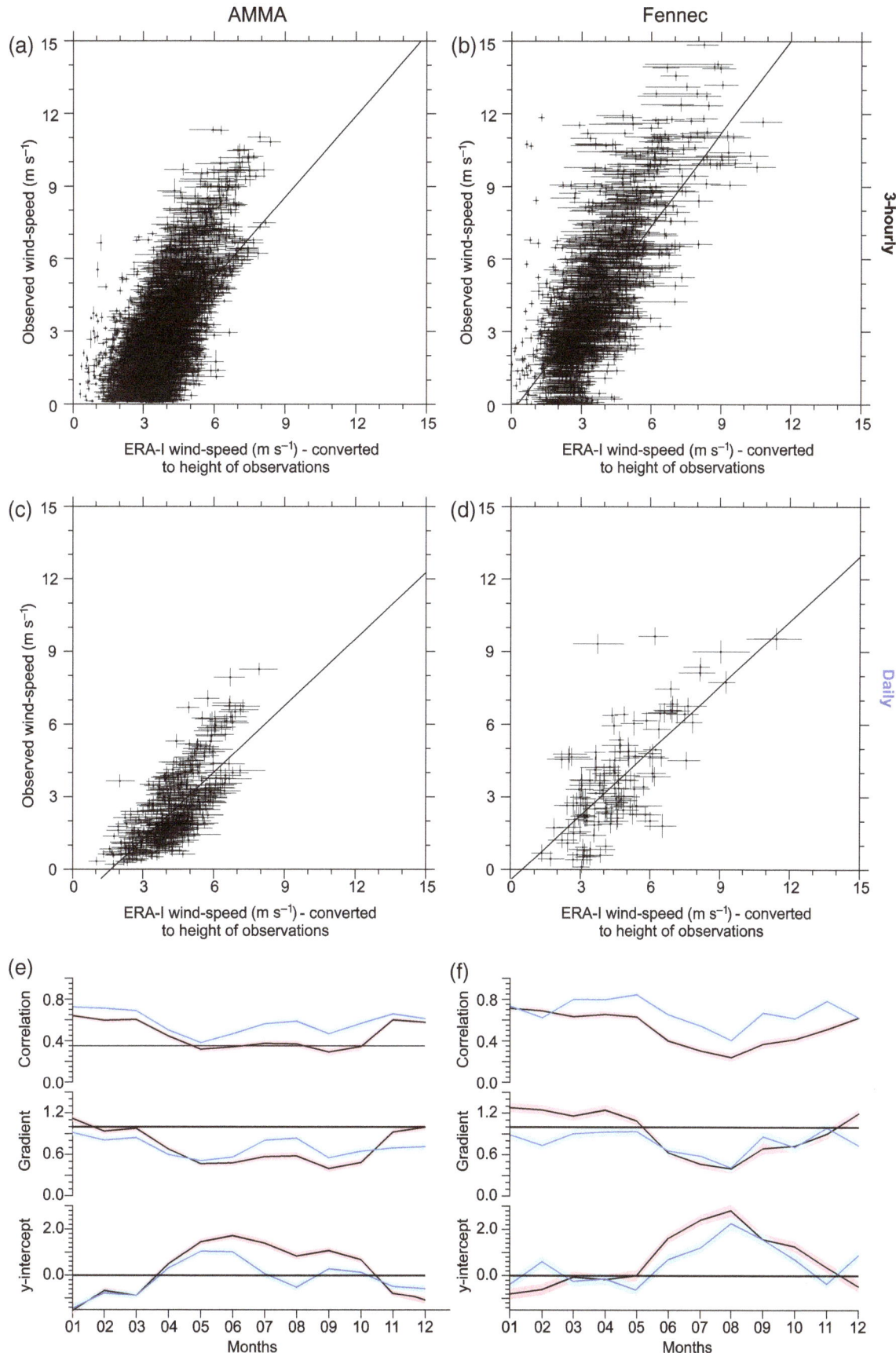

Figure 2. Scatter graphs, AMMA (left) and Fennec (right) for January. (a) and (b) 1-h mean observed against instantaneous ERA-I wind, (3-hourly intervals). (c) and (d) same using daily-mean wind. (e) and (f) seasonal evolution of correlation coefficient and gradient and y-intercept of the best-fit line, 3-hourly (red) and daily (blue). Shading represents ±2 standard deviations of wind uncertainty distribution based on 10 000 iterations of a bootstrapping algorithm.

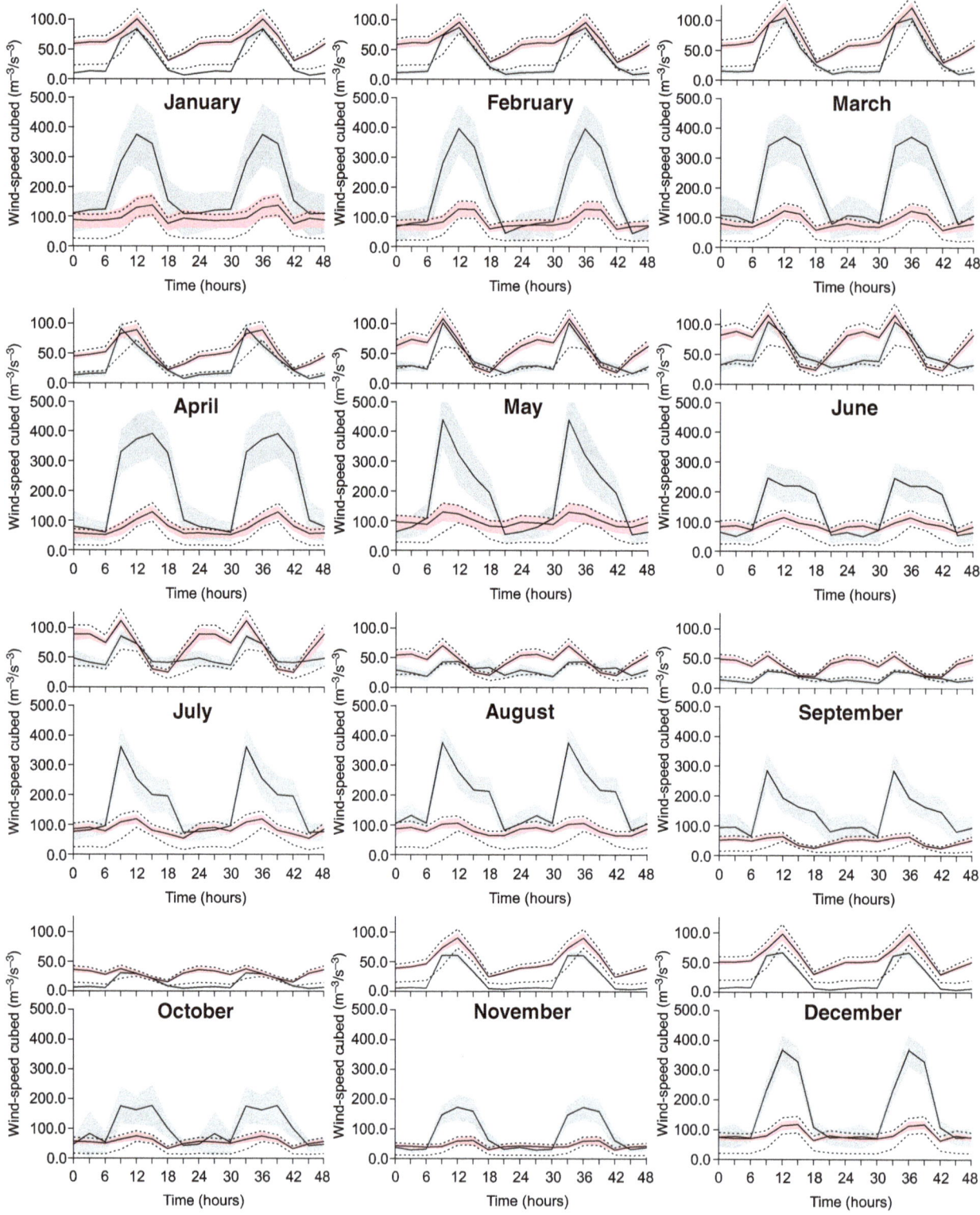

Figure 3. Mean monthly wind-speed cubed diurnal cycle for AMMA (top of panel) and Fennec (bottom), observations (black) and ERA-I (red). Shading represents normalized standard error from monthly diurnal cycle at each station. Dashed red lines indicate the envelope uncertainty associated height conversion of ERA-I winds. Please note the different scales between top and bottom panels.

observed to be regularly affected by moist convection during summer (Marsham *et al.*, 2013a).

ERA-I underestimates the monsoon peak occurring at the AMMA stations (Figure 4(a)), but represents both the variation and magnitude of u^3, during winter.

Observations have greater variability compared to ERA-I at all times of year and increases during the monsoon season for the AMMA stations and at F-136 (Figures 4(a) and (c)). This is consistent with the occurrence of cold pools and their impact on

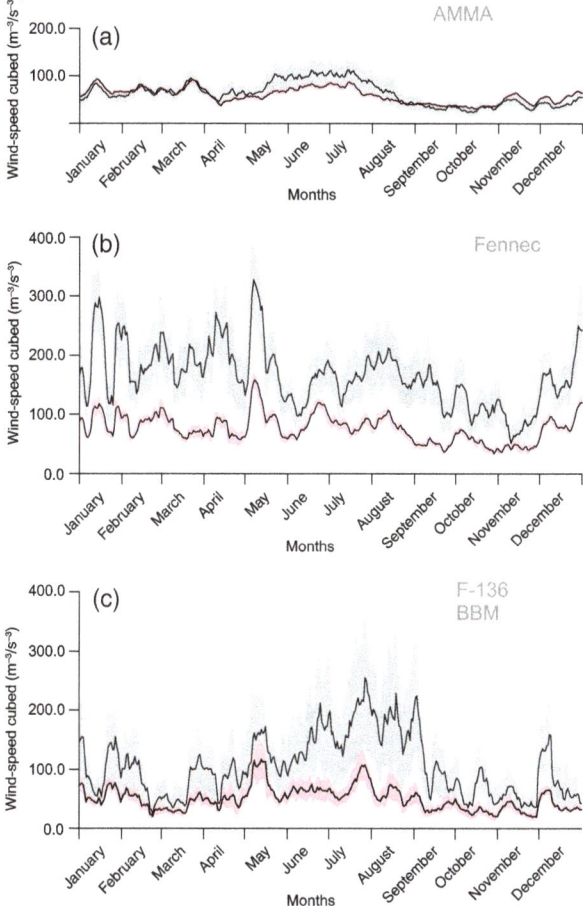

Figure 4. Ten-day running average of daily-mean wind-speed cubed for (a) AMMA, (b) Fennec, and (c) F-136 at BBM showing observations (black) and ERA-I (red). Shading represents normalized standard error of wind-speed cubed from daily-mean for each station.

large-scale meteorology. u^3 values in the Fennec group (Figure 4(b)) are underestimated at all times of year (between 50 and 100 $m^3 s^{-3}$). Despite this, much of the sub-seasonal variation is captured. ERA-I captures the seasonal and synoptic variability at F-136 in winter, but completely misses the summer maximum which is associated with haboobs (Marsham *et al.*, 2013a).

4. Conclusions

This work gives an unprecedented evaluation of Saharan dust-generating winds from within the summertime dust maximum. We compare new Saharan observations from the Fennec project and Sahelian observations from AMMA with ERA-I reanalysis. These data lack the systematic sampling bias found in routine data from the Sahara (Cowie *et al.*, 2014) and capture a full annual cycle, unlike datasets used previously (Marsham *et al.*, 2013a).

The ERA-I distribution of winds misses rare high wind-speed events. ERA-I's wind-power maxima is $3-8 ms^{-1}$, compared to $5-10 ms^{-1}$ in observations. While tuning uplift thresholds can account for

a shortfall in emission, it cannot address important process errors.

For all stations, correlations decrease during the approach of the monsoon front ($\sim$0.8 to $\sim$0.4), highlighting the poor monsoon representation as a key problem. Three-hourly correlations (0.24–0.71) are worse than those for daily data (0.41–0.84), implicating processes on sub-daily time-scales as being improperly represented by ERA-I (e.g. NLLJs and haboobs).

The diurnal cycle in dust-generating winds is too weak in ERA-I, consistent with artificially high nocturnal mixing in ERA-I. ERA-I also misses stronger day-time winds in the Sahara. The diurnal timing of strongest winds in ERA-I is correct in the Sahel, but too late in the Sahara, likely due to misrepresentation of the NLLJ. There is an increase in afternoon/evening winds in the summer in the Sahel in observations, but not in ERA-I, coinciding with the occurrence haboobs.

In the remote Sahara, ERA-I captures much of the synoptic and seasonal variability and there is no clear summertime maximum in u^3. In the Sahel and southern Sahara the summertime maximum in u^3 is missed by ERA-I.

Climate models fail to capture the central Saharan summertime maxima in dust that is close to BBM (Figure 2 in Todd and Cavazos-Guerra, 2016 and Figures 2 and 6 in Heinold *et al.*, 2016). This key failure of climate models is consistent with the new result shown here that even the observationally constrained ERA-I analyses do not capture the summertime peak in dust-generating winds in the region of the maximum dust loads.

The misrepresentation of uplift processes provides one reason for the lack of reliability of dust in climate models (Evan *et al.*, 2014). It supplies motivation for work focussed on a better understanding of the interactions of multiscale processes within the West African Monsoon, which, due to the inherently complex and chaotic nature of convective systems and the strong horizontal gradients are still not adequately represented in any simulations. Alongside improved representation of the monsoon there is a need for further research into parameterization of haboobs (Pantillon *et al.*, 2016). This work also suggests that caution should be applied when studying dust uplift using analysed winds (Evan *et al.*, 2016). It highlights that climate models (technically similar ERA-I) miss key uplift processes. Therefore, climate predictions of dust are likely to have spatial and temporal errors in uplift, potentially leading to unrealistic dust loadings and transport.

Acknowledgements

This study was funded by the Natural Environment Research Council (NERC) Saharan-West African Monsoon Multiscale Analysis (SWAMMA) project (NE/L005352/1). Fennec AWS network developed, tested and installed as part of Fennec project (NE/G017166/1). AMMA-CATCH system was funded by the French Ministry of Research and National Institute for Earth Sciences and Astronomy. Data from two AMF sites is provided by ARM Climate Research Facility.

The authors acknowledge Richard Washington (Oxford University) as the PI of the Fennec consortium, the work of Sebastian Engelstaedter and Christopher Allen (Oxford University) and all the staff of the Offices National de la Météorologie in Algeria and Mauritania who worked to deploy instruments for Fennec. The authors thank ECMWF for ERA-I reanalysis data.

The authors thank two anonymous reviewers for their insight and useful suggestions.

Supporting information

The following supporting information is available:

Appendix S1. Information detailing problems and limitations of observed near surface winds, and seasonal cycles of wind speed cubed plotted for individual observation stations.

References

Agustí-Panareda A, Beljaars A, Cardinali C, Genkova I, Thorncroft C. 2010. Impacts of assimilating AMMA soundings on ECMWF analyses and forecasts. *Weather Forecasting* 25: 1142–1160.

Cowie SM, Knippertz P, Marsham JM. 2014. A climatology of dust emission events from northern Africa using long-term surface observations. *Atmospheric Chemistry and Physics* 14: 8579–8597.

Cowie SM, Marsham JH, Knippertz P. 2015. The importance of rare, high-wind events for dust uplift in northern Africa. *Geophysical Research Letters* 42: 8208–8215.

Dee DP, Uppala SM, Simmons AJ, Berrisford P, Poli P, Kobayashi S, Andrae U, Balmaseda MA, Balsamo G, Bauer P, Bechtold P, Beljaars ACM, van de Berg L, Bidlot J, Bormann N, Delsol C, Dragani R, Fuentes M, Geer AJ, Haimberger HSB, Hersbach H, Hólm EV, Isaksen L, Kållberg P, Köhler M, Matricardi M, McNally AP, Monge-Sanz BM, Morcrette J-J, Park B-K, Peubey C, de Rosnay P, Tavolato C, Thépaut J-N, Vitart F. 2011. The ERA-interim reanalysis: configuration and performance of the data assimilation system. *Quarterly Journal of the Royal Meteorological Society* 137: 553–597.

Evan AT, Flamant C, Fiedler S, Doherty O. 2014. An analysis of aeolian dust in climate models. *Geophysical Research Letters* 41: 5996–6001.

Evan AT, Flamant C, Gaetani M, Guichard F. 2016. The past, present and future of African dust. *Nature* 531: 493–495.

Fiedler S, Schepanski K, Heinold B, Knippertz P, Tegen I. 2013. Climatology of nocturnal low-level jets over North Africa and implications for modeling mineral dust emission. *Journal of Geophysical Research: Atmospheres* 118: 6100–6121.

Garcia-Carreras L, Marsham JH, Parker DJ, Bain CL, Milton S, Saci A, Salah-Ferroudj M, Ouchene B, Washington R. 2013. The impact of convective cold pool outflows on model biases in the Sahara. *Geophysical Research Letters* 40: 1647–1652.

Garcia-Carreras L, Parker DJ, Marsham JH, Rosenberg PD, Brooks IM, Lock AP, Marenco F, McQuaid JB, Hobby M. 2015. The turbulent structure and diurnal growth of the Saharan atmospheric boundary layer. *Journal of Atmospheric Science* 72: 693–713.

Heinold B, Knippertz P, Marsham JH, Fiedler S, Dixon NS, Schepanski K, Laurent B, Tegen I. 2013. The role of deep convection and nocturnal low-level jets for dust emission in summertime West Africa: estimates from convection-permitting simulations. *Journal of Geophysical Research: Atmospheres* 118: 4385–4400.

Heinold B, Knippertz P, Beare RJ. 2015. Idealized large-eddy simulations of nocturnal low-level jets over subtropical desert regions and implications for dust-generating winds. *Quarterly Journal of the Royal Meteorological Society* 141: 1740–1752.

Heinold B, Tegen I, Schepanski K, Banks JR. 2016. New developments in the representation of Saharan dust sources in the aerosol-climate model ECHAM6-HAM2. *Geoscientific Model Development* 9: 765–777.

Hobby M, Gascoyne M, Marsham JH, Bart M, Allen C, Engelstaedter S, Fadel DM, Gandega A, Lane R, McQuaid JB, Ouchene B, Ouladichir A, Parker DJ, Rosenberg PD, Ferroudj MS, Saci A, Seddik F, Todd M, Walker D, Washington R. 2013. The fennec automatic weather station (AWS) network: monitoring the Saharan climate system. *Journal of Atmospheric and Oceanic Technology* 30: 709–724.

Knippertz P, Todd MC. 2012. Mineral dust aerosols over the Sahara: meteorological controls on emission and transport and implications for modeling. *Reviews of Geophysics* 50: RG1007. https://doi.org/RG1007.

Knippertz P, Stuut J-B. 2014. Chapter 6. Meteorological aspects of dust storms. In *Mineral Dust: A key Player in the Earth System*. Springer: Dordrecht, Netherlands.

Largeron Y, Guichard F, Bouniol D, Couvreux F, Kergoat L, Marticorena B. 2015. Can we use surface wind fields from meteorological reanalyses for Sahelian dust emission simulations? *Geophysical Research Letters* 42: 2490–2499.

Lebel T, Parker DJ, Flamant C, Bourlès B, Marticorena B, Mougin E, Peugeot C, Diedhiou A, Haywood JM, Ngamini JB, Polcher J, Redelsperger JL, Thorncroft CD. 2010. The AMMA field campaigns: multiscale and multidisciplinary observations in the west African region. *Quarterly Journal of the Royal Meteorological Society* 136: 8–33.

Marsham JH, Knippertz P, Dixon NS, Parker DJ, Lister GMS. 2011. The importance of the representation of deep convection for modeled dust-generating winds over West Africa during summer. *Geophysical Research Letters* 38: L16,803.

Marsham JH, Hobby M, Allen CJT, Banks JR, Bart M, Brooks BJ, Cavazos-Guerra C, Engelstaedter S, Gascoyne M, Lima AR, Martins JV, McQuaid JB, O'Leary A, Ouchene B, Ouladichir A, Parker DJ, Saci A, Salah-Ferroudj M, Todd MC, Washington R. 2013a. Meteorology and dust in the central Sahara: observations from fennec supersite-1 during the June 2011 intensive observation period. *Journal of Geophysical Research: Atmospheres* 118: 4069–4089.

Marsham JH, Dixon NS, Garcia-Carreras L, Lister GMS, Parker DJ, Knippertz P, Birch CE. 2013b. The role of moist convection in the west African monsoon system: insights from continental-scale convection-permitting simulations. *Geophysical Research Letters* 40: 1843–1849.

Pantillon F, Knippertz P, Marsham JH, Birch CE. 2015. A parameterization of convective dust storms for models with mass-flux convection schemes. *Journal of Atmospheric Science* 72: 2545–2561.

Pantillon F, Knippertz P, Marsham JH, Panitz H-J, Bischoff-Gauss I. 2016. Modeling haboob dust storms in large-scale weather and climate models. *Journal of Geophysical Research: Atmospheres* 121: 2090–2109.

Persson A, Grazzini F. 2007. User Guide to ECMWF forecast products. *Meteorological Bulletin* M3.2 ECMWF

Roberts AJ, Knippertz P. 2014. The formation of a large summertime Saharan dust plume: convective and synoptic-scale analysis. *Journal of Geophysical Research: Atmospheres* 119: 1766–1785.

Roberts AJ, Marsham JH, Knippertz P. 2015. Disagreements in low-level moisture between (re)analyses over summertime West Africa. *Monthly Weather Review* 143: 1193–1211.

Sandu I, Beljaars A, Bechtold P, Mauritsen T, Balsamo G. 2013. Why is it so difficult to represent stably stratified conditions in numerical weather prediction (NWP) models? *Journal of Advances in Modeling Earth Systems* 5: 117–133.

Schepanski K, Knippertz P, Fiedler S, Timouk F, Demarty J. 2015. The sensitivity of nocturnal low-level jets and near-surface winds over the Sahel to model resolution, initial conditions and boundary-layer set-up. *Quarterly Journal of the Royal Meteorological Society* 141: 1442–1456.

Todd MC, Cavazos-Guerra C. 2016. Dust aerosol emission over the Sahara during summertime from cloud-aerosol lidar with orthogonal polarization (CALIOP) observations. *Atmospheric Environment* 128: 147–157.

Touma JS. 1977. Dependence of the wind profile power law on stability for various locations. *Journal of the Air Pollution Control Association* 27: 863–866.

Washington R, Flamant C, Parker DJ, Marsham JH, McQuaid J, Brindley H, Todd MC, Highwood EJ, Chaboureau J-P, Kocha C, Bechir M, Saci A. 2012, 2012. Fennec – the Saharan climate system. *CLIVAR Exchanges* 60: 31–33.

15

Investigation of an extreme Koshava wind episode of 30 January–4 February 2014

Djordje Romanić,[1]* Mladjen Ćurić,[2] Miroljub Zarić,[1] Miloš Lompar[1] and Ilija Jovičić[1]

[1] Department of Meteorology, Republic Hydrometerological Service of Serbia, Belgrade, Serbia
[2] Institute of Meteorology, Faculty of Physics, University of Belgrade, Belgrade, Serbia

*Correspondence to:
Djordje Romanić, Department of
Meteorology, Republic
Hydrometerological Service of
Serbia, Belgrade, Serbia. E-mail:
dromanic@uwo.ca

Abstract

An extreme Koshava episode (EKE) from 30 January to 4 February 2014 has been studied. Koshava is a local windstorm in Southeast Europe. EKE was characterized by wind gusts above $45\,m\,s^{-1}$ and deep snowdrifts. Strong Eurasian anticyclone in combination with large temperature gradient between the anticyclone region and the Mediterranean caused the EKE. The anticyclone had the probability of occurrence of 0.1%. Koshava layer was either statically stable or adiabatic. The event was numerically modelled using two mesoscale models: WRF-NMM and NMMB. Wind directions were forecasted more accurately than the mean speed and gusts.

Keywords: Koshava wind; extreme weather; blizzard; WRF-NMM; NMMB; numerical modelling; stable atmosphere

1. Introduction

From 30 January to 4 February 2014, large parts of Serbia were affected by an extremely severe windstorm. A strong wind, named Koshava, occasionally had hurricane velocities. In addition to extreme wind speeds, Koshava created snowdrifts several meters deep. This weather disaster impacted everyday life across the country – trees were toppled, many buildings and cars were damaged, and rail and air transports in northern Serbia were shut down. Furthermore, between 5000 and 10 000 people were trapped in their vehicles on the snow-covered roads and it was necessary to rescue them from cold. Fortunately, no human casualties were reported. Interestingly, this extreme Koshava episode (EKE) had a positive effect on the air quality in cities. Measurements in Belgrade showed that the atosphere after the EKE was two times cleaner compared with the air quality before the event.

Koshava is a strong windstorm that blows from southeast directions over most regions of Serbia. The southeasterly winds need to last for at least 2 days and possess the mean hourly wind speeds above $2\,m\,s^{-1}$ to be considered as Koshava (Romanić et al., 2015a). Romanić et al. (2015b) demonstrated that the Eurasian anticyclones and Mediterranean cyclones, together with the orography of the Balkan region are the main Koshava drivers. The pressure gradients created by these two pressure systems move air masses from Ukraine and Moldavia towards the Adriatic Sea. However, the South Carpathian and Balkan Mountains redirect that air towards the Iron Gates (Figure 1). The diverging jets downstream from the Iron Gates are known as the Koshava wind.

Koshava is strongest in the Vršac (VR) region. Figure 2 shows that the VR station has $10–15\,m\,s^{-1}$ stronger Koshava gusts than the Belgrade (BG) station. The strongest Koshava gust ever recorded was measured in VR on 11 February 1987 [$48\,m\,s^{-1}$ at 10 m above ground (a.g)]. The strongest Koshava speed at the BG station was $35.9\,m\,s^{-1}$ at 24 m a.g. on 17 October 1976. The measured velocity at 650 m a.g. was $45\,m\,s^{-1}$. These two weather stations mostly recorded different storms as being the most extreme. Exceptions are the storms that occurred in February 1979 and February 2014. These results indicate that EKEs are mostly localized. An absence of extreme Koshava winds in the period from the late 1980s to 2014 is in accordance with the results by Romanić et al. (2015a), which state that Koshava was most active in the period from the 1970s to 1980s.

This article investigates meteorological factors that caused the EKE in late January to early February of 2014. This EKE is the most extreme wind event in Serbia that has been peer-reviewed analysed. The following questions are addressed: (1) what was the synoptic situation which brought EKE, (2) how well two mesoscale models with different configurations forecasted the EKE, and most importantly, (3) what caused such high Koshava speeds to occur? Hereafter, when refereeing to dates related to the EKE, the year (2014) will be omitted for simplicity.

2. Methods and materials

Wind and radiosonde measurements were acquired from the Republic Hydrometeorological Service of

Figure 1. Schematics of a typical Koshava flow. The across-mountain pressure gradient between the Wallachia Valley and the Pannonian Plane drives the flow through mountain passes and gorges. Weather stations are indicated with red dots.

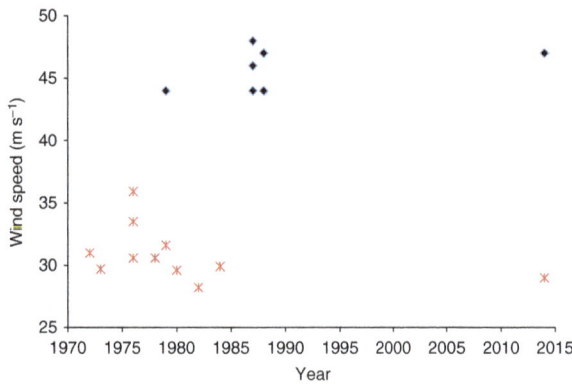

Figure 2. Past records of the extreme Koshava gusts at Vršac station (blue diamonds) and Belgrade station (red stars)

Serbia (RHMSS). The mean sea level pressure (MSLP) and 2-m air temperature fields were obtained from the European Center for Medium-Range Weather Forecasting's (ECMWF) Meteorological Archival and Retrieval System database.

This study uses the Weather Research and Forecasting (WRF) model for the numerical simulation of the EKE. The employed solver is the Nonhydrostatic

Mesoscale Model (NMM; version 3.6.1), which solves the compressible, nonhydrostatic Navier–Stokes equations on Arakawa-E grid. The vertical coordinate is a hybrid sigma-pressure coordinate. Two different configurations of the NMM model were tested (Table 1). In addition, two more numerical simulations were performed using a modified version of the NMM model on Arakawa-B grid, called NMMB. In total, four simulations with different models' configurations are performed – two using NMM and two using NMMB. The forecasts (mean hourly wind speed, wind gust, and wind direction) were compared against measurements from four weather stations situated in the Koshava region (BG, Novi Sad (NS), VR, and Veliko Gradište (VG)). The results are also benchmarked against the ECMWF's Integrated Forecast System (IFS) analysis.

Models' configurations are summarized in Table 1. NMM, NMMB_1, and NMMB_2 setups are operationally used at the RHMSS. Due stably stratified planetary boundary layer (PBL; see Section 3.1), the NMM_QNSE configuration uses the Quasi-Normal Scale Elimination scheme (QNSE; Sukoriansky *et al.*, 2005) for PBL parametrization. The NMM and NMMB simulations have similar physics, but the forecasts are based on different initial and boundary conditions.

Table 1. Overview physical packages, initial and boundary conditions, and computational domain characteristics used in numerical modelling of the EKE.

	NMM	NMMB_1	NMMB_2	NMM_QNSE
Microphysics	Ferrier (Ferrier et al., 2002)			
Longwave radiation	GFDL (Schwarzkopf and Fels, 1991)			RRTMG (Iacono et al., 2008)
Shortwave radiation	GFDL (Lacis and Hansen, 1974)			RRTMG (Iacono et al., 2008)
Surface layer	Similarity scheme			QNSE (Sukoriansky et al., 2005)
Land surface	Noah-MP (Ek et al., 2003)	LISS		Noah-MP (Niu et al., 2011)
PBL	MYJ			QNSE (Sukoriansky et al., 2005)
Cumulus convection	BMJ (Janjić, 1994)		–	
Initial and boundary conditions	IFS	Nested in parent NMMB	Global NMMB	IFS
Horizontal resolution (km)	~2.2	~3	~9	~2.2
Domain centre	LON 17.5°E	LON 18.8°E	LON 8.0°E	LON 17.5°E
	LAT 44.5°N	LAT 44.2°N	LAT 48.5°N	LAT 44.5°N
Grid points	400 × 450 × 45	326 × 287 × 64	394 × 356 × 64	400 × 450 × 45
Time step (s)	5	6.67	20	5
Nest ratio	–	3 : 1	–	–

Namely, the quality of forcing data can have a pronounced influence on the accuracy of forecasts (Talbot *et al.*, 2012).

Numerical modelling of surface and near-surface variables under stable atmospheric conditions is more challenging compared with unstable stratification (Shin and Hong, 2011). This difficulty partially arises because turbulent eddies in stable atmosphere are much smaller compared with unstable (convective) conditions (Beare *et al.*, 2006). For that reason, the NMM_QNSE simulation relies on the QNSE PBL scheme. This scheme is particularly designed for stable stratifications. Namely, QNSE is a turbulent kinetic energy prediction scheme based on a quasi-Gaussian spectral closure model with the surface layer parameters derived from large eddy simulations of stably stratified atmosphere (Sukoriansky *et al.*, 2005).

3. Results and discussion

3.1. Synoptics and dynamics of the EKE

Animation S1 (Supporting Information) shows the ridge of a deep anticyclone stretching across the Balkan Peninsula on 27 January. On 30 January (Figure 3a), the pressure difference between the Eurasian anticyclone and Mediterranean cyclone was 55 mb, and the temperature difference between these two regions was approximately 50 °C. The distance between the centres of these two pressure systems was about 2500 km. The pressure gradients above Serbia were extremely large – approximately 5 mb/100 km. This synoptic situation was favourable for the development of a strong Koshava wind.

EKE started in the afternoon of 30 January. The central pressure in the anticyclone that caused the EKE was 1055 mb, which according to Romanić *et al.* (2015b) has an exceedance probability of occurrence of 0.1%. The Mediterranean cyclone, situated in the Gulf of Genoa, was not particularly strong as its central pressure was 3 mb below the average central pressure of the

cyclones which typically generate Koshava (Romanić *et al.*, 2015b). Therefore, the strong anticyclone was the main trigger for the EKE. Animation S1 further shows that while the centre of the anticyclone was slowly moving eastward from the Koshava region during the EKE, its high-pressure ridge was constantly positioned above Balkan Peninsula and therefore maintained strong pressure and temperature gradients above the region. The disappearance of the anticyclone (Figure 3b) led to the cessation of the EKE.

The Synoptic Koshava Index (SKI; Romanić *et al.* (2015b)), defined as the normalized area-averaged MSLP difference between the anticyclone and cyclone regions is presented in Figure 4a. The SKI values above 25 are noticeable during EKE while the values above 35 can be observed for the days with the strongest Koshava wind. The high values of SKI demonstrate a strong pressure difference between the anticyclone and cyclone regions, which indubitably generated the strong Koshava velocities. Romanić *et al.*'s (2015b) results show that the SKI values above 35 have a probability of occurrence below 1% (less than 0.3% chance of the SKIs above 40). This case study suggests that the SKI is a reliable prognostic Koshava parameter.

The major Koshava contributors on the mesoscales are the across-mountain MSLP and potential temperature differences between west parts of the Wallachia Valley and the Koshava region (Romanić *et al.*, 2015b). The across-mountain MSLP difference between Drobeta-Turnu Severin and VG reached 10 mb during the EKE, whereas the across-mountain potential temperature difference between these two stations was below −7 °C. The probability of occurrence of such a large pressure difference is 0.01%, while the joint probability of these values is practically zero (0.0012%; Romanić *et al.*, 2015b).

Due to these large pressure and temperature differences, northern and northeastern parts of Serbia experienced a blizzard that caused significant damage to infrastructure and posed a threat to human safety. A strong and cold Koshava wind with daily mean speeds of 6–$12\,\mathrm{m\,s^{-1}}$ and occasional gusts between

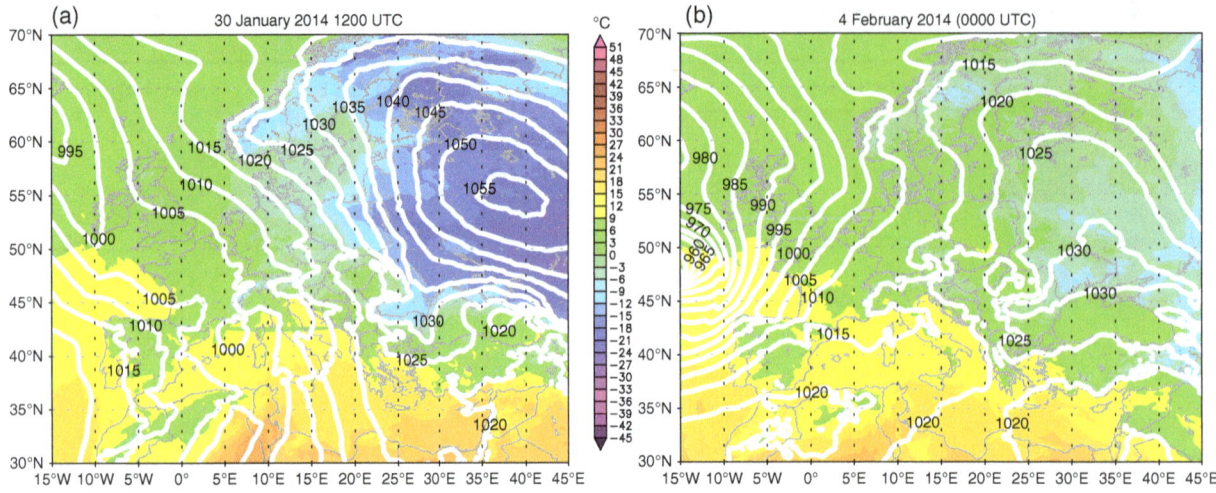

Figure 3. The MSLP in mb (contours) overlying the surface air temperature map. The EKE started on 30 January (a) and lasted until 4 February (b).

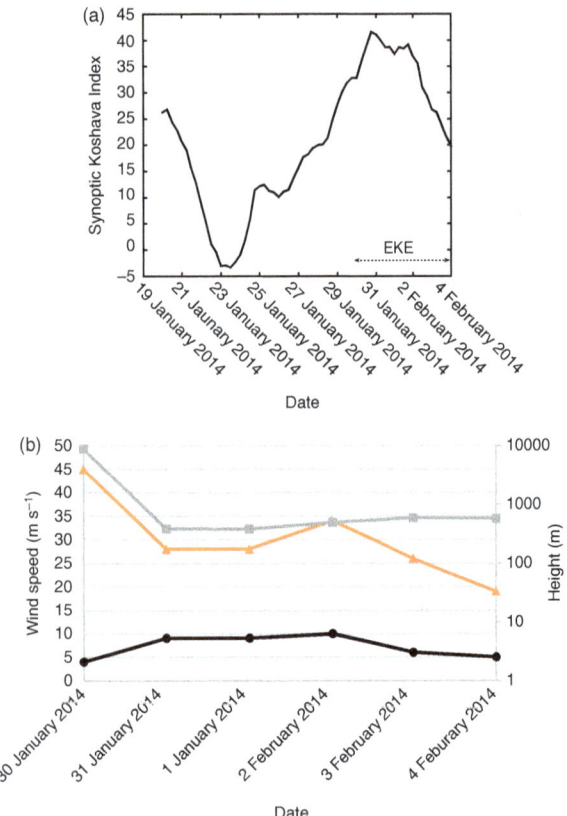

Figure 4. (a) The SKI before and during the EKE. (b) Daily mean (blue line with circles) and maximum (orange line with triangles) Koshava speeds during the EKE at the BG station (primary y-axis). The grey line with squares represents the height of the maximum Koshava speed (logarithmic secondary y-axis).

12 and 25 m s^{-1} was blowing during the EKE. In the early morning of 1 February, Koshava gusts in BG reached 29 m s^{-1}. Radiosonde measurements in BG conducted on 2 February (00 UTC) recorded the maximum Koshava speed of 34 m s^{-1} at a height of 497 m a.g (Figure 4b). The strongest Koshava speed, however, was measured in VR on 1 February, when a gust at 10 m

a.g. reached 47 m s^{-1}. This value is the second largest Koshava gust ever recorded. The 2014 EKE was the first severe Koshava event after more than 25 years without any similar storm (Figure 2).

Emagrams in Figure 5 indicate that the Koshava layer at the BG station was considerably deeper than at the VR station. The Koshava winds above BG occurred in a layer stretching from the surface up to 350 mb. The Koshava layer at the VR station was much shallower; reaching only up to 900 mb. Despite the differences in thickness, the maximum Koshava speeds occurred at similar heights (950–900 mb) at both stations. Figure 4 shows that the maximum Koshava speed at the BG station was measured in a layer between 389 (31 January) and 594 m a.g. (3 February). This observation is in accordance with the finding by Vukmirović (1985) that Koshava is strongest in the layer between 500 and 600 m a.g. The strongest Koshava speed in BG on 2 February occurred at the bottom of an elevated inversion, which was also previously noticed by Unkašević et al. (1999). Koshava layer was characterized either with a deep surface inversion (Figure 5b) or an elevated inversion layer above adiabatic and conditionally stable layers close to the surface (Figure 5a). The elevated inversion was due to the advection of the warm air in the front of the Mediterranean cyclone (Animation S1). Atmospheric stability indices in emagrams show the absence of the atmospheric instability during the EKE. However, both environmental helicity (EH) and storm relative environmental helicity (SREH) had very high values. The helicity represents a measure of the vertical transfer of energy by the shear of the horizontal wind vector. The SREH values at the VR station were around 370 m^2 s^{-2}, which is comparable with the figures favourable for the development of strong tornadoes.

3.2. Numerical modelling of the EKE

Figure 6 and Table 2 show the models' forecasted wind directions (*D*) more precisely than the mean hourly

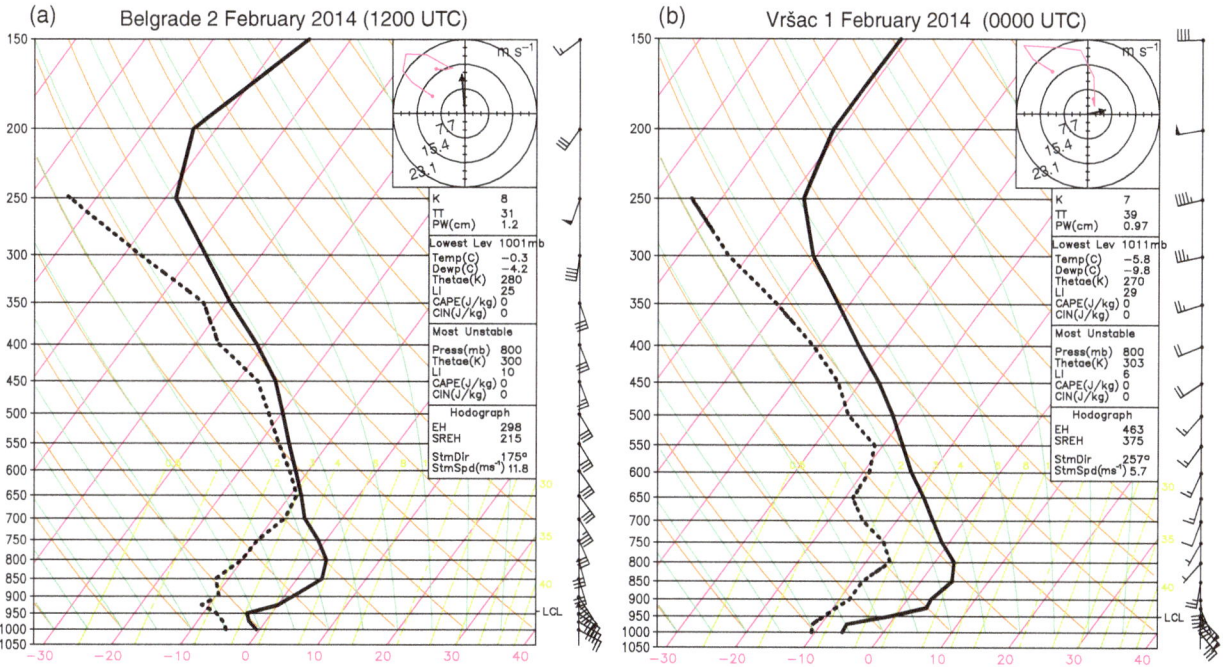

Figure 5. Emagrams for BG (a) and VR (b). Parameters in box as follows: K, K index in °C; TT, total totals index in °C; PW, precipitable water for the entire sounding in centimetres; Temp, temperature on ground in °C; Dewp − dewpoint on ground in °C; Thetae, equivalent potential temperature in K; LI, lifted index in °C; CAPE, convective available potential energy in J kg^{-1}; CIN, convective inhibition in J kg^{-1}; EH, environmental helicity in m^2 s^{-2}; SREH, storm relative environmental helicity in m^{2}s^{-2}; StrmDir, storm direction in degrees; StrmSpd, storm speed in m s^{-1}. For further explanation of the parameters see Doswell III and Schultz (2006).

wind speeds ($\overline{V}$) and wind gusts ($\widehat{V}$). The most accurate wind direction forecasts were at the VR station. A constant 20–25° offset between the forecasted (~130°) and the observed (~110°) wind directions at the VG and NS stations are evident. The forecasts at BG were the least accurate, probably due to the complex urban environment around the BG station. The higher accuracy of wind direction forecasts compared with wind speed forecasts is a result of the strong pressure gradients that drove Koshava unidirectionally. The discrepancies between modelled and observed wind directions are most likely caused by poor resolution of small-scale orography and lack of proper representation of urban environments in the models.

The models were not fully capable of capturing high fluctuations of the observed wind speeds and gusts. The most precise $\overline{V}$ forecasts were at the BG and NS stations, i.e. the stations that are further downstream from the mountain regions of central and eastern Serbia. A general tendency of NMM_QNSE to predict higher $\overline{V}$ and $\widehat{V}$ values than the other four models is noticeable. Small variations between NMM and NMMB simulations indicate that initial and boundary conditions, as well as the grid resolution, had little impact on the forecasts' accuracy. The largest discrepancies between modelled and observed values of $\overline{V}$ are found at the VR station, where the observed $\overline{V}$s were approximately two times higher than the modelled values in the first 90 h of simulation. The largest deviations between NMM_QNSE's outputs and forecasts of

the other four models were in cities (the BG and NS stations). NMM produced the most accurate forecasts of $\overline{V}$ at the VG station. The fluctuations of mean wind speed were best modelled at the BG and VR stations (high CC values). The superiority of the QNSE scheme for numerical modelling of stable conditions, however, has not been confirmed in this study. It can be concluded that numerical modelling of stable PBLs is not at a satisfactory level and more research is required.

The wind gusts are not directly calculated by WRF, but computed in post-processing. The Unified Post Processor (UPP; version 2.2) is used to post-process NMM and NMMB outputs. UPP computes surface wind gusts ($\widehat{V}$) by mixing down momentum from the PBL height (z_{PBL}) level (Mankin, 2015):

$$\widehat{V} = V_{sfc} + \Delta V$$

where

$$\Delta V = \left(V_{PBL} - V_{sfc}\right)\Delta\left(1 - \min\left(0.5, \frac{z_{PBL}}{2000}\right)\right).$$

Here, V_{PBL} is the wind speed at z_{PBL} and V_{sfc} is the surface wind speed. The assumption behind this method is that the gusts are caused by air parcels brought down from the top of the PBL by turbulent eddies. The reliability of calculated gusts, therefore, depends on the accuracy of calculated turbulent kinetic energy and other PBL features. Applicability of the method is particularly questionable for the stable PBL (hence small turbulent eddies) and extremely large helicities, both observed during the EKE. However, analysis of

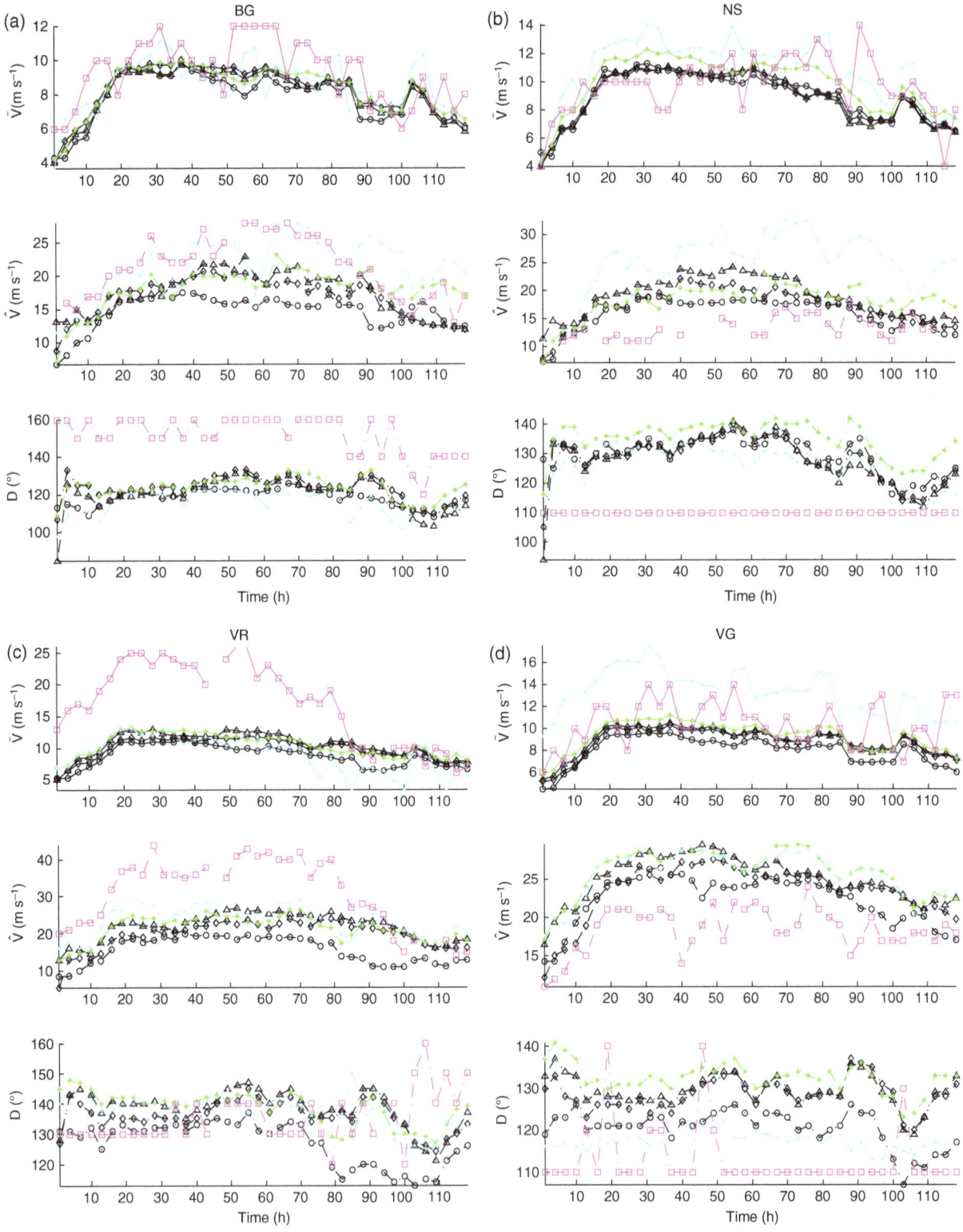

Figure 6. Comparison of simulated mean hourly wind speed (V), gust (V), and wind direction (D) time series against observations (magneta lines with squares) the BG, NS, VR, and VG stations. The utilized models are summarized in Table 1: NMM (green lines with stars), NMMB_1 (blue lines with triangles), NMMB_2 (red lines with diamonds), NMM_QNSE (cyan lines with dots) and IFS (black lines with circles). See Table 2 for for the verification statistics.

different methods for calculations of wind gusts is beyond the scope of this study (c.f. Sheridan, 2011).

RMSDs of estimated gusts are as high as $12.9 \, \mathrm{m \, s^{-1}}$ (NMM_QNSE's forecasts at the VR stations). The discrepancies between observed and simulated gusts are smallest at the BG station were NMM_QNSE captured gusts fairly well. In other cases, gusts were largely either over-predicted (NS and VG) or under-predicted (VR). The presented results demonstrate that more research is needed in the field of wind gust modelling.

Table 2. Bias (i.e. absolute difference), mean absolute difference (MAD), root mean square difference (RMSD) and correlation coefficient (CC) between forecasts and observations during the EKE. The corresponding time series are portrayed in Figure 6.

		NMM			NMMB_I			NMMB_2			NMM_QNSE			IFS		
		$\overline{V}$	$\hat{V}$	D	$\overline{V}$	$\hat{V}$	D	$\overline{V}$	$\hat{V}$	D	$\overline{V}$	$\hat{V}$	D	$\overline{V}$	$\hat{V}$	D
BG	Bias	−1	−3	−28	−1.3	−4.1	−32	−1	−4.5	−28	−0.8	0.2	−37	−1.5	−6.5	−34
	MAD	1.4	3.8	28	1.6	4.1	32	1.4	4.5	28	1.5	3.2	37	1.7	6.6	34
	RMSD	1.7	4.5	30	1.8	4.7	34	1.6	5.1	30	1.8	4.1	38	2	7.3	35
	CC	0.7	0.7	NaN	0.7	0.9	NaN	0.7	0.9	NaN	0.6	0.5	NaN	0.7	0.6	NaN
NS	Bias	0.1	7.8	25	−0.8	5.5	18	−0.8	4.1	18	1.2	12.1	16	−0.8	2.9	19
	MAD	1.5	7.8	25	1.7	5.5	18	1.6	4.2	18	1.8	12.1	16	1.6	3	19
	RMSD	1.9	8.3	26	2.2	6.6	20	2.1	4.9	17	2.3	12.9	17	2.1	3.9	20
	CC	0.5	0.4	NaN	0.5	0.1	NaN	0.5	0.2	NaN	0.6	0.5	NaN	0.5	0.2	NaN
VR	Bias	−6.9	−9.8	3	−6.8	−9.2	3	−7.4	−11.6	1	−8.2	−7.4	2	−8.5	−15.8	−8
	MAD	7.3	10.8	10	7.2	10.1	10	7.7	12.1	8	8.5	8.9	10	8.7	15.8	10
	RMSD	8.5	12	12	8.4	11.5	13	9	13.6	11	9.3	10.1	12	9.9	17	15
	CC	0.8	0.7	NaN	0.8	0.8	NaN	0.7	0.7	NaN	0.8	0.7	NaN	0.8	0.9	NaN
VG	Bias	−1.1	7.5	20	−1.5	6.6	17	−1.6	5	15	2.8	7.2	4	−2.4	3.9	7
	MAD	1.6	7.5	21	1.8	6.6	18	1.9	5	17	3.2	7.2	8	2.6	4	10
	RMSD	2.2	7.9	21	2.4	7.1	19	2.5	5.6	18	3.6	7.6	9	3.1	4.6	12
	CC	0.4	0.6	NaN	0.5	0.6	NaN	0.6	0.7	NaN	0.4	0.6	0.1	0.4	0.7	NaN

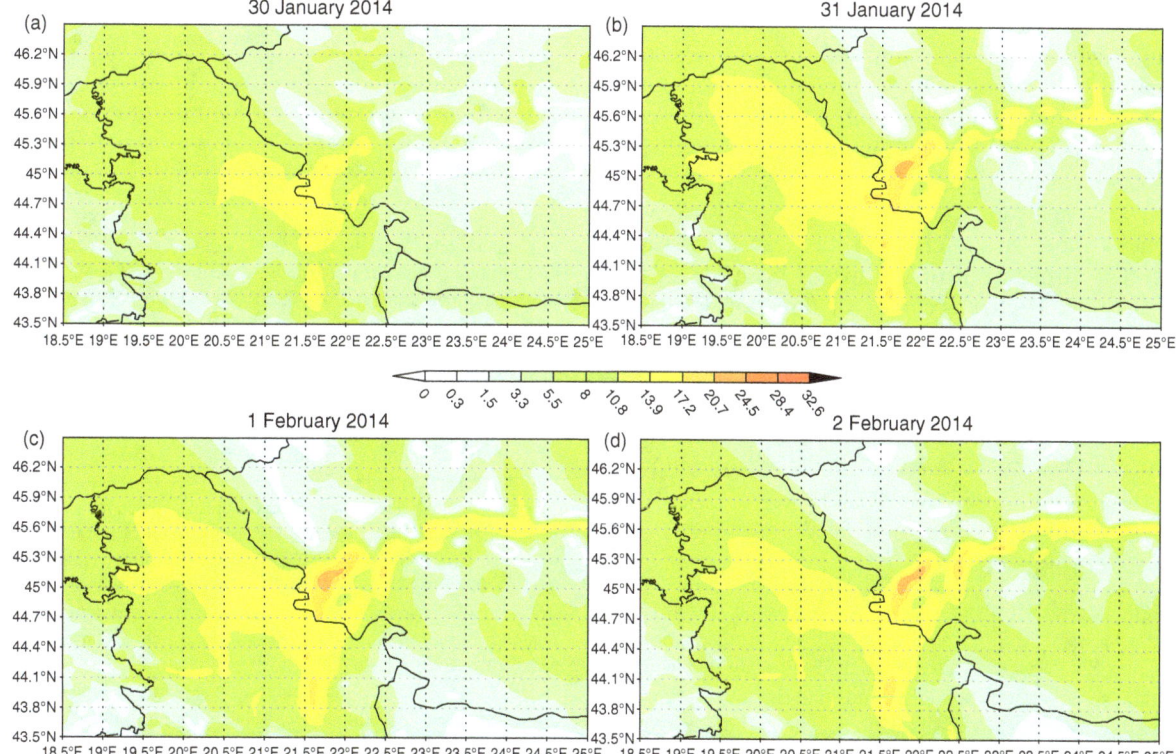

Figure 7. NMM forecasts of the mean daily wind speeds during the EKE. Model start at 30 January (00 h UTC).

Table 2 further shows that the NMM forecast of $\overline{V}$ have the smallest RMSDs. Since RMSD is the absolute measure of the model's accuracy, the model with the lowest RMSD is the most reliable model for forecasts of extreme Koshava speeds. Figure 7 is spatial distribution of the mean daily wind speeds in the Koshava region based on the NMM simulations. The strongest Koshava speeds ($>17\,\text{m s}^{-1}$) occurred in east Serbia (the VR region) and west Romania (the region around Caransebes). High wind speeds were also found in the Morava River basin ($14–21\,\text{m s}^{-1}$). The high intensity of the Koshava wind in the above-mentioned areas is due to the wind channelling effect created by the local orography (Romanić *et al.*, 2015b). Strong winds were also blowing along the gorges in the South Carpathians. Koshava weakened moving downstream from these regions, but strong winds were still present in around BG and NS. Extremely high Koshava speeds were occurring from 31 January to 2 February, when the mean daily wind speeds above the whole Pannonian Plane were greater than $10\,\text{m s}^{-1}$. Figure 7, however, should be interpreted cautiously as the RMSDs in

Table 2 demonstrate that NMM greatly under-predicted $\overline{V}$s in the VR region.

4. Conclusions

This article is a case study of an EKE from 30 January to 4 February, 2014. The mean daily Koshava speeds during the EKE were generally above $10\,\mathrm{m\,s}^{-1}$ with gusts exceeding $30\,\mathrm{m\,s}^{-1}$. Similar Koshava events had not occurred during the past 25 years. Meteorological factors leading to this EKE were: (1) large pressure gradients across the Balkan Peninsula caused by an extremely deep Eurasian anticyclone with the central pressure above 1055 mb combined with (2) large temperature gradients between the anticyclone and cyclone regions – all of which resulted in (3) extremely large across-mountain pressure and potential temperature differences between the Wallachia Valley and the Koshava region. The extreme wind speeds occurred together with deep snowdrifts, thus causing a blizzard. The atmosphere in the Koshava layer was either stable with surface inversion or adiabatic with capping elevated inversion.

Four numerical simulations of the EKE were performed (two using WRF-NMM and two using NMMB) and benchmarked against the observations and the global IFS' analysis. The models captured Koshava's directions more accurately than its mean speeds and gusts. The strongest Koshava speeds occurred in the Vršac region. Higher accuracy of direction forecasts is due to the strong pressure gradients that drove the wind. The low accuracy of modelled speeds and gusts is probably due to the lack of planetary boundary layer schemes capable of properly replicating stable stratifications. More research is needed in the field of numerical modelling of stable boundary layers and wind gusts.

Acknowledgements

The authors whish to thank Dan Parvu, Karishma Hosein, and Jameera Mohamed for their technical help and proof-reading of the manuscript. We would also like to thank Predrag Petrović for providing some of the data for this study and Bojan Kašić for his help regarding the NMMB simulations. The authors acknowledge the two reviewers; their suggestions greatly improved the manuscript.

Supporting information

The following supporting information is available:

Animation S1

References

Beare RJ, Macvean MK, Holtslag AAM, Cuxart J, Esau I, Golaz J-C, Jimenez MA, Khairoutdinov M, Kosovic B, Lewellen D, Lund TS, Lundquist JK, Mccabe A, Moene AF, Noh Y, Raasch S, Sullivan P. 2006. An Intercomparison of Large-Eddy Simulations of the Stable Boundary Layer. *Boundary-Layer Meteorology* **118**(2): 247–272, doi: 10.1007/s10546-004-2820-6.

Doswell CA III, Schultz DM. 2006. On the use of indices and parameters in forecasting severe storms. *E-Journal of Severe Storms Meteorology* **1**(3): 1–22.

Ek MB, Mitchell KE, Lin Y, Rogers E, Grunmann P, Koren V, Gayno G, Tarpley JD. 2003. Implementation of Noah land surface model advances in the National Centers for Environmental Prediction operational mesoscale Eta model. *Journal of Geophysical Research, [Atmospheres]* **108**(D22): 8851, doi: 10.1029/2002JD003296.

Ferrier BS, Springs C, Jin Y, Lin Y, Black T, Rogers E, DiMego G. 2002. Implementation of a new grid-scale cloud and precipitation scheme in the NCEP Eta model. 15th Conference on Numerical Weather Prediction. American Meteorological Society: San Antonio, TX, 280–283.

Iacono MJ, Delamere JS, Mlawer EJ, Shephard MW, Clough SA, Collins WD. 2008. Radiative forcing by long-lived greenhouse gases: calculations with the AER radiative transfer models. *Journal of Geophysical Research, [Atmospheres]* **113**(D13): D13103, doi: 10.1029/2008JD009944.

Janjić ZI. 1994. The Step-Mountain Eta Coordinate Model: further developments of the Convection, Viscous Sublayer, and Turbulence Closure Schemes. *Monthly Weather Review* **122**(5): 927–945, doi: 10.1175/1520-0493(1994)122<0927:TSMECM>2.0.CO;2.

Lacis AA, Hansen J. 1974. A Parameterization for the Absorption of Solar Radiation in the Earth's Atmosphere. *Journal of the Atmospheric Sciences* **31**(1): 118–133, doi: 10.1175/1520-0469(1974)031<0118:APFTAO>2.0.CO;2.

Mankin G. 2015. *The Unified Post Processor: Subprogram CALGUST. en.* Developmental Testbed Center: Boulder, CO.

Niu G-Y, Yang Z-L, Mitchell KE, Chen F, Ek MB, Barlage M, Kumar A, Manning K, Niyogi D, Rosero E, Tewari M, Xia Y. 2011. The community Noah land surface model with multiparameterization options (Noah-MP): 1. Model description and evaluation with local-scale measurements. *Journal of Geophysical Research: Atmospheres* **116**(D12): D12109, doi: 10.1029/2010JD015139.

Romanić D, Ćurić M, Jovičić I, Lompar M. 2015a. Long-term trends of the "Koshava" wind during the period 1949–2010. *International Journal of Climatology* **35**(2): 288–302, doi: 10.1002/joc.3981.

Romanić D, Ćurić M, Lompar M, Jovičić I. 2015b. Contributing factors to Koshava wind characteristics. *International Journal of Climatology*, doi: 10.1002/joc.4397.

Schwarzkopf MD, Fels SB. 1991. The simplified exchange method revisited: An accurate, rapid method for computation of infrared cooling rates and fluxes. *Journal of Geophysical Research, [Atmospheres]* **96**(D5): 9075–9096, doi: 10.1029/89JD01598.

Sheridan P. 2011. *Review of Techniques and Research for Gust Forecasting and Parameterisation.* Forecasting Research Technical Report. Met Office: Exeter, UK; 21.

Shin HH, Hong S-Y. 2011. Intercomparison of planetary boundary-layer parametrizations in the WRF model for a single day from CASES-99. *Boundary-Layer Meteorology* **139**(2): 261–281, doi: 10.1007/s10546-010-9583-z.

Sukoriansky S, Galperin B, Perov V. 2005. Application of a new spectral theory of stably stratified turbulence to the atmospheric boundary layer over sea ice. *Boundary-Layer Meteorology* **117**(2): 231–257, doi: 10.1007/s10546-004-6848-4.

Talbot C, Bou-Zeid E, Smith J. 2012. Nested mesoscale large-eddy simulations with WRF: performance in real test cases. *Journal of Hydrometeorology* **13**(5): 1421–1441, doi: 10.1175/JHM-D-11-048.1.

Unkašević M, Mališic J, Tošić I. 1999. Some aspects of the wind "Koshava" in the lower troposphere over Belgrade. *Meteorological Applications* **6**(1): 69–79, doi: 10.1017/S1350482799000997.

Vukmirović D. 1985. The spatial structure of the koshava wind. Paper presented at the 11-th International Conference for Alpine meteorology 14–16 September, 1983. Hungarian Meteorological Service: Budapest, 10–15.

16

A grid refinement study of trade wind cumuli simulated by a Lagrangian cloud microphysical model: The super-droplet method

Yousuke Sato,[1,2*†] Shin-ichiro Shima[2,1] and Hirofumi Tomita[1]

[1] RIKEN Advanced Institute for Computational Science, Kobe, Japan
[2] Graduated School of Simulation Study, University of Hyogo, Kobe, Japan

*Correspondence to:
Y. Sato, RIKEN Advanced
Institute for Computational
Science, 7-1-26,
Minatojima-Minami Machi,
Chuo-ku, Kobe, Hyogo, Japan.
E-mail: yousuke.sato@riken.jp

†Department of Applied Energy,
Graduate School of Engineering,
Nagoya University, Chikusa-ku,
Nagoya, Aichi, Japan. E-mail:
y-sato@energy.nagoya-u.ac.jp

Abstract

The impact of spatial resolution on the simulation of trade wind cumuli was investigated. The super-droplet method, an efficient stochastic Lagrangian cloud microphysical model, was used to reduce uncertainties due to the empirical parameterisation of cloud microphysics and numerical diffusion for advection, which is inevitable in an Eulerian cloud microphysical model. We showed for the first time that cloud cover numerically converged with a grid resolution of 12.5 m. Our grid refinement analysis elucidated a significant contribution of small cumulus clouds to total cloud cover, as such cumuli are generated by small-scale updrafts that can be resolved only at a fine resolution.

Keywords: Lagrangian cloud microphysical model; grid refinement; shallow cumulus

1. Introduction

Shallow clouds are one of the main sources of uncertainty in climate prediction (IPCC AR5; Stocker *et al.*, 2013). General circulation models (GCMs), which have been used for climate prediction usually express clouds by parameterisations (e.g. Tiedtke, 1993; Considine *et al.*, 1997; Kain, 2004), but they cannot simulate shallow clouds explicitly. Cloud-resolving models based on the large-eddy simulation (LES) technique are powerful tools for reproducing small-scale phenomena associated with clouds and have often been used to improve cloud parameterisation (e.g. Golaz *et al.*, 2007; Bretherton and Park, 2009; Kogan, 2013).

Several factors affect the reliability of LES models for providing a reference solution to GCMs: the spatial grid resolution (e.g. Lewellen and Lewellen, 1998; Stevens and Bretherton, 1999; Cheng *et al.*, 2010; Seifert and Heus, 2013; Seifert and Onishi, 2016), the cloud microphysical scheme (e.g. van Zanten *et al.*, 2011; Sato *et al.*, 2015), the grid aspect ratio (e.g. Nishizawa *et al.*, 2015; Pedersen *et al.*, 2016), and the domain size (e.g. Stevens *et al.*, 2002; de Roode *et al.*, 2004).

Among these issues, the sensitivity to resolution is a key issue of LES. It is still not clear how small a grid must be to obtain an accurate numerical solution of shallow cumuli. For example, in terms of cloud microphysics, Grabowski and Jarecka (2015) determined that a vertical grid resolution of less than 10 m is required

for correctly simulating supersaturation and condensation growth at the cloud base. As well as the effects of the grid resolution on cloud microphysics, Matheou and Chung (2014) investigated the effects on the turbulent structure within shallow cumuli. They indicated that 90% of the total turbulent kinetic energy (TKE) should be resolved to obtain a numerically converged mean profile of trade wind cumulus and 20-m horizontal grid spacing is required. Brown (1999) indicated that the ensemble-mean statistics in shallow cumulus are sensitive not to grid spacing, but to the size of individual clouds, which is directly related to cloud cover and has a strong dependency on grid resolution. Matheou *et al.* (2011) showed that the grid convergence of the mean liquid water profile was achieved with their finest resolution, namely 20-m grid spacing, for a non-precipitating case. However, they could not confirm the convergence for cloud cover. Because cloud cover is a crucial quantity for radiative transfer calculations in GCMs, it is an important quantity, as a reference solution, to construct parameterisation for GCMs.

The cloud microphysical scheme is also crucial for simulated clouds. Bulk schemes and spectral bin schemes have been widely used in LES models. These models are Eulerian cloud models (ECMs), in which prognostic variables associated with cloud are defined on grid points. This treatment has uncertainty caused by numerical diffusion, which is artificially added or implicitly included for model stability and is inevitable in ECMs. (Henceforth, we refer to this as numerical diffusion).

Recently, several Lagrangian cloud microphysical models (LCMs) have been developed (e.g. Andrejczuk *et al.*, 2008; Shima *et al.*, 2009; Sölch and Kärcher, 2010; Riechelmann *et al.*, 2012). Because LCMs explicitly solve the microscopic time evolution equations of particles, they can reduce the uncertainties derived from empirical parameterisations and assumptions related to several cloud microphysical processes (e.g. autoconversion, accretion, saturation adjustment and the size-distribution function of cloud particles). Furthermore, the cloud microphysical variables in LCMs are defined not as Eulerian grid variables but as Lagrangian particles over the calculation space. Due to this particle-based representation, LCMs can also reduce the error derived from the numerical diffusion of cloud microphysics.

To understand the influence of the grid spacing of LES, the grid refinement approach using LCMs is useful because it can reduce uncertainties in cloud microphysics. Arabas and Shima (2013) investigated the influence of LES resolution on shallow trade wind cumuli using the super-droplet method (SDM) (Shima *et al.*, 2009), which is a computationally efficient stochastic LCM. They used a 25-m grid spacing, but the grid convergence on the cloud microphysical properties could not be fully confirmed.

Considering the background information presented above, the aim of this study was to clarify the grid spacing required to produce a well-converged numerical solution of shallow cumuli using an LES model coupled with the SDM. We systematically investigated the influence of resolution on shallow non-precipitating trade wind cumuli, especially their cloud cover, through the grid refinement approach.

2. Model and experimental setup

The LES model used in this study was constructed from the Scalable Computing for Advanced Library and Environment (SCALE) library (Nishizawa *et al.*, 2015; Sato *et al.*, 2015). The SDM (Shima *et al.*, 2009), an LCM using a Monte Carlo scheme for the stochastic coalescence process, was implemented into SCALE (an Eulerian dynamical core). In the SDM, aerosol/cloud/precipitation particles were represented by Lagrangian particles called super-droplets (SDs). In this study, gravitational settling, activation/deactivation, condensation/evaporation and coalescence processes were considered. Nucleation scavenging and rain out of aerosol particles were considered, i.e. the aerosol number concentration changed over time. The influence of sub-grid-scale (SGS) turbulence on the SDs was ignored.

The experimental setup was based on an intercomparison study called the Barbados Oceanographic and Meteorological Experiment (BOMEX) (Siebesma *et al.*, 2003). The domain size of the model was 6.4

Table 1. Spatial resolution, time step of the dynamics and time step of the microphysics used for the sensitivity experiment at each resolution.

Name	Ex6	Ex12	Ex25	Ex50	Ex100
$\Delta x = \Delta y$ (m)[a]	6.25	12.5	25	50	100
Δz (m) [a]	5	10	20	40	80
Δt (dynamics)[b] (s)	0.008	0.0125	0.025	0.05	0.1
Δt (microphysics)[c] (s)	0.08	0.1	0.1	0.1	0.1
Δt (movement of SD) (s)	0.08	0.0125	0.025	0.05	0.1
Total SD at initial (#)	2.1×10^{10}	2.6×10^{9}	3.3×10^{8}	4.0×10^{7}	5.1×10^{6}
(SD number per one grid)	(30)	(30)	(30)	(30)	(30)

[a]The grid aspect ratio of horizontal to vertical grid spacing was kept at 0.8 because a large aspect ratio results in artificial error in the turbulence structure (Nishizawa *et al.*, 2015).
[b]The time step of the dynamics was determined by the Courant–Friedrichs–Lewy (CFL) condition for an acoustic wave.
[c]Microphysics includes aerosol activation/deactivation, condensation, evaporation and coalescence processes.

$\times$ 7.2 $\times$ 3.5 km^3, which was almost the same as that used by Siebesma *et al.*, (2003). To investigate the effects of resolution on cloud microphysical properties and turbulence structure, we conducted sensitivity experiments for spatial resolution. Both horizontal and vertical grid resolution were varied without changing the grid aspect ratio. A summary of the resolution and time step of each experiment is shown in Table 1. To maintain model stability, numerical diffusion with a fourth-order Laplacian operator was added to the Eulerian prognostic variables (density, three-dimensional momentum, density-weighted potential temperature and vapour mixing ratio). The diffusion coefficient was given non-dimensionally as $\gamma = 1 \times 10^{-3}$, based on a sensitivity experiment of shallow cumuli (Sato *et al.*, 2015). For all cases, the number concentration of SDs was fixed as 30 per grid box at the initial time, which was sufficient to simulate cumuli using the SDM (Arabas and Shima, 2013). The total number of SDs used is also summarised in Table 1. The aerosol size distribution at the initial time was set by another intercomparison study, which targeted shallow cumuli (van Zanten *et al.*, 2011).

3. Results and discussion

Figure 1 shows the temporal evolution of cloud cover and the liquid water path (LWP) averaged over the whole calculation domain, together with the results of a previous intercomparison study (Siebesma *et al.*, 2003). Figure 1 also shows the LWP averaged over whole cloudy grid. Cloud cover was defined as the proportion of cloudy columns in the total column that had more than one grid with a cloud water mixing ratio greater than 0.01 g kg^{-1}. Cloud cover with a coarse

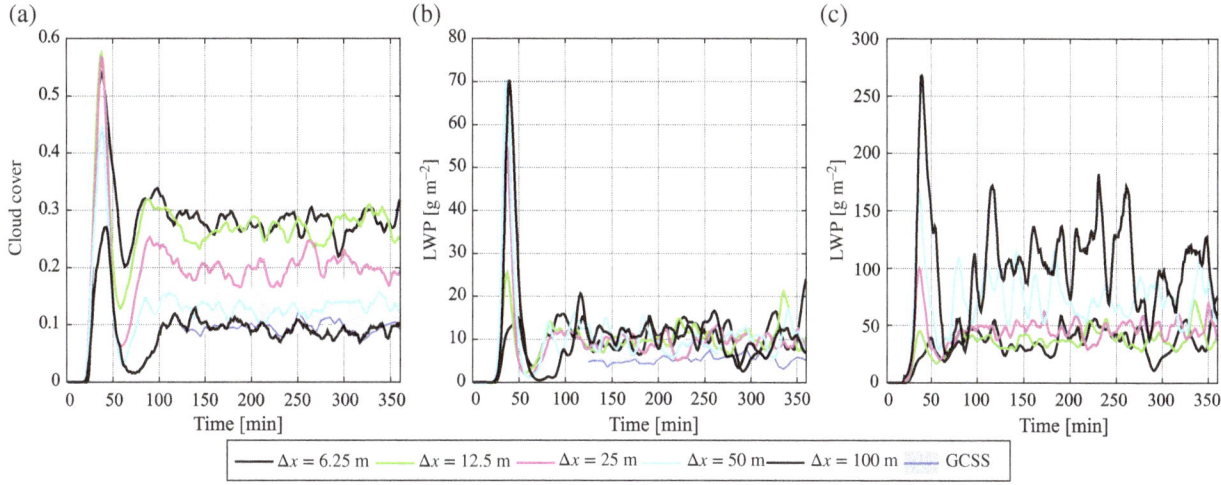

Figure 1. Temporal evolution of (a) cloud cover and (b) the liquid water path (LWP) averaged over the whole calculation domain and (c) the LWP averaged over the whole cloudy grid. The red, green, pink, sky blue and black lines indicate the results with $\Delta x = 6.25$, 12.5, 25, 50 and 100 m, respectively. The grey shade and blue line indicate the range between the maximum and minimum, and the median of the results of the previous intercomparison study conducted by the global energy and water cycle exchanges (GEWEX) cloud system studies (GCSS) (Siebesma *et al.*, 2003).

grid resolution ($\Delta x = 50$ and 100 m), which was close to that of the previous study, was mostly within the range of the previous study. However, the cloud cover increased as the resolution increased. This dependency of cloud cover on resolution was similar to that reported in a previous study targeting shallow cumulus (Matheou *et al.*, 2011) using an ECM. The authors reported that, for the non-precipitating case, the convergence of cloud cover was not achieved with their finest resolution although LWP was converged. The results of the present study, with the finest resolution of 6.25 m, indicate that cloud cover was also converged for the 12.5 m resolution. In contrast to cloud cover, resolution had a small effect on LWP (Figure 1(b)), even though LWP averaged over whole cloudy grids has dependency on the resolution (Figure 1(c)). This indicates that the total mass of cloud water generated by the convection triggered by the surface flux was the same regardless of resolution, whereas the three-dimensional spatial distribution of liquid water was sensitive to resolution.

Figures 2(a)–(e) shows the spatial distribution of the LWP simulated at each resolution. The cloud patterns gradually changed as the resolution increased. As Brown (1999) indicated, the size of an individual cloud is sensitive to resolution. Figure 3(a) shows the frequency histogram $N(S)$ of the horizontal area S of a cloud using uniform bins on the $\log(S)$ axis. $N(S)$ can be fitted by a power law $S^{-1/2}$, which agrees fairly well with the previously reported fractal properties of fair weather cumuli analysed from Landsat satellite data (Cahalan and Joseph, 1989). From this power law, the contribution of clouds in the size range $\log(S)$ to $\log(S) + \Delta\log(S)$ to cloud cover can be evaluated as $SN(S) \sim S^{1/2}$. Because $S^{1/2}$ is a slowly increasing function of S, our analysis suggests that the largest clouds contribute the most to total cloud cover, but smaller clouds also make a notable contribution. Cloud cover values originated from clouds with a small (large) area averaged over the

last 1 h, whose S was smaller (larger) than 0.1 km^2, were 0.061, 0.063, 0.047, 0.029 and 0.016 (0.220, 0.220, 0.150, 0.109 and 0.0726) for $\Delta x = 6.25$, 12.5, 25, 50 and 100 m, respectively. These results support our argument that there is a notable contribution (about 20%) of small clouds to cloud cover. As well as small clouds, the cloud cover from both small and large clouds numerically converged, with a grid resolution of $\Delta x = 12.5$ m.

The dependency of the spatial distribution of clouds on resolution originated from differences in the structure of turbulence, which was visualised by the vertical velocity field (w) near the surface ($z = 50$ m; Figures 2(f)–(i)). The w field had a striped pattern, with a narrow and strong upward velocity in fine-resolution runs ($\Delta x \le 12.5$ m), which is a characteristic of shear-driven turbulence (Moeng and Sullivan, 1994). On the other hand, a broader hexagonal shape with weaker upward velocity, which is typical in buoyancy-driven turbulence (Moeng and Sullivan, 1994), appeared in coarse-resolution runs ($\Delta x \ge 50$ m). The dependency of the spatial pattern of LWP and w is seen during last 4 h of the simulation (see Videos S1–S10, Supporting information).

These differences in the turbulence structure can be explained through an analysis of the vertical profile of the liquid water mixing ratio (q_l) (Figure 3(b)), vertical flux of total water (Figure 3(c)), and skewness of the vertical velocity of grid-resolved turbulence (w'; Figure 3(d)). With a coarse resolution, a positive value of q_l was reached at the surface, corresponding to surface precipitation. The surface precipitation amounts averaged over the last 4 h were 1.12, 0.35, 0.03, 0.002 and 0.016 mm day^{-1} with $\Delta x = 100$, 50, 25, 12.5 and 6.25 m, respectively. Although the precipitation with $\Delta x = 12.5$ m was smaller than that with $\Delta x = 6.25$ m, the difference (0.014 mm day^{-1}) was small enough that the precipitation amounts could be regarded as

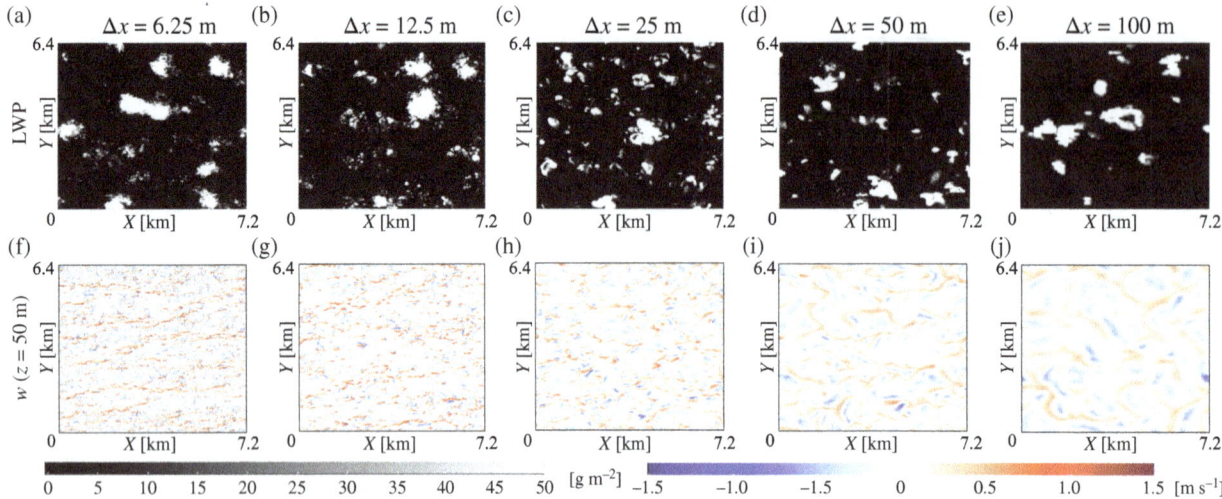

Figure 2. Spatial distribution of (a–e) the liquid water path (LWP) and (f–j) vertical velocity at $z = 50$ m and $t = 14\,400$ s simulated with (a, f) $\Delta x = 6.25$ m, (b, g) $\Delta x = 12.5$ m, (c, h) $\Delta x = 25$ m, (d, i) $\Delta x = 50$ m and (e, j) $\Delta x = 100$ m, respectively.

equal. The surface precipitation with $\Delta x = 12.5$ and 6.25 m was regarded as the non-precipitating. Thus, the surface precipitation gradually decreased as the resolution became finer, and the non-precipitating cumuli of the BOMEX case could be reproduced with $\Delta x = 12.5$ and 6.25 m. The hexagonal structure of w shown in Figures 2(i) and (j) could be produced due to the lack of grid resolution (Piotrowski *et al.*, 2009). The strong upward flux of total water below the cloud layer, shown in Figure 3(c), produced strong precipitation. The strong surface precipitation triggered the mass divergence in the precipitating area near the surface. Consequently, the circulation enhanced the strong convection and precipitation via positive feedback, as suggested by Feingold *et al.* (2010). This circulation enhanced the strong convection and precipitation via positive feedback. The circulation triggered by the surface precipitation resulted in a strong upward flux of total water below the cloud layer, as shown in Figure 3(c). The strong surface precipitation with coarse grid resolution would reduce after implementing the influence of SGS turbulence on the SDs.

On the other hand, the striped pattern of w was constructed by a narrow and strong upward velocity that appeared in the high-resolution runs, which generated small clouds. The skewness of w' clearly shows this distinctive feature of the w field. Although the skewness was included among the results of the previous study, it was found to increase as the resolution became finer. Its numerical convergence was also achieved with $\Delta x = 12.5$ m (Figure 3(d)). These convergences of the vertical velocity field contributed to the convergence of cloud cover. From these results, we can conclude that the fine-resolution simulation ($\Delta x \leq 12.5$ m) accurately captured the detailed structure of the vertical velocity field induced by turbulence. This led to changes in cloud structure and precipitation properties.

Figure 4 shows the turbulence kinetic energy spectra $E(k)$ at three altitudes, $z = 250$, 700 and 1200 m. These spectra were calculated from the velocity profile ($u(x,y)$,

$v(x,y)$, $w(x,y)$) at each altitude. Let $E\left(k_x, k_y\right) = \tilde{u}^2 + \tilde{v}^2 + \tilde{w}^2$, where $\left(\tilde{u}\left(k_x,k_y\right), \tilde{v}\left(k_x,k_y\right), \tilde{w}\left(k_x,k_y\right)\right)$ the Fourier image of the velocity profile. $E(k)$ was obtained by integrating $E(k_x,k_y)$ along a circle of radius $k = \sqrt{k_x^2 + k_y^2}$ and then averaged over the last half-hour of the simulation (i.e. $t = 5.5$ to 6 h) using snapshots taken every minute. The slope of $E(k)$ in the cloud layer is flatter than Kolmogorov's $-5/3$ power law, which is characterised by a constant energy cascade in wave number space (Kolmogorov, 1941a, 1941b). Based on Kolmogorov's theory, this flatter energy spectrum suggests that the energy is not cascading constantly in the wave number space, but there exist input and output of energy at various scales. We can speculate that the cumuli population is maintaining the turbulence because they also have a self-similar size distribution as shown in Figure 3(a). More detailed analyses to understand the mechanism will be conducted in the future.

4. Concluding remarks

In this study, the influence of spatial resolution was investigated through a grid refinement approach using three-dimensional LES simulations. For the cloud microphysics, we used the SDM, a stochastic LCM, to reduce the uncertainty of cloud microphysics and numerical diffusion as much as possible. The results of the sensitivity experiment indicated that a numerical convergence for cloud cover of trade wind cumuli was achieved with a horizontal and vertical grid spacing of 12.5 and 10 m. The detailed structure of turbulence, which can only be resolved with fine spatial resolution, played a notable role in simulating cloud cover.

As suggested by previous studies (Matheou *et al.*, 2011), the grid size required for accurate simulation can differ for precipitating cumulus and stratocumulus. Further investigation targeting the properties of turbulence when using LCMs is required to improve our

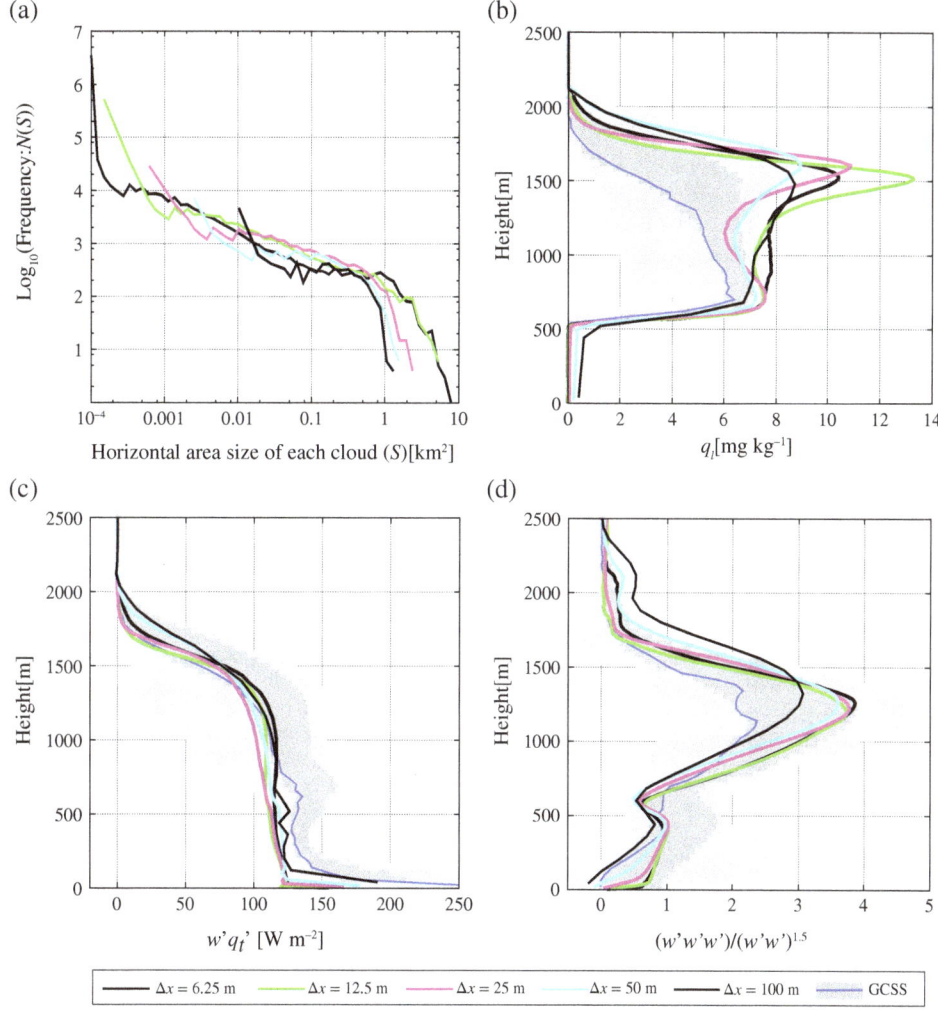

Figure 3. (a) Frequency distribution of the area (S) of each cloud accumulated during the last 4 h (i.e. $t = 2\,h \sim t = 6\,h$) of the calculation, and (b–d) domain averaged vertical profile of the (b) liquid water mixing ratio, (c) mass flux of total water ($w'q_t'$) and (d) skewness of the vertical component of grid-resolved turbulence normalised by the variance averaged during the last 3 h (i.e. $t = 3 \sim 6\,h$) of the calculation. The horizontal axis of (a) was separated into 50 bins, with logarithmically (logarithmic base 10) uniform width. S is defined as $\Delta x \times \Delta y \times$ (the number of grids in each closed interval of cloudy grids). (a) was created by accumulating the results of every minute of the last 4 h of the calculation. The legend is the same as for Figure 2.

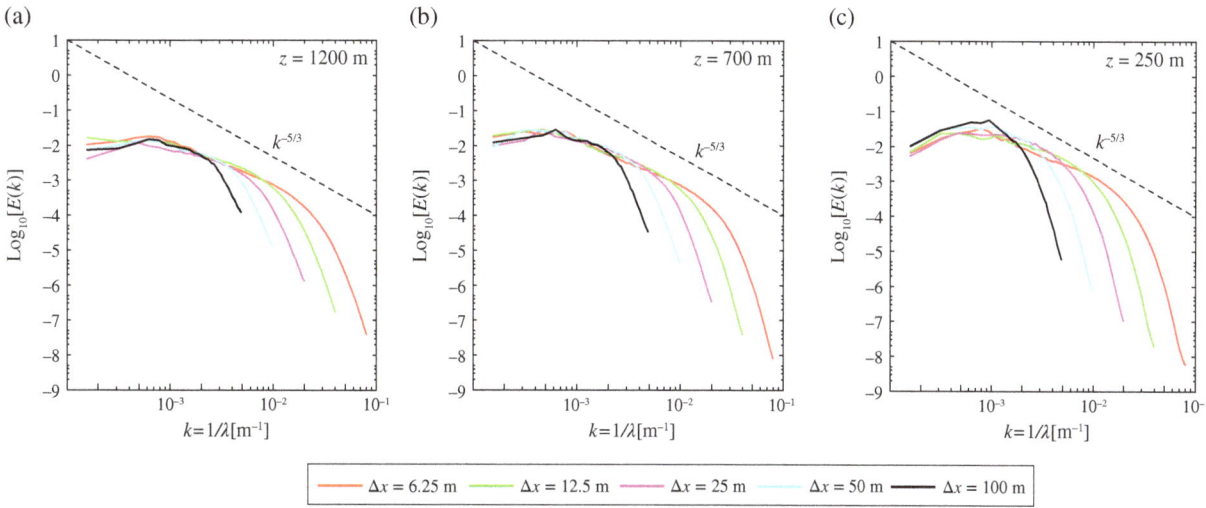

Figure 4. Turbulence kinetic energy spectrum of three-dimensional velocity at a height of (a) $z = 1200\,m$ (middle of the cloud layer), (b) $z = 700\,m$ (lower layer of the cloud), and (c) $z = 250\,m$ (sub-cloud layer) averaged over the last 30 min of the calculation (i.e. from $t = 5.5$ to 6 h). The red, green, pink, sky blue and black lines indicate the results of $\Delta x = 6.25$, 12.5, 25, 50 and 100 m, respectively. The dotted line indicates the linear slopes of a $-5/3$ power law.

understanding of the impact of spatial resolution on the detailed structure of turbulence. These analyses also enable a more comprehensive understanding of the origin of the flatter turbulence energy spectrum (Figure 4). Comparing the grid size sensitivity between LCMs and ECMs is another important issue for future studies. Investigating all these issues should provide useful information to improve the cloud models used in mesoscale and global-scale models.

Acknowledgements

S.S. is grateful to L.-P. Wang for informative discussions. This research used computational resources of the K computer and FX10 of the HPCI system provided by the RIKEN Advanced Institute for Computational Science (AICS) and Kyushu University through the HPCI System Research Project (Project ID: hp140094, hp150153). This research is supported by JSPS KAKENHI Grant-in-Aid for Scientific Research(B): (Proposal number: 26 286 089), the JSPS Grant-in-Aid for Young Scientists (B) (Proposal number: 15 K17766), the Center for Cooperative Work on Computational Science, University of Hyogo, and RIKEN special postdoctoral researcher program. SCALE library was developed by Team-SCALE of RIKEN AICS (http:// scale.aics.riken.jp/). Part of figures were created by tools developed by GFD-Dennou Club (http://www.gfd-dennou.org/index .html.en) and Ncview (http://meteora.ucsd.edu/~pierce/ncview_ home_page.html).

Supporting information

The following supporting information is available:

Video S1. Movie of the spatial distribution of the liquid water path (LWP) from $t = 4$ h to $t = 6$ h simulated with $\Delta x = 6.25$ m.

Video S2. Movie of the spatial distribution of the liquid water path (LWP) from $t = 4$ h to $t = 6$ h simulated with $\Delta x = 12.5$ m.

Video S3. Movie of the spatial distribution of the liquid water path (LWP) from $t = 4$ h to $t = 6$ h simulated with $\Delta x = 25$ m.

Video S4. Movie of the spatial distribution of the liquid water path (LWP) from $t = 4$ h to $t = 6$ h simulated with $\Delta x = 50$ m.

Video S5. Movie of the spatial distribution of the liquid water path (LWP) from $t = 4$ h to $t = 6$ h simulated with $\Delta x = 100$ m.

Video S6. Movie of the spatial distribution of vertical velocity at $z = 50$ m from $t = 4$ h to $t = 6$ h simulated with $\Delta x = 6.25$ m.

Video S7. Movie of the spatial distribution of vertical velocity at $z = 50$ m from $t = 4$ h to $t = 6$ h simulated with $\Delta x = 12.5$ m.

Video S8. Movie of the spatial distribution of vertical velocity at $z = 50$ m from $t = 4$ h to $t = 6$ h simulated with $\Delta x = 25$ m.

Video S9. Movie of the spatial distribution of vertical velocity at $z = 50$ m from $t = 4$ h to $t = 6$ h simulated with $\Delta x = 50$ m.

Video S10. Movie of the spatial distribution of vertical velocity at $z = 50$ m from $t = 4$ h to $t = 6$ h simulated with $\Delta x = 100$ m.

References

Andrejczuk M, Reisner JM, Henson B, Dubey MK, Jeffery CA. 2008. The potential impacts of pollution on a nondrizzling stratus deck: does aerosol number matter more than type? *Journal of Geophysical Research* **113**: D19204. https://doi.org/10.1029/2007JD009445.

Arabas S, Shima S. 2013. Large-eddy simulations of trade wind cumuli using particle-based microphysics with Monte Carlo coalescence. *Journal of the Atmospheric Sciences* **70**: 2768–2777.

Bretherton CS, Park S. 2009. A new moist turbulence parameterization in the community atmosphere model. *Journal of Climate* **22**: 3422–3448.

Brown AR. 1999. The sensitivity of large-eddy simulations of shallow cumulus convection to resolution and subgrid model. *Quarterly Journal of the Royal Meteorological Society* **125**: 469–482.

Cahalan RF, Joseph JH. 1989. Fractal statistics of cloud fields. *Monthly Weather Review* **117**: 261–272.

Cheng A, Xu K-M, Stevens B. 2010. Effects of resolution on the simulation of boundary-layer clouds and the partition of kinetic energy to subgrid scales. *Journal of Advances in Modeling Earth Systems* **2**: 3.

Considine G, Curry JA, Wielicki B. 1997. Modeling cloud fraction and horizontal variability in marine boundary layer clouds. *Journal of Geophysical Research* **102**: 13517–13525.

Feingold G, Koren I, Wang H, Xue H, Brewer WA. 2010. Precipitation-generated oscillations in open cellular cloud fields. *Nature* **466**: 849–852.

Golaz J-C, Larson VE, Hansen JA, Schanen DP, Griffin BM. 2007. Elucidating model inadequacies in a cloud parameterization by use of an ensemble-based calibration framework. *Monthly Weather Review* **135**: 4077–4096.

Grabowski WW, Jarecka D. 2015. Modeling condensation in shallow nonprecipitating convection. *Journal of the Atmospheric Sciences* **72**: 4661–4679.

Kain JS. 2004. The Kain–Fritsch convective parameterization: an update. *Journal of Applied Meteorology* **43**: 170–181.

Kogan Y. 2013. A cumulus cloud microphysics parameterization for cloud-resolving models. *Journal of the Atmospheric Sciences* **70**: 1423–1436.

Kolmogorov A. 1941a. On degeneration (decay) of isotropic turbulence in an incompressible viscous liquid. *Doklady Akademii Nauk SSSR* **31**: 538–540.

Kolmogorov A. 1941b. The local structure of turbulence in incompressible viscous fluid for very large Reynolds' numbers. *Doklady Akademii Nauk SSSR* **30**: 301–305.

Lewellen DC, Lewellen WS. 1998. Large-eddy boundary layer entrainment. *Journal of the Atmospheric Sciences* **55**: 2645–2665.

Matheou G, Chung D. 2014. Large-eddy simulation of stratified turbulence. Part II: application of the stretched-vortex model to the atmospheric boundary layer. *Journal of the Atmospheric Sciences* **71**: 4439–4460.

Matheou G, Chung D, Nuijens L, Stevens B, Teixeira J. 2011. On the fidelity of large-eddy simulation of shallow precipitating cumulus convection. *Monthly Weather Review* **139**: 2918–2939.

Moeng C-H, Sullivan PP. 1994. A comparison of shear- and buoyancy-driven planetary boundary layer flows. *Journal of the Atmospheric Sciences* **51**: 999–1022.

Nishizawa S, Yashiro H, Sato Y, Miyamoto Y, Tomita H. 2015. Influence of grid aspect ratio on planetary boundary layer turbulence in large-eddy simulations. *Geoscientific Model Development* **8**: 3393–3419.

Pedersen JG, Malinowski SP, Grabowski WW. 2016. Resolution and domain-size sensitivity in implicit large-eddy simulation of the stratocumulus-topped boundary layer. *Journal of Advances in Modeling Earth Systems* **8**: 885–903.

Piotrowski ZP, Smolarkiewicz PK, Malinowski SP, Wyszogrodzki AA. 2009. On numerical realizability of thermal convection. *Journal of Computational Physics* **228**: 6268–6290. https://doi.org/10. 1016/j.jcp.2009.05.023.

Riechelmann T, Noh Y, Raasch S. 2012. A new method for large-eddy simulations of clouds with Lagrangian droplets including the effects of turbulent collision. *New Journal of Physics* **14**: 65008. http:// iopscience.iop.org/article/10.1088/1367-2630/14/6/065008/meta.

de Roode SR, Duynkerke PG, Jonker HJJ. 2004. Large-eddy simulation: how large is large enough? *Journal of the Atmospheric Sciences* **61**: 403–421.

Sato Y, Nishizawa S, Yashiro H, Miyamoto Y, Kajikawa Y, Tomita H. 2015. Impacts of cloud microphysics on trade wind cumulus: which cloud microphysics processes contribute to the diversity in a large eddy simulation? *Progress in Earth and Planetary Science* **2**: 23. https://doi.org/10.1186/s40645-015-0053-6.

Seifert A, Heus T. 2013. Large-eddy simulation of organized precipitating trade wind cumulus clouds. *Atmospheric Chemistry and Physics* **13**: 5631–5645.

Seifert A, Onishi R. 2016. Turbulence effects on warm-rain formation in precipitating shallow convection revisited. *Atmospheric Chemistry and Physics* **16**: 12127–12141.

Shima S, Kusano K, Kawano A, Sugiyama T, Kawahara S. 2009. The super-droplet method for the numerical simulation of clouds and precipitation: a particle-based and probabilistic microphysics model coupled with a non-hydrostatic model. *Quarterly Journal of the Royal Meteorological Society* **135**: 1307–1320.

Siebesma AP, Bretherton CS, Brown A, Chlond A, Cuxart J, Duynkerke PG, Jiang H, Khairoutdinov M, Lewellen D, Moeng C-H, Sanchez E, Stevens B, Stevens DE. 2003. A large eddy simulation intercomparison study of shallow cumulus convection. *Journal of the Atmospheric Sciences* **60**: 1201–1219.

Sölch I, Kärcher B. 2010. A large-eddy model for cirrus clouds with explicit aerosol and ice microphysics and Lagrangian ice particle tracking. *Quarterly Journal of the Royal Meteorological Society* **136**: 2074–2093.

Stevens DE, Bretherton CS. 1999. Effects of resolution on the simulation of stratocumulus entrainment. *Quarterly Journal of the Royal Meteorological Society* **125**: 425–439.

Stevens DE, Ackerman AS, Bretherton CS. 2002. Effects of domain size and numerical resolution on the simulation of shallow cumulus convection. *Journal of the Atmospheric Sciences* **59**: 3285–3301.

Stocker TF, Qin D, Plattner G-K, Tignor M, Allen SK, Boschung J, Nauels A, Xia Y, Bex V, Midgley PM. 2013. *IPCC, 2013: Climate Change 2013: The Physical Science Basis*. Cambridge University Press: Cambridge and New York, NY.

Tiedtke M. 1993. Representation of clouds in large-scale models. *Monthly Weather Review* **121**: 3040–3061.

van Zanten MC, Stevens B, Nuijens L, Siebesma AP, Ackerman AS, Burnet F, Cheng A, Couvreux F, Jiang H, Khairoutdinov M, Kogan Y, Lewellen DC, Mechem D, Nakamura K, Noda A, Shipway BJ, Slawinska J, Wang S, Wyszogrodzki A. 2011. Controls on precipitation and cloudiness in simulations of trade-wind cumulus as observed during RICO. *Journal of Advances in Modeling Earth Systems* **3**: M06001.

17

Using highly resolved maximum gust speed as predictor for forest storm damage caused by the high-impact winter storm Lothar in Southwest Germany

Dirk Schindler,* Christopher Jung and Alexander Buchholz

Environmental Meteorology, Albert-Ludwigs-University of Freiburg, Freiburg, Germany

*Correspondence to:
D. Schindler, Environmental
Meteorology,
Albert-Ludwigs-University of
Freiburg, Werthmannstrasse 10,
D-79085 Freiburg, Germany.
E-mail:
dirk.schindler@meteo.uni-freiburg.de*

Abstract

Results from a newly available empirical maximum gust speed model were evaluated for their predictive power of forest storm damage caused by the high-impact winter storm 'Lothar' in the German federal state of Baden-Wuerttemberg. In this state, Lothar was the most severe storm event of the last decades, causing nearly 30 million m³ of damaged timber. By applying a least squares boosting procedure, daily maximum gust speed values measured at 28 meteorological stations were used to empirically model highly resolved (50 × 50 m) near-surface gust speed fields associated with Lothar. Gust speed was modelled using terrain- and roughness-related variables as predictors. The modelled gust speed fields were then used as input to an empirical forest storm damage model. To build the damage model, the machine learning method random forests was applied. Results from this study demonstrate that the empirically modelled maximum gust speed field associated with Lothar was the most important predictor variable for forest storm damage at the landscape scale. Modelled maximum gust speed was nearly twice as important as all other available predictor variables. The medians of classified proportions of storm-damaged timber quasi-linearly increased with increasing maximum gust speed.

Keywords: natural hazard; high-impact winter storm; gust speed model; empirical modelling; random forests

1. Introduction

In the period 1950–2000 windstorms caused 18.6 million m³ of damaged timber in European forests per year (Schelhaas *et al.*, 2003). Not less than 65% of the total damage was caused by high-impact winter storms over Europe in the months from November to January (Gardiner *et al.*, 2010). An exceptional winter storm that caused catastrophic damage in France, Germany, Switzerland and Italy on 26 December 1999 was 'Lothar' (Mayer and Schindler, 2002). In terms of the amount of damaged timber, Lothar was the most damaging storm event over the last decades in the southwestern German federal state of Baden-Wuerttemberg (Jung *et al.*, 2016) and Switzerland (Usbeck, 2015). After the passage of Lothar about 44 million m³ of storm-damaged timber had to be salvaged in these areas and the loss in Baden-Wuerttemberg alone was estimated at 770 million Euros (Kohnle *et al.*, 2003).

The factors that affect storm damage formation in forests are typically divided into weather conditions, orography, human influence, soil conditions and stand conditions (Schindler *et al.*, 2012a) with high-impact gusts being the initial cause for damage occurrence (Mayer, 1987).

In previous studies (Schindler *et al.*, 2009, 2012b; Albrecht *et al.*, 2012, 2013; Pasztor *et al.*, 2015), however, the predictive power of modelled gust speed was low. The coarse spatial resolution (often 1 × 1 km) did not realistically represent the scales of gusts causing damage. In combination with the coarsely resolved orography, the insufficient resolution of gusts often resulted in a weak association with forest storm damage.

In a recent study, Jung *et al.* (2016) demonstrated that empirically modelled gust speed fields (resolution: 50 × 50 m) can be used as useful predictor variable for storm damage in forests. Based on their modelling approach, Jung *et al.* (2016) were able (1) separating endemic damage, i.e. damage resulting at regular intervals from recurring severe winds, from catastrophic damage, which is related to infrequently recurring high-impact storms and (2) estimating endemic storm damage risk.

In the period 1979–2008, the catastrophic storm events 'Vivian'/'Wiebke' (Schüepp *et al.*, 1994) and Lothar accounted for as many as 75% of the total amount of storm-damaged timber in Baden-Wuerttemberg. Owing to the (1) limited availability of useful long-term forest damage data, (2) infrequent recurrence of high-impact storm events and (3) limited knowledge about storm-related gust speed fields, it has not been possible to estimate catastrophic forest damage risk at the landscape scale in Baden-Wuerttemberg by empirical modelling approaches until today. However, in forest management there is need for information on the statistical estimation of the return levels of exceptional damage-causing storm events because forest plans set out management objectives for decades. Thus, the goals of this study are (1) statistically classifying Lothar into the extreme gust climatology in Baden-Wuerttemberg to provide an estimation of the return level of Lothar, (2) empirically modelling a representative near-surface gust speed field associated with Lothar at high-spatial resolution (50×50 m) and (3) building an empirical storm damage model that uses the modelled gust speed field as informative input allowing for a gust speed driven estimation of Lothar-induced forest damage.

2. Data and methods

2.1. Study area and data

The study area (size: 35.752 km^2) was the German federal state Baden-Wuerttemberg (Figure 1). It borders on Switzerland in the south and on France in the west. The orography within the study area is very complex because it includes the flat, broad Rhine Valley as well as the low mountain ranges Black Forest and Swabian Alb. In the study area, the elevation varies between 85 and 1493 m (Jung and Schindler, 2015). According to data available from the CORINE Land Cover (CLC) project with a resolution of 100×100 m (Keil *et al.*, 2005), agricultural land covers 51%, forests cover 38% and artificial surfaces (urban areas, roads) cover 9% of the study area. In 45% of the forests, conifers dominate the tree species composition.

The near-surface gust speed field associated with Lothar was modelled and analysed based on daily maximum gust speed values ($U_{max,emp}$) measured on 26 December 1999 at 28 meteorological stations (DS1) operated by the German Weather Service (DWD). A subset of 18 stations, where $U_{max,emp}$-time series (DS2) were available for the period 1979–2013 (data availability greater than 85%), was used for extreme value analysis. Gust speed data were prepared for further analysis according to the methodology described in Jung and Schindler (2016) including measurement height correction, homogenisation and detrending.

Forest storm damage data, which were available for 15.871 compartments (mean size ~ 20 ha) of the public forest, served as a basis for the analysis of the damage pattern caused by Lothar at the landscape scale. The damage pattern was reconstructed based on annual booking records provided by the Forest Research Institute of Baden-Wuerttemberg as described in Jung *et al.* (2016).

The orographic features relative elevation (Φ) (Jung, 2016), orographic sheltering (τ) (Wilson, 1984; Quine and White, 1998), curvature (φ) and aspect (η) were computed from a digital terrain model with a resolution of 50×50 m using the ArcGIS$^\circledR$ 10.2 software.

For analysis and model building all available data were mapped from their original resolution to a 50×50 m resolution grid.

2.2. Extreme value analysis

The extreme value analysis was an application of the block-maxima method with the time-limited blocks being single years. The empirical cumulative distribution function (CDF$_{emp}$) is defined as: $F(U_{max}) = P(X \le U_{max,emp})$ with X as random variable which denotes an annual gust speed maximum. The $U_{max,emp}$-time series included in DS2 were fitted to seven theoretical cumulative (CDF) extreme value distributions (2- and 3-parameter (P) Frechet, 3-P Generalized Extreme Value, 2-P Gumbel, 5-P Wakeby, 2- and 3-P Weibull).

The goodness-of-fit (GoF) between CDF and CDF$_{emp}$ was evaluated by the Kolmogorov–Smirnov test statistic, Anderson–Darling test statistic and chi-square statistic. For the CDFs, GoF was tested using the Easyfit software (MathWave Technologies, Dnepropetrovsk, Ukraine, 215 version 5.5). From the results of the GoF-evaluation (Table S1, Supporting Information) it was found that the Wakeby distribution is the most appropriate CDF for representing CDF$_{emp}$.

The Wakeby distribution is defined by its quantile function (Houghton, 1978; Jung and Schindler, 2015, 2016; Jung, 2016)

$$U_{max}(F) = \varepsilon + \frac{\alpha}{\beta} \cdot \left[1 - (1-F)^{\beta} \right]$$
$$- \frac{\gamma}{\delta} \cdot \left[1 - (1-F)^{-\delta} \right] \quad (1)$$

where F is the non-exceedance probability and $U_{max}(F)$ is the associated gust speed value. The parameters α, β, γ and δ are distribution parameters and ε is the location parameter.

Station-related annual return level (T) values associated with F were calculated according to

$$T = -\frac{1}{F-1} \quad (2)$$

To represent regional annual T-patterns in the study area, station-related T-values were interpolated between the DS2-stations using the spline tool implemented in Spatial Analyst Toolbox of the ArcGIS 10.2 software.

2.3. Gust speed model building

The empirical gust speed model used as input to the forest storm damage model was built according to the methodology of Jung and Schindler (2016). Here, its application includes (1) predictor selection and building, (2) using DS1 for least squares boosting (LSBoost) according to Friedman (2001) to model maximum gust speed related to the passage of Lothar ($U_{max,mod}$) and (3) mapping $U_{max,mod}$ at every grid point in the study area.

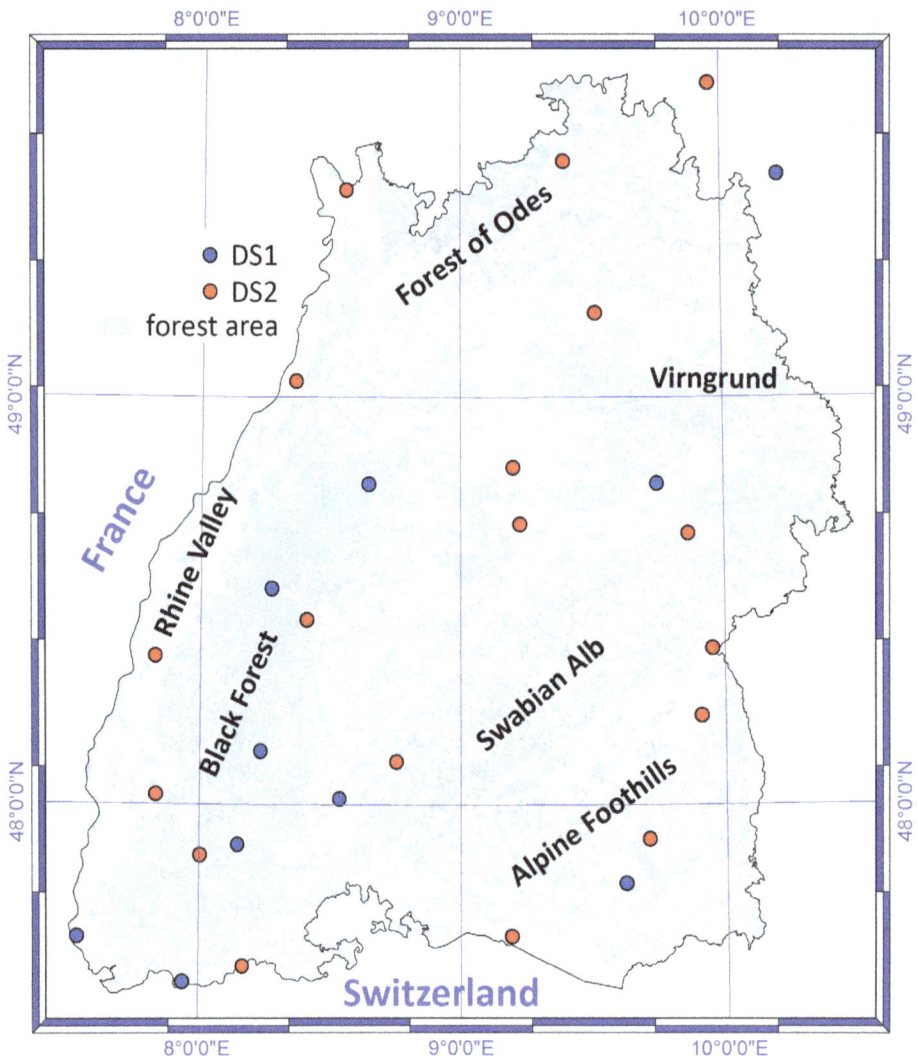

Figure 1. Meteorological stations included in DS1 and DS2 (DS2 ⊆ DS1) as well as forests in the study area.

The predictor variables that were tested for their predictive power to model maximum gust speed related to the passage of Lothar are listed in Table S2. Latitude (λ) and longitude (μ) were used as proxy variables for the synoptic storm field. Roughness length was derived from CLC-data and assigned to land cover types according to Jung and Schindler (2015) yielding the local roughness length ($z_{0,l}$). Roughness length and orographic sheltering were calculated for various sectors and distances around each grid point; Φ was calculated within circles of different radii around each grid point.

After testing for collinearity according to Belsley *et al.* (2001), various combinations of predictor variables were evaluated for their power to predict maximum gust speed related to the passage of Lothar at the DS1-stations. Starting with one predictor variable, further predictor variables were sequentially added to the gust speed model and retained when the model error decreased. During gust speed model building, it turned out that there is no unique best predictor variable combination for maximum gust speed related to the passage of Lothar. Out of 22 meaningful evaluated predictor variable combinations, 6 combinations

(supplied in Table S3) yielded modelling results that represented the measured gust speed data with very high accuracy (coefficient of determination $R^2 > 0.99$). Thus, each of the six $U_{\text{max,mod}}$-fields was used as input for six damage models which form a small damage model ensemble.

2.4. Storm damage model building

For damage modelling, forest compartment-related data on storm-damaged timber (DAM_{emp}) available from Jung *et al.* (2016) were used. The damage data was exclusively related to Lothar, i.e. amounts of endemically damaged timber were subtracted from the total amount of storm-damaged timber. To model DAM_{emp}, the machine learning method random forests (Breiman, 1996) was applied. Owing to the availability of six $U_{\text{max,mod}}$-fields, six damage models were built. Based on the findings of Jung *et al.* (2016), soil type, soil moisture, soil substrate, soil depth, soil acidification, geology, groundwater-affected soils, slope, forest type were also included as predictor variables. Contributions of the predictor variables to the model outputs were

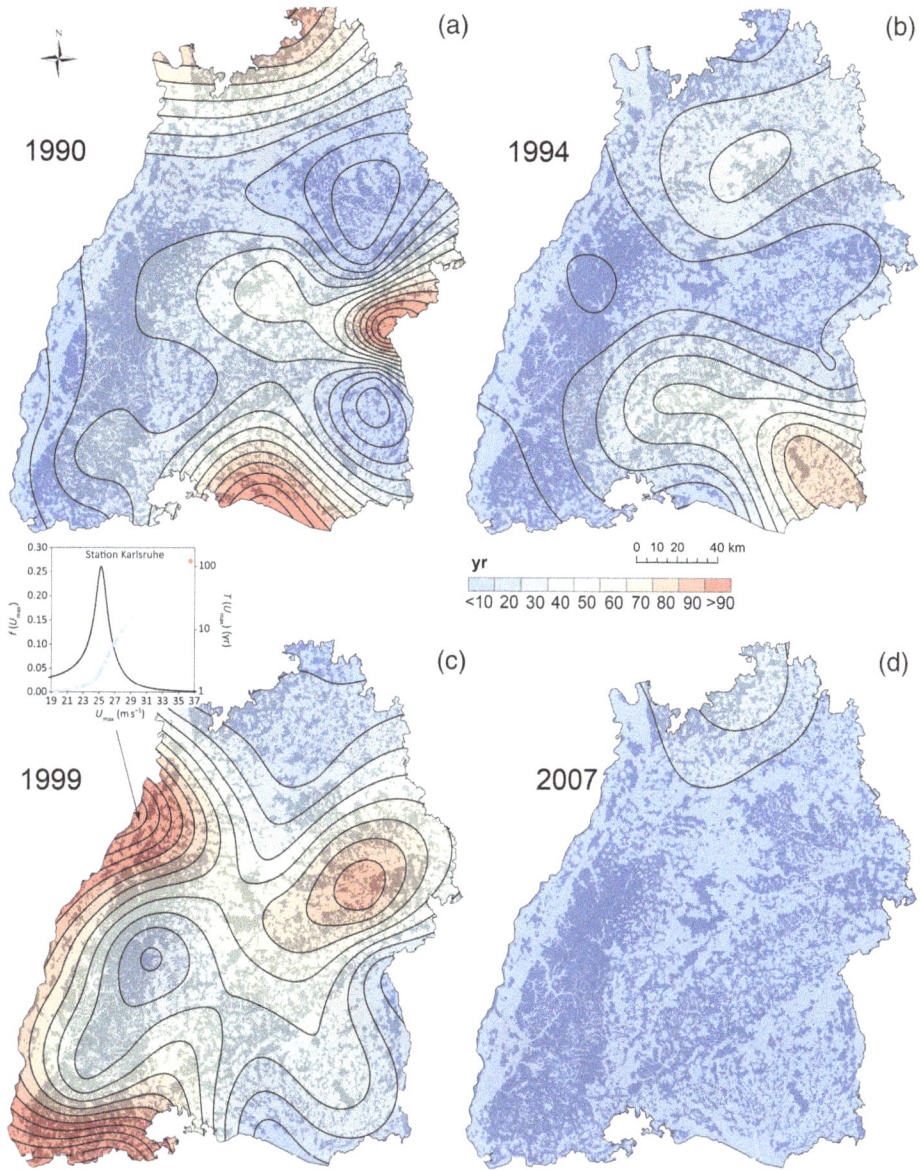

Figure 2. Spatial pattern of T for the years (a) 1990, (b) 1994, (c) 1999 and (d) 2007. The inset of (c) shows the probability of $U_{max}(f(U_{max}))$ and T-values (dots) for station Karlsruhe. The red dot indicates the T-value associated with the year 1999.

evaluated by calculating predictor importance (PI) according to Breiman (1996). The output of the final damage model (DAM_{mod}) was the average of the six individual storm damage models.

3. Results and discussion

3.1. Extreme value analysis

At 5, 2 and 9 of the 18 DS2-stations highest T-values were calculated for 1990, 1994 and 1999, respectively. In these years the winter storms Vivian/Wiebke, 'Lore' (Hofherr and Kunz, 2010) and Lothar, respectively, passed over the study area. Spatial patterns of T-values from these years as well as from the year 2007, in which high-impact storm 'Kyrill' (Fink *et al.*, 2009) passed over Central Europe, are shown in Figure 2. It is obvious that the T-patterns are remarkably different

in the study area. For example, in 1990, maximum T-values occurred in the north (Forest of Odes) and the southeast (parts of Alpine Foothills) whereas in 1999 maximum T-values were calculated in the west (lowlands of the Rhine Valley) and in central eastern parts (Virngrund) of the study area.

In 1999, the range of T-values was 5–135 years. The inset of Figure 2(c) shows the T-values calculated for the DWD-station Karlsruhe which is located in the Rhine Valley. At that station the T-value related to 1999 equals 123 years (red dot).

The storm Kyrill, which occurred in 2007, caused large amounts of damage in central parts of Germany (Klaus *et al.*, 2011). However, it only scraped the study area in the north. Therefore, except for one station, T-values related to 2007 were below 10 years.

The comparison of the T-fields demonstrates the uniqueness of maximum gust speed fields in individual

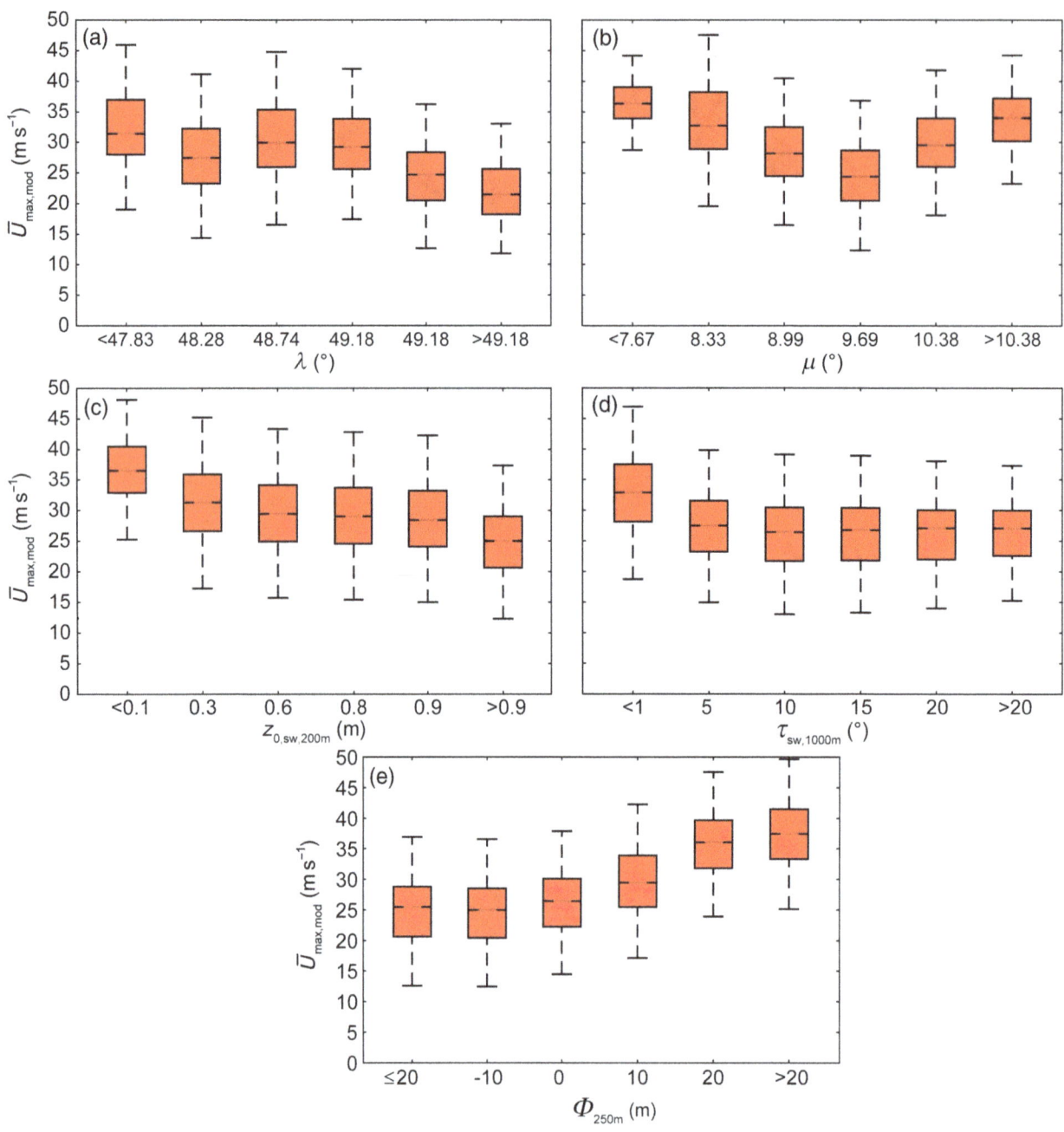

Figure 3. Boxplots of $\bar{U}_{max,mod}$ as a function of (a) latitude (λ), (b) longitude (μ), (c) effective roughness length in the southwest direction up to a distance of 200 m ($z_{0,sw,200m}$), (d) orographic sheltering in the southwest direction up to a distance of 1000 m ($\tau_{sw,1000m}$) and (e) relative elevation within a radius of 250 m (Φ_{250m}) around each grid point.

years during the investigation period. Owing to unique paths and near-surface characteristics, severe storms stroke different parts of the study area. From this it follows that the spatial variability of maximum gust speed fields associated with single severe storm events currently prevents the calculation of catastrophic storm damage risk based on these storm events.

3.2. Gust speed and damage modelling

To present the variation of the six $U_{max,mod}$-fields in the study area, they were averaged yielding the modelled mean maximum gust speed field ($\bar{U}_{max,mod}$). In Figure 3, $\bar{U}_{max,mod}$ is shown as a function of the most

important predictor variables. While no linear relationship can be identified between (1) $\bar{U}_{max,mod}$ and λ, and (2) $\bar{U}_{max,mod}$ and μ, $\bar{U}_{max,mod}$ decreases with increasing effective roughness ($z_{0,sw,200m}$) and increasing orographic sheltering ($\tau_{sw,200m}$) in southwestern direction, which was the main track direction of Lothar. The most pronounced change of $\bar{U}_{max,mod}$ occurred as a function of relative elevation determined within a radius of 250 m around each grid point (Φ_{250m}). During the predictor selection process, it turned out that the absolute elevation (ψ) was not informative for gust speed modelling.

Results from model evaluation showed that about 71% of the modelling error is in the range between −0.3

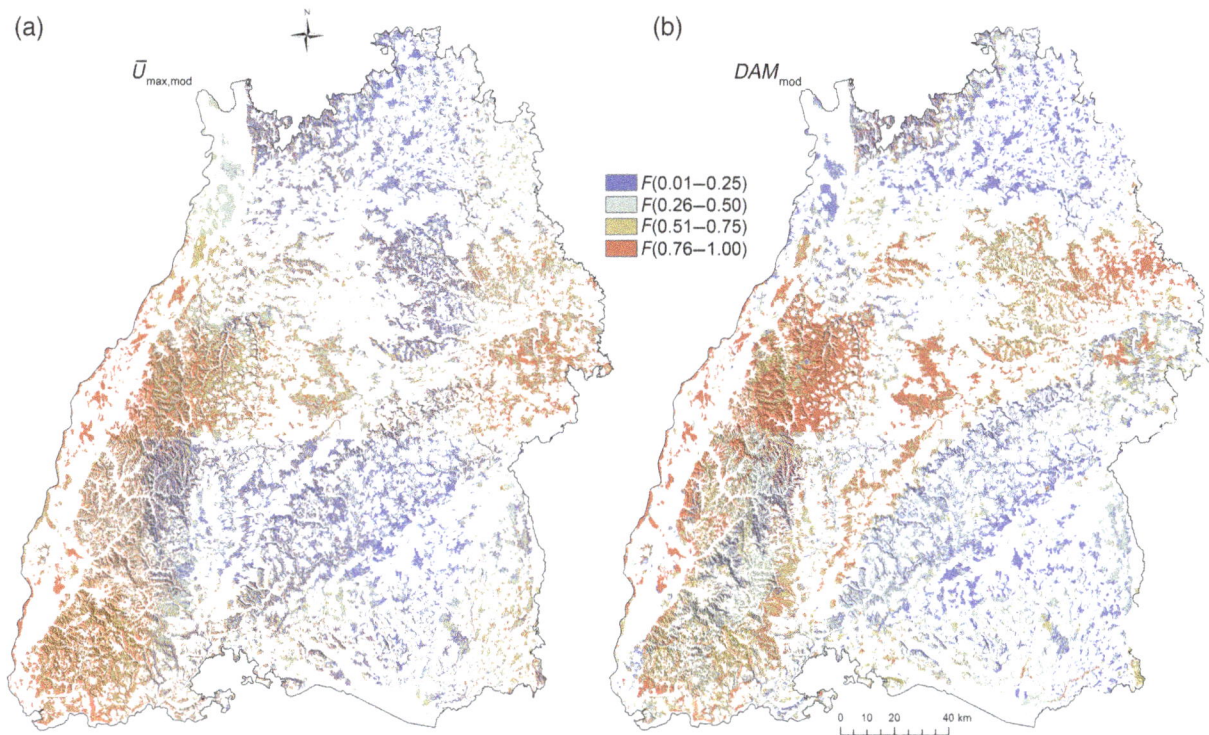

Figure 4. (a) Daily maximum gust speed ($\bar{U}_{\text{max,mod}}$) associated with the passage of storm Lothar, (b) proportions of storm-damaged timber (DAM_{mod}) caused in the forested area by storm Lothar.

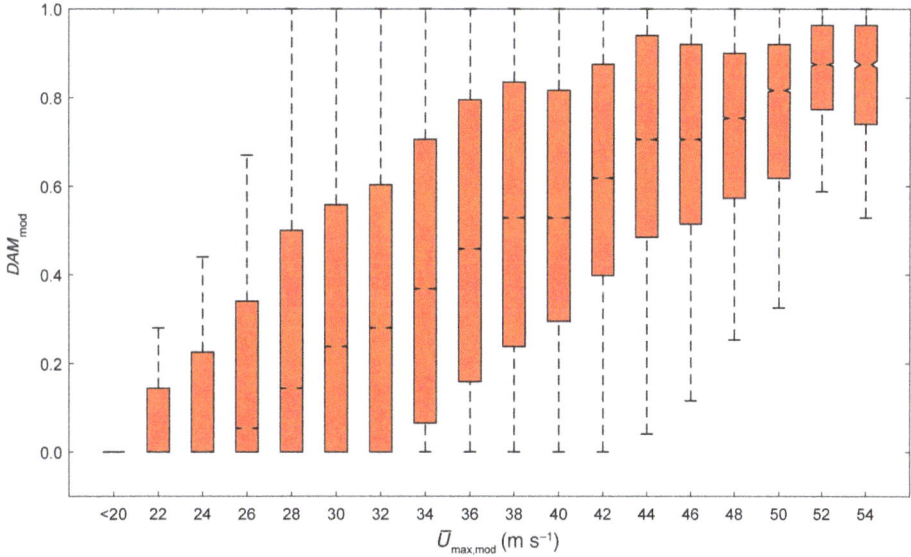

Figure 5. Dependence of DAM_{mod} upon mean daily maximum gust speed $\bar{U}_{\text{max,mod}}$ associated with the passage of storm Lothar.

and $0.3\ \text{m s}^{-1}$. The histogram of the modelling error of $\bar{U}_{\text{max,mod}}$ is shown in Figure S1.

The Lothar-related $\bar{U}_{\text{max,mod}}$-values varied in the range $11\text{–}59\ \text{m s}^{-1}$. The classified $\bar{U}_{\text{max,mod}}$-pattern was closely related to the synoptic storm pattern with highest $\bar{U}_{\text{max,mod}}$-values occurring along the Rhine Valley and in the Virngrund area (Figure 4(a)). The $\bar{U}_{\text{max,mod}}$-mean value within the forested area equalled to $29\ \text{m s}^{-1}$. In 20% of the forested area $\bar{U}_{\text{max,mod}}$ exceeded $35\ \text{m s}^{-1}$, which has been demonstrated to be a critical gust speed at which forest damage is

likely to occur (Schindler *et al.*, 2009, 2012b). Lowest $\bar{U}_{\text{max,mod}}$-values were calculated for the northern parts of the study area and the Swabian Alb.

Figure 4(b) displays classified DAM_{mod} caused by Lothar. The classified DAM_{mod}-pattern is very similar to the classified $\bar{U}_{\text{max,mod}}$-pattern and indicates that the modelled forest storm damage pattern at the landscape scale is mainly induced by Lothar-related gust speed.

The great importance of $\bar{U}_{\text{max,mod}}$ for forest storm damage modelling is confirmed by *PI*-evaluation. The most important predictor variables ($PI \geq 2$) for DAM_{mod}

were $\bar{U}_{\text{max,mod}}$ ($PI = 4.7$), geology ($PI = 2.7$), soil moisture ($PI = 2.7$) and forest type ($PI = 2.4$) with $\bar{U}_{\text{max,mod}}$ being nearly twice as important for DAM_{mod} as all other evaluated predictor variables.

Results from model evaluation demonstrated that about 58% of the modelling error is in the range -25 to 25%. The histogram of the modelling error of DAM_{mod} is shown in Figure S2.

Figure 5 shows DAM_{mod} as a function of $\bar{U}_{\text{max,mod}}$-classes. Related to the class median, DAM_{mod} quasi-linearly increases with increasing $\bar{U}_{\text{max,mod}}$. Below $\bar{U}_{\text{max,mod}} = 20\ \text{m s}^{-1}$, DAM_{mod} is very small. The proportions of storm-damaged timber were virtually zero. At $\bar{U}_{\text{max,mod}} = 38\ \text{m s}^{-1}$, DAM_{mod} is greater than 0.5. In all areas where $\bar{U}_{\text{max,mod}}$ exceeded $50\ \text{m s}^{-1}$, DAM_{mod} reached values around 0.9.

Because this study focused on an individual storm event, the dependence of proportions of storm-damaged timber on maximum gust speed is much stronger than in the study of Jung *et al.* (2016). In that study the proportions of storm-damaged timber included both endemic and catastrophic damage for the period 1999–2003.

The results also demonstrate that for very high maximum gust speed values the proportions of storm-damaged timber is exceptionally high, regardless of other factors. This is in contrast to results presented in previous studies (Schindler *et al.*, 2009, 2012b; Schmidt *et al.*, 2010; Albrecht *et al.*, 2012, 2013) for the study area. In all these studies no useful gust speed field was available for forest storm damage modelling. Thus, proxy variables for maximum gust speed such as topographic exposure or absolute elevation were used as surrogate data leading to biased interpretation of (1) the importance of predictor variables and (2) modelling results.

4. Conclusions

Results from this study demonstrate that it is possible to include an airflow variable – such as maximum gust speed – as informative predictor variable into empirical forest storm damage model building. This enables a better representation of the storm damage formation process compared with previous modelling studies as high-impact airflow is always the initial cause of damage formation. In many previous studies, airflow-related variables explained only little variation of observed storm damage patterns. Given the condition that the influence of extreme wind loads on storm damage formation is now reasonably estimated, the evaluation of the influence of orography, soil and stand conditions on forest storm damage formation can be improved.

Acknowledgement

The authors would like to thank the Forest Research Institute Baden-Wuerttemberg for providing forest damage data analysed in this study.

Supporting information

The following supporting information is available:

Table S1. Results of Kolmogorov–Smirnov (D_{KS}), Anderson–Darling (D_{AD}) and chi-square (χ^2) fit evaluation. Numbers indicate the absolute frequency of CDF ranked best after being fitted to CDF_{emp}.

Table S2. List of predictor variables available for maximum gust speed ($U_{\text{max,mod}}$) models building. For detailed information on predictor variables see Jung and Schindler (2015, 2016), Jung (2016).

Table S3. List of predictor variable combinations ($P_{i,1}$–$P_{i,7}$; $i = 1, \ldots, 6$) used to model the maximum gust speed field associated with storm Lothar ($U_{\text{max,mod},i}$).

Figure S1. Histogram showing the error ($U_{\text{max,emp}} - \bar{U}_{\text{max,mod}}$) of modelled mean daily maximum gust speed ($\bar{U}_{\text{max,mod}}$) at 28 meteorological stations on 26 December 1999.

Figure S2. Histograms showing the error ($DAM_{\text{emp}} - DAM_{\text{mod}}$) of modelled proportions of storm-damaged timber caused by storm Lothar in the public forest in the study area.

References

Albrecht A, Hanewinkel M, Bauhus J, Kohnle U. 2012. How does silviculture affect storm damage in forests of south-western Germany? Results from empirical modeling based on long-term observations. *European Journal of Forest Research* **131**: 229–247.

Albrecht A, Kohnle U, Hanewinkel M, Bauhus J. 2013. Storm damage of Douglas-fir unexpectedly high compared to Norway spruce. *Annals of Forest Science* **70**: 195–207.

Belsley DA, Kuh E, Welsh RE. 2001. *Regression Diagnostics*. John Wiley & Sons: New York, NY.

Breiman L. 1996. Bagging predictors. *Machine Learning* **26**: 123–140.

Fink AH, Brücher T, Ermert V, Krüger A, Pinto JG. 2009. The European storm Kyrill 2009: synoptic evolution, meteorological impacts and some considerations with respect to climate change. *Natural Hazards and Earth System Sciences* **9**: 405–423.

Friedman J. 2001. Greedy function approximation: a gradient boosting machine. *Annals of Statistics* **29**: 1189–1232.

Gardiner B, Blennow K, Carbus J-M, Fleischer P, Ingemarson F, Landmann G, Lindner M, Marzano, M, Nicoll B, Orazio C. et al. 2010. Destructive storms in European forests: past and forthcoming impacts. Final Report to European Commission – DG Environment. European Forest Institute. Joensuu, Finland; 138 pp.

Hofherr T, Kunz M. 2010. Extreme wind climatology of winter storms in Germany. *Climate Research* **41**: 105–123.

Houghton JC. 1978. Birth of a parent: the Wakeby distribution for modeling flood flows. *Water Resources Research* **14**: 1105–1109.

Jung C. 2016. High spatial resolution simulation of annual wind energy yield using near-surface wind speed time series. *Energies* **9**: 344.

Jung C, Schindler D. 2015. Statistical modeling of near-surface wind speed: a case study from Baden-Wuerttemberg (Southwest Germany). *Austin Journal of Earth Science* **2**: id1006.

Jung C, Schindler D. 2016. Modelling monthly near-surface maximum daily gust speed distributions in Southwest Germany. *International Journal of Climatology*, doi: 10.1002/joc.4617.

Jung C, Schindler D, Albrecht AT, Buchholz A. 2016. The role of highly-resolved gust speed in simulations of storm damage in forests at the landscape scale: a case study from Southwest Germany. *Atmosphere* **7**: 7.

Keil M, Kiefl R, Strunz G. 2005. CORINE Land Cover 2000 – Germany. Final Report. German Aerospace Center: Germany, Wessling; 72 pp.

Klaus M, Holsten A, Hostert P, Kropp JP. 2011. Integrated methodology to assess windthrow impacts on forest stands under climate change. *Forest Ecology and Management* **261**: 1799–1810.

Kohnle U, Gauckler S, Risse FJ, Stahl S. 2003. Orkan Lothar im Spiegel der Be-triebsinventur und Einschlagbuchführung: Auswirkungen auf einen Forstbezirk im Randbereich des Sturms. *AFZ-DerWald* **58**: 1203–1207 (In German).

Mayer H. 1987. Wind-induced tree sway. *Trees* **1**: 95–106.

Mayer H, Schindler D. 2002. Forstmeteorologische Grundlagen zur Auslösung von Sturmschäden im Wald im Zusammenhang mit dem Orkan "Lothar". *Allgemeine Forst- und Jagdzeitung* **173**: 200–208 (In German).

Pasztor F, Matulla C, Zuvela-Aloise M, Rammer W, Lexer MJ. 2015. Developing predictive models of wind damage in Austrian forests. *Annals of Forest Science* **72**: 289–301.

Quine CP, White IMS. 1998. The potential of distance-limited topex in the prediction of site windiness. *Forestry* **71**: 325–332.

Schelhaas M-J, Nabuurs G-J, Schuck A. 2003. Natural disturbances in the European forests in the 19th and 20th centuries. *Global Change Biology* **9**: 1620–1633.

Schindler D, Grebhan K, Albrecht A, Schönborn J. 2009. Modelling the wind damage probability in forests in Southwestern Germany for the 1999 winter storm 'Lothar'. *International Journal of Biometeorology* **53**: 543–554.

Schindler D, Bauhus J, Mayer H. 2012a. Wind effects on trees. *European Journal of Forest Research* **131**: 159–163.

Schindler D, Grebhan K, Albrecht A, Schönborn J, Kohnle U. 2012b. GIS-based estimation of the winter storm damage probability in forests: a case study from Baden-Wuerttemberg (Southwest Germany). *International Journal of Biometeorology* **56**: 57–69.

Schmidt M, Hanewinkel M, Kändler G, Kublin E, Kohnle U. 2010. An inventory-based approach for modeling singletree storm damage – experiences with the winter storm of 1999 in southwestern Germany. *Canadian Journal of Forest Research* **40**: 1636–1652.

Schüepp M, Schiesser HH, Huntrieser H, Scherrer HU, Schmidtke H. 1994. The winterstorm "Vivian" of 27 February 1990: About the meteorological development, wind forces and damage situation in the forests of Switzerland. *Theoretical and Applied Climatology* **49**: 183–200.

Usbeck T. 2015. Wintersturmschäden im Schweizer Wald von 1865 bis 2014. *Schweizerische Zeitschrift für Forstwesen* **166**: 184–190 (In German).

Wilson JD. 1984. Determining a topex score. *Scottish Forestry* **38**: 251–256.

18

Prediction and attribution of quiescent tropical cyclone activity in the early summer of 2016: Case study of lingering effects by preceding strong El Niño events

Yuhei Takaya,[1,2]* ⓘ Yutaro Kubo,[1,2] Shuhei Maeda[2] and Shoji Hirahara[1,2] ⓘ

[1] Meteorological Research Institute, Japan Meteorological Agency, Ibaraki, Japan
[2] Climate Prediction Division, Japan Meteorological Agency, Tokyo, Japan

*Correspondence to:
Y. Takaya, Meteorological
Research Institute, Japan
Meteorological Agency, 1-1
Nagamine, Tsukuba, Ibaraki
305-0052, Japan.
E-mail:
yuhei.takaya@mri-jma.go.jp

Abstract

We investigated mechanisms contributing to the quiescent tropical cyclone (TC) activity in the western North Pacific (WNP) during the early summer (May–July) of 2016 by conducting and analysing seasonal predictions and sensitivity experiments with an atmosphere–ocean coupled model. In the seasonal prediction experiment, the model successfully predicted the inactive TC condition. Sensitivity experiment simulations, in which the warmer-than-normal sea surface temperature (SST) in the Indian Ocean (IO) was restored to the climatology, represented a weakened lower-tropospheric anticyclonic anomaly and near-normal TC activity over the WNP. These results suggest that the quiescent TC activity is attributable to the warm IO SST anomalies induced by the preceding 2015/2016 El Niño. Verification and analysis of reforecasts indicated that the TC count in early summer is highly predictable due to IO warming, a lingering effect of preceding El Niño events.

Keywords: tropical cyclone; seasonal prediction; Indian Ocean; El Niño

1. Introduction

The western North Pacific (WNP) is the most active basin for tropical cyclones (TCs), and its TC activity has a profound socio-economic influence on Asian countries (Zhang et al., 2009; Weinkle et al., 2012). Accurate seasonal prediction of TCs is expected to contribute to mitigate the risk of human and economic losses caused by TCs.

Much effort has been made in developing and improving seasonal TC predictions (Vitart, 2006; Camargo et al., 2007; Takaya et al., 2010; Vecchi et al., 2014; Camp et al., 2015). Some operational and research institutes currently issue seasonal TC predictions (Camargo et al., 2007). A better understanding of the interannual variability of the TC activity and improving the predictive performance of seasonal prediction systems are important in facilitating the improved use of forecast information.

Recent studies provide a promising perspective on the seasonal prediction of WNP monsoon and TC activity in boreal summer on the basis of the potential predictability originating from lingering effects of preceding strong El Niño events (Xie et al., 2009; Kosaka et al., 2013; Wang et al., 2013; Xie et al., 2016). Two mechanisms have been proposed to explain these lingering effects on the WNP monsoon (Wang et al., 2003; Xie et al., 2009). This study focuses on one of the mechanisms: the 'Indian Ocean capacitor effect' put forward by Xie et al. (2009). Strong El Niño events trigger sea surface temperature (SST)

warming in the Indian Ocean (IO) during their mature and decay phases (boreal winter–spring) through oceanic Rossby wave and wind–evaporation feedback mechanisms (Xie et al., 2002; Du et al., 2009), and the warmed SSTs and enhanced convections in the IO in turn induce an anticyclonic anomaly in the lower troposphere and suppressed convections over the WNP. These conditions often persist through the summer.

Moreover, a few studies have investigated impacts of preceding El Niño events on TCs in the WNP. Such work has found that the IO warming by preceding strong El Niño events inhibits WNP TC activity in the following summer and pointed out a possible IO contribution to the high predictability of seasonal TC activity (Du et al., 2011, hereafter DYX11, Zhan et al., 2011a, hereafter ZWL11, Zhan et al., 2011b). DYX11 examined the relationship between TC activity and SST in the IO basin during July–September and showed that TC activity is suppressed in summers following El Niño events. DYX11 reported that the warm tropical IO induced by a preceding El Niño causes an inactive TC condition during July–September. ZWL11 examined the relationship between the eastern IO SST and TC activity during June–October.

The early summer (May–July) of 2016 exhibited profound quiescent TC activity. The year also marked the second latest record of the first typhoon genesis date of the year since 1951 (the first named Typhoon "Nepartak" formed on 3 July 2016). From the results of previous studies (DYX11, ZWL11), the quiescent

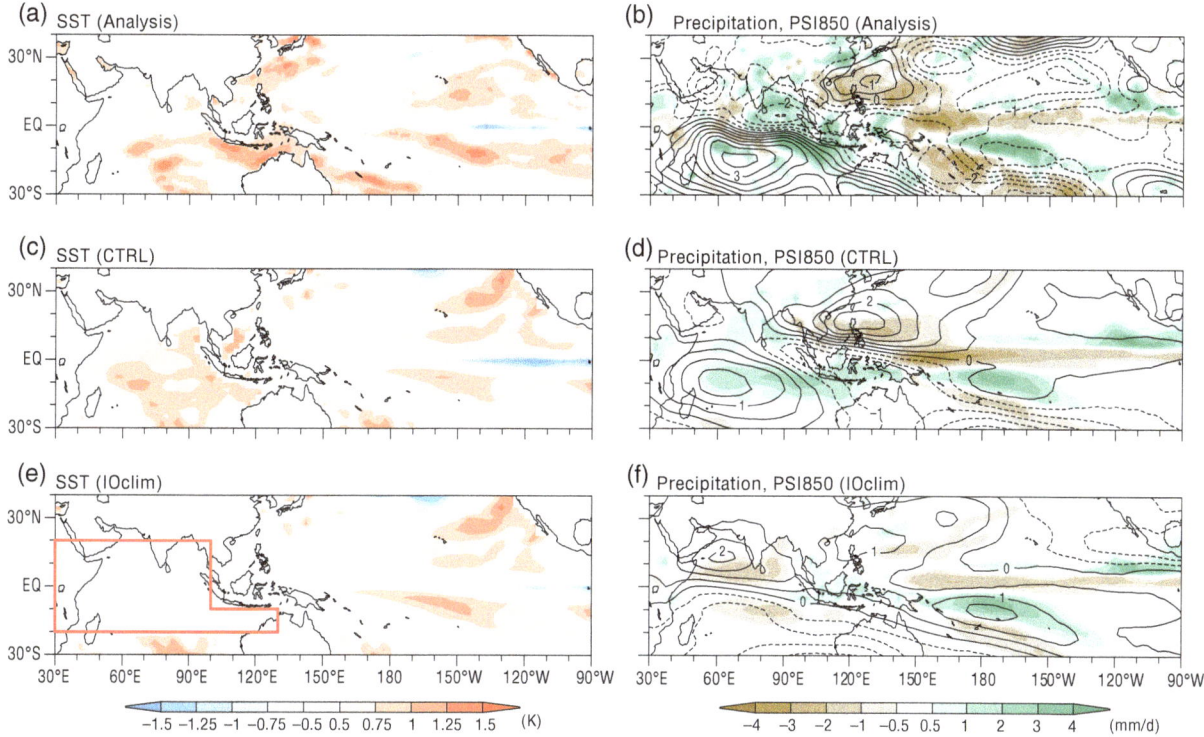

Figure 1. May–July averaged anomalies of (a,c,e) SST, and (b,d,f) precipitation (colours) and 850-hPa stream-function (contours, contour interval is $10^6\,s^{-3}$) of (a,b) analyses, (c,d) CTRL, and (e,f) IOclim experiments. The anomalies are given relative to the analysis and reforecast climatology during 1981–2010. An ocean region surrounded by a red line illustrates the area where SST was restored to the model climatology (see Section 2 for more details).

TC activity was likely associated with a warmed IO induced by the preceding El Niño.

As reviewed above, a few studies have focused on the relationship between TC activity and IO SST; however, to our best knowledge, the feasibility of the seasonal TC prediction has not been studied in this context. Moreover, the seasonal dependence of seasonal TC predictability and its link to the IO have not been reported. Here we analyse a set of numerical simulations for the early summer of 2016 using a state-of-the-art atmosphere–ocean coupled model with a special focus on the early summer. Furthermore, we analyse reforecasts and discuss underlying mechanisms and the seasonal dependence of seasonal TC predictability.

2. Data and methods

We conducted seasonal prediction and sensitivity experiments using the Japan Meteorological Agency/Meteorological Research Institute-Coupled Prediction System version 2 (JMA/MRI-CPS2, Takaya *et al.*, 2017b), which is the latest operational seasonal prediction system of JMA. The system adopts an atmosphere–ocean–land–sea-ice coupled model with an atmospheric resolution of 110 km and ocean resolution of $1° \times 0.5°$. Further details are described in Takaya *et al.* (2017b). We verified the performance of the system in predicting seasonal

TCs and found that it is comparable to that of its precedent system (JMA/MRI-CPS1, Takaya *et al.*, 2017a, 2010).

In the control seasonal prediction experiment (CTRL), a total of 52-member ensembles were integrated from initial dates of 11, 16, 21, and 26 April 2016 (13 members from each initial date). In the sensitivity experiment (IOclim), the experimental settings were almost the same as those of CTRL, except that SST in the IO basin (illustrated in Figure 1(e)) was strongly nudged to the model climatology of the same system; specifically, the Newtonian nudging term with a coefficient of $2400\,W\,m^{-2}\,K^{-1}$ was added to the surface layer so that SST was sufficiently restored to the climatology of reforecasts during 1981–2010 (c.f. Morioka *et al.*, 2014).

In this study, we analysed TCs classified in categories stronger than the tropical storm, which have sustained winds exceeding 34 knots ($17.2\,m\,s^{-1}$). We detected simulated TCs using an objective algorithm (Takaya *et al.*, 2010, Appendix A). We set parameters in the detection algorithm so that the model produces a realistic climatological number of TC genesis. Validation of the algorithm using the JRA-25 analysis (Onogi *et al.*, 2007) showed a reasonable agreement between the detected TCs and best track data (Takaya *et al.*, 2010).

Analysis data used in this study were the JRA-55 reanalysis (Kobayashi *et al.*, 2015) for atmospheric analysis, CPC Merged Analysis Precipitation (CMAP,

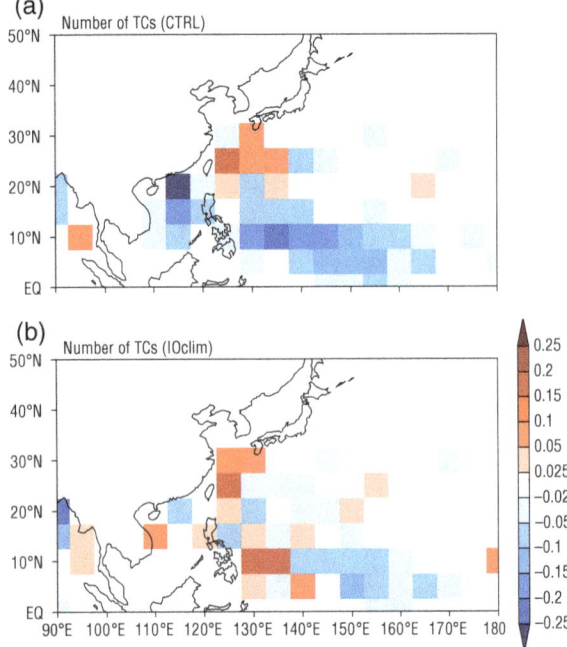

Figure 2. May–July averaged anomalies of TC counts in 5° × 5° grids for the (a) CTRL and (b) IOclim experiments. The anomalies are given relative to the reforecast climatology of each dataset during 1981–2010.

Xie and Arkin, 1997) for monthly precipitation analysis, COBE SST (Ishii *et al.*, 2005) for SST analysis, and RSMC Tokyo best track data for the TC analysis.

3. Results

The pronounced quiescent TC activity during May–July 2016 was associated with large-scale atmospheric and oceanic conditions in the Indo-Pacific basin. Figure 1 shows the observed and predicted SST, precipitation, and 850-hPa stream function anomalies during May–July 2016. From the analyses (Figures 1(a) and (b)), warm SSTs and enhanced convections dominated in the central to eastern tropical IO. In the WNP, a lower-tropospheric anticyclonic anomaly and deficit precipitation were observed as a forced atmospheric response to the IO (Watanabe and Jin, 2003). The prediction (CTRL) successfully captured these conditions with a 0-month lead time (Figures 1(c) and (d)).

Figure 2(a) illustrates anomalies of the TC count in 5° × 5° boxes in CTRL during the period May–July 2016. The negative anomalies spread broadly in the WNP in response to the above-mentioned weak WNP monsoon conditions is consistent with ZWL11.

Figure 3 shows the observed and predicted time-series of the TC count in the WNP (0°–60°N, 100°–180°E) during May–July. During May–July 2016, the observed TC activity was extremely weak compared with the climatology and only four TCs were determined in the RSMC Tokyo analysis. This inactive TC condition was well captured by CTRL

with an ensemble mean TC count of 4.3. The interannual variability of TC genesis is well captured by the 10-member reforecasts, in particular, the inactive TC years following strong El Niño events such as 1998 and 2010, indicating strong lingering effects of the preceding El Niño events. The correlation coefficient of the total TC counts between the best track analysis and the prediction is 0.69 during 1981–2010, which is higher than that of the other summer periods (Table 1). These results indicate that the TC count is highly predictable especially during May–July.

The inactive early TC season (May–July 2016) is likely to have resulted from the IO warming (Figure 1(a)), which was plausibly forced by the tropical teleconnection of the preceding El Niño event in the 2015/2016 winter (Klein *et al.*, 1999; Lau and Nath, 2000). To attribute the inactive early-summer TC to the warm IO condition, we conducted an idealized experiment (IOclim). Figures 1(e) and (f) display the simulated SST, precipitation anomalies and 850-hPa stream function anomalies in IOclim. The SST anomalies in the IO were close to zero, indicating that the SST in the IO is sufficiently constrained by the model climatology. Below-normal precipitation is seen in the IO in contrast to CTRL, and the anticyclonic anomaly in the WNP is rather weak compared with that of CTRL.

Figure 2 compares anomalies of the TC counts in CTRL and IOclim. As expected from the large-scale atmospheric conditions over the WNP, CTRL presents broad negative anomalies of TC genesis over the WNP, whereas IOclim presents near-normal TC genesis. Correspondingly, the total TC count was close to the climatology in IOclim in contrast to the remarkably small TC count in CTRL (Figure 3). Our results presented herein suggest that the quiescent TC activity observed in the 2016 early TC season is mostly attributable to the IO warming.

4. Conclusions and discussion

We have presented results of seasonal TC simulations for the early summer (May–July) of 2016. The latest JMA seasonal prediction system well predicted the quiescent TC activity in the season. Our verification (Table 1) suggested that the predictive skill of the TC count for the early TC season is high (correlation coefficient of 0.69), and it is higher than that for the other consecutive 3-month periods during the TC season. Kosaka *et al.* (2013) presented the lagged correlation between the NINO3 (5°N–5°S, 150°–90°W) SST during December–February and the monthly TC counts, which shows the largest negative correlation in June (Figure 1 in their paper). Together with their finding, our results suggest that the lingering effect of an El Niño on the TC activity is stronger in the early summer and TC activity is more predictable during the early summer than other 3-month periods during the TC season. We note that JMA/MRI-CPS2 has a high predictive skill (correlation coefficient of >0.8) for the TC

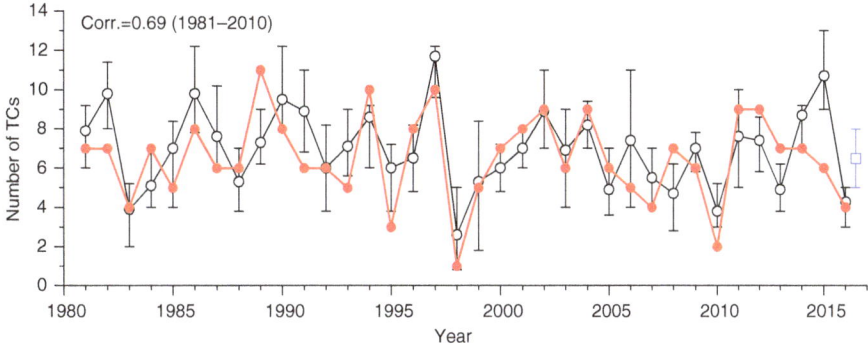

Figure 3. Time-series of total TC counts during May–July. Red line indicates the best track analysis and the black line indicates the predictions with a 0-month lead time. Whiskers represent the maximum and minimum ranges of 10 member ensembles, except for those of 52 members from the real-time operational predictions for 2015 and 2016 (CTRL experiment). The blue dot and whiskers show the result of the IOclim experiment.

Table 1. Correlations between TC counts for 3-month periods in the best track analysis and those of the predictions with a 0-month lead time. The 10-member ensemble mean predictions were verified for a 30-year period (1981–2010).

Initial month	May	June	July	August
Target months	May–July	June–August	July–September	August–October
Correlation	0.69	0.53	0.42	0.17

count in the southeast region (17.5°–0°N, 140°–180°E) of the WNP especially during late summer to autumn (August–November) reflecting a strong influence of SST over the tropical central North Pacific associated with El Niño-Southern Oscillation (ENSO) on the inter-annual TC variability (Mei *et al.*, 2015) and high predictive skill for ENSO. On the other hand, the model has relatively low skill for the total TC count in the WNP during late summer to autumn (Table 1). More research is needed to elucidate the reasons for the seasonality of the TC predictive skill and predictability.

We also elucidated the attribution of the quiescent TC activity in the early summer of 2016 by conducting and analysing the prediction and sensitivity experiment. The prediction experiment (CTRL) represented a broad decrease in TC genesis over the WNP in contrast to the sensitivity experiment, in which the IO SST was restored to the model climatology. The anomaly pattern in CTRL differs from that expected for an El Niño/La Niña response in summer, which is characterized by a northwest–southeast shift of genesis location (Wang and Chan, 2002; Takaya *et al.*, 2010). Our results are consistent with the results of ZWL11, who pointed out that a warm eastern IO SST reduces the total number of TCs in the WNP during the TC season (June–October).

It is worth comparing the 2016 case with other cases following El Niño events. Figure 4 shows composites of the anomalies of TC counts in the best track analysis and predictions during the years following El Niño events (1983, 1988, 1992, 1998, 2003, and 2010) based on the JMA's definition (http://ds.data.jma.go.jp/tcc/tcc/products/elnino/ensoevents.html). The year 1987 was excluded since the 1987 El Niño event lasted from 1986 to 1988. In general, suppressed activity in the

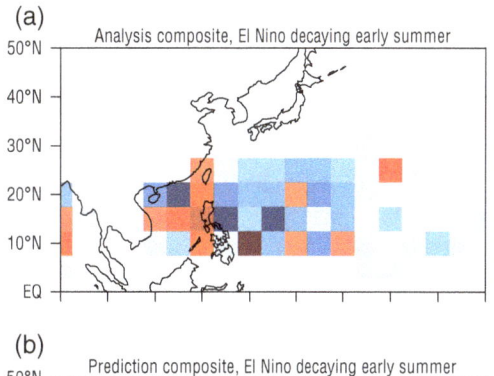

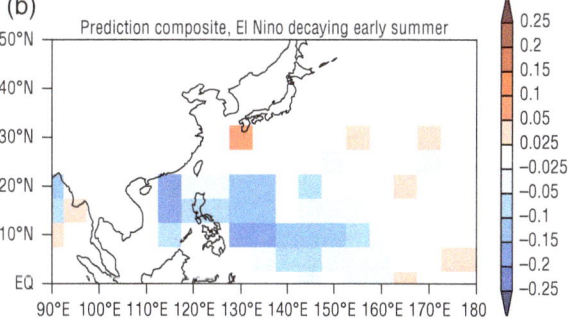

Figure 4. Composites of anomalies of the TC count in 5° × 5° grids during May–July for (a) the best track analysis and (b) predictions during El Niño decaying summers (1983, 1988, 1992, 1998, 2003, and 2010). The anomalies are given relative to the reforecast climatology of each dataset during 1981–2010.

early summer following El Niño events was observed in the historical record, as seen in Figure 3. Nakazawa (2001) pointed out that inactive TC genesis in the WNP in 1998 was related to anomalous lower-tropospheric anticyclonic circulation and subsidence over the WNP, which was induced by the anomalous IO convective activity and ascending motion, associated with the 1997–1998 El Niño. This is a recursive situation in summers following El Niño events (DYX11, ZWL11). The observed inactive TC condition in the early summer following El Niño events was well reproduced in the reforecasts of JMA/MRI-CPS2, supporting the hypothesis that the preceding El Niño and lingering IO warming are sources of the predictably of early summer TC activity in the WNP. The results presented in this paper indicate high predictably of seasonal TC activity and feasibility of operational seasonal TC prediction during early summer.

Appendix A: TC detection algorithm

The TC detection algorithm used in this study is briefly described here. TCs in the model simulations are objectively detected using model outputs at $1.5° \times 1.5°$ resolution according to the following conditions and criteria.

1. A grid point with a local sea level pressure (SLP) minimum within $6° \times 6°$ over the ocean between the equator and 30°N is searched to detect the center of a candidate TC.
2. At any point within $3° \times 3°$ of the center of the TC, the relative vorticity at 850-hPa is below $5 \times 10^{-5}\,\text{s}^{-1}$.
3. A warm core exists near the local SLP minimum. The thickness between 200 and 500 hPa geopotential height at the center of the TC is 7 gpm higher than the average thickness within grids surrounding the center of the TC within a $9° \times 9°$ box, excluding the center grid (24 grids).
4. At the center of the TC, the wind speed at 200 hPa is lower than that at 850 hPa.
5. The four conditions listed above continue for at least 12 h.

The parameters were chosen so that the number of detected TCs was reasonably close to that for analysed TCs (maximum wind exceeding $17.2\,\text{m s}^{-1}$) in the RSMC Tokyo best track analysis.

Acknowledgements

The authors would acknowledge C. Matsukawa and H. Sugimoto at CPD/JMA for their support on conducting numerical experiments. This work was supported by the Eighth Precipitation Measuring Mission Science Research Announcement of the Japan Aerospace Exploration Agency (JAXA) and JSPS KAKENHI Grant Number JP17K14395. The CMAP Precipitation data provided by the NOAA/OAR/ESRL PSD, Boulder, Colorado, USA, from their website at http://www.esrl.noaa.gov/psd/.

References

Camargo S, Barnston A, Klotzbach PJ, Landsea CW. 2007. Seasonal tropical cyclone forecasts. *WMO Bulletin* **56**: 297–309.

Camp J, Roberts M, MacLachlan C, Wallace E, Hermanson L, Brookshaw A, Arribas A, Scaife AA. 2015. Seasonal forecasting of tropical storms using the Met Office GloSea5 seasonal forecast system. *Quarterly Journal of the Royal Meteorological Society* **141**: 2206–2219. https://doi.org/10.1002/qj.2516.

Du Y, Xie SP, Huang G, Hu KM. 2009. Role of air–sea interaction in the long persistence of El Nino-induced North Indian Ocean warming. *Journal of Climate* **22**: 2023–2038. https://doi.org/10.1175/2008JCLI2590.1.

Du Y, Yang L, Xie SP. 2011. Tropical Indian Ocean influence on Northwest Pacific tropical cyclones in summer following strong El Niño. *Journal of Climate* **24**: 315–322. https://doi.org/10.1175/2010JCLI3890.1.

Ishii M, Shouji A, Sugimoto S, Matsumoto T. 2005. Objective analyses of sea-surface temperature and marine meteorological variables for the 20th century using ICOADS and the Kobe collection. *International Journal of Climatology* **25**: 865–879. https://doi.org/10.1002/joc.1169.

Klein SA, Soden BJ, Lau NC. 1999. Remote sea surface temperature variations during ENSO: evidence for a tropical atmospheric bridge. *Journal of Climate* **12**: 917–932. https://doi.org/10.1175/1520-0442(1999)012<0917:RSSTVD>2.0.CO;2.

Kobayashi S, Ota Y, Harada Y, Ebita A, Moriya M, Onoda H, Onogi K, Kamahori H, Kobayashi C, Endo H, Miyaoka K, Takahashi K. 2015. The JRA-55 reanalysis: general specifications and basic characteristics. *Journal of the Meteorological Society of Japan* **93**: 5–48. https://doi.org/10.2151/jmsj.2015-001.

Kosaka Y, Xie SP, Lau NC, Vecchi GA. 2013. Origin of seasonal predictability for summer climate over the Northwestern Pacific. *Proceeding of National Academy of Science of the United States of America* **110**: 7574–7579. https://doi.org/10.1073/pnas.1215582110.

Lau NC, Nath MJ. 2000. Impact of ENSO on the variability of the Asian-Australian monsoons as simulated in GCM experiments. *Journal of Climate* **13**: 4287–4309. https://doi.org/10.1175/1520-0442(2000)013<4287:IOEOTV>2.0.CO;2.

Mei W, Xie SP, Zhao M, Wang Y. 2015. Forced and internal variability of tropical cyclone track density in the western North Pacific. *Journal of Climate* **28**: 143–167. https://doi.org/10.1175/JCLI-D-14-00164.1.

Morioka Y, Masson S, Terray P, Prodhomme C, Behera SK, Masumoto Y. 2014. Role of tropical SST variability on the formation of subtropical dipoles. *Journal of Climate* **27**: 4486–4507. https://doi.org/10.1175/JCLI-D-13-00506.1.

Nakazawa T. 2001. Suppressed tropical cyclone formation over the western North Pacific in 1998. *Journal of the Meteorological Society of Japan* **79**: 173–183. https://doi.org/10.2151/jmsj.79.173.

Onogi K, Tsutsui J, Koide H, Sakamoto M, Kobayashi S, Hatsushika H, Matsumoto T, Yamazaki N, Kamahori H, Takahashi K, Kadokura S, Wada K, Kato K, Oyama R, Ose T, Mannoji N, Taira R. 2007. The JRA-25 reanalysis. *Journal of the Meteorological Society of Japan* **85**: 369–432. https://doi.org/10.2151/jmsj.85.369.

Takaya Y, Yasuda T, Ose T, Nakaegawa T. 2010. Predictability of the mean location of typhoon formation in a seasonal prediction experiment with a coupled general circulation model. *Journal of the Meteorological Society of Japan* **88**: 799–812. https://doi.org/10.2151/jmsj.2010-502.

Takaya Y, Yasuda T, Fujii Y, Matsumoto S, Soga T, Mori H, Hirai M, Ishikawa I, Sato H, Shimpo A, Kamachi M, Ose T. 2017a. Japan Meteorological Agency/Meteorological Research Institute-Coupled Prediction System version 1 (JMA/MRI-CPS1) for operational seasonal forecasting. *Climate Dynamics* **48**: 313–333. https://doi.org/10.1007/s00382-016-3076-9.

Takaya Y, Hirahara S, Yasuda T, Matsueda S, Toyoda T, Fujii Y, Sugimoto H, Matsukawa C, Ishikawa I, Mori H, Nagasawa R, Kubo Y, Adachi N, Yamanaka G, Kuragano T. 2017b. Japan Meteorological Agency/Meteorological Research Institute-Coupled Prediction System version 2 (JMA/MRI-CPS2): atmosphere-land-ocean-sea ice coupled prediction system for operational seasonal forecasting. *Climate Dynamics*. https://doi.org/10.1007/s00382-017-3638-5.

Vecchi GA, Delworth T, Gudgel R, Kapnick S, Rosati A, Wittenberg AT, Zeng F, Anderson W, Balaji V, Dixon K, Jia L, Kim HS, Krishnamurthy L, Msadek R, Stern WF, Underwood SD, Villarini G, Yang X, Zhang S. 2014. On the seasonal forecasting of regional tropical cyclone activity. *Journal of Climate* **27**: 7994–8016. https://doi.org/10.1007/s00382-017-3638-5.

Vitart F. 2006. Seasonal forecasting of tropical storm frequency using a multi-model ensemble. *Quarterly Journal of the Royal Meteorological Society* **132**: 647–666. https://doi.org/10.1256/qj.05.65.

Wang B, Chan JCL. 2002. How strong ENSO events affect tropical storm activity over the western north Pacific. *Journal of Climate* **15**: 1643–1658. https://doi.org/10.1175/1520-0442(2002)015<1643:HSEEAT>2.0.CO;2.

Wang B, Xiang B, Lee JY. 2013. Subtropical high predictability establishes a promising way for monsoon and tropical storm predictions. *Proceeding of National Academy of Science of the United States of America* **110**: 2718–2722. https://doi.org/10.1073/pnas.1214626110.

Watanabe M, Jin FF. 2003. A moist linear baroclinic model: coupled dynamical–convective response to El Niño. *Journal of Climate* **16**: 1121–1140. https://doi.org/10.1175/1520-0442(2003)161121:AMLBMC2.0.CO;2.

Weinkle J, Maue R, Pielke R. 2012. Historical global tropical cyclone landfalls. *Journal of Climate* **25**: 4729–4735. https://doi.org/10.1175/JCLI-D-11-00719.1.

Xie P, Arkin PA. 1997. Global precipitation: a 17-year monthly analysis based on gauge observations, satellite estimates, and numerical model outputs. *Bulletin of the American Meteorological Society* **78**: 2539–2558. https://doi.org/10.1175/1520-0477(1997)0782539:GPAYMA2.0.CO;2.

Xie SP, Annamalai H, Schott F, McCreary JP. 2002. Structure and mechanisms of South Indian Ocean climate variability. *Journal of Climate* **15**: 864–878. https://doi.org/10.1175/1520-0442(2002)0150864:SAMOSI2.0.CO;2.

Xie SP, Hu K, Hafner J, Tokinaga H, Du Y, Huang G, Sampe T. 2009. Indian Ocean capacitor effect on Indo–western Pacific climate during the summer following El Niño. *Journal of Climate* **22**: 730–747. https://doi.org/10.1175/2008JCLI2544.1.

Xie SP, Kosaka Y, Du Y, Hu K, Chowdary JS, Huang G. 2016. Indo-western Pacific ocean capacitor and coherent climate anomalies in post-ENSO summer: a review. *Advances in Atmospheric Sciences* **33**: 411–432. https://doi.org/10.1007/s00376-015-5192-6.

Zhan R, Wang Y, Lei X. 2011a. Contributions of ENSO and East Indian Ocean SSTA to the interannual variability of Northwest Pacific tropical cyclone frequency. *Journal of Climate* **24**: 509–521. https://doi.org/10.1175/2010JCLI3808.1.

Zhan R, Wang Y, Wu CC. 2011b. Impact of SSTA in the East Indian Ocean on the frequency of Northwest Pacific tropical cyclones: a regional atmospheric model study. *Journal of Climate* **24**: 6227–6242. https://doi.org/10.1175/JCLI-D-10-05014.1.

Zhang Q, Liu Q, Wu L. 2009. Tropical cyclone damages in China 1983–2006. *Bulletin of the American Meteorological Society* **90**: 489–495. https://doi.org/10.1175/2008bams2631.1.

19

Quantifying the extremity of windstorms for regions featuring infrequent events

Michael A. Walz,[1,2]* Tim Kruschke,[2,3] Henning W. Rust,[2] Uwe Ulbrich[2] and Gregor C. Leckebusch[1]

[1] School of Geography, Earth and Environmental Science, University of Birmingham, UK
[2] Institut fuer Meteorologie, Freie Universität Berlin, Germany
[3] Department of Ocean Circulation and Climate Dynamics, Research Unit Marine Meteorology, GEOMAR Helmholtz Centre for Ocean Research Kiel, Germany

*Correspondence to:
M. A. Walz, School of Geography, Earth and Environmental Science, University of Birmingham, Birmingham B15 2TT, UK.
E-mail: maw526@bham.ac.uk

Abstract

This paper introduces the Distribution-Independent Storm Severity Index (DI-SSI). The DI-SSI represents an approach to quantify the severity of exceptional surface wind speeds of large scale windstorms that is complementary to the SSI introduced by Leckebusch *et al.* While the SSI approaches the extremeness of a storm from a meteorological and potential loss (impact) perspective, the DI-SSI defines the severity in a more climatological perspective. The idea is to assign equal index values to wind speeds of the same singularity (e.g. the 99th percentile) under consideration of the shape of the tail of the local wind speed climatology. Especially in regions at the edge of the classical storm track, the DI-SSI shows more equitable severity estimates, e.g. for the extra-tropical cyclone Klaus. In order to compare the indices, their relation with the North Atlantic Oscillation is studied, which is one of the main large scale drivers for the intensity of European windstorms.

Keywords: windstorms; quantification of extremity; North Atlantic Oscillation; extra-tropical cyclones; generalized Pareto distribution

1. Introduction and motivation

1.1. Background

Winter windstorms are among the biggest natural hazards occurring in the mid-latitudes causing human casualties as well as economic losses up to billions of Euros each year. According to SwissRe, the winter storm Kyrill, which strongly affected Central Europe on 18, 19 January 2007 caused an economic insured loss of about \$6.1 billion and casualties of 54 people (SwissRe, 2016). An approach to objectively quantify the meteorological hazard is represented by the Storm Severity Index (SSI) introduced by Leckebusch *et al.* (2008). The SSI is widely used (e.g. Osinski *et al.*, 2016) for assessing the severity of windstorms within the actuarial sector by linking extreme surface winds (i.e. exceedances of the 98th percentile of local 6-hourly wind speeds) to potential loss on buildings. Furthermore, the 98th percentile is used as a criterion for identifying extreme windstorms in a wind tracking algorithm by Leckebusch *et al.* (2008) and further developed by Kruschke (2015). Equation (1) shows the mathematical definition of the SSI. The index t represents the time step, k represents the grid cell and A_k represents the area of the associated cell divided by a reference cell at the equator:

$$SSI_{T,K} = \sum_{t=1}^{T} \sum_{k=1}^{K} \left[\left(\max \left(0, \frac{v_{k,t} - v_{98,k}}{v_{98,k}} \right) \right)^3 \cdot A_k \right] \tag{1}$$

The $v_{98,k}$ refers to the local 98th percentile of the kth grid cell which is the minimum wind speed at which damage on housing or nature is to be expected. This relationship was established based on real damage experience (Klawa and Ulbrich, 2003) which proved the assumption of Palutikof and Skellern (1991) who assumed storm damages to occur at about 2% of all days.

For the further development of the index, the focus will be on the Meteorological Contribution Γ to the SSI defined by Equation (2) which shifts and scales wind speeds by the 98th percentile:

$$\Gamma = \frac{v_{k,t} - v_{98,k}}{v_{98,k}} \tag{2}$$

1.2. Motivation for a supplementary severity index (DI-SSI)

Technically, the SSI is based on excesses over a fixed quantile (percentile) for a given wind speed distribution; however, the SSI does not take into account the shape of the distribution of these excesses, i.e. the tail behaviour. This property becomes particularly obvious in areas with little storm occurrence. The top panel of Figure 1 illustrates this effect: The two panels depict histograms of Γ (scaled and shifted wind speeds, cf. Equation (2)) of a grid cell south of Iceland (called Iceland hereafter) and on the Isle of Corsica. The coloured lines mark the 98th (red) and the 99th (green) percentile of the shifted and scaled distribution. As known from various

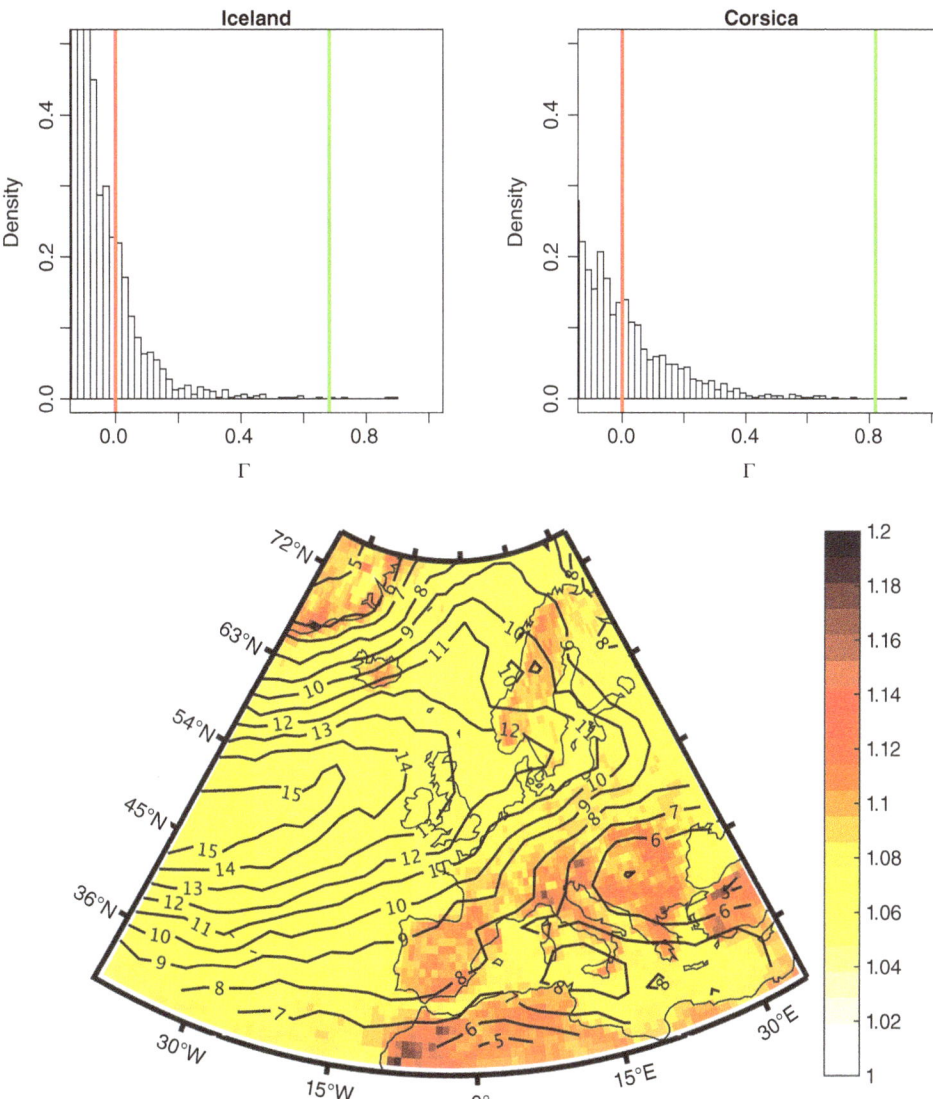

Figure 1. Top: Histograms of the Meteorological contribution (Γ) as defined in Equation (2) for Iceland (left) and Corsica (right). The distribution of the events looks visibly different. The red line marks the 98th percentile, thus the cut off threshold. The green line marks the 99th percentile, thus it illuminates the larger difference between the two percentiles for Corsica in comparison with Iceland. Bottom: Quotient between the 99th and the 98th percentile of local wind speeds for Europe in colour. Large values indicate a large difference between the percentiles. The contours depict the average trackdensity per winter (ONDJFM; average number of tracks within a 500 km radius around a given grid point). The quotient is clearly larger in areas with a reduced windstorm frequency.

other studies (e.g. Leckebusch *et al.*, 2008 or Klawa and Ulbrich, 2003) the area south of Iceland is located within the corridor of extra-tropical cyclones (ETCs), whereas Corsica is less affected by ETCs (cf. bottom panel of Figure 1).

By definition of the SSI, the red line is equal to 0. The histogram for Iceland resembles a light-tailed distribution, whereas the histogram for Corsica shows features of a heavy-tailed distribution. Accordingly, the gap between the 99th and 98th percentile is substantially different (0.68 for Iceland and 0.82 for Corsica). Note that due to the cubic relationship of the SSI (Equation (1)), Γ is taken to the third power; for Corsica this is around six times larger than Γ^3 for Iceland (cf. Table 1). From a probabilistic perspective, however, both wind speeds are equally likely. As the

SSI consists of spatially and temporally accumulated Γ^3, this implies that a potential storm over Corsica with wind speeds exceeding the 98th percentile will result in a much larger integral SSI value than a storm over Iceland with the same exceptional wind speeds (c.f. examples for storms Klaus and Martin in Table 2). The gap in Γ^3 is a direct result of the local wind speed climatology and this in turn is a consequence of the scarcity of very extreme winds at the edge of the main storm track. Thus, within the storm track there are larger 'background' winds with a higher number of extreme events, whereas the edges of the storm track feature lower 'background' winds and comparatively few extreme events. The exceedance compared with the background wind in relative terms is larger for the edge of the storm track.

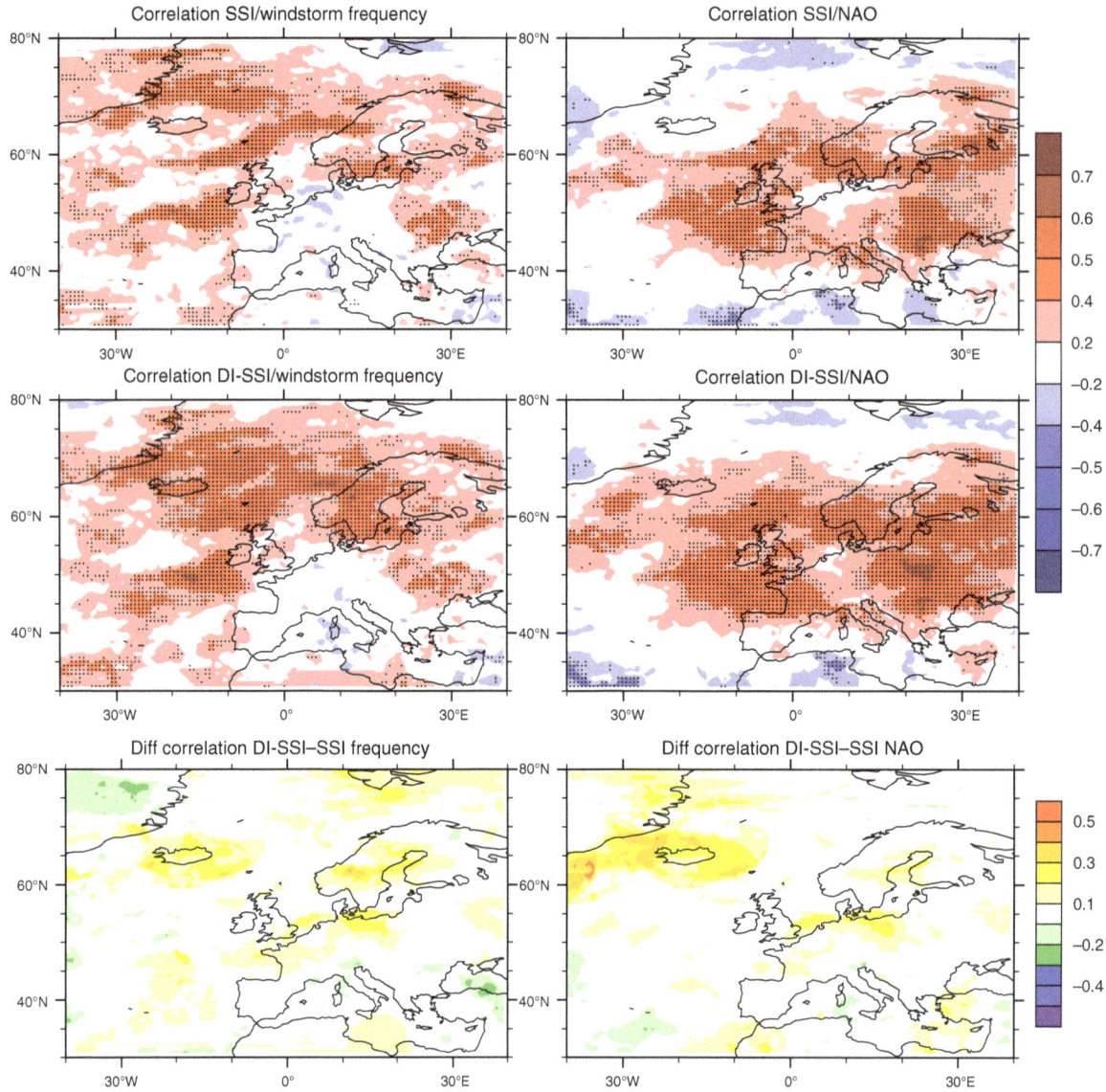

Figure 2. Left panel: Correlation between the storm frequency and storm intensity (SSI on the top; DI-SSI in the middle) for each grid cell. Correlation coefficients significant at the 5% level are stippled. There is a significant link between more storms and more intense storms for much of the North Atlantic and Scandinavia. Right panel: Correlation coefficients between the yearly NAO time series and the yearly windstorm intensity on grid cell level (SSI on the top; DI-SSI in the middle). Again correlation coefficients significant at the 5% level are tagged. Bottom row: Differences between the respective correlations. Positive values represent areas where the correlation for the DI-SSI is larger compared to the SSI. This is the case for most of the North Atlantic Domain.

Table 1. Meteorological and DI-SSI contributions of the two example grid cells. Note that the contribution of the grid cell in Corsica is more than five times larger, although the wind is of the same singularity in both cases.

	Γ^3 in a single grid cell for a surface wind equal to the 99th percentile	DI-SSI contribution in a single grid cell for a surface wind equal to the 99th percentile
Iceland	2.05×10^{-4}	0.71
Corsica	1.24×10^{-3}	0.67
Theoretical value	–	0.69

The effect of the gaps in the two histograms shown in Figure 1 (top panel) can be shown on grid cell level as well: Figure 1 (bottom panel) depicts the quotient of the local 99th and 98th percentile (as the division by the 98th percentile is part of the calculation of the SSI) and the average storm frequency per grid cell per extended winter season (i.e. how often on average a windstorm track is detected within a 500 km radius of a particular grid cell). Klawa and Ulbrich (2003) calculated the same quotients for station data of wind speeds in various locations in Germany. Their conclusion was that the quotient was sufficiently homogeneous for the entire country. This assumption can be supported and confirmed by Figure 1 (bottom). Values above 1.1, however, indicate grid cells which are potentially affected by large SSI values, thus in particular Southern Europe. These areas coincide with regions of little storminess over the winter period (less than 8–10 windstorm events per year).

Table 2. Integral SSI and DI-SSI values for some prominent European windstorms. The rank of severity for the respective index is denoted in parenthesis. Note that storms which occurred outside of the main storm tracks feature relatively large SSI values (e.g. Klaus, Martin, Xynthia, Torsten) compared to the ones within the main storm track (Daria or Jeanette). This applies especially for the SSI/DI-SSI values per time step. The largest discrepancy in terms of rank of the integral values is observed for Martin and for Vivian/Klaus for time step based values.

Storm	Date	Integral SSI value	Integral DI-SSI value	SSI per timestep	DI-SSI per timestep	References
Daria	23–26 January 1990	26.69 (7)	1940.20 (4)	2.05 (8)	149.25 (5)	Heming (1990)
Vivian	25–28 February 1990	58.52 (2)	4126.34 (1)	3.90 (4)	275.10 (1)	McCallum and Norris (1990)
Anatol	2–4 December 1999	23.57 (8)	1565.67 (6)	1.81 (9)	120.44 (8)	Ulbrich et al. (2001)
Martin	26–28 December 1999	43.81 (4)	1435.09 (8)	5.48 (2)	179.39 (3)	Ulbrich et al. (2001)
Torsten	10–13 November 2001	15.94 (9)	789.95 (9)	2.66 (6)	131.66 (7)	Tripoli et al. (2005)
Jeanette	25–31 October 2002	32.53 (6)	1576.27 (5)	2.32 (7)	112.60 (9)	Parton et al. (2009)
Kyrill	15–24 January 2007	53.03 (3)	2439.57 (2)	4.08 (3)	187.66 (2)	Fink et al. (2009)
Klaus	23–28 January 2009	74.30 (1)	2117.52 (3)	5.72 (1)	162.89 (4)	Liberato et al. (2011)
Xynthia	26 February to 7 March 2010	37.92 (5)	1459.10 (7)	3.45 (5)	132.65 (6)	Lumbroso and Vinet (2011)

Regarding a potential loss in these areas, large integral SSI values are possibly intentional. Due to a lack of severe large scale storms, the infrastructure might not be as adapted to severe wind speeds as it is in regions within the main storm track. This study, however, aims at creating a metric for the extremeness of a storm using an alternative approach. In order to compare the severity of storms independently from their geographical occurrence, a metric is developed that accounts for the wind speed distribution at a given grid cell. Particularly, the shape of the upper tail of the local wind speed distribution is considered. Due to that feature and its resemblance to the original SSI, the index is named Distribution-Independent SSI (DI-SSI). The DI-SSI can be seen as a side development to the original SSI as it represents a complementary method to assess the severity of storms. The DI-SSI is particularly useful when comparing storms occurring in and outside of the main storm corridor. The two indices are correlated with the North Atlantic Oscillation (NAO), as it is currently recognized to be the most prominent driver of the inter annual variability of European storminess (e.g. Donat et al., 2010; Pinto et al., 2009 or Ulbrich and Christoph, 1999). Due to its parametric nature it is expected that the DI-SSI gives a more coherent and distinct link for areas outside of the classic storm track as it is a smoother function compared to the highly variable nonparametric SSI signal.

2. Data and methods

2.1. Data and event identification

The wind speed data used for this work are taken from the ERA Interim reanalysis (Dee et al., 2011) which is managed by the European Centre for Medium Range Forecasts (ECMWF). The spectral resolution of ERA Interim is T255 which corresponds to a grid cell size of $0.7° \times 0.7°$ at the equator. An objective wind-speed-based tracking algorithm (Leckebusch et al., 2008; Kruschke, 2015) was applied to the 6-hourly 10-m wind field of the extended boreal winter

period (October to March) in order to extract windstorm trajectories. ERA Interim has been frequently used in other ETC studies (e.g., Hodges et al., 2011). The NAO time series is obtained as the first principal component of a rotated EOF analysis of monthly (October to March) mean 700 hPa geopotential height anomalies for the North Atlantic domain ($70°W - 40°E$, $30° - 80°N$) as done by Hunter et al. (2016).

2.2. The DI-SSI

The derivation of the DI-SSI is based on the idea that excesses over a sufficiently large threshold can be well approximated by a generalized Pareto distribution (GPD). The approach of modelling excesses is one of the main concepts within extreme value theory (see e.g. Coles, 2001). Modelling excesses of geophysical data with a GPD has been proposed in various other studies in connection with extreme precipitation [e.g. Vrac and Naveau, 2007 or Cooley et al., 2007], wind speeds (Kunz et al., 2010) and also SSI values (Donat et al., 2011 or Held et al., 2013).

The concept of the DI-SSI is to understand the numerator of the SSI equation (Equation (1)) as the exceedance of a threshold (i.e. the 98th percentile). In contrast to the common method of determining a threshold for the GPD, the threshold is fixed at the 98th percentile for every grid cell. The goodness of fit test provides satisfying results for this threshold (see below). A new variable is introduced to which the GPD is applied: v^* is defined as the random variable of the excess wind speeds over the local 98th percentile $v_{98,k}$ at grid cell k and time t:

$$v^* = v_{k,t} - v_{98,k} | v_{k,t} > v_{98,k} \qquad (3)$$

Estimating parameters of the GPD for v^* (using the ismev library in R; Heffernan and Stephenson, 2015) results in a pair of shape (ξ) and scale parameters (σ) for every grid cell. To get an idea of how well the GPD performs in describing v^* in the mid-latitudes, a Kolmogorov-Smirnov test (ks-test) is used to assess the goodness of fit of the GPD distribution at every single grid point. Most grid cells over the North Atlantic and Europe do not show distances larger than the critical value D of the ks-test at the 5% significance level. Between 30° and 70°N, only 6% (2578 grid cells out of 43520) of all grid cells fail the test. A potential spatial dependence of neighbouring grid cells is neglected as each grid cell is considered as an individual contributor to the DI-SSI, although spatial dependence would potentially increase the amount of rejected cells. Being aware of the weaknesses of the ks-test when distributional parameters are estimated from the sample and the multiple-testing setting, we still consider this test as evidence that a GPD represents a sufficiently good model of v^* in our region of interest. To avoid the problem with estimated distributional parameters, one could simulate the distribution of the test-statistic under the Null for every grid point. However, we consider this as too costly for the scope of this study here.

Analogous to the equiprobability transformation to yield the Standardized Precipitation Index (SPI; Lloyd-Hughes and Saunders (2002)), the GPD fitted cumulative probability distribution of v^* is transformed into a standard exponential distribution as the GPD is closely related to the exponential family (rate $\lambda = 1$; cf. Lloyd-Hughes and Saunders (2002), their Figure 1). Equation (4a) defines the transformed value x by equating the GPD for v^* with the exponential probability distribution for x. The resulting Equation (4b) gives the contribution to the DI-SSI (x in Equation (4b)) of a single grid cell where ξ represents the shape and σ the scale parameter of the GPD distribution.

$$1 - \left(1 + \frac{\xi v^*}{\sigma}\right)^{\frac{-1}{\xi}} = 1 - e^{-x} \qquad (4a)$$

$$\frac{1}{\xi} ln \left(1 + \frac{\xi v^*}{\sigma}\right) = x \qquad (4b)$$

The definition of the (integral) DI-SSI in turn is the result of the summation over the entire footprint of a respective storm, equivalent to the definition of the SSI (cf. Equations (1) and (5)).

$$\text{DI} - \text{SSI}_{T,K} = \sum_t^T \sum_k^K \left[\frac{1}{\xi} \ln \left(1 + \frac{\xi v^*}{\sigma}\right) \cdot A_k\right] \qquad (5)$$

3. DI-SSI in practice and in comparison the SSI

The 99th percentile of the original wind speed distribution V_k at grid cell k is equal to the 50th percentile of v^* (as only wind speeds above the 98th percentile are considered). By definition of the standard exponential distribution, its 50th percentile (median) is equal to ln $2 = 0.69$. Thus, a wind speed v_k equal to the 99th percentile results in a DI-SSI contribution of 0.69. In the same way, the 99.9th percentile of V_k is equal to a contribution of $-\ln 0.05 = 3.00$ (cf. Equation (4b)). By using these values, the DI-SSI becomes more readily interpretable: The average integral DI-SSI value per

time step of storm Kyrill is 187.66 (cf. Table 2). This value divided by 3.0 (or ln 2, respectively) represents an equivalent number of 1×1 degree reference grid cells that feature the 99.9th (99th) percentile. Thus for Kyrill this corresponds to around 63 (270) 'virtual' grid cells per time step in which the local 99.9th (99th) percentile was observed. In order to compare the SSI contributions to its DI-SSI equivalents, both were calculated for the grid cells described in Section 1.2. As opposed to the SSI contribution for that particular grid cell, which differs by a factor of almost 20, the DI-SSI contribution is almost equal for the two wind speeds (see Table 1).

Table 2 presents integral values of the SSI and DI-SSI for some of the most prominent European windstorms. As expected storms that occurred on the edges of the classical storm track yield comparatively large SSI values. One of the most striking examples is represented by windstorm Klaus (Liberato et al., 2011) whose SSI value is almost three times as large as the respective value for windstorm Daria (Heming, 1990), whereas their DI-SSI values are almost of the same magnitude. Similar conclusions can be drawn from the storms Klaus and Kyrill (Fink et al., 2009): The DI-SSI is similar for both events; however the SSI is about 1.5 times larger for Klaus. Thus, judging from the SSI it appears that storm Klaus was far more intense than both Daria and Kyrill. The different assessment of severity for storms in different climatic background conditions is even more striking when comparing average SSI/DI-SSI values per time step. Klaus and Martin (Ulbrich et al., 2001) which follow similar tracks across Southern and Central Europe exhibit the largest SSI values per time step whereas the largest DI-SSI per time step can be identified for Vivian and Kyrill. Daria, Klaus and Vivian show the largest difference in rank if assessed by the average value per time step. The largest DI-SSI is associated with the storm Vivian (McCallum and Norris, 1990) which ranks second with regard to the SSI ranking. The large magnitude of the DI-SSI can potentially be explained by very extreme winds observed over the Atlantic Ocean (cf. Figure S1, Supporting information). As shown for storm Klaus in Section 3.1, the DI-SSI contributions over the sea are considerably larger than for the SSI. Thus, a storm with extreme surface winds over the sea is subject to high DI-SSI values as the DI-SSI is purely based on the singularity of wind speeds without any potential impact consideration. The biggest discrepancy between the respective rankings for the integral values of the storms is observed for storm Martin (4th compared with 8th) which is in line with the arguments for storm Klaus. An application of both indices is shown in Figure 2 (left panel) where the correlation of the annual storm intensity (annual sum of all SSI/DI-SSI contributions within a 500 km radius around a grid cell) and the annual storm occurrence per grid cell is presented. Hunter et al. (2016) showed a similar figure (their Figure 4(a)) using the vorticity as a severity metric. The coherent area of significant values over

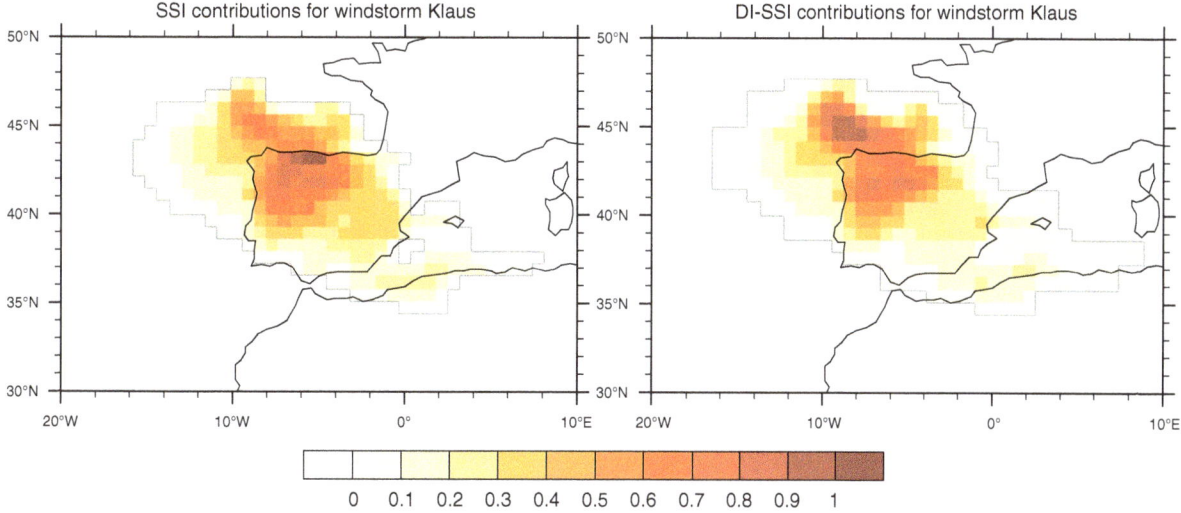

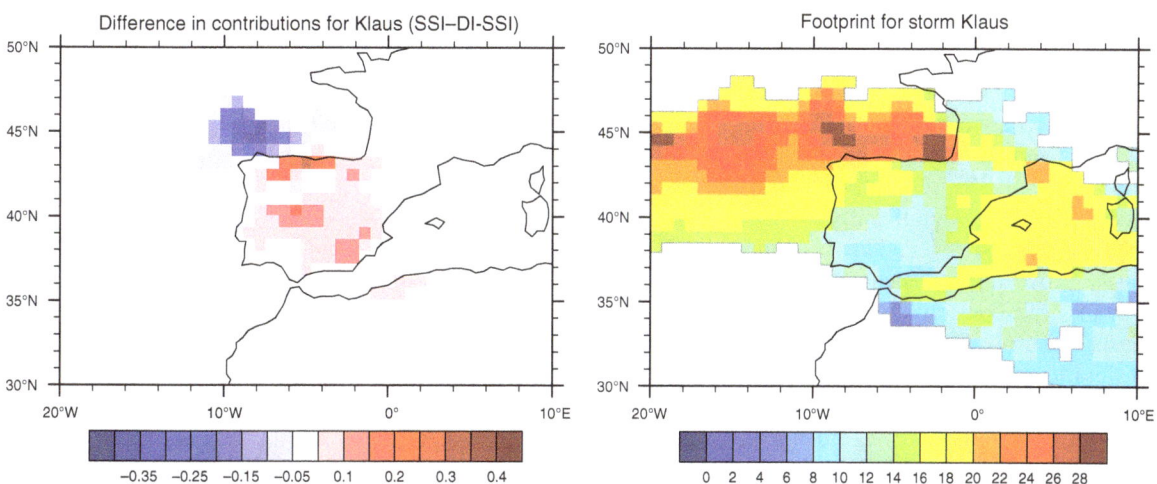

Figure 3. Footprint of storm Klaus on 24 January 2009 with SSI contributions shown on the top left, the DI-SSI contributions on the top right and the differences between the both values on the bottom left. The footprint of maximum wind speeds (m s⁻¹) for the entire storm is shown in the bottom right panel. Both SSI and DI-SSI were standardized for comparison reasons. Positive values indicate grid cells with larger SSI contributions value; negative values indicate larger DI-SSI contributions. There is a distinct separation represented by the coast line of the northern and western coast of the Iberian Peninsula.

Scandinavia is smaller compared with their results. The overall pattern looks fairly similar though, with most of Scandinavia showing positive correlations, implying that seasons with more storms also feature more intense storms.

Especially for the DI-SSI (middle-left panel), there is a large area of significant correlation between occurrence and intensity southwest of the British Isles that is not visible in their figure. This indicates the enhanced capability of the DI-SSI to characterize intense and unusual wind speeds not only over land but also over the sea compared with using vorticity as a severity metric. This is in line with the large DI-SSI value for windstorm Vivian for the DI-SSI is capable of quantifying extreme surface winds regardless of their occurrence. This is also supported by the difference between the two correlations shown in the bottom-left panel of Figure 2 as most areas over the Central Atlantic are positive, thus denoting larger DI-SSI correlations.

3.1. SSI and DI-SSI compared for a European storm example

Figure 3 serves as an example of how the previously discussed differences between the SSI and the DI-SSI arise: The figure shows a snapshot of the footprint of storm Klaus (Liberato *et al.*, 2011) and the footprint of the entire storm in the bottom right panel. The overall footprint of the storm looks exactly the same by definition as the local 98th percentile is used as a detection criterion in the storm tracking algorithm. The geographical intensity distribution however is different for the two indices (both indices are standardized for comparison). Whereas the SSI has its largest contributions over land on the northern coast of the Iberian Peninsula, the DI-SSI has in fact its largest contributions over the sea just north of the northern coast of Spain. This area coincides with the area of the most extreme wind speeds. This difference becomes more obvious when looking at the differences of the contributions of both indices. The coast line of the Iberian Peninsula represents an almost perfect segregation between negative and positive differences. This is according to the expectation regarding the features of the SSI and DI-SSI. The SSI can be used well to assess the potential damage to infrastructure, however judging from this figure it would seem that the wind speeds over the Atlantic do not have the same exceedance probability as they have over land. The DI-SSI draws a different picture: Albeit still showing large values over land, the more extreme values are apparent over the ocean indicating that the wind speeds in that area were even more exceptional with regard to their climatological wind speed distribution. This supports the arguments regarding the large DI-SSI for the storm Vivian in Section 3.

3.2. Intensity indices in connection with the NAO

A more quantitative comparison is supplied in the right panel of Figure 2. These two figures show the correlation coefficients between the annual winter NAO time series and the annual intensity time series per grid cell for the SSI and DI-SSI, respectively. Grid cells with a correlation coefficient significantly different from zero at the 5% level are stippled. This correlation does not necessarily prove any physical evidence; however, it indicates that the correlation was unlikely if the null hypothesis was true. Considering this fact, both maps show a significant link between the NAO and the intensity of European windstorms for most of Europe. However, overall there are more significant grid cells for the correlation using the DI-SSI compared with the SSI. This applies especially to large parts of southwest France, parts of Northern Africa and some areas in northeast Europe, thus regions which are affected less frequently by large scale windstorms. The largest difference in correlation is observed south of Greenland and around Iceland. According to the bottom panel of Figure 1, these areas are also on the edge of the storm track. This is another indication showing that the DI-SSI is a suitable metric to quantify extreme wind speeds outside the main storm track. The correlation pattern within the main storm track (central North Atlantic) is almost equal for both indices This supports the expectation that they behave fairly similar given the amount of storms per grid cell is sufficiently large.. Thus, the DI-SSI is a useful metric to represent the extremeness of wind speeds both in areas with little annual storm activity and also in areas with increased storminess.

4. Summary and discussion

This study introduces the DI-SSI: It serves as a quantification of exceptional surface wind speeds, especially for those high wind speeds occurring outside of the main storm tracks. Due to strongly diverse wind climatologies in different regions, the actual wind speed is an improper metric for the assessment of extremeness. A widely used index, especially in the impact community is the SSI developed by Leckebusch et al. (2008). The SSI is a metric that relates extreme winds to their potential damage on housing or infrastructure, whereas the newly introduced DI-SSI assesses the severity of exceptional wind speeds based on their occurrence probability: The shape of the tail of the local wind climatology is taken into account and used as input regarding the objective severity of an event. Thus, the severity of storms occurring within the main storm track can be compared more accurately to those rarer events, e.g. in the Mediterranean area. SSI/DI-SSI values are presented for nine prominent European winter storms. These values reveal the difference between the two indices. The largest SSI values arise for storms occurring on the edges of the storm track (Klaus, Martin), whereas the DI-SSI ranks storm Vivian and Kyrill as the most severe events.

In connection with the NAO index, the DI-SSI time series shows a more coherent area of significant correlation over Southwest Europe and also the Baltic Sea area compared with the SSI. This proves the capability of the DI-SSI to assess the severity of extreme winds both inside and outside of the main storm track. A larger area of correlation is also apparent for the correlation between frequency and intensity. The overall pattern of correlation for both indices is in agreement with Hunter et al. (2016). The results imply that especially within the main storm track in the North Atlantic and for most parts of Scandinavia, seasons with many storms also tend to feature more intense storms. This is in accordance with Vitolo et al. (2009) who found that serially clustered seasons are likely to spawn more intense storms. Technically, the SSI is easier to compute than the DI-SSI for it only requires wind speed data on grid cell level and no fitting of a statistical model. The DI-SSI requires more processing of the data, however it is a useful additional tool to assess the severity of storms/extreme winds regardless of their geographic occurrence.

Acknowledgements

M.A. Walz has been supported by a National Environmental Research Council (NERC) CENTA PhD scholarship kindly funded by Research Councils UK (RCUK). Fragments of this manuscript were prepared as part of a MSc Thesis at Freie Universität Berlin. H. W. Rust received support by the Freie Universität Berlin within the Excellence Initiative of the German Research Foundation. G.C. Leckebusch is supported for his research by the European Union (EU) FP7-MC-CIG-322208 grant. The authors thank the ECMWF for providing ERA Interim reanalysis data. The authors would also like to thank Ben Youngman and an anonymous reviewer for many helpful comments.

Supporting information

The following supporting information is available:

Figure S1. Footprint of maximum wind speed for storm Vivian and Kyrill. The very extreme wind speeds over the Central Atlantic Ocean for Vivian are responsible for the very large DI-SSI value compared to all the other storms which are compared in Table 2. This is due to the fact that the DI-SSI shows the same magnitude for wind speeds of the same extremeness.

References

Coles SG. 2001. *An Introduction to Statistical Modeling of Extreme Values*, (Vol. 208). Springer: London.

Cooley D, Nychka D, Naveau P. 2007. Bayesian spatial modeling of extreme precipitation return levels. *Journal of the American Statistical Association* **102**(479): 824–840.

Dee D, Uppala S, Simmons A, Berrisford P, Poli P, Kobayashi S, Andrae U, Balmaseda M, Balsamo G, Bauer P. 2011. The ERA-Interim reanalysis: configuration and performance of the data assimilation system. *Quarterly Journal of the Royal Meteorological Society* **137**(656): 553–597.

Donat MG, Leckebusch GC, Pinto JG, Ulbrich U. 2010. Examination of wind storms over Central Europe with respect to circulation weather types and NAO phases. *International Journal of Climatology* **30**(9): 1289–1300.

Donat MG, Pardowitz T, Leckebusch GC, Ulbrich U, Burghoff O. 2011. High-resolution refinement of a storm loss model and estimation of return periods of loss-intensive storms over Germany. *Natural Hazards and Earth System Sciences* **11**(10): 2821–2833.

Fink AH, Brücher T, Ermert V, Krüger A, Pinto JG. 2009. The European storm Kyrill in January 2007: synoptic evolution, meteorological impacts and some considerations with respect to climate change. *Natural Hazards and Earth System Sciences* **9**(2): 405–423.

Heffernan J, Stephenson A. 2015. Ismev: R package version 1.40, viewed.

Held H, Gerstengarbe F-W, Pardowitz T, Pinto J, Ulbrich U, Born K, Donat M, Karremann M, Leckebusch G, Ludwig P, Nissen K, Österle H, Prahl B, Werner P, Befort D, Burghoff O. 2013. Projections of global warming-induced impacts on winter storm losses in the German private household sector. *Climatic Change* **121**(2): 195–207.

Heming J. 1990. The impact of surface and radiosonde observations from two Atlantic ships on a numerical weather prediction model forecast for the storm of 25 January 1990. *Meteorological Magazine* **119**(1421): 249–259.

Hodges KI, Lee R, Bengtsson L. 2011. A comparison of extratropical cyclones in recent reanalyses ERA-Interim, NASA MERRA, NCEP CFSR, and JRA-25. *Journal of Climate* **24**(18): 4888–4906.

Hunter A, Stephenson DB, Economou T, Holland M, Cook I. 2016. New perspectives on the collective risk of extratropical cyclones. *Quarterly Journal of the Royal Meteorological Society* **142**(694): 243–256.

Klawa M, Ulbrich U. 2003. A model for the estimation of storm losses and the identification of severe winter storms in Germany. *Natural Hazards and Earth System Sciences* **3**(6): 725–732.

Kruschke T. 2015. Winter wind storms: Identification, verification of decadal predictions, and regionalization. Dissertation, Freie Universität Berlin, Berlin, Germany.

Kunz M, Mohr S, Rauthe M, Lux R, Kottmeier C. 2010. Assessment of extreme wind speeds from Regional Climate Models – Part 1: estimation of return values and their evaluation. *Natural Hazards and Earth System Sciences* **10**(4): 907–922.

Leckebusch GC, Renggli D, Ulbrich U. 2008. Development and application of an objective storm severity measure for the Northeast Atlantic region. *Meteorologische Zeitschrift* **17**(5): 575–587.

Liberato ML, Pinto JG, Trigo IF, Trigo RM. 2011. Klaus – an exceptional winter storm over northern Iberia and southern France. *Weather* **66**(12): 330–334.

Lloyd-Hughes B, Saunders MA. 2002. A drought climatology for Europe. *International Journal of Climatology* **22**(13): 1571–1592.

Lumbroso D, Vinet F. 2011. A comparison of the causes, effects and aftermaths of the coastal flooding of England in 1953 and France in 2010. *Natural Hazards and Earth System Sciences* **11**(8): 2321–2333.

McCallum E, Norris W. 1990. The storms of January and February 1990. *Meteorological Magazine* **119**(1419): 201–210.

Osinski R, Lorenz P, Kruschke T, Voigt M, Ulbrich U, Leckebusch G, Faust E, Hofherr T, Majewski D. 2016. An approach to build an event set of European windstorms based on ECMWF EPS. *Natural Hazards and Earth System Sciences* **16**(1): 255–268.

Palutikof J, Skellern A. 1991. Storm Severity over Britain. A Report to Commercial Union General Insurance, Climatic Research Unit, School of Environmental Sciences, University of East Anglia, Norwich (UK).

Parton G, Vaughan G, Norton E, Browning K, Clark P. 2009. Wind profiler observations of a sting jet. *Quarterly Journal of the Royal Meteorological Society* **135**(640): 663–680.

Pinto JG, Zacharias S, Fink AH, Leckebusch GC, Ulbrich U. 2009. Factors contributing to the development of extreme North Atlantic cyclones and their relationship with the NAO. *Climate Dynamics* **32**(5): 711–737.

SwissRe. 2016. Natural Catastrophes and Man-Made Disasters in 2015: Asia suffers substantial losses. Sigma No 1/2016. Order number: 270_0116_EN.

Tripoli G, Medaglia C, Dietrich S, Mugnai A. 2005. The 9–10 November 2001 Algerian flood. *Bulletin of the American Meteorological Society* **86**(9): 1229.

Ulbrich U, Christoph M. 1999. A shift of the NAO and increasing storm track activity over Europe due to anthropogenic greenhouse gas forcing. *Climate Dynamics* **15**(7): 551–559.

Ulbrich U, Fink A, Klawa M, Pinto J. 2001. Three extreme storms over Europe in December 1999. *Weather* **56**(3): 70–80.

Vitolo R, Stephenson DB, Cook IM, Mitchell-Wallace K. 2009. Serial clustering of intense European storms. *Meteorologische Zeitschrift* **18**(4): 411–424.

Vrac M, Naveau P. 2007. Stochastic downscaling of precipitation: from dry events to heavy rainfalls. *Water Resources Research* **43**(7): W07402.

20

Characteristics of multiscale vortices in the simulated formation of Typhoon Durian (2001)

Yaping Wang,[1,2] Xiaopeng Cui[1,3] and Yongjie Huang[1,2]*

[1] Key Laboratory of Cloud-Precipitation Physics and Severe Storms (LACS), Institute of Atmospheric Physics, Chinese Academy of Sciences, Beijing, China
[2] University of Chinese Academy of Sciences, Beijing, China
[3] Collaborative Innovation Center on Forecast and Evaluation of Meteorological Disasters, Nanjing University of Information Science & Technology, Nanjing, China

*Correspondence to:
Y. Huang, Key Laboratory of
Cloud-Precipitation Physics and
Severe Storms (LACS),
Institute of Atmospheric
Physics, Chinese Academy of
Sciences, No. 40 Huayanli,
Chaoyang District, Beijing
100029, China.
E-mail: huangyj@mail.iap.ac.cn

Abstract

The formation of Typhoon Durian (2001) was simulated well by the Weather Research and Forecasting model. The vorticity field was separated into three scales: small scale ($L < 30$ km), intermediate scale (30 km $< L < 120$ km), and system scale ($L > 120$ km), where L is wavelength. During the formation, small-scale vorticity anomalies, associated with convection, aggregated radially inward and rotated anticlockwise. The system-scale vorticity field presented a distinct mid-level vortex before genesis, and then a well-organized low-level vortex. The spectral power of the low-level vorticity at the small scale barely changed, while that at the system scale continued to grow steadily. The vorticity growth or spin-up of the near-surface tropical cyclone embryo was mainly due to the convergence of vertical vorticity flux at the planetary boundary layer. The positive (negative) effect of vorticity flux convergence at lower (middle-to-upper) levels appears to cause the shift of system-scale vorticity center from the middle to lower troposphere.

Keywords: tropical cyclogenesis; multiscale vortices; WRF model; Typhoon Durian

1. Introduction

Tropical cyclone (TC) formation, involving interactions among multiple spatial- and temporal-scale systems, is a very important aspect of TC research. Large-scale environmental conditions and physical factors associated with TC formation have been studied for a long time (Palmen, 1948; Riehl, 1948; Gray, 1968; Cheung, 2004; Zhang and Cui, 2013). Gray (1968) and Mcbride and Zehr (1981) proposed six factors favorable for TC formation: sufficient ocean thermal energy with sea surface temperature over 26 °C; enhanced mid-troposphere relative humidity; a deep layer of conditional instability; weak to moderate vertical wind shear; displacement by at least 5° away from the Equator; and cyclonic low-level relative vorticity. Besides, interactions among large-scale circulation, mesoscale convective activities, and convective-scale systems during TC genesis have also received considerable attention (Zhang and Bao, 1996a, 1996b; Bister and Emanuel, 1997; Ritchie and Holland, 1997, 1999; Zhang et al., 2010; Fang and Zhang, 2011; Lu et al., 2012; Park et al., 2015). All of these lines of enquiry can be summed up into a key issue: how mesoscale vortices and small-scale deep convection organize to create a system-scale TC vortex.

In terms of TC development, several theories have been proposed, such as the convective instability of the second kind (Charney and Eliassen, 1964) and wind-induced surface heat exchange (Emanuel, 1986; Rotunno and Emanuel, 1987), both of which require a preexisting initial low-pressure vortex with a certain intensity. More recently, a number of theories describing TC-scale vorticity enhancement have been proposed. Bister and Emanuel (1997) postulated a pathway for the development of Hurricane Guillermo, based on observations and numerical simulation; that is, a mid-level mesoscale convective vortex (MCV) expanded downward and, later, a near-surface cyclone with a warm core developed. This kind of approach is referred to as 'top-down' thinking. In contrast, another hypothesis, following 'bottom-up' thinking, emphasizes the key role of surface-based convection in the genesis process. Several hypotheses on the organizational mechanisms of small-scale deep convection have also been proposed. Montgomery et al. (2006) and Hendricks et al. (2004) suggested that vortical hot towers (VHTs) contribute to system-scale spin-up via the aggregation and merger of VHT-induced vorticity anomalies, as well as via diabatic heating. Lu et al. (2012) and Zhang et al. (2012) suggested that a system-scale vortex forms with the merger and aggregation of VHTs as one of the possible mechanisms, while Fang and Zhang (2011) (FZ11 hereinafter) pointed out that system-scale vorticity grows through diabatic heating-induced convergence of environmental and convectively generated vorticity. Above all, TC genesis remains an open question involving multiscale systems from several kilometers to several hundred kilometers.

TCs forming over the northwest Pacific Ocean, including the South China Sea (SCS), often impose serious and rapidly occurring disasters upon the dense population of south and southeast China. Therefore, examination of TC formation processes is very important. With the development of remotely measured observations including those obtained from radar, satellites, aircraft and surface stations, multiscale vortex features, and their evolutions during TC genesis have begun to be revealed, especially those taking place in the Western Pacific and Atlantic (Reasor et al., 2005; Sippel et al., 2006; Houze et al., 2009; Davis and Ahijevych, 2012). However, observations remain inadequate in the SCS region. Thus, high-resolution numerical simulations are a very useful tool to spatiotemporally resolve the dynamics and thermodynamics of TC formation processes in detail. In this context, this study focused on the characteristics and evolution of multiscale vortices, as well as how they related to the genesis of Typhoon Durian (2001) in the SCS, using numerical simulation. The next section introduces Typhoon Durian (2001) and describes how we verified the simulation using available observations. Section 3 reports the multiscale features during Durian's genesis through the scale-separation method.

2. Simulation of Durian

2.1. Review of TC Durian (2001)

Durian was recorded to initially generate as a tropical depression (TD, with winds of $10.8-17.1\,\mathrm{m\,s^{-1}}$ in China) at ($117°E$, $16°N$) within the monsoon trough at 0600 UTC on 29 June 2001 in the middle region of the SCS. After that, Durian moved northwest and developed into a tropical storm (TS, with winds of $17.2-24.4\,\mathrm{m\,s^{-1}}$ in China) at 0600 UTC on 30 June. Finally, it intensified into a typhoon (TY, with winds of $32.7-41.4\,\mathrm{m\,s^{-1}}$ in China), with a minimum sea level pressure of 970 hPa and maximum wind speed of $35\,\mathrm{m\,s^{-1}}$ (Figure 1(c) and (d)), and landed at Zhanjiang, Guangdong Province, causing tremendous damage in adjacent areas (Zhang et al., 2008).

2.2. Numerical simulation

To examine the genesis process of TC Durian in detail, this study utilized a numerical simulation obtained using the Weather Research and Forecasting (WRF) model, version 3.5.1, driven by National Centers for Environmental Prediction (NCEP) reanalysis data, on $2.5° \times 2.5°$ grids. Based on the work in Zhang et al. (2008), the simulation used four nesting domains (displayed in Figure 1(a)) with horizontal resolutions of 54, 18, 6, and 1.2 km and mesh grids of 112×145, 235×223, 373×289, and 951×776, respectively. The Goddard microphysics scheme and Yonsei University boundary layer scheme were used in each domain, while the Kain–Fritsch cumulus parameterization scheme was only used in the two outermost domains. The model was initiated by analysis fields dominated by a monsoon trough, without a bogus vortex, and integrated for

84 h from 0000 UTC on 28 June to 1200 UTC on 1 July 2001. The two innermost domains started their integration 12 h after the initial time.

The model basically reproduced the large-scale circulation characteristics, including the position and evolution of the monsoon trough and the subtropical high (not shown). The simulated Durian was identified to generate at 0800 UTC on 29 June with the 6 km-resolution model data, when a system-scale low pressure center and the closed cyclonic circulation near the surface both first appeared (Figure 1(b)). The simulated track and intensity of Durian were compared with the best-track analysis of the Shanghai Typhoon Institute (http://tcdata.typhoon.gov.cn/index.html). The simulated track of Durian appeared to agree well with observations in the early stage of 0600 UTC on 29 June to 1200 UTC on 1 July (Figure 1(c)). The simulated genesis moment (0800 UTC on 29 June) of Durian was about 2 h later than observed (0600 UTC on 29 June), and the simulated formation location of Durian was ($116.2°E$, $16°N$), which was about 80 km away from the observed location of ($117°E$, $16°N$). The intensity errors reduced gradually as the circulation intensified (Figure 1(d)). The minimum sea level pressure decreased steadily from 0600 UTC on 29 June to 1200 UTC on 30 June and then decreased quickly from 1200 UTC on 30 June to 1200 UTC on 1 July. The observed Durian stably developed from a TD to a TS before 0600 UTC on 30 June 2001 and then rapidly intensified from a TS to a TY. The simulated maximum wind speed near the storm center was basically larger than the observed before 1800 UTC on 30 June, then the deviation significantly decreased. Even so, the wind speed of simulation and observation underwent a similar evolution trend.

Geostationary meteorological satellite (GMS) cloud images with a resolution of $0.05° \times 0.05°$, and simulated outgoing longwave radiation (OLR) with a resolution of 6 km, are presented in Figure 2. The incipient Durian cloud system evolved from a mesoscale cloud cluster, which was isolated from the north part of a mature mesoscale convective system (MCS, Figure 2). At 1800 UTC on 28 June, the disturbance in the north of the MCS developed into a mesoscale convective cloud (Figure 2(a)). Six hours later (0000 UTC on 29 June), the distribution of observed disturbance clouds became cyclonic (Figure 2(c)). After the moment of Durian's genesis (0600 UTC on 29 June), the cloud clusters grew larger and were better organized (Figure 2(e) and (g)). From the simulated OLR, a mesoscale cloud belt at the south of $15°N$ could be seen along the monsoon trough at 1800 UTC on 28 June. However, the convective cloud clusters had not yet organized, and were instead scattered irregularly at the north of the main cloud belt, possibly because the simulated mid-level vortex has not set up a proper condition to assist the organization of convection (Zhang and Bao, 1996a, 1996b). During the following period, a closed sea level pressure center developed. Meanwhile, the cloud clusters grew larger and generally organized into a tight, cyclonic system.

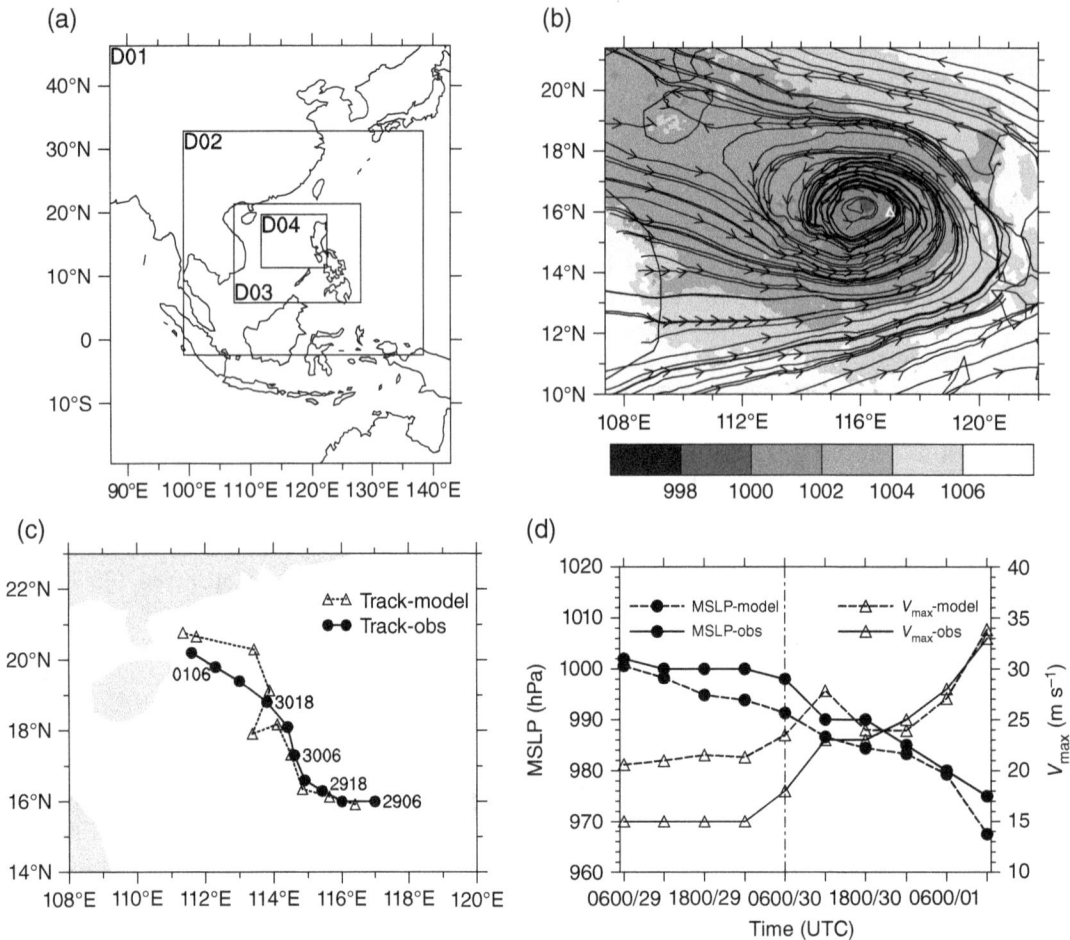

Figure 1. (a) Model domain configuration. (b) Streamlines at 900 hPa and sea level pressure (shaded, units: hPa) obtained from domain 3 at 0800 UTC on 29 June 2001. The triangle indicates the observed genesis location. (c) Track and (d) minimum sea level pressure (units: hPa) and maximum wind speed (units: m s^{-1}) of observed and simulated storms from 0600 UTC on 29 June to 1200 UTC on 1 July 2001. The interval is 6 h, and the vertical dashed line in (d) indicates the observed transition from a TD to a TS.

The success of the simulation, in terms of its reflection of the observed situation, justifies further examination of Durian's genesis process based on its results.

3. Multiscale features of Durian's genesis

3.1. Relative vorticity

Convection and vortices during TC genesis usually possess substantial and different-scale anomalous vorticity, and TC genesis is intimately linked to the enhancement of TC-scale cyclonic vorticity. The relative vorticity of domain 4, with grid spacing of 1.2 km at low [1 km above sea level (ASL)], middle (6 km ASL), and upper (11 km ASL) levels, was decomposed into three scales according to the spectral characteristics of the vorticity field: system scale (wavelengths >120 km), intermediate scale (wavelengths >30 km but <120 km), and small scale (wavelengths <30 km). The threshold values were selected based on the fact that the power spectra of the vorticity field at wavelengths <30 km, between 30 km and 120 km, and >120 km, exhibited different evolutions. The spectral decomposition was obtained by a Fourier transform and filter, following Lin and Zhang (2008) and FZ11. The small-scale (convective-scale) relative vorticity centers are presumably induced by small-scale moist convection. Among them, those deep, rapidly rotating convective cores with horizontal scale of about 10–30 km are defined as VHTs, which usually develop a cyclonic vorticity bias at lower levels and appear as vorticity dipoles at middle–upper levels (Montgomery *et al.*, 2006). The VHTs have been demonstrated to exist in observation and play critical roles in TC genesis by numerical simulation analysis (Hendricks *et al.*, 2004; Montgomery *et al.*, 2006; Zhang *et al.*, 2008; Houze *et al.*, 2009). The intermediate-scale vorticity is regarded as a cluster of convective-scale vorticity, as that in FZ11.

Figure 3 presents the three-dimensional distribution of relative vorticity and its three scale components at 1800 UTC on 28 June (14 h prior to genesis), 0800 UTC on 29 June (the genesis moment), and 1800 UTC on 29 June (10 h after genesis). In general, the intensity of the small-scale vorticity was the strongest and the system-scale vorticity was the weakest (Figure 3). The small-scale vorticity

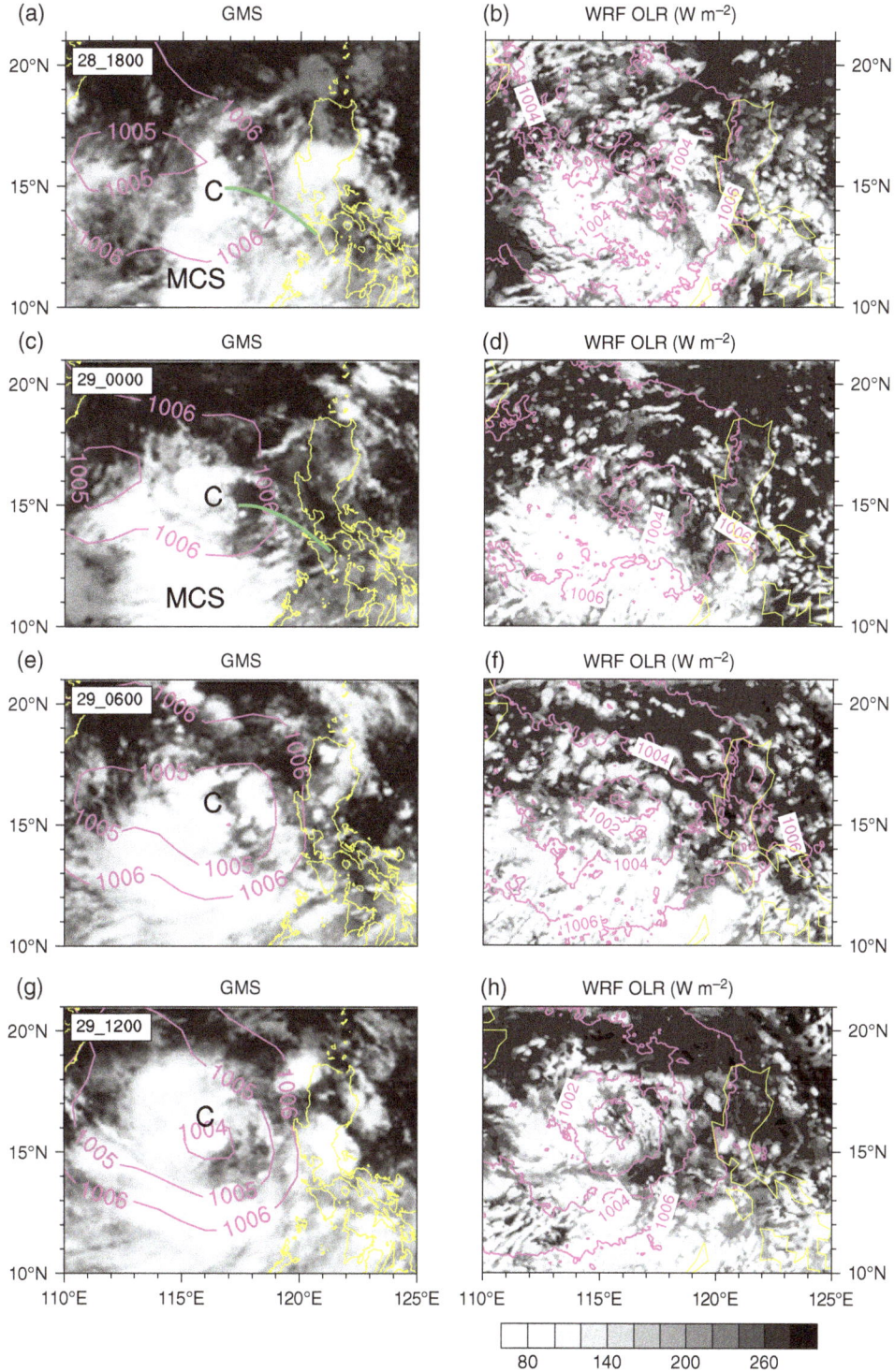

Figure 2. (a, c, e, g) GMS satellite images and mean sea level pressure (contours, units: hPa) from NCEP Final Operational Global Analysis data on a 1.0° × 1.0° grid. (b, d, f, h) Simulated OLR (units: W m^{-2}) and sea level pressure (contours, units: hPa,). (a, b) 1800 UTC on 28 June 2001; (c, d) 0000 UTC on 29 June 2001; (e, f) 0600 UTC on 29 June 2001; (g, h) 1200 UTC on 29 June 2001. MCS denotes mesoscale convective system; C denotes TC-related vortex; and the green line denotes the monsoon trough.

centers distributed irregularly inside and outside the 200 km-radius circular domain at 1800 UTC on 28 June (the fourth column in Figure 3(a)). The most intense small-scale vorticity anomalies were located at high levels near the tropopause. Fourteen hours later, at the genesis moment, low-level small-scale vorticity centers emerged and became stronger, tending to aggregate radially inward in the genesis area and rotate anticlockwise (the fourth column in Figure 3(b)). Meanwhile, fewer small-scale vorticity centers existed at middle–upper levels with the TC's formation. This coincided with a weakening of vertical velocity in the upper troposphere and a downward shift of vertical velocity cores (not shown). At 1800 UTC on 29 June

(10 hours after genesis), low-level small-scale and stronger vorticity centers were basically located within the 200 km-radius circular domain (the fourth column in Figure 3(c)). The aggregation of small-scale vorticity anomalies associated with VHTs and other small-scale convection is consistent with the analysis in Zhang *et al.* (2008); that is, a large number of VHTs occurred, while some of them as well as their residual vorticity aggregated and merged with each other. This phenomenon might be induced by larger-scale convergent or confluent flow, according to Fang and Zhang (2010). The intermediate-scale vorticity shared similar features of aggregation and development as the small-scale vorticity (the third column in Figure 3(a)–(c)). As for system-scale vorticity, a mid-level vortex, identified as the MCV according to Zhang *et al.* (2008), already existed at 1800 UTC on 28 June, and then became slightly weaker at the genesis moment. A relatively weak and well-organized low-level vorticity center, which represented the embryo of the storm, occurred at 0800 UTC on 29 June and subsequently intensified (the second column in Figure 3). In other words, the near-surface TC circulation had not been established until 0800 UTC on 29 June, and then it was able to self-develop.

3.2. Multiscale power spectra of relative vorticity

Figure 4 shows the integrated power spectra of relative vorticity at different levels, by adding up the power spectra at three different-scale ranges (FZ11). The scale separation has been carried out in a rectangular domain (400 × 400 km) moving with the mid-level vortex center (before genesis) and storm center (after genesis). The maximum amplitudes (spectral power) of small-scale and intermediate-scale vorticity were located in the upper troposphere at first, then decreased before 0600 UTC on 29 June and increased after 0600 UTC on 29 June with the genesis process of Durian (Figure 4(a), (b), and (f)). The maximum amplitudes of intermediate-scale vorticity moved downward from the upper troposphere to the middle troposphere. The center of system-scale spectral power existed at the middle level before 0300 UTC on 29 June, and then a low-level center gradually developed (Figure 4(c)). The low-level power spectra analysis shows that the amplitudes of the vorticity at the system scale (larger than 120 km) steadily increased, indicating that the near-surface TC circulation developed throughout Durian's genesis process. The amplitudes of the low-level vorticity at scales below 30 km was saturate and barely changed, while the power spectra of the intermediate-scale vorticity exhibited a slight decrease (Figure 4(d)). Furthermore, those amplitudes of the middle–upper-level vorticity at all scales showed a decreasing trend. Though the saturation of convective-scale vorticity may imply an immediate upscale transfer from the convective to the intermediate scale or system scale (FZ11), how the vorticity grows upscale remains an open question. Whether the transfer

of the maximum vorticity core from the middle troposphere to lower levels indicates a top-down mechanism is also a pending issue. The following section may provide some clues to these aspects via an analysis of the vorticity budget.

3.3. Sources of lower-to-middle tropospheric vorticity

The given analysis shows that the maximum system-scale vorticity gradually moved from the middle troposphere to the near surface during TC Durian's initial genesis process. To further evaluate the sources of different-level vorticity, a vorticity budget equation in a reference frame moving with the storm was solved, and the results are presented in this section. As shown in Davis and Galarneau (2009) and Raymond and López (2011), by integrating a traditional vorticity equation over any closed region and applying Gauss's theorem, the tendency of circulation around the area can be obtained (Raymond and López, 2011; Raymond *et al.*, 2011; Lussier *et al.*, 2014):

$$\frac{\partial \Gamma}{\partial t} = -\oint v_n \zeta_z dl + \oint \zeta_n v_z dl$$
$$+ \oint F_t dl \qquad (1)$$

where Γ is the circulation; the line integrals are taken to be in the anticlockwise direction over the periphery of a horizontal area A; v_n and ζ_n are the horizontal outward normal components of the velocity and relative vorticity; v_z and ζ_z are the vertical components of the velocity and absolute vorticity; and F_t is the component of frictional force **F** tangential to the periphery of A in the direction of the integration. The term on the left hand side virtually represents the total vorticity tendency in area A. The first term on the right-hand side of Equation (1) is the vertical absolute vorticity flux on the periphery of A, representing the convergence of vorticity flux to the area A. The second term is referred to as a 'tilting' term here, which actually combines the tilting term with the vertical advection. Davis and Galarneau (2009) explicitly interpreted the relationship of this term with vortex tilting. The third term represents the spin-down tendency due to friction. The baroclinic generation term is omitted in Equation (1) since it is generally neglected (as in this TC genesis case) except in highly baroclinic environment such as the hurricane eyewall or mid-latitude fronts. More detail regarding the calculation method of these terms can be found in Raymond and López (2011).

Figure 5 represents the temporally averaged vertical profiles of the terms in Equation (1) in a 400 km-long box moving with the storm center. Results are expressed in terms of box-integral vorticity changes. The box contains the circulation centers in the lower-to-middle troposphere. Red lines denoted 'net' in Figure 5 indicate the sum of three terms on the right-hand side of Equation (1). During 1400–2000 UTC on 28 June

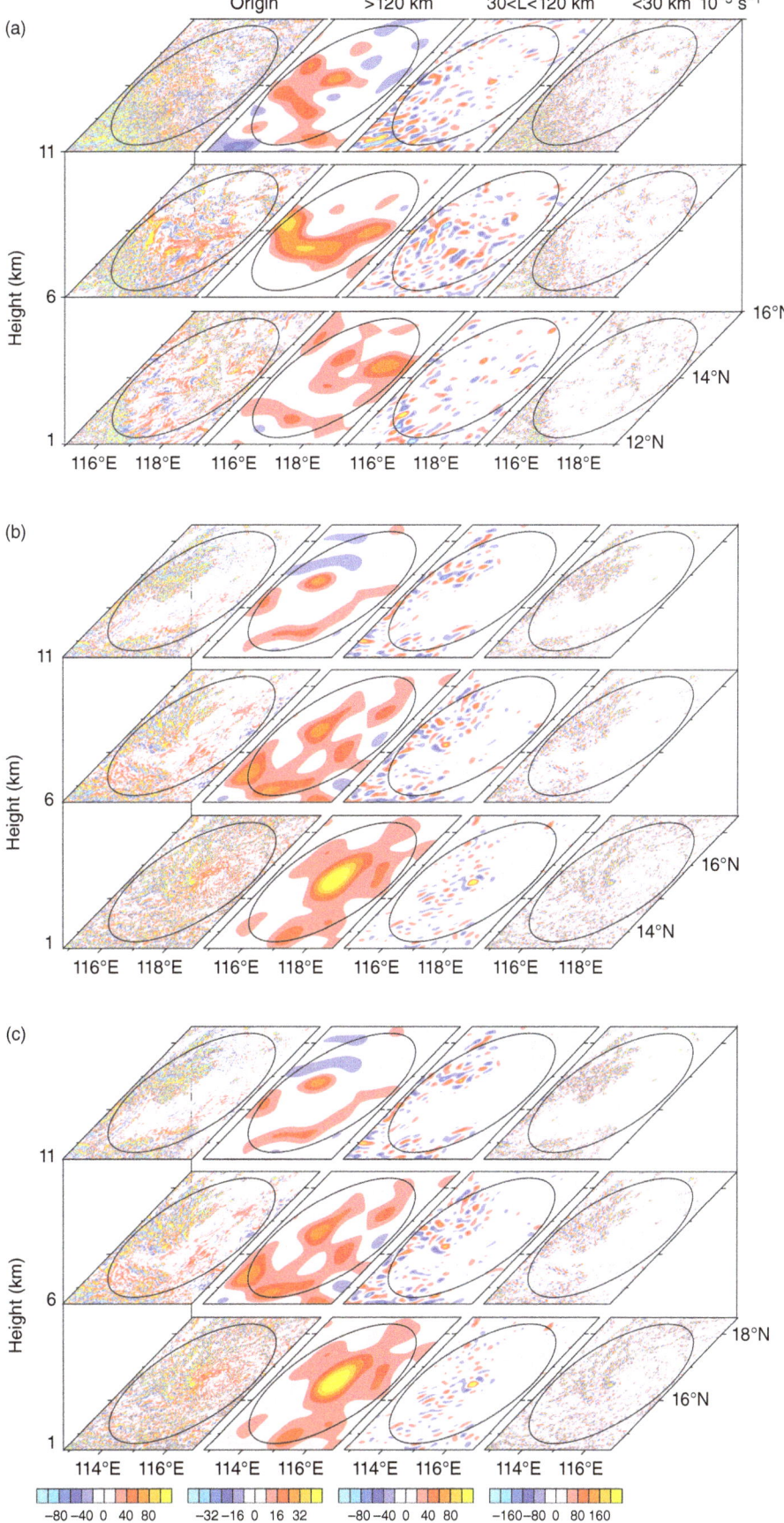

Figure 3. Horizontal distribution of low- ($z = 1$ km), middle- ($z = 6$ km), and upper-level ($z = 11$ km) relative vorticity (shaded) at (a) 1800 UTC on 28 June, (b) 0800 UTC on 29 June, and (c) 1800 UTC on 29 June, of different scales: the original field (first column); scales >120 km (second column); scales between 30 and 120 km (third column); and scales <30 km (fourth column). Units: 10^{-5} s^{-1}. The solid-line circle indicates the 200 km-radius region around the circulation center.

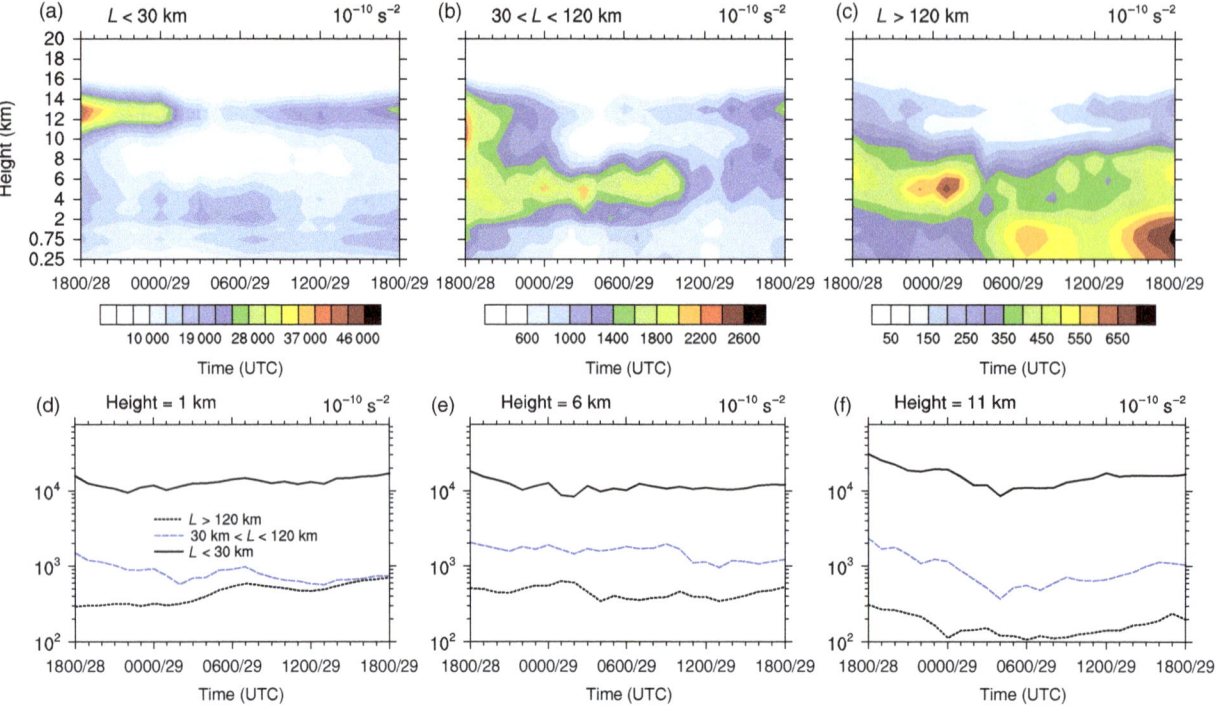

Figure 4. (a) Height–time cross section of the integrated power spectra of relative vorticity at the scale ranges <30 km. Panels (b, c) as in (a), but for the scale ranges between 30 and 120 km, and scales >120 km, respectively. (d) Temporal evolution of the integrated power spectra of relative vorticity at the three scale ranges at $z = 1$ km. Panels (e, f) as in (d), but for $z = 6$ km and $z = 11$ km, respectively. Units: 10^{-10} s^{-2}.

(18–12 h prior to genesis), vorticity increased throughout the whole troposphere. The circulation tendency at the lower-to-middle level was mainly due to the spin-up effect of vorticity flux convergence and the spin-down effect of the tilting term; while at the middle-to-upper level, the two effects were opposite. During 2000 UTC on 28 June to 0200 UTC on 29 June (12–6 h prior to genesis), the positive circulation tendency in the middle troposphere, corresponding to the development of the mid-level MCV, was contributed by both the convergence of vorticity flux and the tilting term. Friction almost offset the positive effect of convergence near the surface. During 0200–0800 UTC on 29 June (genesis moment), the middle-to-upper-level vorticity exhibited a significant decrease, which was induced by the divergence of vorticity flux above the planetary boundary layer (PBL). However, circulation still strengthened near the surface owing to the vorticity flux convergence, resulting in the system-scale vorticity center generally evolving from a mid-level vortex to a near-surface vortex. After Durian's genesis (0800 UTC on 29 June 2001), positive circulation tendency recovered above the PBL due to the significant contribution of vortex tilting.

In general, the convergence of vertical vorticity flux and vortex tilting both played important roles in the circulation tendency. The accumulation of near-surface vorticity and the spin-up of the low-level vortex was mainly induced by the continuous vorticity flux convergence at the PBL. As for the vorticity in the middle-to-upper troposphere, the spin-up effect

of vortex tilting and the spin-down effect of vorticity flux convergence both worked. Meanwhile, the positive effect of vorticity flux convergence at the lower level, as well as the negative effect at the middle-to-upper level appears to have caused the shift of the system-scale vorticity center from the middle troposphere to the near surface.

4. Conclusions and discussion

In this study, TC Durian (2001), forming over the SCS, was simulated using the WRF model, and the model output data were compared with observations. The characteristics of multiscale vortices were then explored by separating the vertical relative vorticity into three scales: wavelength <30 km, wavelength >30 km but <120 km, and wavelength >120 km.

The numerical simulation adequately reproduced the track and intensity of Durian, as well as its genesis process. The formation moment of the simulated TC was just 2 h later than observed, while the deviation of the formation location was about 80 km. The model was also capable of describing the organizational process of mesoscale cloud clusters during the formation.

In general, the smaller-scale vorticity was stronger than the larger-scale vorticity. As the TC generated, small-scale vorticity associated with VHTs and other convective-scale vortices tended to aggregate radially inward and rotate anticlockwise within the 200 km-radius circular domain. Meanwhile,

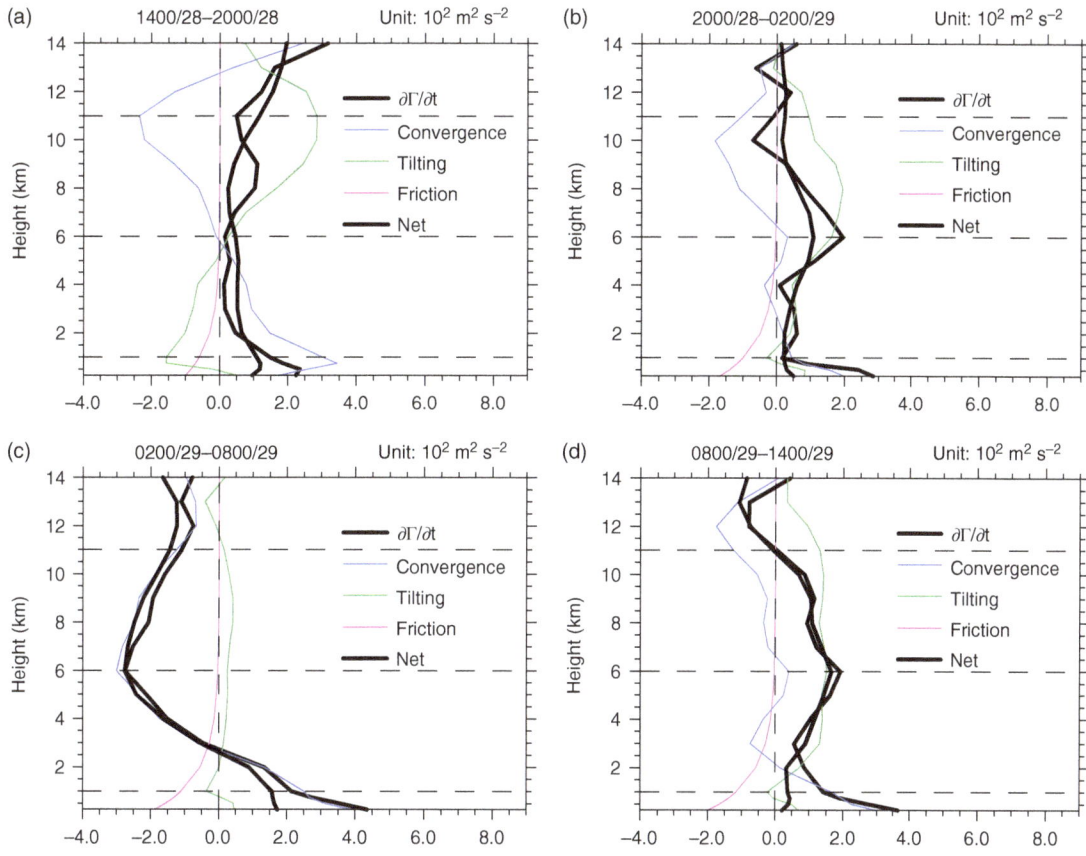

Figure 5. Vertical profiles of circulation budget terms averaged from (a) 1400 UTC on 28 June to 2000 UTC on 28 June, (b) 2000 UTC on 28 June to 0200 UTC on 29 June, (c) 0200 UTC on 29 June to 0800 UTC on 29 June, and (d) 0800 UTC on 29 June to 1400 UTC on 29 June, within a 400 km-long box moving with the storm: circulation tendency (black), contributions to the total circulation tendency (red) due to vorticity convergence (blue), vortex tilting (green), and friction (magenta) of Equation (1), units: 10^2 m^2 s^{-2}. The three horizontal dashed lines denote 1, 6, and 11 km. The vertical dashed line denotes zero.

middle–upper level vorticity gradually weakened while low-level vorticity strengthened. Large-scale vorticity at the middle level before the simulated genesis moment (0800 UTC on 29 June 2001), and that at the low level after genesis, both possessed a well-organized center, representing the mid-level MCV and embryo of the TC, respectively. The power spectra analysis showed that the low-level disturbances at all scales <30 km barely changed, while the amplitudes at scales >120 km steadily increased. Analysis of the vorticity budget suggested that the development of low-level vorticity, or spin-up of near-surface circulation, was mainly caused by the convergence of the vertical vorticity flux. Meanwhile, the positive effect of low-level vorticity flux convergence and the negative effect of middle-to-upper-level convergence appear to have caused the shift of the system-scale vorticity center from the middle troposphere to the near surface.

Acknowledgements

The authors were supported by the National Basic Research Program of China (973 Program) under grant number 2015CB452804. The work was carried out at the National Supercomputer Center in Tianjin, and the calculations were performed on TianHe-1(A). The authors thank the two anonymous reviewers for their helpful comments on the manuscript.

References

Bister M, Emanuel KA. 1997. The genesis of Hurricane Guillermo: TEXMEX analyses and a modeling study. *Monthly Weather Review* **125**: 2662–2682.

Charney J, Eliassen A. 1964. On the growth of the hurricane depression. *Journal of the Atmospheric Sciences* **2**: 68–75.

Cheung KKW. 2004. Large-scale environmental parameters associated with tropical cyclone formations in the western North Pacific. *Journal of Climate* **17**: 466–484.

Davis CA, Ahijevych DA. 2012. Mesoscale structural evolution of three tropical weather systems observed during PREDICT. *Journal of the Atmospheric Sciences* **69**: 1284–1305.

Davis CA, Galarneau TJ. 2009. The vertical structure of mesoscale convective vortices. *Journal of the Atmospheric Sciences* **66**: 686–704.

Emanuel KA. 1986. An air-sea interaction theory for tropical cyclones. Part I: steady-state maintenance. *Journal of the Atmospheric Sciences* **43**: 585–605.

Fang J, Zhang FQ. 2010. Initial development and genesis of Hurricane Dolly (2008). *Journal of the Atmospheric Sciences* **67**: 655–672.

Fang J, Zhang FQ. 2011. Evolution of multiscale vortices in the development of Hurricane Dolly (2008). *Journal of the Atmospheric Sciences* **68**: 103–122.

Gray WM. 1968. Global view of the origin of tropical disturbances and storms. *Monthly Weather Review* **96**: 669–700.

Hendricks EA, Montgomery MT, Davis CA. 2004. The role of "vortical" hot towers in the formation of tropical cyclone Diana (1984). *Journal of the Atmospheric Sciences* **61**: 1209–1232.

Houze RA, Lee WC, Bell MM. 2009. Convective contribution to the genesis of Hurricane Ophelia (2005). *Monthly Weather Review* **137**: 2778–2800.

Lin Y, Zhang F. 2008. Tracking gravity waves in baroclinic jet-front systems. *Journal of the Atmospheric Sciences* **65**: 2402–2415.

Lu XY, Cheung KKW, Duan YH. 2012. Numerical study on the formation of Typhoon Ketsana (2003). Part I: roles of the mesoscale convective systems. *Monthly Weather Review* **140**: 100–120.

Lussier LL, Montgomery MT, Bell MM. 2014. The genesis of Typhoon Nuri as observed during the tropical cyclone structure 2008 (TCS-08) field experiment – part3: dynamics of low-level spin-up during the genesis. *Atmospheric Chemistry and Physics* **14**: 8795–8812.

McBride JL, Zehr R. 1981. Observational analysis of tropical cyclone formation. Part II: comparison of non-developing versus developing systems. *Journal of the Atmospheric Sciences* **38**: 1132–1151.

Montgomery M, Nicholls M, Cram T, Saunders A. 2006. A vortical hot tower route to tropical cyclogenesis. *Journal of the Atmospheric Sciences* **63**: 355–386.

Palmen E. 1948. On the formation and structure of tropical hurricanes. *Geophysica* **3**: 26–38.

Park M-S, Kim H-S, Ho C-H, Elsberry RL, Lee M-I. 2015. Tropical cyclone Mekkhala's (2008) formation over the South China Sea: mesoscale, synoptic-scale, and large-scale contributions. *Monthly Weather Review* **143**: 88–110.

Raymond DJ, López CC. 2011. The vorticity budget of developing Typhoon Nuri (2008). *Atmospheric Chemistry and Physics* **11**: 147–163.

Raymond DJ, Sessions SL, López CC. 2011. Thermodynamics of tropical cyclogenesis in the northwest Pacific. *Journal of Geophysical Research* **116**: D18101.

Reasor PD, Montgomery MT, Bosart LF. 2005. Mesoscale observations of the genesis of Hurricane Dolly (1996). *Journal of Geophysical Research* **62**: 3151–3171.

Riehl H. 1948. On the formation of typhoons. *Journal of Meteorology* **5**: 247–265.

Ritchie EA, Holland GJ. 1997. Scale interactions during the formation of Typhoon Irving. *Monthly Weather Review* **125**: 1377–1396.

Ritchie EA, Holland GJ. 1999. Large-scale patterns associated with tropical cyclogenesis in the western Pacific. *Monthly Weather Review* **127**: 2027–2043.

Rotunno R, Emanuel KA. 1987. An air-sea interaction theory for tropical cyclones. Part II: evolutionary study using a nonhydrostatic axisymmetric numerical model. *Journal of the Atmospheric Sciences* **44**: 542–561.

Sippel JA, Nielsen-Gammon JW, Allen SE. 2006. The multiple-vortex nature of tropical cyclogenesis. *Monthly Weather Review* **134**: 1796–1814.

Zhang D-L, Bao N. 1996a. Oceanic cyclogenesis as induced by a mesoscale convective system moving offshore. Part I: a 90-h real-data simulation. *Monthly Weather Review* **124**: 1449–1469.

Zhang D-L, Bao N. 1996b. Oceanic cyclogenesis as induced by a mesoscale convective system moving offshore. Part II: genesis and thermodynamic transformation. *Monthly Weather Review* **124**: 2206–2226.

Zhang WL, Cui XP. 2013. Review of the studies on tropical cyclone genesis. *Journal of Tropical Meteorology* **29**: 337–346 (in Chinese).

Zhang WL, Wang AS, Cui XP. 2008. The role of the middle tropospheric mesoscale vortex in the genesis of Typhoon Durian (2001) – simulation and verification. *Chinese Journal of Atmospheric Sciences* **32**: 1197–1209 (in Chinese).

Zhang WL, Cui XP, Dong JX. 2010. The role of the middle tropospheric mesoscale vortex in the genesis of Typhoon Durian (2001) – diagnostic analysis of simulated data. *Chinese Journal of Atmospheric Sciences* **34**: 45–57 (in Chinese).

Zhang GP, Cheung KKW, Lu XY. 2012. Numerical study on the formation of Typhoon Ketsana (2003) in the Western North Pacific. In *30th Conference on Hurricanes and Tropical Meteorology*, FL, USA, 15–20 April 2012.

21

Model study of the asymmetry in tropical cyclone-induced positive and negative surges

Benjamin Wong* and Ralf Toumi

Blackett Laboratory, Space and Atmospheric Physics Group, Imperial College London, UK

*Correspondence to:
B. Wong, Blackett Laboratory,
Space and Atmospheric Physics
Group, Imperial College London,
Exhibition Road, London
SW72AZ, UK.
E-mail: benjamin.wong11@
imperial.ac.uk

Abstract

Storm surges pose significant threats to coastal communities, yet negative surges are not as well understood as positive surges. In this study, idealized experiments of a tropical cyclone forcing a 3D ocean model are conducted to investigate the asymmetry of positive and negative surges. Negative surges are larger in magnitude and extend further across the coastline than positive surges. While positive surges are driven by wind blowing onshore, negative surges are largely dominated by alongshore winds, with horizontal divergence as the main mechanism. This asymmetry also increases with decreasing depth and increasing latitude.

Keywords: negative surge; sea level; asymmetry

1. Introduction

With 44% of the world's population living within 150 km from the coast (Resio and Westerink, 2008), storm surges are a substantial threat to human lives and activities. Positive surges are widely studied due to the high impacts of coastal flooding but negative surges are less well understood. Some impacts of negative surges include ship grounding and draining of coastal aquifers. Ship grounding can lead to hull damage, subsequent collisions and oil spill disasters. Draining of coastal aquifers can lead to the depletion of drinking water supply and is detrimental to coastal communities dependent on it (Pousa *et al.*, 2013). Negative surges can also destroy coastal aquacultures, which are major economic contributors to countries such as Bangladesh (AsSalek, 1997).

AsSalek (1997) showed that negative surges are affected by factors such as the cyclone's inflow angle, central pressure, radius of maximum winds, speed of translation, propagation path, angle of coastal crossing and interaction with astronomical tides. However, the study was specific to selected points on the unique coastal geometry of the Meghna estuary, making it difficult to apply the same conclusions to other coastal cases. Peng *et al.* (2006) studied positive and negative surges using an idealized coastal setup, investigating the sensitivity of the surges to the cyclone's inflow angle, radius of maximum winds and the speed of translation. However, this idealized study was limited to a one-dimensional (1D) analysis where the positive and negative sea surface height (SSH) response was investigated at a single point. Modelling results of the surge at the Orissa coast of India in 1982 showed a positive surge to the right of the track and a negative surge to the left of the track and this was attributed to winds blowing onshore and offshore respectively (Pugh, 1987). However, the relatively larger negative surges could not simply be accounted for by 1D advection from the onshore and offshore winds. The idealized study performed here provides a basic framework for understanding the asymmetry between the negative and positive surges. We show, for the first time, the importance of the 2D wind field and the important role of the alongshore wind for negative surges.

2. Methods

An idealized tropical cyclone is set up using the wind field as described by Chan and Williams (1987). The domain spans $5° - 35°N$ and $120° - 180°E$ at a horizontal resolution of 15 km. The axisymmetric vortex is initialized to the right of the domain at 20°N, 165°E and translated westwards at the speed of 15 km h^{-1} for a total of 8 days. The maximum wind speed is 77 m s^{-1}. The temperature, pressure, humidity, long- and short-wave radiation fields are prescribed to be spatially constant and time-invariant. Increasing the atmospheric pressure has been known to decrease the SSH by 1 cm per mbar (Roden and Rossby, 1999). However, this is a systematic and symmetric change localized near the cyclone centre (runs not shown here). For clarity, the inverse barometer effect, together with other environmental conditions that can affect SSH are not considered here in order to isolate the effect of wind on the asymmetry of storm surges.

The 3D ocean model used is the Regional Ocean Modelling System (ROMS v4.3). ROMS is a free-surface, hydrostatic, primitive equation ocean model that uses stretched, terrain-following coordinates in the vertical and orthogonal curvilinear

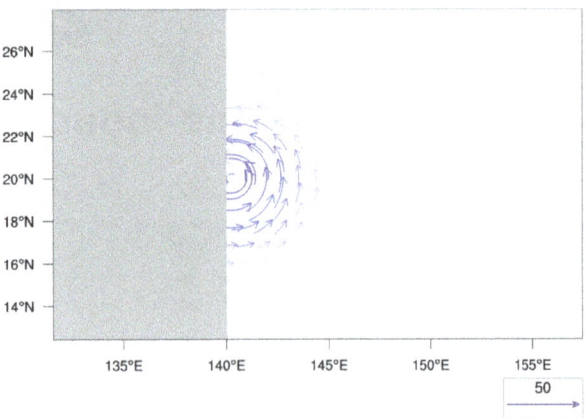 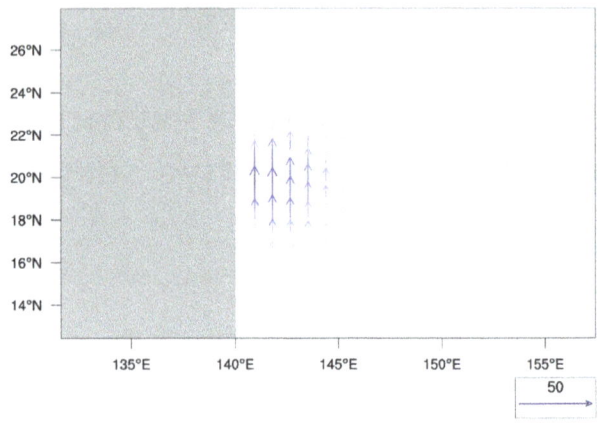

Figure 1. Surface stress (N m^{-2}) response to the total wind stress (a) and to only the alongshore wind stress (b).

coordinates in the horizontal, with Shchepetkin and McWilliams (2005) and Warner *et al.* (2008) outlining the computational algorithms used. The model has 21 vertical levels and a horizontal resolution of 15 km. The initial ocean state has a surface temperature of 28 °C, decreasing linearly to a constant 22 °C from the depth of 500 m and below. The salinity is uniform at 35 psu. The boundaries are closed, with the land mask specified at the western boundary up to 140°E to simulate the coastline.

The 'control' ocean domain has a spatial extent identical to the atmospheric domain (5° – 35°N, 120° – 180°E). The bathymetry is uniform at 100-m deep. The 'higher latitude' case is shifted northwards by 10° (15° – 45°N, 120° – 180°E). The cyclone wind field used to force the ocean model is unchanged, now translating along 30°N instead. The bathymetry is identical to the control setup, uniform at 100 m. A 'sloping bathymetry' case with the same spatial extent as the control case was also studied. The bathymetry slopes linearly eastwards from 50 m near the coast to a maximum of 954 m. The chosen ratio of 1:5000 is an average of the range of coastal slopes studied by Irish *et al.* (2008).

3. Results

Figure 1 shows the surface stress at 105 h when the cyclone crosses the coastline at 140°E. At this time, the alongshore surface stress near the coastline is directed northwards. The cross-shore surface stress (not shown) is westwards to the north of 20°N and eastwards to the south of 20°N. Figure 2 shows the SSH response and the corresponding surface currents at 105 h from initialization. This particular time is chosen to show the maximum magnitude of the positive and negative surges during the event. Not all the cases exhibit their maximum surges at the same time and 105 h was selected for consistency. A later analysis (Figure 3) shows that 105 h is a good representation of the peak surges. The figures are zoomed into 13° – 28°N and 132° – 158°E to display details at the coastline. In the control run (Figure 2(a)),

an anticlockwise circulation centering 20°N 142°E is generated with a decrease in SSH at the circulation centre. Along the coastline at 140°E, the magnitude and the alongshore extent of the negative surge are much larger compared to the positive surge (Figure 3). The positive surge is up to +5.3 m and the negative surge is up to −13.1 m. Along the coast, the surface currents flow in the southeastward direction between 18° and 20°N and in the northeastward direction between 13° and 17°N. Figure 2(b) shows the SSH response to only the cross-shore component of the wind stress, with the alongshore component removed. Similar to Figure 2(a), an anticlockwise circulation is generated, but it is elongated zonally and the circulation centre is shifted eastwards to 143°E instead of 142°E. The surge pattern is substantially more symmetrical with a positive surge of up to +4.0 m and a negative surge of only up to −4.5 m.

We next explore the role of the Coriolis parameter in Figure 2(c). A 10° northward shift in latitude shows a greater rightward deflection in the surface currents compared to the control run in Figure 2(a). There is a larger decrease in SSH at the anticlockwise circulation centred at 30°N 142°E. The alongshore extent of the negative surge is much larger compared to the positive surge (Figure 3). The positive surge is decreased from the control case of +5.3 to +4.4 m, but the negative surge is enhanced from the control case of −13.1 to −14.6 m. Without the alongshore wind stress (Figure 2(d)), the positive surge is decreased slightly to +3.9 m but the negative surge is weakened substantially to −5.0 m.

With a sloping bathymetry (Figure 2(e)), there is an overall weaker response in SSH and the ocean currents are less regular. A negative SSH can still be observed at the centre of the anticlockwise circulation. The magnitude and alongshore extent of the negative surge is also much larger compared to the positive surge. The positive surge is up to +4.0 m and the negative surge is up to −6.7 m. In Figure 2(f), the absence of the alongshore wind stress generated a much weaker response in SSH compared to all the other cases. The positive and negative surges are more symmetrical with similar magnitudes of up to 2.8 m.

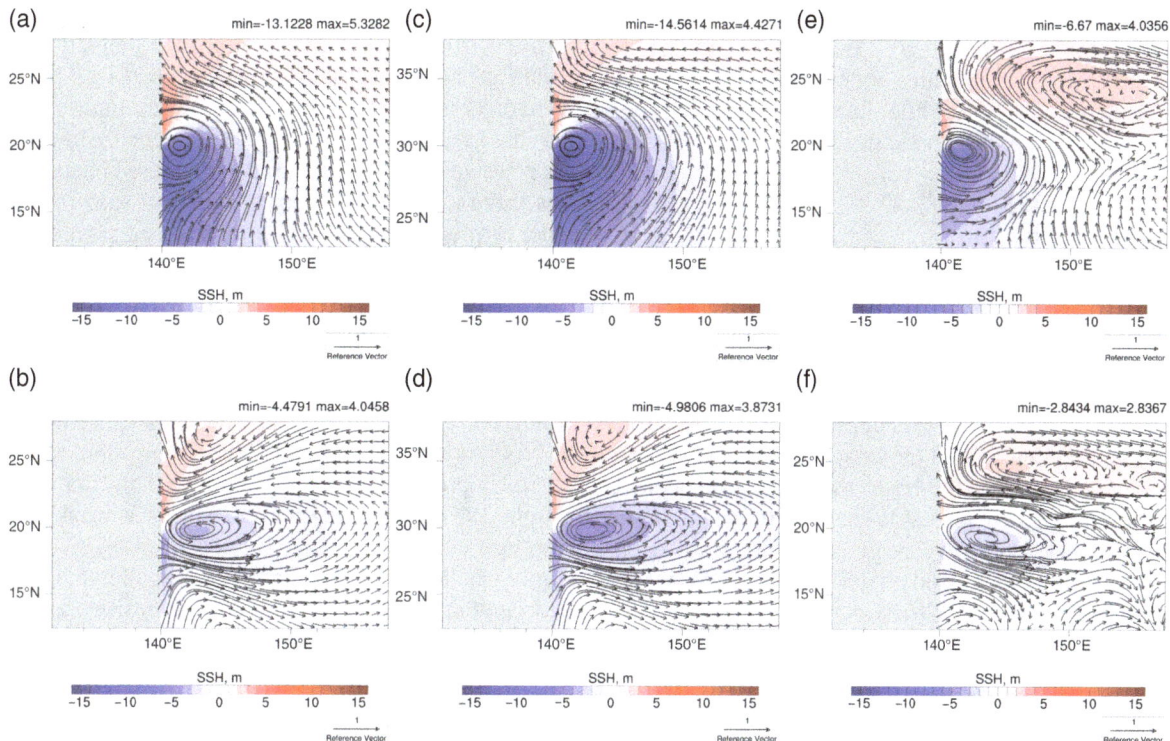

Figure 2. Sea surface height (m) response to the total wind stresses (a, c, e) and to only the cross-shore wind stress (b, d, f): for the control case (a, b), higher latitude case (c, d) and sloping bathymetry case (e, f) at 105 h into the simulation. The reference vector for the sea surface currents is 1 m s⁻¹.

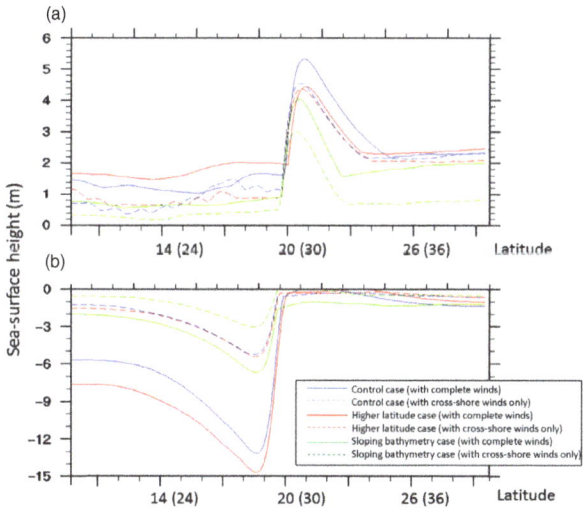

Figure 3. The maximum magnitude of the positive surge (a) and negative surge (b) in meters along the coast line during the simulation for 10 – 30°N (control case) and 20 – 40°N (higher latitude case).

We next examine the peak surges in Figure 3, showing the worst case scenario that the coastline experiences for this event. The three cases with both the cross-shore and alongshore wind stresses show a larger magnitude in the negative surge compared to the positive surge. For the control case, the surge ranges from −13.1 to +5.3 m. The higher latitude case ranges from −14.6 to +4.5 m. The sloping bathymetry case ranges from −6.7 to +4.0 m. The three cases without the alongshore wind stress

have comparable positive and negative surges. For the control case, the surge ranges from −5.3 to +4.6 m. The higher latitude case ranges from −5.4 to +4.4 m. The sloping bathymetry case ranges from −3.0 to +3.0 m. These results also confirm the choice of 105 h to map the horizontal extent of the surges for all cases (Figure 2) as the surge at 105 h and the simulation maximum in Figure 3 are very similar.

Figure 3 also shows the horizontal extent that the positive and negative surges affect the coastline. In cases with both the cross-shore and alongshore wind stresses, the positive surge is narrowly distributed along the coast compared to the negative surge which extends far to the south. The cases without the alongshore wind stress show comparably narrow distributions for both the positive and negative surges. As an illustration, the horizontal coastal extent (control case) affected by a positive surge greater than +4 m is 2°, while the region affected by a negative surge larger than −4 m is 20° or more. Without the alongshore wind stress, the coastal region with a positive surge greater than +4 m is halved to about 1°, while the region with a negative surge larger than −4 m is reduced by a factor of 10 to about 2°. Shifting the latitude northwards by 10° reduces the extent of the positive surge to 1° while the extent of the negative surge is increased much further to 25° or more. Without the alongshore wind stress, the region with a positive surge greater than 4 m remains at 1°, while the region with a negative surge larger than −4 m is reduced by a factor of 13 to only 2°. For the sloping bathymetry case, the coastal region impacted by a positive surge greater than 2 m is 2.5°, while the region with a negative surge larger than −2 m is 10°. Without the alongshore wind stress, the region with a positive surge greater than 2 m is reduced to 1.5°, while the region impacted by a negative surge larger than −2 m is only 2.5°.

4. Discussion

A mechanism has been proposed by Peng *et al.* (2006) to account for the 1D asymmetry in the sea level response, where the pressure gradient force required to balance the wind is a function of $(h + \zeta)d\zeta/dx$ (where h, ζ, x are the undisturbed water depth, the sea surface elevation and distance from the coast, respectively). This 1D analysis explains the basic asymmetry in positive and negative surges, since it is easier to move the lower water mass for a negative surge, given the same wind. However, observations and analyses described by Pugh (1987) and Pousa *et al.* (2013) showed horizontal surge features that a 1D mechanism cannot fully represent quantitatively.

The idealized study here provides a basic spatial framework for the understanding of cyclone-driven surges and the asymmetry between positive and negative surges. In the absence of the alongshore wind stress (Figure 2(b)), the coastal surge pattern is more symmetrical along 20°N. The anticlockwise circulation is elongated along 20°N since only the cross-shore wind stress is present. The centre of the anticlockwise circulation is shifted eastwards with the western land-mask acting as a barrier, shifting the elongated water mass eastwards. The northward alongshore wind stress generates a northward advection of water mass that increases the positive surge and decreases the negative surge. In addition, Ekman transfer of momentum deflects the northward alongshore flow rightwards, decreasing the overall SSH at the coast. Without the alongshore wind stress, the two effects oppose in the case of the positive surge, showing an overall decrease from +5.3 to +4.0 m. For the negative surge, the two effects add up and resulted in a substantial decrease in the magnitude of the negative surge from −13.1 to −4.5 m. This substantial additional decrease demonstrates that the alongshore component of the wind stress plays a very important role in creating the horizontal divergence that is the main driver of the negative surge. Surges simply generated by onshore and offshore winds alone generate fairly symmetrical surge patterns as shown in Figure 2(b). In the case of a southern hemisphere scenario, the tropical cyclone will be rotating clockwise, creating a southward alongshore component in the same domain setup. The results will be inverted but conceptually the same.

When the ocean domain is shifted northwards by 10° (comparing Figures 2(a) and 2(c)), the ocean currents are deflected more towards the right with the enhanced Coriolis force. The increased rightward deflection of the northward alongshore surface stress decreases both the positive and negative surges. Comparing Figures 2(b) and 2(d), both the positive and negative surges decrease in magnitude with the weakened cross-shore stress that resulted from the enhanced rightward deflection. The enhanced rightward deflection also increases the outward divergence of the anticlockwise circulation, resulting in a greater decrease in the SSH here. Overall, both the positive and negative surge decreases. That the Coriolis force affects the extent of negative storm surge is not a surprise since the storm negative surges are driven by horizontal divergence. This cannot be accounted for by 1D mechanisms.

While the previous two cases represent a near-shore or continental shelf scenario, the sloping bathymetry case (Figures 2(e) and 2(f)) shows a different scenario where the cyclone approaches the coastline from deep waters. As the response of the SSH to wind stress is inversely proportional to the depth of the water column (Pugh, 1987), the deeper water shows a weaker SSH response compared to the shallower cases. While the sloping bathymetry case produced smaller magnitude and range in asymmetry between the positive and negative surges, it delivers the same qualitative message as the control case. The asymmetry in the magnitude and horizontal extent of the positive and negative surges is still substantial, with horizontal divergence as the main driver of this asymmetry. The absence of the alongshore wind stress component generates a fairly symmetrical surge response that is expected from having onshore and offshore winds alone.

The surge asymmetry can be characterized by the ratio defined as $\frac{\zeta^-}{\zeta^+}$ (where ζ^- and ζ^+ are the magnitudes of the minimum and the maximum SSH at the coast, respectively). In the control case, the alongshore wind stress increases the ratio from 1.2 to 2.5, demonstrating the significance of the alongshore wind stress in generating the large asymmetry. With a 10° northward shift in latitude, the asymmetry increases to 3.2. This increase is expected with horizontal divergence as the main mechanism for driving the negative surges. A deeper and sloping bathymetry lowers the ratio to 1.7, illustrating the effect of the deeper waters weakening the SSH response. The surge ratios for the storms at the Orissa coast in 1982 (Pugh, 1987) and the Argentinian coast in 1984 (Pousa *et al.*, 2013) are 2.7 and 1.9, respectively, both of which are well within the range estimated in this study.

The asymmetry in the horizontal extent of the positive and negative surges along the coastline has not been noted in literature. Comparing the horizontal extent of the positive and negative surges in Figure 3 shows that the alongshore wind stress is the main contributor to this large asymmetry. In the control case, the negative surge affected coastal regions by a factor of 10 compared to the positive surge. Without the alongshore wind stress, the asymmetry decreases by a factor of 2. Increasing the latitude by 10° northwards raises this asymmetry to 25 but in the absence of the alongshore component, the asymmetry decreases to 2. For the sloping bathymetry case, the extent of the negative surge is four times larger than the positive surge. Without the alongshore wind stress, the coastal extent affected by the negative surge is only a factor of 1.7 larger than the positive surge. This comparison in horizontal extents gave a similar conclusion as the surge ratios discussed previously. The simulation results of Pugh (1987) for the storm hitting the Orissa coast of India also show the negative surge affecting a greater coastal extent than the positive surge.

5. Conclusion

Storm surges can result in high economic consequences and even the loss of lives. While the impacts and occurrences of positive surges are widely investigated, negative surges are less understood. Existing studies on negative surges have been limited to either specific real case studies with specific inferences, or idealized studies that do not consider the horizontal extent of the SSH response. In this study, the asymmetry of positive and negative surges in the magnitude and spatial extent is investigated using idealized experiments. Three cases have been examined. While the occurrence of positive and negative surges has previously been understood as being driven by winds blowing onshore and offshore, respectively, the wind stress experiments here reveal that the alongshore component is the main cause of asymmetry in the magnitude and spatial scale of the surge. We further examine the sensitivity of the asymmetry to latitude and bathymetry. The asymmetry increases with increasing Coriolis force. A northward shift in latitude increases the asymmetry. The third case investigates the SSH response in a deeper, sloping ocean floor. The overall SSH response for this case is weaker than the shallower cases and the surge asymmetry decreases. The alongshore wind stress is also the main contributor to the large asymmetry in the length of coastline affected by the positive and negative surges. This study shows that positive and negative surges have some similarities but also different causes and properties. Simplified storm surge models that only incorporate cross-shore winds to analyse the positive and negative surges will not capture the large asymmetry shown here.

References

AsSalek JA. 1997. Negative surges in the Meghna estuary in Bangladesh. *Monthly Weather Review* **125**(7): 1638–1648, doi: 10.1175/1520-0493(1997)125<1638:Nsitme>2.0.Co;2.

Chan JCL, Williams RT. 1987. Analytical and Numerical-Studies of the Beta-Effect in Tropical Cyclone Motion. Part 1: zero mean flow. *Journal of the Atmospheric Sciences* **44**(9): 1257–1265, doi: 10.1175/1520-0469(1987)044<1257:Aansot>2.0.Co;2.

Irish JL, Resio DT, Ratcliff JJ. 2008. The influence of storm size on hurricane surge. *Journal of Physical Oceanography* **38**(9): 2003–2013, doi: 10.1175/2008jpo3727.1.

Peng MC, Xie L, Pietrafesa LJ. 2006. Tropical cyclone induced asymmetry of sea level surge and fall and its presentation in a storm surge model with parametric wind fields. *Ocean Model* **14**(1–2): 81–101, doi: 10.1016/j.ocemod.2006.03.004.

Pousa JL, D'Onofrio EE, Fiore MME, Kruse EE. 2013. Environmental impacts and simultaneity of positive and negative storm surges on the coast of the Province of Buenos Aires, Argentina. *Environmental Earth Sciences* **68**(8): 2325–2335, doi: 10.1007/s12665-012-1911-9.

Pugh DT. 1987. *Tides, Surges and Mean Sea-Level*. John Wiley and Sons: Chichester, UK.

Resio DT, Westerink JJ. 2008. Modeling the physics of storm surges. *Physics Today* **61**(9): 33–38, doi: 10.1063/1.2982120.

Roden GI, Rossby HT. 1999. Early Swedish contribution to oceanography: Nils Gissler (1715–71) and the inverted barometer effect. *Bulletin of the American Meteorological Society* **80**(4): 675–682, doi: 10.1175/1520-0477(1999)080<0675:Escton>2.0.Co;2.

Shchepetkin AF, McWilliams JC. 2005. The regional oceanic modeling system (ROMS): a split-explicit, free-surface, topography-following-coordinate oceanic model. *Ocean Model* **9**(4): 347–404, doi: 10.1016/j.ocemod.2004.08.002.

Warner JC, Sherwood CR, Signell RP, Harris CK, Arango HG. 2008. Development of a three-dimensional, regional, coupled wave, current, and sediment-transport model. *Computers & Geosciences* **34**(10): 1284–1306, doi: 10.1016/j.cageo.2008.02.012.

22

Extreme Arctic cyclone in August 2016

Akio Yamagami,[1]* Mio Matsueda[1,2]◉ and Hiroshi L. Tanaka[1]

[1]Center for Computational Sciences, University of Tsukuba, Japan
[2]Department of Physics, University of Oxford, UK

*Correspondence to:
A. Yamagami, Center for
Computational Sciences,
University of Tsukuba, 1-1-1
Tennodai, Tsukuba, Ibaraki
305-8577, Japan.
E-mail:
yamagami@ccs.tsukuba.ac.jp

Abstract

An extremely strong Arctic cyclone (AC) developed in August 2016. The AC exhibited a minimum sea level pressure (SLP) of 967.2 hPa and covered the entire Pacific sector of the Arctic Ocean on 16 August. At this time, the AC was comparable to the strong AC observed in August 2012, in terms of horizontal extent, position, and intensity as measured by SLP. Two processes contributed to the explosive development of the AC: growth due to baroclinic instability, similar to extratropical cyclones, during the early phase of the development stage, and later nonlinear development via the merging of upper warm cores. The AC was maintained for more than 1 month through multiple mergings with cyclones both generated in the Arctic and migrating northward from lower latitudes, as a result of the high cyclone activity in summer 2016.

Keywords: Arctic cyclone; warm core; baroclinic instability; merging

1. Introduction

An extremely strong Arctic cyclone (AC) developed over the Arctic on 16 August 2016, spanning the Pacific sector of the Arctic Ocean. Cyclones were quite active over the Arctic in general in the summer of 2016. Many cyclones migrated into the Arctic, and these cyclones maintained the AC for a prolonged period.

The AC that is known as 'The Great Arctic Cyclone of August 2012' is well analyzed as a remarkable summer AC (Simmonds and Rudeva, 2012). They showed that this AC had the lowest central pressure and largest size of any summer AC from 1979 to 2012. Zhang et al. (2004) showed that, in general, summer ACs have longer lifetimes, are more numerous, and are weaker than those in winter. On the other hand, Simmonds et al. (2008) found a greater number of cyclones in winter than in summer, as a result of 'open depression' systems (Murray and Simmonds, 1991). Most cyclones in the Arctic during summer are generated over the Arctic Ocean, and most of the remainder is generated over the northern Eurasian continent (Brümmer et al., 2000; Serreze and Barrett, 2008). In the context of these previous studies, the intensification of AC in August 2016 (AC16) was unusual in summer.

ACs can impact the wider Arctic climate system through fields, such as seawater temperature (Inoue and Hori, 2011). It has been reported that the AC in August 2012 (AC12) contributed greatly to the record-low sea-ice extent of that summer (Parkinson and Comiso, 2013; Zhang et al. 2013) and influenced biological activities in Arctic Ocean (Zhang et al., 2014).

ACs have warm (cold) core at upper (lower) level and barotropic vorticity in the troposphere (Tanaka et al., 2012). Previous studies showed that the baroclinicity over the Arctic frontal zone was one of the main factors for generation and intensification of ACs (e.g. Serreze and Barrett, 2008). Recently, Crawford and Serreze (2016) indicated that the baroclinicity affected only on an intensification of ACs. The coupling with lower and upper cyclones was also important for the development of ACs (Simmonds and Rudeva, 2012, 2014).

This study investigates the features and mechanisms behind the development of the AC16.

2. Data and methods

ERA-Interim (Dee et al., 2011) data were used in this analysis. Specifically, we used temperature (T), geopotential height, horizontal wind ($V = (u, v)$), relative vorticity, and sea level pressure (SLP) at 6-hourly intervals, and surface sensible and latent heat fluxes at 12-hourly intervals. The sensible and latent heat fluxes were converted from 12-hourly accumulated values to 12-hourly average rates. The horizontal resolution of all variables was $1.25° \times 1.25°$.

The method developed by Aizawa and Tanaka (2016) was used to identify cyclone centers at each timestep. The method uses an SLP field interpolated from a regular longitude–latitude grid to an equal-distance grid centered on the North Pole. The equal-distance grid used in this study has 200 grid points in the x and y directions with a spacing of 40 km. The SLP at each grid point is then compared with the SLP averaged over all grid points between 500 and 550 km from the target grid point. If the SLP at the target grid point is lower than the area-averaged SLP, the grid point is regarded as a candidate for a cyclone center. The horizontal resolution of ERA-Interim did not affect the cyclone center detection substantially. Cyclone tracks were then formed based on the nearest-neighbor method

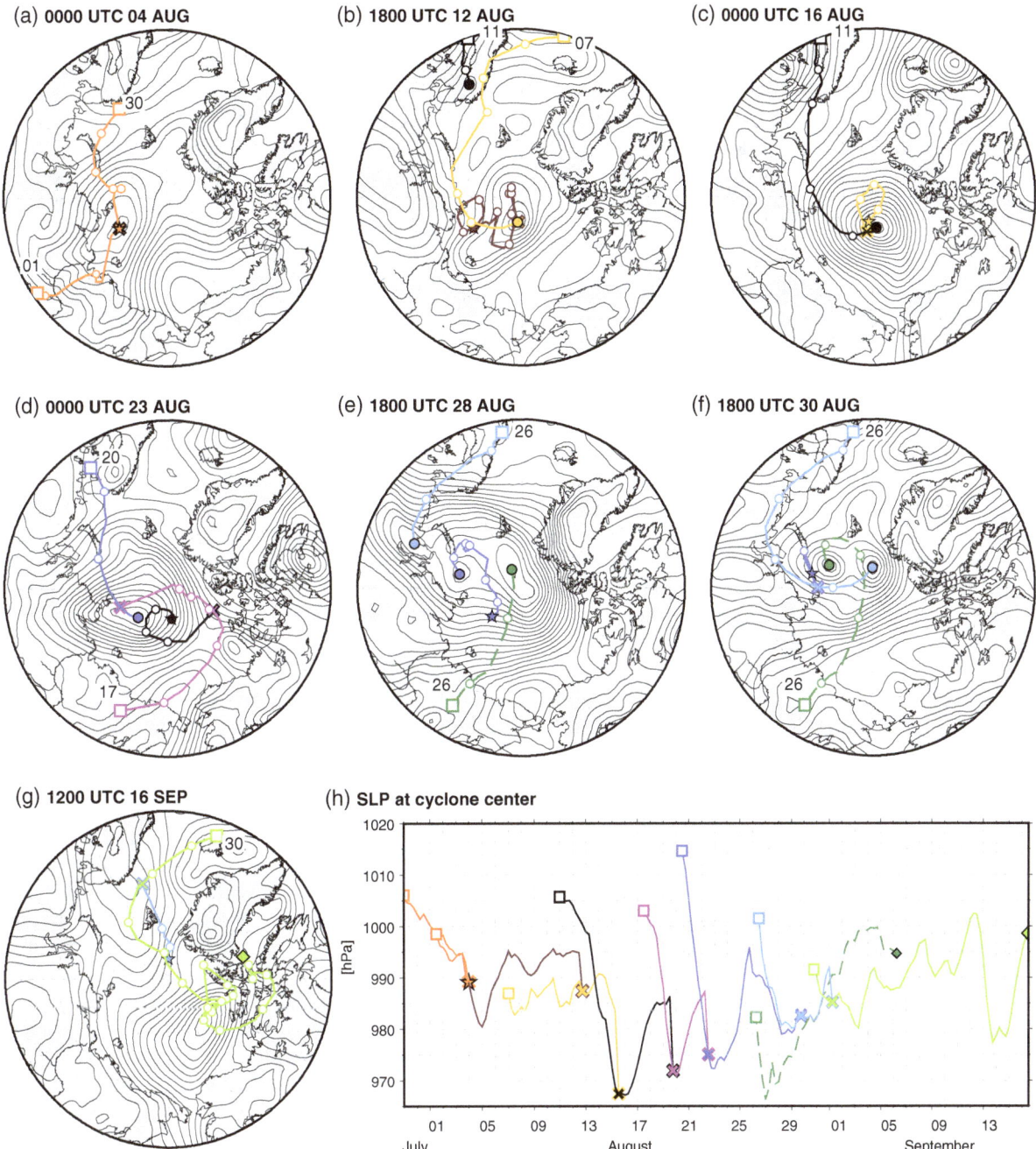

Figure 1. (a–g) Cyclone tracks related to the AC16 (colored lines) up to (a) 0000 UTC on 4 August, (b) 1800 UTC on 12 August, (c) 0000 UTC on 16 August, (d) 0000 UTC on 23 August, (e) 1800 UTC on 28 August, (f) 1800 UTC on 30 August, and (g) 1200 UTC on 16 September 2016. The star in (a) represents the start position of the AC16 track and that in (b)–(g) represents the last position of AC16 in the previous panel. The filled crosses with colored edge indicate the merging position. The white squares and colored numbers show the position and the day at which each cyclone was first identified. Open circles mark the position at 0000 UTC. The rhombus in (g) represents the position at which the AC was last identified. Contour and filled circles show SLP and the position of each AC or cyclone at a given time above each panel, respectively. The contour interval is 2 hPa. (h) Time series of cyclone center SLP, with colors, squares, crosses, and rhombuses corresponding to (a)–(g).

(Serreze, 1995), except in the cases of merging and splitting cyclones, which were tracked manually. The radius of ACs was calculated based on outermost closed SLP contour with 1 hPa intervals including single cyclone center.

The magnitude of the temperature gradient at 850 hPa, and the Eady growth rate (EGR) (Simmonds and Lim, 2009) at 700 hPa were also used, in order to identify fronts and measure baroclinicity. They were calculated as follows:

$$|\nabla T| = \sqrt{\left(\frac{\partial T}{\partial x}\right)^2 + \left(\frac{\partial T}{\partial y}\right)^2}$$

$$\mathrm{EGR} = 0.3098 \frac{|f|\left|\frac{\partial V}{\partial z}\right|}{N}$$

where f is the Coriolis parameter; x, y, and z are the longitudinal, latitudinal, and vertical coordinates, respectively; and N is the Brunt–Väisälä frequency. The vertical derivatives in EGR were calculated using centered finite differences at the 750 and 650 hPa levels. The heat budget was also calculated as follows:

$$\frac{\partial \theta'}{\partial t} = \frac{\partial (u\theta)'}{\partial x} + \frac{\partial (v\theta)'}{\partial y} + \frac{\partial (\omega\theta)'}{\partial p} + F'$$

where θ is the potential temperature, ω is the vertical p-velocity, p is the pressure, and F is the diabatic heating evaluated as the residual. The prime indicates an anomaly from the 6-hourly climatology.

In some studies, the term 'AC' was applied to both cyclones coming from mid-latitude to the Arctic and those generated over the Arctic. In this study, the cyclones having upper warm core in the upper troposphere were called 'AC' according to Tanaka *et al.* (2012). The cyclones coming from mid-latitude were called 'cyclone.' The term 'merging' was used based on the definition of Hanley and Caballero (2012). However, the cyclone center detection in this study can identify open depression systems. Therefore, 'merging' in this study includes the following case: A cyclone center exists as an open depression system within 1000 km of the other cyclone center at a timestep and then these two SLP minima merge into a single SLP minimum at the next timestep. Most mergings in this study were accompanied by a vorticity merging.

3. Results

The AC16 began with a merging of two cyclones from the Barents Sea and the northeast Siberia (orange in Figure 1(a)) over the Laptev Sea on 4 August (orange star). The AC16 wandered over the Arctic Ocean for 8.5 days (brown in Figure 1(b)). After that, the AC16 merged with a cyclone from the North Atlantic (yellow) on 13 August. Meanwhile, a cyclone formed over the Scandinavian Peninsula (red). This cyclone developed rapidly, and its central pressure dropped by ~30 hPa during 13–15 August (red in Figure 1(h)). The cyclone merged with the AC16 at 1200 UTC on 15 August, and the lowest minimum central pressure of 967.2 hPa was recorded at 0000 UTC on 16 August (Figure 1(c)). As seen from the SLP, the whole Pacific sector of the Arctic Ocean was covered by the AC. The AC16 was quite similar to the AC12 in terms of the SLP values measured, horizontal extent, and position. The central pressure of the AC12 was 964.1 hPa. The radius and the center for the AC16 (AC12) were ~1028 km (~1035 km) and at 187.60°E, 84.56°N (188.53°E, 82.73°N), respectively.

During the time period 19–22 August, the AC16 merged with cyclones that originated over northeast Siberia (purple cross in Figure 1(d)) and the Scandinavian Peninsula (dark blue cross). After these mergings, the AC16 exhibited a minimum pressure of 972.3 hPa and covered the Laptev, Kara, and Arctic oceans (Figure 1(d)). At 1200 UTC on 28 August, the AC16

(dark blue in Figure 1(e)) and cyclones moving from the Atlantic (light blue) and from the Sea of Okhotsk (dark green) formed a huge multicenter cyclone that covered the whole Arctic Ocean. After the merging with the cyclone on 29 August (light blue cross in Figure 1(f)), the resulting AC16 center (light blue) orbited the remaining cyclone center (dark green) according to the Fujiwhara effect (Fujiwhara, 1923). The AC16 (light blue) merged with the cyclone from the North Atlantic (light green cross in Figure 1(g)) on 1 September. The merged AC16 wandered for 15.5 days and dissipated on 16 September over the Canadian Arctic Archipelago. The AC was thus maintained for more than 1 month through repeated cyclone mergings.

The structure of each cyclone was changed after merging with the AC16. The temperature anomalies from 6-hourly climatology (1981–2010) were near-zero or negative at the formation stage of the cyclones (Figure 2(b)). Each cyclone had an upper-level warm core of ~6 K during its mature stage. The change from cold to warm core is considered to mark the change from an extratropical cyclone to an AC (Aizawa and Tanaka, 2016).

In particular, the fourth cyclone (red) had a weak warm core of 1 K and a strong cold core of −8 K at 250 and 850 hPa, respectively, in its formation stage. As the cyclone developed on 13 August, the cold core at the lower level weakened to near-zero. These suggest that the shift in position from a highly stable environment to the climatological environment contributed to the beginning of the cyclone's development. The warm core at the upper level strengthened during the early phase of the rapid development stage from 1200 UTC on 13 August to 1800 UTC on 14 August. The cold core at the lower level also strengthened at this time. From Figures 3(d)–(f) and Figures 4(c)–(f), the warm and cold cores accompanied by the cyclone developed on the rear of the center. Although both cores continued to develop during the later phase of the development stage, the warm core reached its peak value, 6 K, faster than did the cold core.

The relative vorticity also strengthened at upper and lower levels during its development and mature stages (Figures 2(d) and (e)). These also indicated that the AC16 has barotropic vorticity. The vorticity at 850 hPa represents that the vorticity accompanied by a cyclone was as high as that accompanied by the AC16 when the cyclone merged with the AC16. The horizontal structure of the vorticity merging is shown in Figures 3(a)–(c). The maximum relative vorticity at 850 hPa for AC16 (3.04×10^{-5} s^{-1}) was also very similar to that for AC12 (3.05×10^{-5} s^{-1}).

The cyclone featured relatively large surface heat fluxes, especially latent heat flux, on 11 and 12 August (Figure 2f). These large fluxes occurred when the cyclone was located over the Scandinavian Peninsula. LeDrew (1984) indicated that the surface enthalpy flux contributed only for cyclones over the Laptev Sea in end of fall significantly and for cyclone in summer insignificantly. These surface heat fluxes seem insignificant

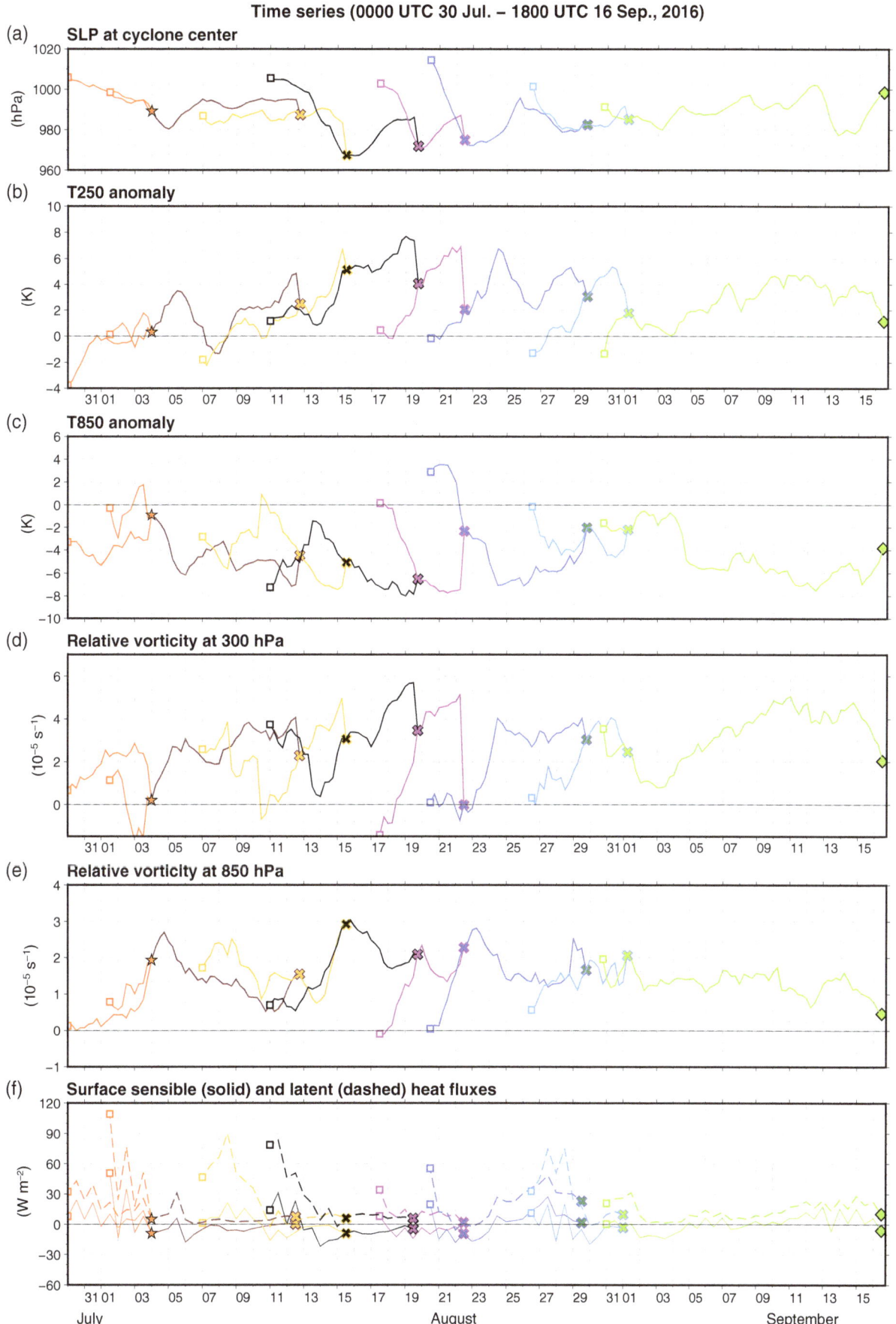

Figure 2. (a) As in Figure 1(h). (b–f) Cyclone temperature anomalies at (b) 250 hPa and (c) 850 hPa, defined as departures from 6-hourly climatology, relative vorticity at (d) 300 hPa and (e) 850 hPa, and (f) surface sensible (solid lines) and latent (dashed lines) heat fluxes, averaged over a circle of radius 1000 km centered at the cyclone center. Colors and symbols are as in Figure 1.

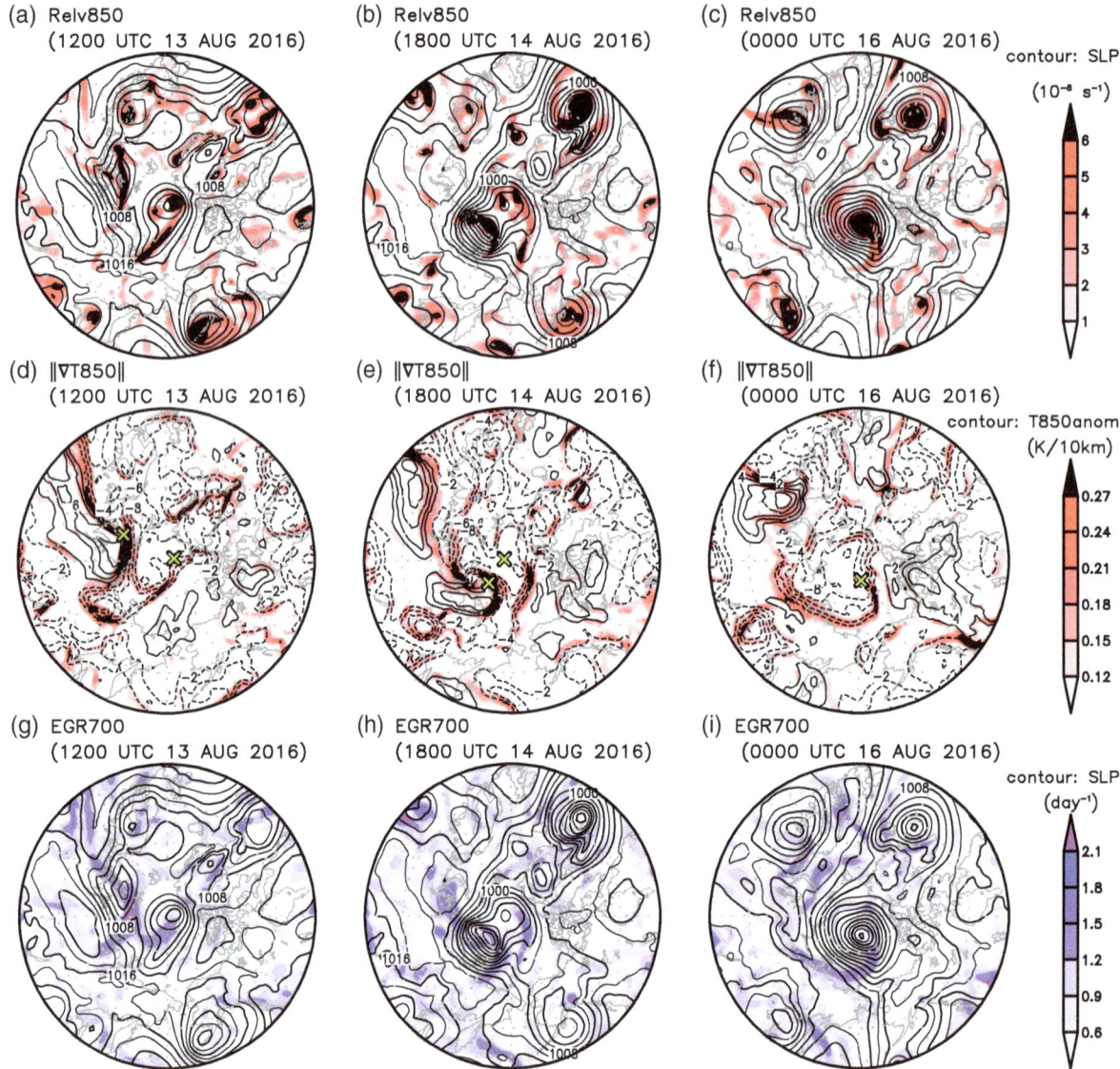

Figure 3. (a–c) Relative vorticity at 850 hPa (shading) and SLP (contour), (d–f) magnitude of the temperature gradient at 850 hPa (shading) and the temperature anomaly at 850 hPa, and (g–i) Eady growth rate at 700 hPa (shading) and SLP (contour), at (a, d, g) 1200 UTC on 13 August, (b, e, h) 1800 UTC on 14 August, and (c, f, i) 0000 UTC on 16 August 2016. Contours intervals are (a–c, g–i) 3 hPa for SLP and (d–f) 2 K for temperature anomaly at 850 hPa. (d–f) Green crosses mark the minimum central pressure of each cyclone.

for the development of the AC16, since values were at most 100 W m^{-2} and were comparable to monthly mean value in August (Serreze and Barry, 2014).

On 13 August, the AC16 (yellow in Figure 1(c)) did not have any remarkable frontal structure in its mature stage (Figure 3(d)). The cyclone over Novaya Zemlya (red in Figure 1(c)) possessed cold and warm fronts. On 14 August, the Novaya Zemlya cyclone had moved northeastward to the Kara Sea (Figure 3(e)). The cyclone's fronts formed a T-bone structure, indicating that the cyclone had reached its mature stage. The surrounding environment of the cyclone was unstable on 13 August (Figure 3(g)). The unstable area expanded into the northern part of the cyclone (Figure 3(h)), corresponding to the direction of motion of the cyclone. On the other hand, EGR around the AC16 was small on August 13 and 14 (Figures 3(g) and (h)). Therefore, the

cyclone over the northern coast of Eurasia developed via baroclinic instability, as do extratropical cyclones, during the early phase of its rapid development. After the merging of the cyclone and the AC16 on 16 August, the fronts became separated from the AC16 center (Figure 3(f)) and the strong instability disappeared (Figure 3(i)). The results suggest that baroclinicity was not the primary energy source in the later phase of the rapid development stage of the AC.

Figure 4(a) shows that a polar vortex with a warm core was located over the Arctic Ocean on 11 August and was connected to the AC16 (brown in Figure 1(b)). Deep and shallow troughs extended over the Scandinavian Peninsula and the Kara Sea, respectively. Cyclones (yellow and red in Figure 1) were located to the east of these troughs. The deep trough moved eastward; accordingly, the surface cyclone (red) also traveled

Z300 and T250 anomaly

contour: Z300 (cint: 70m), shade: T250 anomaly

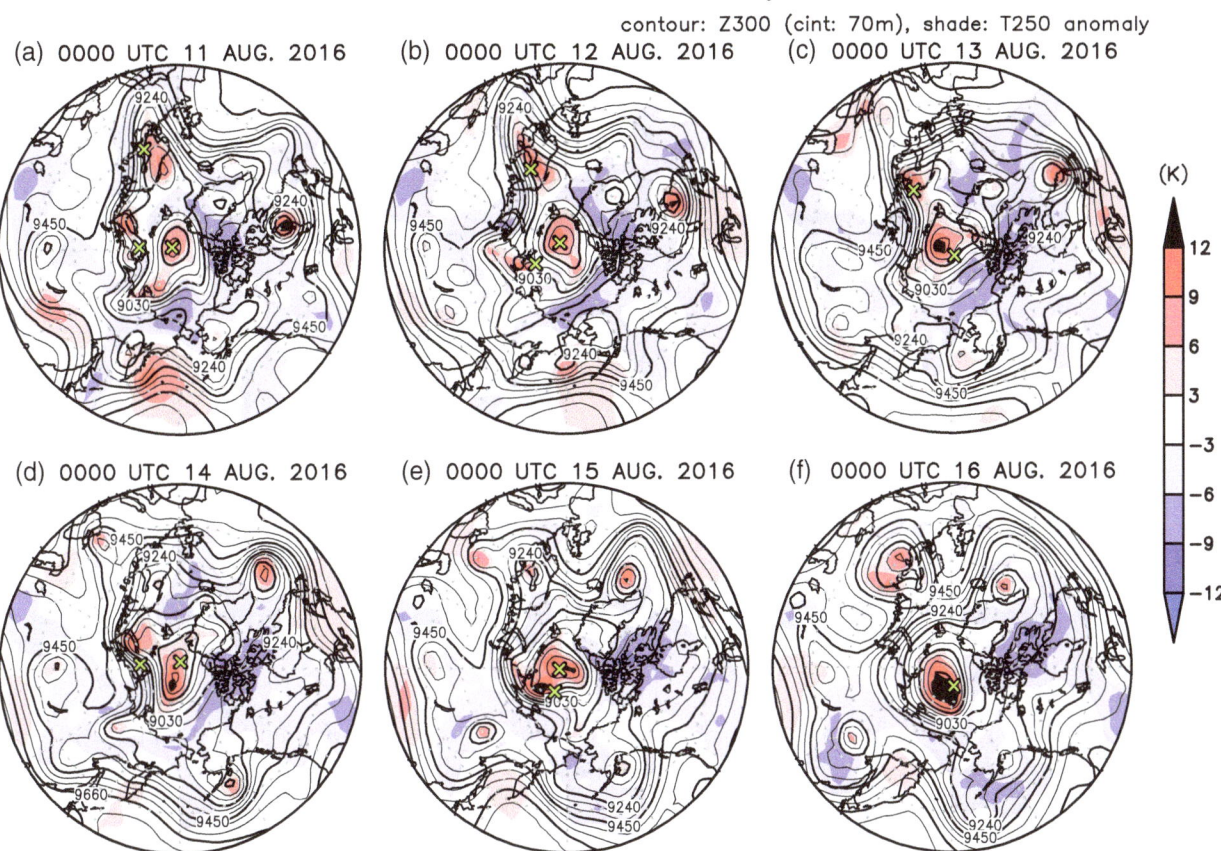

Figure 4. Geopotential height at 300 hPa (contours) and temperature anomaly at 250 hPa (shading) at 0000 UTC on 11–16 August 2016. The contour interval is 70 m. Green crosses mark the minimum central pressure of each cyclone.

eastward to the Kara Sea on 14 August. The cyclone center was located east of the trough, suggesting that the cyclone was developing in the same manner as extratropical cyclones in their formation stage. The trough and surface cyclone then traveled farther eastward, and the cyclone gradually progressed to its mature stage. In contrast, although the AC16 merged with the cyclone that was accompanied by the shallow trough on 13 August (Figure 4(c)), the AC16 remained over the Arctic Ocean from 11 to 14 August (Figure 4(d)). From 14 to 16 August, the warm core in the polar vortex merged with the warm core in the trough over the Kara Sea (Figure 4(e)). The merger resulted in the warm core strengthening and expanding over the AC16, marking the later phase of the rapid development stage. The AC16 center and the upper vortex exhibited a barotropic structure on 16 August (Figure 4(f)).

For the development of the upper-level warm core accompanied by the cyclone (red in Figure 1(c)), the horizontal flux convergence (dotted line in Figure 5(b)) was a dominant component during the whole period. The horizontal component was relatively small for the generation of the cyclone before 1200 UTC on 11 August. The sum of the horizontal and the vertical components (solid line in Figure 5(b)) was ~0.8 K $(6\,h)^{-1}$ at 1800 UTC on 11 August, for the beginning of the cyclone's development. The sum first peaked at

1800 UTC on 13 August with ~1.9 K $(6\,h)^{-1}$, corresponding to the early phase of the rapid development stage (Figure 5(a)). The peak indicates that the warm core in the upper trough caught up with the surface cyclone (Figures 4(d) and (e)). A second peak appeared on 15 August, corresponding to the later phase of the rapid development stage (Figure 5(a)). The second peak indicates that the warm cores are merging (Figure 4(e)). Although the sum during the second peak is weaker than that during the first peak, the heat budget at upper levels also indicates that two processes have contributed to the development of the warm core over the cyclone, and as a result, the cyclone has developed to an extreme AC. The upper warm core was mainly maintained by the horizontal flux convergence after 17 August. Although the eddy component of the vertical flux convergence was small, the background component of that was ~0.4 K $(6\,h)^{-1}$. It indicates that an upper warm core accompanied by ACs was maintained by both merging of two cyclones and the background downward flow.

4. Summary and conclusions

In this study, the features and mechanisms behind the development of the extreme AC of August 2016 were investigated. The AC16 occurred over the Laptev Sea on 4 August and was maintained for more than 1 month

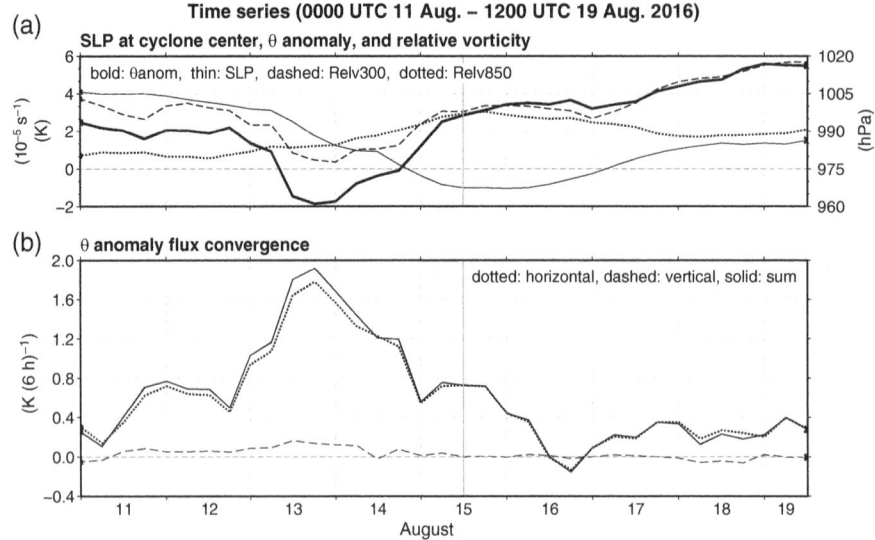

Figure 5. Time series for the cyclone featuring the lowest minimum SLP (marked in red in Figure 1), for: (a) potential temperature anomaly averaged over the cylinder with radius 1000 km centered on the cyclone center at heights between 300 and 100 hPa (bold solid), SLP at the cyclone center (thin solid) and relative vorticity at 300 hPa (dashed) and 850 hPa (dotted); (b) horizontal (dotted) and vertical (dashed) potential temperature anomaly flux convergences through the surface of the same cylinder and their sum (solid). The gray lines represent the time of the merging.

through repeated mergings with other cyclones. The AC16 recorded a minimum SLP of 967.2 hPa and covered the entire Pacific sector of the Arctic Ocean. In addition, the AC16 experienced two notable periods of development after its initial development.

On 15 August, the AC16 merged with a cyclone that originated to the west of the trough over the Scandinavian Peninsula on 11 August. The combined cyclone moved along the northern coast of Eurasia and developed rapidly from 13 to 16 August, with a decrease in central pressure of ~30 hPa. The extreme development of the cyclone occurred via two processes: a baroclinic process, as occurs in extratropical cyclones, in the early phase of the development stage (from 13 to 14 August), and a nonlinear process caused by the merging of the upper-level warm cores in the later phase of the development stage (on 15 August). Simmonds and Rudeva (2012) concluded that not only baroclinicity but also the establishment of a connection with the tropopause polar vortex were important to the development of the AC12. Both processes were also seen during the rapid development of the AC16. Furthermore, our results confirm that a merging of warm cores accelerates the development of the AC16.

The lifetime of the AC16 was much longer than that of the AC12 due to multiple merging events. The merging process is essential to ACs, and it may correspond to the connection between an upper polar vortex and a surface vortex. However, when the cyclones were as strong as the AC16, two vortices were merged with in some cases (purple cross in Figure 1(d)) and not in the other cases (light blue and dark green in Figure 1(f)). Thus, it is suggested that the occurrence of merging for ACs is not determined by only length scale or strength of cyclones.

In this study, we focused only on the cyclone with the lowest minimum SLP; however, many other interesting events occurred during the lifetime of the AC. The mechanisms of merging and the AC in terms of atmosphere–ocean–sea-ice interactions were also interesting.

Acknowledgements

The authors are grateful to Dr Takuro Aizawa of the University of Tsukuba for providing the cyclone detection algorithm and for useful discussions. The authors thank ECMWF for providing the ERA-Interim reanalysis dataset. Part of this research was supported by the Arctic Challenge for Sustainability (ArCS) Project.

References

Aizawa T, Tanaka HL. 2016. Axisymmetric structure of the long lasting summer Arctic cyclones. *Polar Science* **10**: 192–198.

Brümmer B, Thiemann S, Kirchgäßner A. 2000. A cyclone statistics for the Arctic based on European centre re-analysis data. *Meteorology and Atmospheric Physics* **75**: 233–250. https://doi.org/10.1007/s007030070006.

Crawford AD, Serreze MC. 2016. Does the summer Arctic frontal zone influence Arctic Ocean cyclone activity? *Journal of Climate* **29**: 4977–4993. https://doi.org/10.1175/JCLI-D-15-0755.1.

Dee DP, Uppala SM, Simmons AJ, Berrisford P, Poli P, Kobayashi S, Andrae U, Balmaseda MA, Balsamo G, Bauer P, Bechtold P, Beljaars ACM, van de Berg L, Bidlot J, Bormann N, Delsol C, Dragani R, Fuentesn M, Geer AJ, Haimberger L, Healy SB, Hersbach H, H'olm EV, Isaksen L, Kållberg P, Köhler M, Matricardi M, McNally AP, Monge-Sanz BM, Morcrette JJ, Park BK, Peubey C, de Rosnay P, Tavolato C, Thépaut JN, Vitart F. 2011. The ERA-Interim reanalysis: configuration and performance of the data assimilation system. *Quarterly Journal of the Royal Meteorological Society* **137**: 553–597.

Fujiwhara S. 1923. On the growth and decay of vortical systems. *Quarterly Journal of the Royal Meteorological Society* **49**: 75–104. https://doi.org/10.1002/qj.49704920602.

Hanley J, Caballero R. 2012. Objective identification and tracking of multicenter cyclones in the ERA-Interim reanalysis dataset. *Quarterly Journal of the Royal Meteorological Society* **138**: 612–625.

Inoue J, Hori ME. 2011. Arctic cyclogenesis at the marginal ice zone: a contributory mechanism for the temperature amplification? *Geophysical Research Letters*. **38**: L12502. https://doi.org/10.1029/2011GL047696.

LeDrew EF. 1984. The role of local heat sources in synoptic activity within the polar basin. *Atmosphere-Ocean* **22**: 309–327.

Murray RJ, Simmonds I. 1991. A numerical scheme for tracking cyclone centers from digital data. Part I: development and operation of the scheme. *Australian Meteorological Magazine* **39**: 155–166.

Parkinson CL, Comiso JC. 2013. On the 2012 record low Arctic sea ice cover: combined impact of preconditioning and an August storm. *Geophysical Research Letters* **40**: 1356–1361. https://doi.org/10.1002/grl.50349.

Serreze MC. 1995. Climatological aspects of cyclone development and decay in the Arctic. *Atmosphere-Ocean* **33**: 1–23.

Serreze MC, Barrett AP. 2008. The summer cyclone maximum over the central Arctic Ocean. *Journal of Climate* **21**: 1048–1065.

Serreze MC, Barry RG. 2014. *The Arctic Climate System*, 2nd ed. Cambridge University Press: New York, NY.

Simmonds I, Lim EP. 2009. Biases in the calculation of Southern Hemisphere mean baroclinic eddy growth rate. *Geophysical Research Letters*. **36**: L01707. https://doi.org/10.1029/2008GL036320.

Simmonds I, Rudeva I. 2012. The great Arctic cyclone of August 2012. *Geophysical Research Letters* **39**: L23709. https://doi.org/10.1029/2012GL054259.

Simmonds I, Rudeva I. 2014. A comparison of tracking methods for extreme cyclones in the Arctic basin. *Tellus A: Dynamic Meteorology and Oceanography* **66**: 25252. https://doi.org/10.3402/tellusa.v66.25252.

Simmonds I, Burke C, Keay K. 2008. Arctic climate change as manifest in cyclone behavior. *Journal of Climate* **21**: 5777–5796. https://doi.org/10.1175/2008JCLI2366.1.

Tanaka HL, Yamagami A, Takahashi S. 2012. The structure and behavior of the Arctic cyclone analyzed by the JRA-25/JCDAS data. *Polar Science* **6**: 54–69.

Zhang X, Walsh JE, Zhang J, Bhatt US, Ikeda M. 2004. Climatology and interannual variability of Arctic cyclone activity: 1948-2002. *Journal of Climate* **17**(12): 2300–2317.

Zhang J, Lindsay R, Schweiger A, Steele M. 2013. The impact of an intense summer cyclone on 2012 Arctic sea ice retreat. *Geophysical Research Letters* **40**: 720–726. https://doi.org/10.1002/grl.50190.

Zhang J, Ashjian C, Campbell R, Hill V, Spitz YH, Steele M. 2014. The great 2012 Arctic Ocean summer cyclone enhanced biological productivity on the shelves. *Journal of Geophysical Research, Oceans* **119**: 297–312. https://doi.org/10.1002/2013JC009301.

23

A comparison between thermal tropopauses derived from mandatory and significant levels for the Indian subcontinent upper-air network

Adrián E. Yuchechen*[iD] and Pablo O. Canziani

Universidad Tecnológica Nacional, Facultad Regional Buenos Aires, Consejo Nacional de Investigaciones Científicas y Técnicas (CONICET), Unidad de Investigación y Desarrollo de las Ingenierías, Buenos Aires, Argentina

*Correspondence to:
A. E. Yuchechen, Unidad de
Investigación y Desarrollo de las
Ingenierías, Facultad Regional
Buenos Aires, Universidad
Tecnológica Nacional, Mozart
2300 – C1407IVT, Ciudad
Autónoma de Buenos Aires,
Argentina.
E-mail:
aeyuchechen@frba.utn.edu.ar*

Abstract

Differences between lapse rate tropopauses (LRTs) and LRT-like tropopauses retrieved from mandatory levels (LRTMs) were studied for height, pressure and temperature at 37 locations of the Indian subcontinent on a long-term annual, summer and winter basis covering the period 1973–2015. LRTM is usually found below LRT and statistical distinctions hinder the use of the former tropopause as a replacement for the latter one, yet significant positive Spearman's correlations show a relationship through a monotonic increasing function that enable the estimation of LRT variables from the corresponding LRTM ones. The slope and the intercept for a linear function relating corresponding variables were obtained at each location.

Keywords: thermal tropopause; mandatory levels; Indian subcontinent

1. Introduction

The tropopause is the layer that separates the troposphere from the stratosphere. Temperature in the former (latter) layer decreases (increases) with height. The lapse rate tropopause (LRT) determines the region at which the lapse rate changes its sign. The World Meteorological Organization (WMO) defines LRT as 'the lowest level at which the lapse rate decreases to $2 \,°C\,km^{-1}$ or less, provided that the average lapse rate between this level and all higher levels within 2 km does not exceed $2 \,°C\,km^{-1}$' (WMO, 1992). The last condition avoids labeling low-level inversions as tropopauses. The definition permits the direct calculation of LRTs from radiosonde ascents, but they should be obtained from significant levels since it is at them where abrupt changes of thermodynamic variables occur.

Radiosondes at most upper-air stations show no recent issues with the report of significant levels (Simmons, 2011) but past records may lack them, hindering the operational retrieval of LRTs. One way to overcome this is to derive LRTs from mandatory levels, i.e. pressure levels at which temperature and wind data must be reported (WMO, 1992). Biases between LRTs and lapse rate tropopauses calculated from mandatory levels (LRTMs) are latitude-dependent (Zängl and Hoinka, 2001; Reichler *et al.*, 2003). Considering the advantages of the tropopause for assessing anthropogenic effects (Santer *et al.*, 2003), a direct application for LRTMs is their employment in tropopause trend studies, including the recovery and analysis of old radiosonde records registering only mandatory levels. Even at radiosonde void regions, long tropopause records can be operationally generated from historic reanalyses (e.g. The Twentieth Century Reanalysis, Compo *et al.*, 2011). It is therefore important for such purposes to know how LRTM relates to LRT.

Intra- and inter-seasonal variability in the Indian subcontinent, which ranks among the highest variabilities in the World, prompted the study of the region. The goal of the paper is to compare LRTM with LRT in the long-term annual and seasonal means for upper-air stations that span tropical and subtropical latitudes in the region in order to establish the differences and whether the LRTM is a suitable replacement for LRT. Given that LRTMs can be easily obtained from a variety of products with their output at fixed levels (e.g. reanalyses and circulation models), ways to estimate LRT from LRTM are also presented.

2. Data and methodology

The study was carried out using radiosonde ascents from the University of Wyoming's worldwide radiosonde database (http://weather.uwyo.edu/upperair/sounding.html). Radiosondes from 39 upper-air stations in India and one in Bangladesh were used. They are shown in Table S1, Supporting information along with the period spanned at each location: most of them extend from January 1973 to December 2015. Average

A comparison between thermal tropopauses derived from mandatory and significant levels for the Indian...

163

radiosonde availability is above two thirds of the total at half of the stations. Owing to relocations, the time series at three stations were amalgamated to form a single record (see Table S1). Hence, the net number of locations in this study is 37. The location of these stations appear in Figure S1.

Individual radiosondes include both mandatory and significant levels. The number of either level varies spatially and temporally. Figure S2 shows the average annual percentages of significant levels for the study region, evidencing a net increase in the recent years. Figure S2 also shows the average annual number of significant levels between 250 hPa and 50 hPa. Mandatory and significant levels were separated in order to calculate LRTMs and LRTs. Following the WMO definition, LRTs were obtained from significant levels with an algorithm that had the ability to calculate up to five different tropopauses from each ascent. Table S2 shows the percentages of detected LRTs and the breakdown into single and double events. Single tropopauses represent most of the cases at all sites, with the largest occurrence of double tropopauses taking place in north-western India. Events having three or more tropopauses were found at some locations. They were not included due to the very small percentages they represent, the largest of them also occurring in the aforementioned region. This is the reason some percentages in Table S2 do not add up to 100.

LRT height, pressure, temperature, wind direction and speed, and potential temperature were used in a selection procedure aimed at detecting outliers. A long-term climatology was carried out to obtain monthly means and standard deviations for the variables at each different tropopause. If each variable exceeded the grand-time mean by more than two standard deviation units it was considered an outlier and that tropopause was excluded from the analysis. Table S2 shows the percentages of rejections: on average, they are around 10% and 30% for single and double tropopauses, respectively. Composite time series were built for relocated stations. Single LRT and double LRT's lower tropopause time series were consolidated at each location.

One way to characterize the temperature profile in the vicinities of the upper troposphere/lower stratosphere is the LRT sharpness. It was calculated from significant levels as the difference of the mean temperature gradients between the height of the first detected LRT and 500 m above and below (Wirth, 2000). Given this definition, it might be argued that the number of significant levels restrains the sharpness values. However, the annual values of the Spearman's correlation coefficient (r_S) between the number of these levels and the sharpness do not show a clear pattern either with all the locations considered altogether (Figure S2) or individually (not shown).

Figure S3 shows the annual pressure distribution from LRT at two stations with distinct regimes. Bimodality occurs at Patiala (station 42101; 30°19′48″N, 76°27′36″E), with a primary (secondary) maximum at

100 hPa (220 hPa). If only single events are considered the percentages for the secondary maximum decreases in favor of the primary ones. At Thiruvananthapuram (station 43371; 8°28′48″N, 76°56′60″E) the distribution is unimodal irrespective of the inclusion of double LRT events, suggesting single LRTs are by far most frequent at lower latitudes. Unimodality may also be due partly to the inability of the WMO definition to capture the intricate structures of multiple tropopauses in the Tropics (Mehta *et al.*, 2011) and partly due to early balloon bursts before reaching the second tropopause (Seidel *et al.*, 2001; Yuchechen *et al.*, 2010). Separately, LRTM height, pressure and temperature were obtained from mandatory levels (Appendix S1). The aforementioned selection procedure was carried out on these three LRTM variables. In general, detection percentages are quite similar to the single LRT ones (Table S2).

3. Results

Annual, summer and winter LRT mean pressures are shown in Figure S4. Annual values south of 25°N are slightly above the 100-hPa reference for the tropical tropopause (Holton *et al.*, 1995); north of this latitude pressures increase, with highest values (i.e. lowest tropopauses) occurring at the northwest portion of the study region. The wet (summer) and dry (winter) seasons span from June to September (Ding and Sikka, 2006) and from November to February (Chang *et al.*, 2006), respectively. LRTs are higher in the north during summer. Given that the tropopause can be broadly associated to the temperature of the troposphere underneath (Schneider, 2007), the surface temperature in particular (Thuburn and Craig, 1997), results are in agreement with a reversal of the temperature gradient (i.e. directed southwards) created by the large heating of the Tibetan Plateau in during the summer season (Ding and Sikka, 2006). In winter, LRT pressure is lower than (higher than) the annual mean in the southern (northern) portion of the domain.

The LRTM method finds a single tropopause that may correspond either to the single LRT or to the lowest one in multiple LRT events (cf. Figure S3(a)). The differences between LRTM and LRT were calculated for height (z), pressure (P) and temperature (T) during the entire year as well as for the summer and winter seasons. Significances were evaluated using the test for differences of mean for paired samples (Wilks, 2006) with a level of confidence of 95%. Annual, summer and winter long-term means and standard deviations (SDs) for $z_{LRTM} - z_{LRT}$ are shown in Figure 1. Annual means are significant and negative except at Srinagar where $z_{LRTM} > z_{LRT}$. Summer and winter replicate the annual picture, with a statistically indistinguishable difference at Srinagar in summer. As to SD, it increases northwards in all three cases, with a maximum meridional gradient in winter. This may be attributable to the greater baroclinicity impinged by the westerly

$$Z_{LRTM} - Z_{LRT}$$

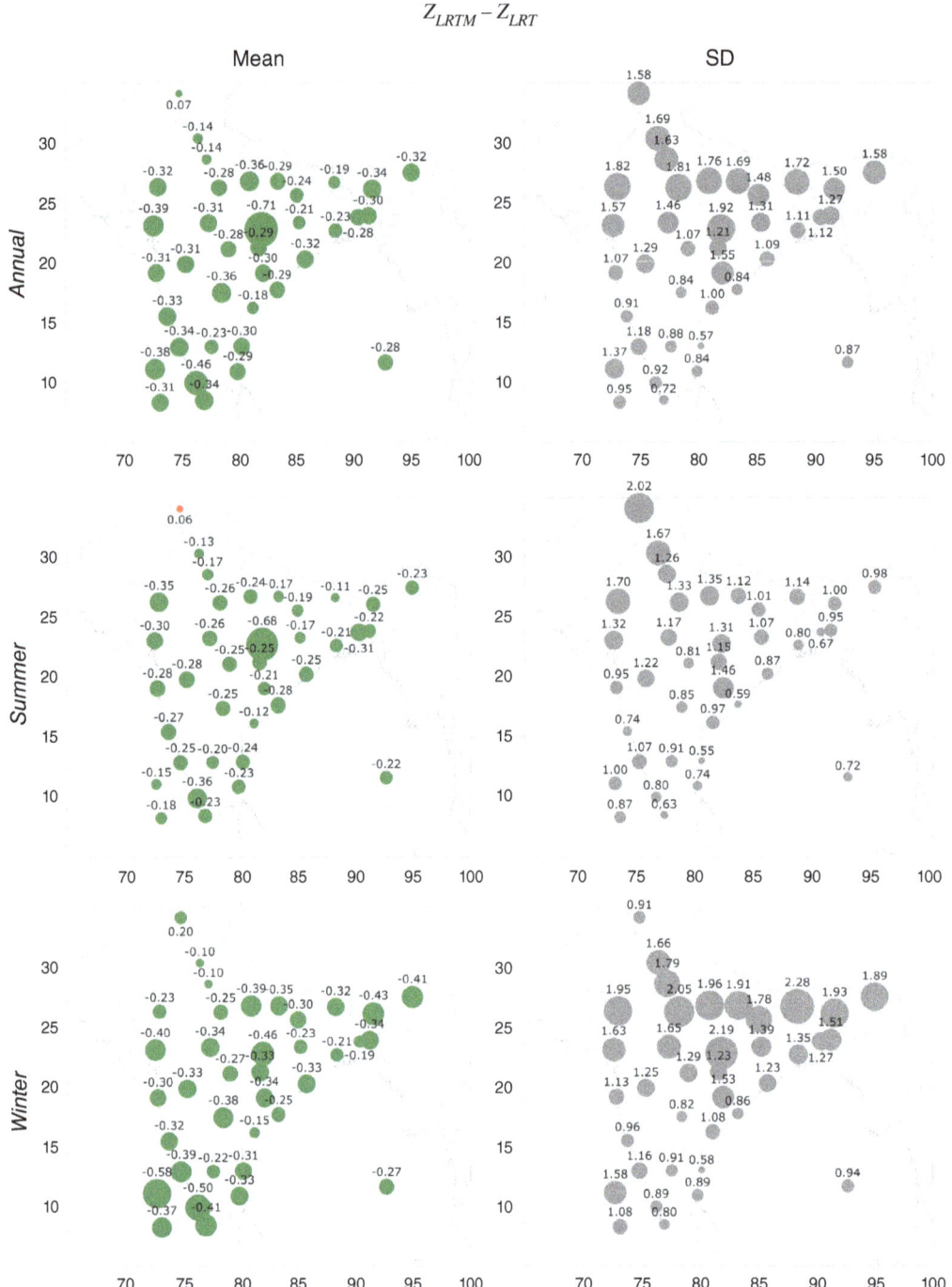

Figure I. Long-term annual, summer, and winter means and standard deviations (SDs) for $z_{LRTM} - z_{LRT}$ (in km). Green (red) dots marks significance (no significance); the size is proportional to the variable's absolute value.

subtropical jet (STJ). In summer it migrates to higher latitudes (Yanai and Wu, 2006). It is in this season when Srinagar experiences the largest SD (above 2 km) between the compared seasons. Significance implies LRTM and LRT averaged values are statistically distinguishable. Notwithstanding, the large SDs at Srinagar indicate that no significance does not necessarily mean individual values are close to one another at any particular time.

Annual, summer and winter means for $P_{LRTM} - P_{LRT}$ and $T_{LRTM} - T_{LRT}$ are shown in Figure 2. When $P_{LRTM} - P_{LRT}$ annual values are considered a significant

negative difference occurs at Srinagar only; at the rest of the stations the value is positive, with both variables being statistically indistinguishable at Patiala and Siliguri only. Discrepancies are generally smaller in summer at most of the stations; $P_{LRT} > P_{LRTM}$ just at Srinagar, with a separation between the two variables greater than in the annual case, no significance occurring at seven locations, and significance taking place elsewhere. Most stations experience the largest discrepancies in winter, in agreement with Reichler *et al.* (2003). Like in the annual and summer cases, Srinagar is the only location that has a significant negative difference. Six

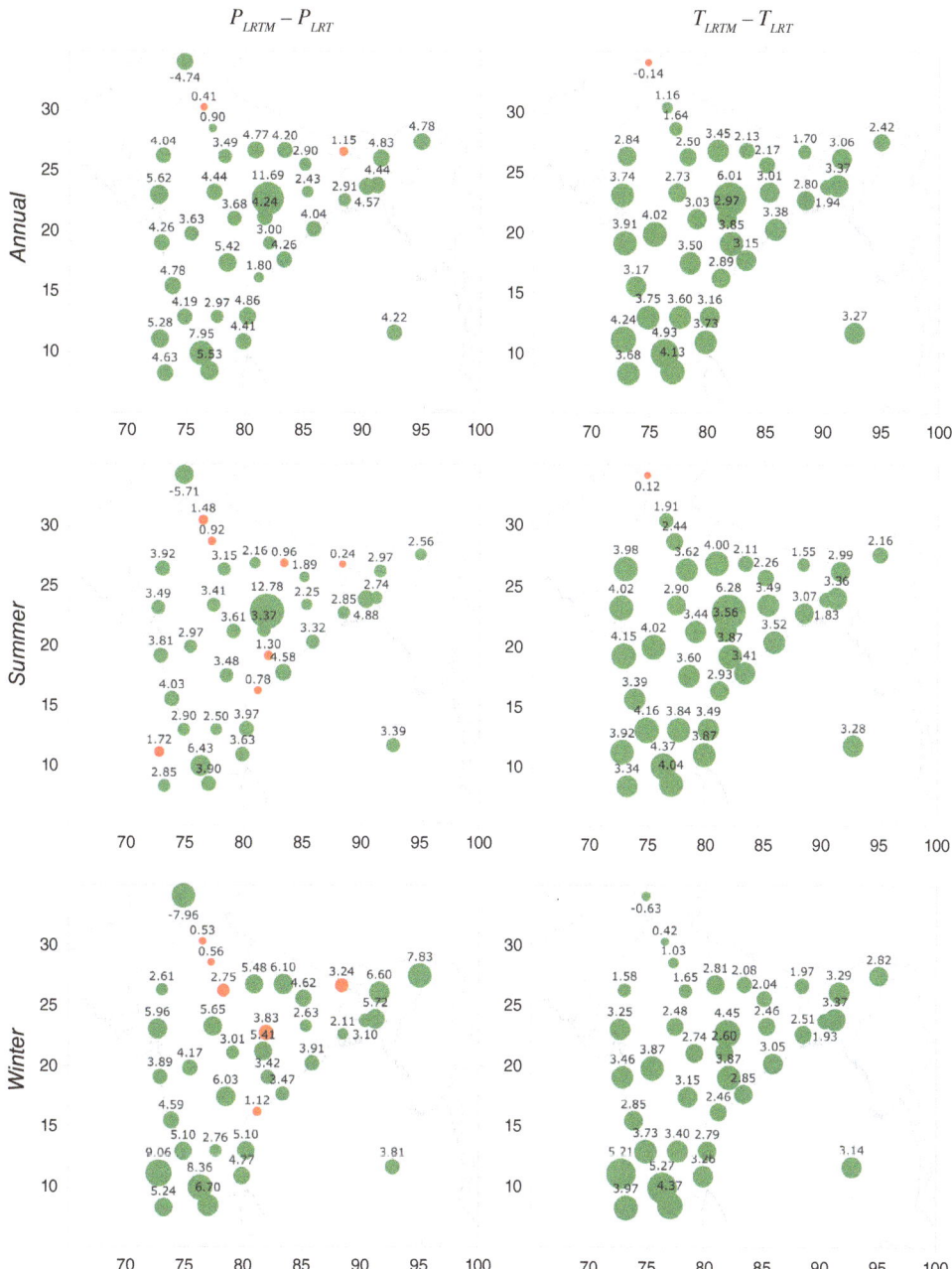

Figure 2. Long-term annual, summer and winter means for $P_{LRTM} - P_{LRT}$ (in hPa) and $T_{LRTM} - T_{LRT}$ (in °C). Size is proportional to the variable's absolute value. Color code as in Figure 1.

stations have their pressures indistinguishable from each other, and the rest of the sites have a significant positive difference. Overall, results match the inverse relation between pressure and height. The average summer and winter difference SDs are 27 and 36 hPa, respectively, with mean individual values increasing northwards in both seasons but being no greater than 67 hPa anywhere (results not shown). These figures are also in line with those of Reichler *et al.* (2003) for the subtropics.

The annual means for $T_{LRTM} - T_{LRT}$ reveal a warm bias at all but one of the stations, with largest values in the southernmost tip of the subcontinent. Similarly, all but one of the values are significant. Summer shows a warm bias too but at all of the locations, with an

overall increase of the values with respect to the annual case. Regarding significance, the summer configuration parallels the annual one. Concerning winter, all values are significant, with Srinagar distinguishing itself from the rest of the sites with its cold bias. In general, differences are smaller than in the annual case at all but the southernmost stations, whose values exhibit the largest discrepancies between the compared seasons.

Since remarkable differences exist between the two tropopauses, LRTM cannot be directly used as a replacement for LRT (to fill missing data, for instance), except at a reduced number of stations, during specific seasons, depending on the variable considered and, most importantly, on its variability. The long-term annual, summer and winter values of r_S were calculated

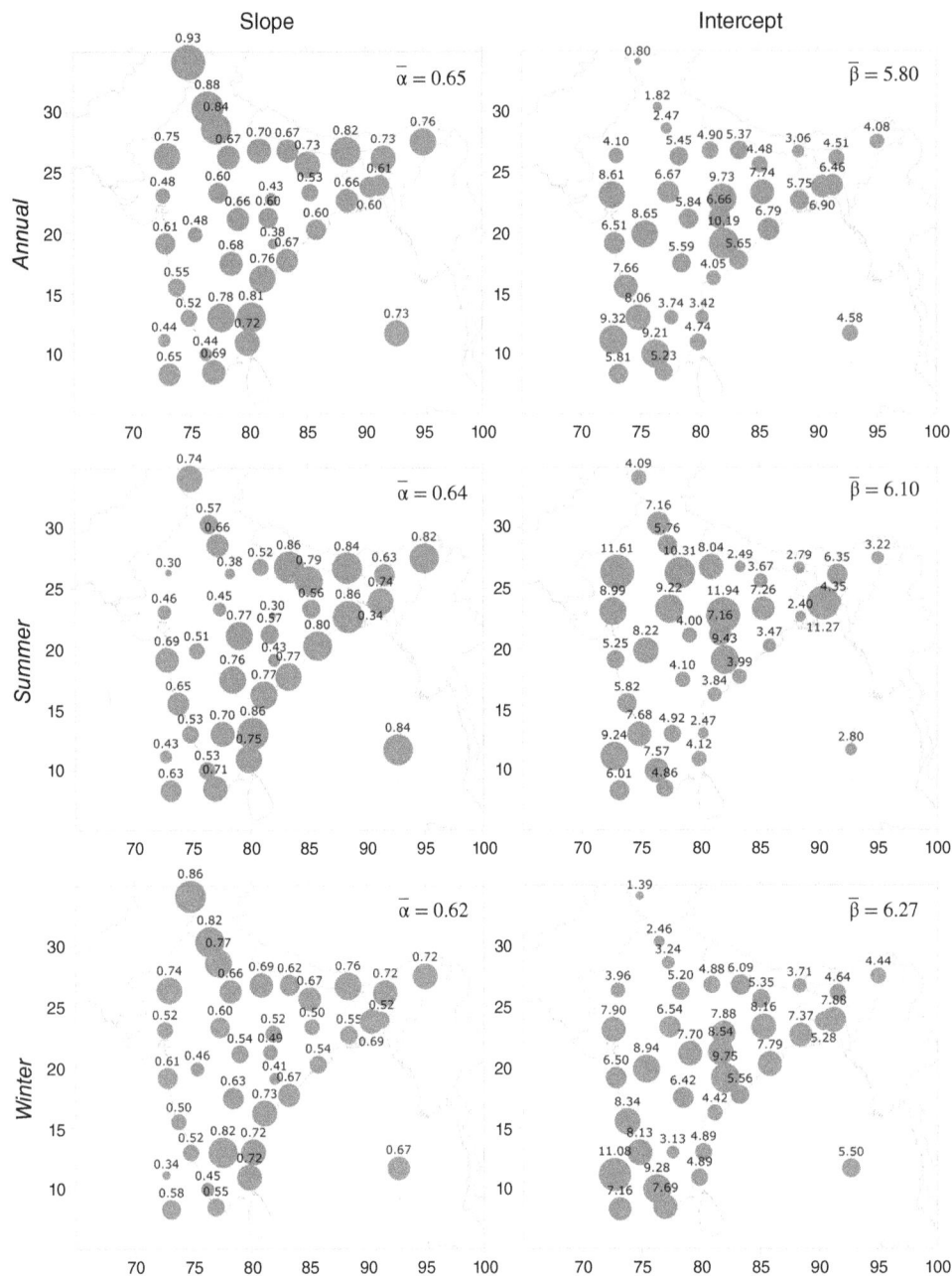

Figure 3. Slope (α) and intercept (β) for the retrieval of the tropopause height from z_{LRTM} with a linear function. Size is proportional to the variable's value. The mean values over all the stations are shown in the upper right corner of each panel.

between each variable's two time series. Given that r_S operates onto the ranks of the series (Wilks, 2006), it is preferred, for the purposes of the paper, over the Pearson's correlation coefficient, as joint behaviors are better represented. Figure S5 shows the long-term annual, summer and winter values of r_S for P. All of them are positive and 95% significant. Results are similar for z and T (not shown). Considering the positive values of r_S, the use of any monotonic function that relates LRT variables with the corresponding LRTM ones is warranted. A linear function $x_{\mathrm{LRT}} = \alpha\, x_{\mathrm{LRTM}} + \beta$, where x denotes any of the analyzed variables, is preferred for its simplicity, and used in a point-by-point fitting of the data to obtain the dimensionless slope α and the intercept β that correct the value of the LRTM

variable to the LRT one. Positive r_S values indicate a direct relationship between the correlated variables, so $\alpha \geq 0$. With all the pairs considered altogether, no bias exists between corresponding LRTM and LRT variables if $\alpha = 1$ and $\beta = 0$, whereas $\alpha = 0$ represents the maximum uncorrelatedness. These two extreme situations help interpreting the results, since the greater the value of α the smaller the uncorrelatedness given that the independent LRTM variable is given more weight.

The values of α and β for z for the annual, summer and winter cases are shown in Figure 3. In all cases α is positive and less than unity and its value is inversely related to β. Maximum values for α occur in the north-western portion of the study region, the region

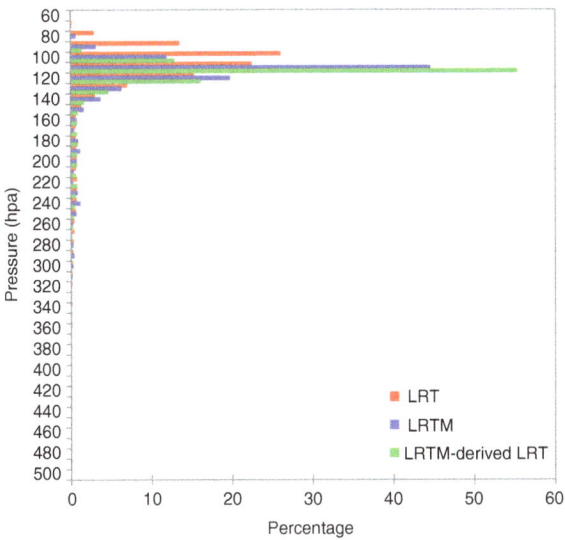

Figure 4. Annual pressure distributions from LRT, LRTM and LRTM-derived LRT over all the stations.

with maximum r_S values for the variable (not shown). Both α and β being positive is in agreement with the condition $z_{LRT} > z_{LRTM}$ at all stations, even at Srinagar where year-round averages show the inverse condition, possibly related to the number of double LRT events at the site, i.e. a greater variability for the difference due to cases with the lower tropopause being considerably low. Summer differs from the annual conditions in that most of the stations have a greater (smaller) $\alpha(\beta)$, in compliance with a reduction of both the mean difference and the variability (cf. Figure 1). The big picture reverses in winter, when both the difference and the SD are, in general, the largest between the compared seasons.

Comparisons between different variables can be made through α given its dimensionless character. Generally speaking α year-round values for P (Table S3) are slightly smaller than for z, thus increasing the uncorrelatednesses. For summer, α generally takes on greater (smaller) values than for the entire year at most of the stations south (north) of 20°N, thus increasing (decreasing) the correlatedness with respect to the annual case. Approximately the inverse situation takes place during winter. Regarding T, α annual values are the smallest between the analyzed variables and they generally decrease during summer (Table S4), making summer uncorrelatednesses greater than the annual ones. Changes in α for winter are not as noticeable as with the other two variables: they are close to the annual values at most of the stations. In a winter versus summer comparison α generally decreases south of 20°N. All α values are significant at a 95% confidence level. Annual, summer and winter β values for P and T are also included in Tables S3 and S4, respectively. The annual pressure distribution from LRT, LRTM and LRTM-derived LRT is shown in Figure 4.

Figure 5 shows the root mean square (RMS) for the discrepancy between z_{LRT} and the one linearly calculated using z_{LRTM} for the annual, summer, and winter cases. Assuming that there exists a dependence between

$z_{LRTM} - z_{LRT}$ and the LRT sharpness, it also shows the values of r_S for the correlation between both variables. RMSs generally increase northwards, year-round as well as in summer and winter, whereas the values of r_S show that the LRT sharpness dependence decreases with latitude. During summer, RMSs are larger in the north-westernmost stations, in coincidence with a monsoon trough located there. Notwithstanding, the largest RMSs occur during winter in the 25°–30°N belt, in agreement with both the southernmost position of the STJ and the northwest-by-southeast ridge corridor in the region (Ding and Sikka, 2006). The behaviors for P and T (Figures S6 and S7, respectively) parallel the one described for z, with T having the strongest inverse dependence with latitude in winter.

4. Summary and concluding remarks

Differences between tropopauses calculated from mandatory levels (LRTMs) and WMO-defined lapse rate tropopauses (LRTs) at 37 upper-air stations spanning the Indian subcontinent's tropical and subtropical latitudes have been analyzed for height (z), pressure (P) and temperature (T), focusing on a 42-year annual, summer and winter means. Results indicate that LRTM is lower than LRT at all the stations with the exception of Srinagar, which is located in a region where the occurrence of multiple LRTs is relatively high all year long. In general, $|z_{LRTM} - z_{LRT}|$ shows a decrease in summer with respect to annual conditions at all the stations, whereas the separation between both tropopauses increases during winter. P behaves in much the same way. As for $T_{LRTM} - T_{LRT}$, there is an overall increase in summer with respect to annual conditions; in winter, lesser values occur with the exception of the southernmost stations, whose differences are even greater than in summer. Significance for the differences were also addressed.

Results indicate that the use of LRTM as a direct replacement for LRT is discouraged. However, significant positive Spearman's correlations show that they are related so LRT variables can be estimated from LRTM ones. The station-specific slope and intercept for a linear function linking the variables were obtained for the annual, summer and winter cases. To the best of our knowledge, results here are the first ones including double tropopauses in the comparisons [they were not considered in Reichler et al. (2003) and Zängl and Hoinka (2001)]. Largest RMSs occur in winter and seem to be connected to anticyclonic anomalies. However, the absolute values of the winter correlations between the LRT/LRTM discrepancy and the LRT sharpness (Figures 4, S6, and S7) are the lowest ones, an indication that the sharpness may not play a fundamental role in establishing the differences. This is a matter of future investigation, as also is the possibility of calculating multiple tropopauses from the LRTM method. One of the utilities of this work is the direct retrieval of LRTM variables from fixed pressure levels

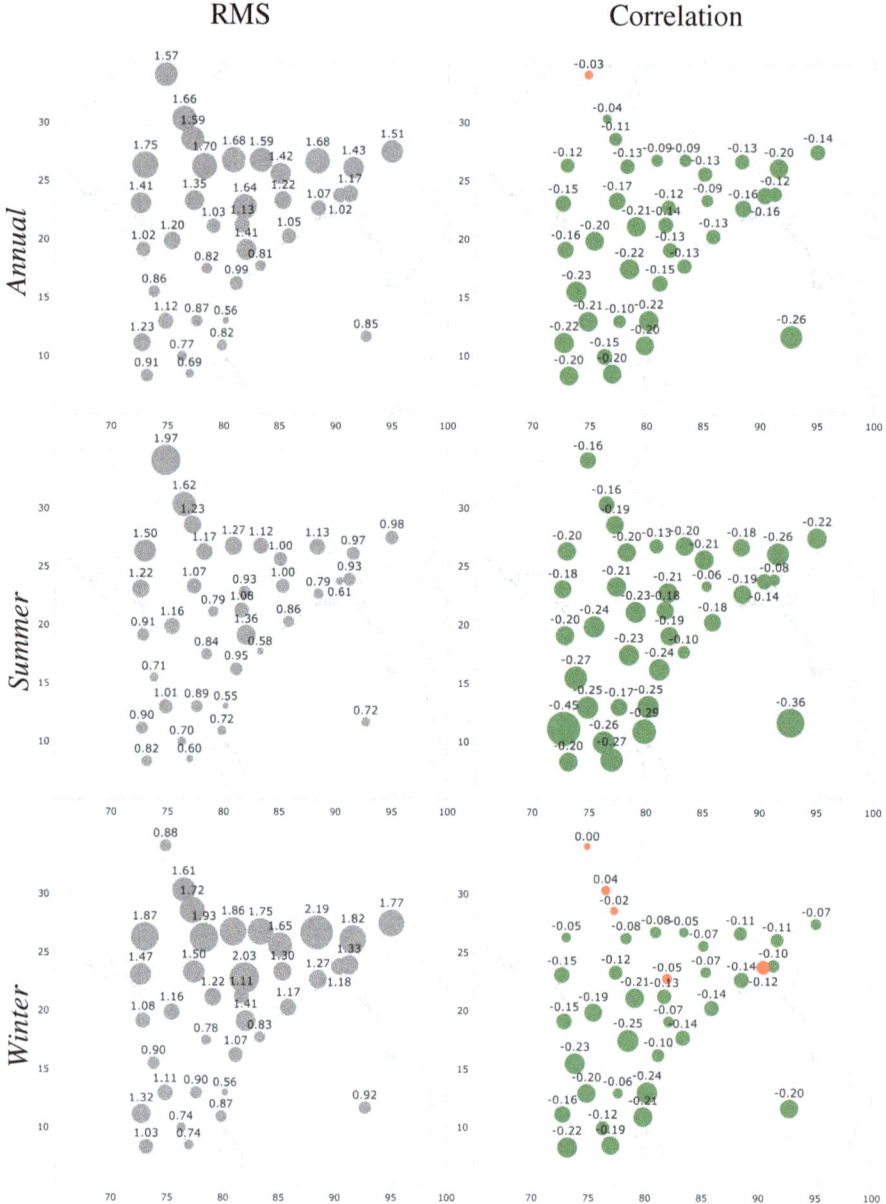

Figure 5. Root mean squares (RMS) for the difference between z_{LRT} and the tropopause height calculated from z_{LRTM}, and Spearman's correlations between $z_{LRTM} - z_{LRT}$ and the LRT sharpness. RMS values in km. Color code for correlations as in Figure I.

for worldwide regions where radiosonde launchings are scant and, on the other hand, to use them as a proxy for LRT variables in a variety of applications, such as variability studies or anthropogenic effects in the troposphere through trends. It is worth emphasizing that the only requirement for the method in order to make comparisons between the tropopauses retrieved with it is that it should be used on fixed pressure levels, i.e. mandatory levels, or model levels whenever possible.

Acknowledgements

We gratefully acknowledge two anonymous reviewers for their comments and suggestions. This research was partly supported by CONICET PIP 2012–14 0075, Ministerio de Defensa PID-DEF 26/2014 and Agencia Nacional de Promoción Científica y Tecnológica PICT-2012-2927 grants.

Supporting information

The following supporting information is available:

Appendix S1. Calculation of thermal tropopauses from mandatory levels and supplemental figures and tables.

References

Chang C-P, Wang Z, Hendon H. 2006. *The Asian Monsoon.* Praxis Publishing: Chichester; 89–127.

Compo GP, Whitaker JS, Sardeskmukh PD, Matsui N, Allan RJ, Yin X, Gleason BE Jr, Vose RS, Rutledge G, Bessemoulin P, Brönnimann, Brunet M, Crouthamel RI, Grant AN, Groisman PY, Jones PD, Kruk MC, Kruger AC, Marsahll GJ, Maugeri M, Mok HY, Nordli Ø, Ross TF, Trigo RM, Wang XL, Woodruff SD, Worley SJ. 2011. The twentieth century reanalysis project. *Quarterly Journal of the Royal Meteorological Society* **137**: 1–28.

Ding Y, Sikka DR. 2006. *The Asian Monsoon*. Praxis Publishing: Chichester; 131–201.

Holton JR, Haynes PH, McIntyre ME, Douglass AR, Rood RB, Pfister L. 1995. Stratosphere-troposphere exchange. *Reviews of Geophysics* **33**: 403–439.

Mehta SK, Ratnam MV, Krishna Murthy BV. 2011. Multiple tropopauses in the Tropics: a cold point approach. *Journal of Geophysical Research* **116**: D20105.

Reichler T, Dameris M, Sausen R. 2003. Determining the tropopause height from gridded data. *Geophysical Research Letters* **30**(20): 2042.

Santer BD, Sausen R, Wigley TML, Boyle JS, AchutaRao K, Doutriaux C, Hansen JE, Meehl GA, Roeckner E, Ruedy R, Schmidt G, Taylor KE. 2003. Behavior of tropopause height and atmospheric temperature in models, reanalyses, and observations: decadal changes. *Journal of Geophysical Research* **108**(D41): 4002.

Schneider T. 2007. *The Global Circulation of the Atmosphere*. Princeton University Press: Princeton, NJ; 47–77.

Seidel DJ, Ross RJ, Angell JK, Reid GC. 2001. Climatological characteristics of the tropical tropopause as revealed by radiosondes. *Journal of Geophysical Research* **106**(D8): 7857–7878.

Simmons A. 2011. From observations to service delivery: challenges and opportunities. *WMO Bulletin* **60**: 96–107.

Thuburn J, Craig GC. 1997. GCM tests of theories for the height of the tropopause. *Journal of the Atmospheric Sciences* **54**: 869–882.

Wilks DS. 2006. *Statistical Methods in the Atmospheric Sciences*, 2nd ed. Academic Press: Burlington, MA; 627.

Wirth V. 2000. Thermal versus dynamical tropopause in upper-tropospheric balanced flow anomalies. *Quarterly Journal of the Royal Meteorological Society* **126**: 299–317.

World Meteorological Organization (WMO). 1992. International Meteorological Vocabulary, WMO/OMM/BMO – No. 182, Secretariat of the WMO, Geneva, Switzerland. 784 pp.

Yanai M, Wu G-X. 2006. *The Asian Monsoon*. Praxis Publishing: Chichester; 513–549.

Yuchechen AE, Bischoff SA, Canziani PO. 2010. Latitudinal height couplings between single tropopause and 500 and 100 hPa within the Southern Hemisphere. *International Journal of Climatology* **30**: 492–508.

Zängl G, Hoinka KP. 2001. The tropopause in the polar regions. *Journal of Climate* **14**: 3117–3139.

24

Changes of probabilities in different wind grades induced by land use and cover change in Eastern China Plain during 1980–2011

Jinlin Zha,[1] Jian Wu[1]* and Deming Zhao[2]

[1] Department of Atmospheric Science, Yunnan University, Kunming, China
[2] Key Laboratory of Regional Climate-Environment for Temperate East Asia, Institute of Atmospheric Physics, Chinese Academy of Sciences, Beijing, P. R. China

*Correspondence to:
J. Wu, Department of
Atmospheric Science, Yunnan
University, Kunming 650091,
China.
E-mail: wujian@ynu.edu.cn

Abstract

The differences of the near-surface wind speed (SWS) between the frictional wind model (FWM) and the observation are used to reflect the impacts of land use and cover change (LUCC) on SWS in Eastern China Plain (ECP). Results show that LUCC makes the range of SWS narrow, which had significantly weakening effect to stronger wind than weaker wind. In addition, the decrease of probabilities for observed gentle breeze (GB), moderate breeze (MB), and wind speed greater than or equal to $8\,\mathrm{m\,s^{-1}}$ (WSGE8) are more significant in large cities than that in small cities.

Keywords: near-surface wind speed; frictional wind model; LUCC; probabilities; wind grade

1. Introduction

Long-term change of the observed near-surface wind speed (SWS) can be affected by both of land use and cover change (LUCC) and variation of atmospheric circulation strength, which has relation with anthropogenic activities and climate change, respectively. SWS declined has been reported in recent decades in different regions, such as North America (Greene *et al.*, 2012), East Asia, and Central Asia and Europe (Vautard *et al.*, 2010). A slowdown of SWS was also reported in China (Guo *et al.*, 2011). Possible reasons causing SWS decrease include the weakening of large-scale circulations (Sušelj *et al.*, 2010), the variability of the pressure gradient force (PGF) (Klink, 2007), and the weakening of the monsoon driving force (Xu *et al.*, 2006).

However, some researches tend to deduce the cause of SWS decrease to the rise of surface roughness, which induced mainly by anthropogenic LUCC and natural factors, such as urbanization (Zhang *et al.*, 2010) and vegetation recovery (Vautard *et al.*, 2010). Li *et al.* (2008) showed the significant decrease of SWS was induced by urbanization and other types LUCC over China in recent 40 years. The SWS in urban areas was also found to be lower than suburban regions in some megacities of China (Li *et al.*, 2011). In any case, LUCC can affect SWS changes, but it's difficult to isolate the impacts of LUCC on SWS and quantify its effects on SWS. In our former research, the frictional wind model (FWM) was used to separate the effects of PGF and LUCC on the long-term changes of SWS, in which the balance among the PGF, the Coriolis force, and the surface drag force was supposed (Wu *et al.*, 2016).

To isolate the influences of LUCC on SWS, the drag coefficient was derived using FWM with observed SWS and PGF calculated from observed air pressure from 1980 to 2011. Then, FWM was used to calculate the wind speed when the drag coefficient was held constant at its value of the year with less LUCC for each station, such as 1980, the beginning of the period used in the article, and this wind speed was called the model wind speed (MWS). Obviously, the MWS included the effect from temporal change of PGF and excluded the influence of drag coefficient change induced by LUCC, because the constant drag coefficient was used in the calculation of MWS. Finally, the difference between MWS and observed SWS at each station could quantify the influence of LUCC on SWS (Wu *et al.*, 2016).

The changes of probabilities for SWS are another way to recognize the long-term variations of SWS besides the speed value itself. Meanwhile, the knowledge of probability distribution of SWS is essential for surface flux estimation, wind power estimation, and wind risk assessments (He *et al.*, 2010). The increase of the drag coefficient and associated slowdown in SWS over ECP have been demonstrated by Wu *et al.* (2016), but the impacts of LUCC on changes in probabilities for different wind grades are remain unclear. In this article, temporal characteristics in probabilities of different wind grades induced by LUCC are investigated further based on our former FWM.

2. Data and Methods

ECP region is selected as research region. Daily mean wind speed data from 93 meteorological stations on the

Table 1. Six wind grades criteria.

Grade	Name	Wind speed (m s⁻¹)
1	Calm	0–0.2
2	Light air (LA)	0.3–1.5
3	Light breeze (LB)	1.6–3.3
4	Gentle breeze (GB)	3.4–5.4
5	Moderate breeze (MB)	5.5–7.9
6	Wind speed greater than or equal to 8 m s⁻¹ (WSGE8)	≥8.0

ECP during 1980–2011 are used. The 93 stations are selected according to the following criteria: (1) the elevation of the station is below 200 m above sea level, (2) it is a national standard station, and (3) the missing data account for less than 1% of the total period studies. The observed SWS data was operated, provided, and quality tested by the China Meteorological Administration (CMA, 2003), and that which passed the homogeneity test and were therefore regarded as the credible dataset in China (Feng *et al.*, 2004). Typhoon track data from the Joint Typhoon Warning Center for 1980–2011 were used to remove wind speed observation data influenced by typhoons at stations located within a circle with a radius of 2° in latitude and longitude centered on the middle of each typhoon. Population data is obtained from the National Bureau of Statistics of China, which is available only for the year 2005. We define the large and small city as having a population of more than 1 million, and less than 0.5 million (Wu *et al.*, 2012), and there are 38 large cities and 23 small cities in the ECP region.

According to the criteria of CMA, SWS can be divided into 13 grades (CMA, 2003), which includes five and eight grades for SWS lower and higher than 8 m s⁻¹ respectively. The 13-grade criterion is usually used to describe the daily change of SWS under different weathers. The probabilities of wind speeds more than 8 m s⁻¹ are very low in China, which are usually induced by tropical cyclones. When this 13-grade criterion is used to depict the long-term change of SWS, it can be found the probabilities of SWS higher than 8 m s⁻¹ are too low to indicate clear climatic meaning. Therefore, the wind speed greater than or equal to 8 m s⁻¹ (WSGE8) is merged into one grade in the long-term wind speed change studies. Hence, the SWS is divided into six grades, which include calm, light air (LA), light breeze (LB), gentle breeze (GB), moderate breeze (MB), and WSGE8. The criteria for these wind grades are summarized in Table 1.

In this article, we discuss the probability distribution of wind speed and the influences of LUCC on six-grade wind speed based on the results of FWM. The probabilities in six wind grades $P_{l,i}$ is computed based on Equation (1).

$$P_{l,i} = \left(f(l, i) / n_i \right) \cdot 100\% \qquad (1)$$

where, l represents six wind grades respectively, i denotes the years from 1980 to 2011. $f(l, i)$ denotes the

days of l wind grade in i year. n_i is the total observation days in i year. To fit the distribution of probability of wind speed, the kernel probability estimator is used, which is given in Equation (2) (Bowman and Azzalini, 1997).

$$\text{PDF}(X) = \frac{1}{m} \sum_{i=1}^{m} w\left(X - X'_i; h\right) \qquad (2)$$

where, w is a probability density, which is symmetric with mean 0, and the variance of w is controlled by the parameter h, which is called smoothing parameter. The detail information about w and h can be found in Bowman and Azzalini (1997). $X = \{X_1, \dots, X_m\}$ represents wind speed, X'_i denotes the center value of the interval in which X_i falls, and m denotes the integer part of S, which is defined by Equation (3).

$$S = (\max(X) - \min(X)) / 0.1; \quad (m = int(S)) \qquad (3)$$

Additionally, Student's t-test is used to determine the significance of the data, and the linear trend coefficient is computed using the least-squares method.

3. Results

3.1. Probability distribution of observed SWS and MWS

Distinct decrease of observed SWS in China has been demonstrated (Xu *et al.*, 2006), and similar changes are also found in ECP. Figure 1 shows that observed SWS is bigger in inshore region than in inland region, which is also bigger in northern ECP than that in southern ECP with the highest value in Yangtze River Delta Region and the average of 2.3 m s⁻¹ in whole ECP. On the other hand, the average of MWS reaches 2.8 m s⁻¹, which is distinctly higher than SWS. Comparing MWS with observed SWS, it can be found that the fine regional difference is fewer in MWS than that in observed SWS, which means that more local characteristics were enclosed in observed SWS than MWS. Meanwhile, observed SWS shows a pronounced decreasing trend with the average linear trend of −0.19 m s⁻¹ (10 year)⁻¹, which can pass the significant t-test at 95% level, but MWS presents an increasing trend with an average of 0.11 m s⁻¹ (10 year)⁻¹ in ECP. The correlation coefficient between MWS and SWS is −0.35, which fail to pass significant t-test at 99% level. These results show that spatial distribution and long-term changes of observed SWS are inconsistent with that of MWS, because the influence of LUCC on MWS is excluded, while this influence is included in observed SWS.

Probabilities of MWS and observed SWS were calculated using Equation (1) and fitted using Equation (2). The maximum of the probability density for observed SWS and MWS is 0.830 and 0.376, respectively, and the corresponding wind speed is 2.1 m s⁻¹ and 2.0 m s⁻¹, respectively. At the same time, the probability close to 2 m s⁻¹ has been doubled by the influence of LUCC. The enclosed area under the probability

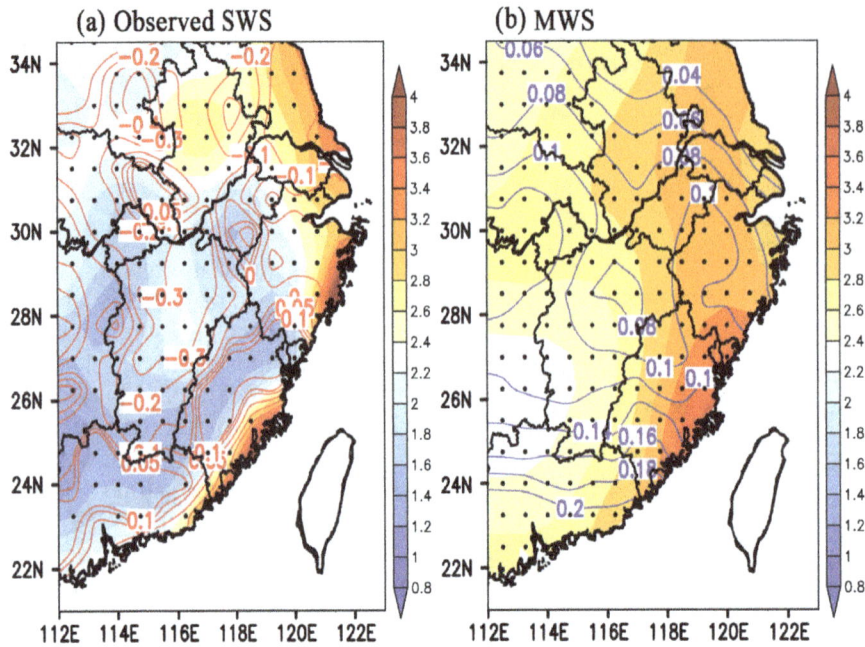

Figure 1. Spatial distribution of observed SWS (a) and MWS (b) (shaded), as well as their linear trend coefficients (contour) (The dot in (a), (b) denotes the trend coefficients can pass significant *t*-test at 95% level).

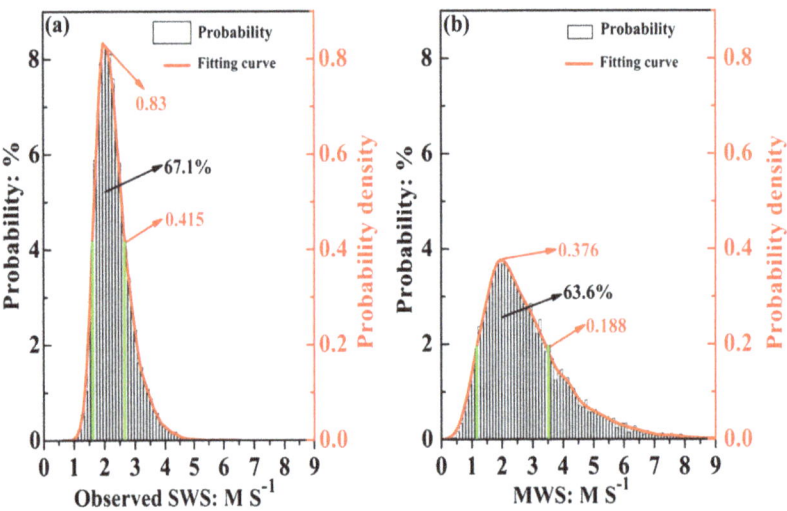

Figure 2. Probability distribution of observed SWS (a) and MWS (b) (the red line represents the PDF curve, and the green lines show the positions of half maximums).

density function (PDF) curve of observed SWS and between the two half maximums, with the corresponding wind speed of 1.6 and 2.6 m s^{-1}, respectively, reaches 67.1% of the total area under the PDF curve (Figure 2(a)). While the wind speed corresponding to the two half maximums of the PDF curve of MWS is 1.2 and 3.5 m s^{-1} respectively, and the enclosed area between the two half maximums reaches 63.6% of the total area under the curve (Figure 2(b)). It is obvious that LUCC made the PDF curve of observed SWS higher and narrower than that of MWS, at the same time wind speeds corresponding to the maximum of PDF curves are almost equal. Additionally, the probability of wind speed beyond 3.8 m s^{-1} is 1.8 and 20.6% for observed SWS and MWS, respectively,

and such distinct difference indicates that LUCC had significantly weakening effect to stronger wind than weaker wind, which implies observed SWS decrease was mainly induced by the decrease of strong wind episodes in ECP. Vautard *et al.* (2010) also addressed that LUCC decreased the strong wind more significantly than the weak wind over almost the northern hemisphere.

3.2. Temporal changes of probabilities in six wind grades

Temporal changes of probabilities of observed SWS in six wind grades are shown in Figure 3(a). The probabilities of observed LA and LB increase in recent

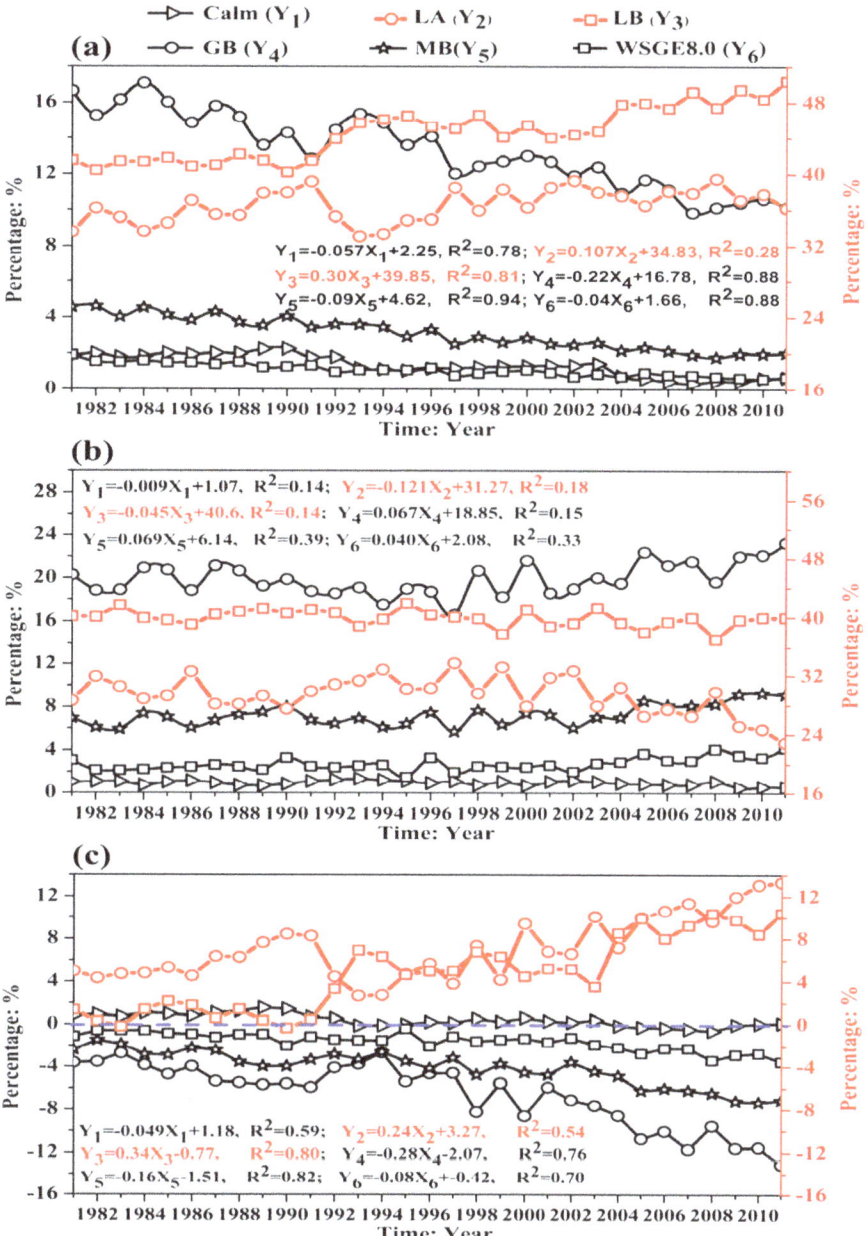

Figure 3. Temporal changes of probabilities in six wind grades ((a), (b) represents probabilities of observed SWS, and MWS respectively, and (c) denotes probability difference between observed SWS and MWS. R^2 is correlation coefficient and the threshold of 99% confidence level is 0.21).

30 years with the linear trend of 1.1% $(10\,\text{year})^{-1}$ and 3.0% $(10\,\text{year})^{-1}$, respectively, and the probabilities of observed calm, GB, MB, and WSGE8 show decrease trends with linear trend of -0.57% $(10\,\text{year})^{-1}$, -2.2% $(10\,\text{year})^{-1}$, -0.9% $(10\,\text{year})^{-1}$, and -0.4% $(10\,\text{year})^{-1}$, respectively. All of these linear trends are significant at 99% confidence level. Figure 3(b) indicates the probabilities of calm, LA, and LB of MWS that show weak decreases at rates of -0.09% $(10\,\text{year})^{-1}$, -1.2% $(10\,\text{year})^{-1}$, and -0.45% $(10\,\text{year})^{-1}$, respectively, which are indistinctive at 99% confidence level. The probabilities of GB, MB, WSGE8 of MWS have increasing trends, in which MB and WSGE8 passed significant t-test at 99% level. These characteristics are inconsistent with that of observed SWS, which means

that the wind speed excluding the influence of LUCC has indistinctive linear trends in weak wind grades, and has evident increase in strong wind grades. The temporal changes of the probability differences in the same wind grades between observed SWS and MWS (PDs) are shown in Figure 3(c), which was induced by LUCC purely, because the impact of LUCC on probability variations of MWS was excluded, and this influence was included in probability variations of observed SWS. The PDs of LA and LB show striking increases with linear rates of 2.4% $(10\,\text{year})^{-1}$ and 3.4% $(10\,\text{year})^{-1}$ respectively, which is bigger than the increase trends of LA and LB in observed SWS, respectively. On the contrary, the PDs of GB, MB, WSGE8 show decreasing trends of -2.8% $(10\,\text{year})^{-1}$,

Table 2. Linear trend coefficients (unit: % $(10 \, \text{year})^{-1}$) of probability differences between SWS and MWS for six wind grades respectively, as well as their mean values in large and small cities.

		Clam	LA	LB	GB	MB	WSGE8
Linear trend	Large cities	−0.05[a]	0.27[a]	0.32[a]	−0.35[a]	−0.16[a]	−0.05[a]
	Small cities	−0.02	0.28[a]	0.21[a]	−0.30[a]	−0.13[a]	−0.04[a]
Mean value	Large cities	0.39%	4.57%	9.12%	−6.30%	−5.43%	−2.48%
	Small cities	0.45%	5.98%	6.05%	−5.89%	−4.49%	−2.11%

[a]Denotes the trend coefficients passed significant t-test at 99% level.

-1.6% $(10 \, \text{year})^{-1}$, and -0.8% $(10 \, \text{year})^{-1}$, respectively. These long-term changes of PDs are similar to Figure 3(a), therefore, the influences of LUCC on the probabilities in six wind grades were mainly embodied by the pronounced increases of LA and LB, and the significant decreases of GB, MB, and WSGE8.

3.3. Probability changes in large and small cities

Because the changes in probabilities for six grades were induced by LUCC, there should exist probability changes of different extents in large cities relative to small cities, therefore, the PDs between MWS and observed SWS in different scale cities are investigated. Table 2 presents that the long-term changes of PDs for GB, MB, and WSGE8 had downward trends in large and small cities both, and that the decrease trends of PDs for GB, MB, WSGE8 in large cities were stronger than that in small cities. The average of PDs for GB, MB, WSGE8 in large cities is -6.30%, -5.43%, -2.48%, respectively, which is more significant than -5.89%, -4.49%, -2.11% in small cities, respectively. The main reasons include that the urban canopy layer in large cities is taller and denser than in small cities, so the surface roughness and drag force in large cities are bigger than in small cities (Coceal and Belcher, 2005). Coceal and Belcher (2005) have reported an increase in canopy density and height was usually associated with a decrease in the mean wind speed. The linear trends and average of PDs for LA and LB are positive both in large cities and in small cities because the drag induced by LUCC is increasing in recent 30 years (Wu et al., 2016), which induced the transfer of probabilities from stronger wind speed grades to LA and LB. Table 2 also reveals the decrease trend of PDs for calm in large cities is more prominent than that in small cities. Liu and Tian (2010) reported that China has experienced enormous urbanization after 1980. There was more than 20% expansion in urban area in ECP and southeastern China, and the largest increase in urban areas was found in the ECP region. Additionally, other studies have also reported that the urban land expansion was significant over the ECP region in recent 30 years, which was regarded as a main driver of LUCC (Asselen and Verburg, 2013). Coceal and Belcher (2005) revealed that based on the rapid urbanization, the higher momentum air at the top of the taller canopy is mixed down through the canopy, thus increasing the wind speed at the lower levels, which is in favor of the decrease of

calm. Observed SWS in large cities is smaller than that in small cities, which has been demonstrated by the previous studies (Guo et al., 2011), according to the above analysis, the fact is induced by the combination effects that the probabilities decrease of GB, MB, WSGE8 and the probabilities increase of LB in large cities are more distinct than that in small cities.

4. Conclusion

In this article, the differences between FWM wind speed and observed SWS are used to reflect the impacts of LUCC on SWS. The changes of probabilities of six wind grades are investigated, in which the influences of LUCC on these probabilities are revealed, and the main results are as follows:

1. The observed SWS corresponding to the maximum probability is close to $2.0 \, \text{m s}^{-1}$, and the probability close to $2.0 \, \text{m s}^{-1}$ was increased more than twice from 0.376 to 0.830 by the influence of LUCC purely. Meanwhile, LUCC made the PDF curve of observed SWS much higher and narrower than that of MWS. The probability of wind speed beyond $3.8 \, \text{m s}^{-1}$ is 1.8 and 20.6% for observed SWS and MWS, respectively, and such distinct difference shows that LUCC has significantly weakening effect to stronger wind than weaker wind, which means that observed SWS decrease was mainly induced by the decrease of strong wind episodes in ECP.

2. The most distinct characteristic for the changes of wind speed probabilities induced by LUCC included two aspects: one was the decreases of GB, MB, WSGE8 and the increases of LA and LB significantly; the other one was the decrease of the probability of calm. The decrease trends of probabilities of GB, MB, and WSGE8 in large cities were more pronounced than that in small cities. Additionally, when excluding the influences of LUCC, the probabilities of weak wind speed grades should show insignificant linear trends, and the probabilities of MB and WSGE8 should have an evident increase rate of 0.69 % $(10 \, \text{year})^{-1}$ and 0.40% $(10 \, \text{year})^{-1}$, respectively.

This article mainly analysis the spatio-temporal characteristics of wind speed probabilities in six grades. However, some limitations and drawbacks should be mentioned. The turbulent vertical mixing and blocking effect of buildings are ignored in the FWM, so the physical mechanism of changes in probabilities for different wind grades should be further studied from the view of considering both dynamical and thermodynamic effects. LUCC in ECP region includes urbanization, farmland irrigation, returning farmland to forest, and other types, so it is necessary to quantify the influences of different type LUCC on SWS, but the effects of different LUCC type are hard to be distinguished in diagnostic analysis, and therefore this issue should be simulated by regional climate models in the near future.

Acknowledgements

This study was sponsored by Chinese Natural Science Foundation (41275162) and Chinese Jiangsu Collaborative Innovation Center for Climate Change.

References

Asselen SV, Verburg PH. 2013. Land cover change or land-use intensification: simulating land system change with a global-scale land change model. *Global Change Biology* **19**: 3648–3667.

Bowman AW, Azzalini A. 1997. *Applied Smoothing Techniques for Data Analysis: The Kernel Approach with S-Plus Illustrations: The Kernel Approach with S-Plus illustrations*. Oxford University Press: England; 1–205.

China Meteorological Administration (CMA). 2003. *Ground Surface Meteorological Observation*. China Meteorological Press: Beijing, China; 157.

Coceal O, Belcher ES. 2005. Mean winds through an inhomogeneous urban canopy. *Boundary-Layer Meteorology* **115**: 47–68.

Feng S, Hu Q, Qian WH. 2004. Quality control of daily meteorological data in China, 1951–2000: a new dataset. *International Journal of Climatology* **24**: 853–870.

Greene JS, Chatelain M, Morrissey M, Stadler S. 2012. Estimated changes in wind speed and wind power density over the western High Plains, 1971–2000. *Theoretical and Applied Climatology* **109**: 507–518.

Guo H, Xu M, Hu Q. 2011. Change in near-surface wind speed in China: 1969–2005. *International Journal of Climatology* **31**: 349–358.

He YP, Monahan AH, Jones CG, Dai A, Biner S, Caya D, Winger K. 2010. Probability distributions of land surface wind speeds over North America. *Journal of Geophysical Research, [Atmospheres]* **115**: D04103, doi: 10.1029/2008JD010708.

Klink K. 2007. Atmospheric Circulation effects on the wind speed variability at Turbine Height. *Journal of Applied Meteorology and Climatology* **46**: 445–456.

Li Y, Wang Y, Chu HY, Tang JP. 2008. The climate influence of anthropogenic land-use changes on near-surface wind energy potential in China. *The Chinese Science Bulletin* **53**(18): 2859–2866.

Li Z, Yan ZW, Tu K, Liu WD, Wang YC. 2011. Changes in wind speed and extremes in Beijing during 1960–2008 based on homogenized observations. *Advances in Atmospheric Sciences* **28**(2): 408–420.

Liu ML, Tian HQ. 2010. China's land cover and land use change from 1700 to 2005: estimations form high-resolution satellite data and historical archives. *Global Change Biology* **24**: GB3003.

Sušelj K, Sood A, Heinemann D. 2010. North Sea near-surface wind climate and its relation to the large-scale circulation patterns. *Theoretical and Applied Climatology* **99**: 403–419.

Vautard R, Cattiaux J, Yiou P, Thepaut JN, Ciais P. 2010. Northern Hemisphere atmospheric stilling partly attributed to an increase in surface roughness. *Nature Geoscience* **3**: 756–761, doi: 10.1038/ngeo979.

Wu J, Fu CB, Zhang LY, Tang JP. 2012. Trend of visibility on sunny days in China in the resent 50 years. *Atmospheric Environment* **55**: 339–346.

Wu J, Zha JL, Zhao DM. 2016. Estimating the impacts of the changes in land use and cover on the surface wind speed over the East China Plain during the period 1980–2011. *Climate Dynamics* **46**: 847–863, doi: 10.1007/s00382-015-2616-z.

Xu M, Chang CP, Fu CB, Qi Y, Robock A, Robinson D, Zhang HM. 2006. Steady decline of East Asian monsoon winds, 1969–2000: evidence from direct ground measurements of wind speed. *Journal of Geophysical Research, [Atmospheres]* **111**: D24111.

Zhang N, Gao ZQ, Wang XM, Chen Y. 2010. Modeling the impact of urbanization on the local and regional climate in Yangtze River Delta, China. *Theoretical and Applied Climatology* **102**: 331–342.

25

Assessment of medium-range ensemble forecasts of heat waves

Hyun-Ju Lee, Woo-Seop Lee* and Jin Ho Yoo

Climate Research Department, APEC Climate Center, Busan, Republic of Korea

*Correspondence to:
Dr W.-S. Lee, Climate Research
Department, APEC Climate
Center, 12 Centum 7-ro,
Haeundae-gu, Busan 612-020,
Republic of Korea.
E-mail: wslee@apcc21.org*

Abstract

This study assesses the predictability of heat wave occurrence over Korea applying The Observing system Research and Predictability Experiment Interactive Grand Global Ensemble (TIGGE) data for the Heat Wave Index (HWI) associated with large-scale circulation. HWI shows higher predictability of heat waves than those of maximum temperature (TMAX) and bias corrected TMAX. The verification scores using HWI are compared to other cases of forecasts. It is concluded that the proposed HWI can be useful for heat wave forecasting, which may reduce the health impacts from heat waves through appropriate and timely mitigation efforts.

Keywords: heat wave; medium-range forecast; large scale circulation

1. Introduction

Extreme weather events such as heat waves, droughts, heavy rain/snow and storms/floods have become more frequent and severe all around the world. In the future, high temperatures or heat waves are expected be more intense and last longer due to global warming (Meehl and Tebaldi, 2004).

The daily mean temperature across most of East Asia was much higher than usual during the summer of 2013. South Korea observed its second warmest summer since records began in 1973. As a result, the government of Korea issued a power shortage warning, as electricity consumption rose sharply due to a persistent heat wave (Min *et al.*, 2014). Japan set a new national record for daily maximum temperature of 41.0 °C in Kochi and more than 143 observation stations exceeded that threshold during this period. China also experienced the strongest heat wave associated with high pressure patterns since 1995. During this period, most of central-eastern China was 3.0 °C warmer than the climatological normals. Such heat waves can have devastating impacts on health, agriculture and economy in East Asian countries.

In spite of the significant enhancement of the forecasting capability for climate/regional model and establishment of emergency preparedness, there have occurred many natural hazards, which impact on health, economy and infrastructure in recent years. During the summer of 2003, a severe heat wave in Europe caused more 35 000 heat-related mortalities in many countries and costed above 13 billion dollars. In 2010, western Russian experienced an intense heat wave that lasted for 33 consecutive days with temperatures above 30 °C. Total economic losses for Russia are estimated at 15 billion dollars due to wild fire and drought (Alexander, 2011). Papanastasious *et al.* (2014) have examined the relation of heat waves to air quality in Athens, Greece.

A better understanding of the large-scale teleconnection processes is essential to improve the forecasting predictability of high impact weather event such as heat waves (Pezza *et al.*, 2012). Grotjahn (2011) showed that the hottest days are associated with large-scale subsidence and offshore flow in the Central valley of California, USA.

Matsueda (2011) showed that the ensemble forecast derived from multiple meteorological centres may actually perform better compared to an ensemble derived from a single meteorological model in advance of the potential occurrence of high-impact weather. Boer (1984) suggested that the forecast error growth for small scales is much faster than for large scales.

The purpose of this study is to assess the predictability of Korean heat wave occurrence, by applying operational medium-range ensemble forecast datasets available from The Observing system Research and Predictability Experiment Interactive Grand Global Ensemble (TIGGE) portal for Heat Wave Index (HWI) associated with large-scale circulations.

2. Data and methodology

2.1. TIGGE data and observation data

Model data used for this research come from the Interactive Grand Global Ensemble (TIGGE) of the Observing system Research and Predictability Experiment (THORPEX), available through the European Centre for Medium-Range Weather Forecasts (ECMWF) portal. THORPEX, a 10-year World Weather Research Programme organized by the World Meteorological Organisation (WMO) that started in

Table 1. Configurations of TIGGE models for NCEP, CMC, KMA, UKMO and ECMWF.

Country/region	Centre	Ensemble members	Forecast length	Initial perturbation	Model resolution
USA	National Centres for Environmental Prediction (NCEP)	20	15 days	Ensemble transform with rescaling	T126
Canada	Canadian Meteorological Centre (CMC)	20	15 days	Ensemle Kalman filter	TL213
Korea	Korea Meteorological Administration (KMA)	23	10 days	Ensemble transform Kalman filter	N320
UK	UK Meteorological Office (UKMO)	23	15 days	Ensemble transform Kalman filter	N214
Europe	European Centre for Medium-Range Weather Forecasts (ECMWF)	50	10 days	Singular vectors	TL399

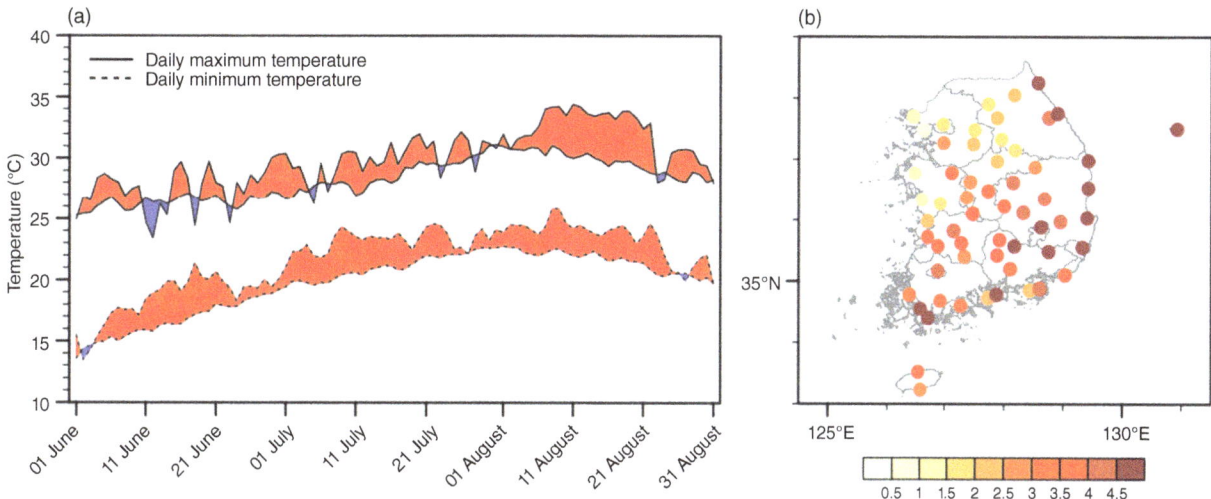

Figure 1. (a) Time series of the anomaly of the daily maximum temperature (solid line) and the daily minimum temperature (dashed line) averaged over 60 stations in Korea for summer 2013 relative to the 1981–2010 climatological average. (b) Anomaly of the daily maximum temperature at South Korean stations for 6–23 August 2013 relative to the 1981–2010 climatological average.

2008, was created in order to accelerate improvement in the accuracy of the 1–2weeks high-impact weather forecasts. We have performed our research based on the ensemble prediction data for zonal wind at 200 hPa, meridional wind at 200 hPa and daily maximum temperature in summer from five numerical weather prediction (NWP) centres including the National Centres for Environmental Prediction (NCEP), the Canadian Meteorological Centre (CMC), the Korea Meteorological Administration (KMA), the United Kingdom Meteorological Office (UKMO) and the ECMWF (Table 1).

The daily maximum temperature data from 60 Korean weather stations were used in order to analyse the weather patterns of summer 2013. The NCEP and National Centre for Atmospheric Research (NCAR) reanalysis version 1 data were used here to characterize the atmospheric circulation field (Kalnay *et al.*, 1996). The database includes daily maximum temperature, zonal wind at 200 hPa, meridional wind at 200 hPa and geopotential height at 500 hPa.

Although there is no standardized definition for heat waves, the general approach can be used to define them as a prolonged period of excessive threshold (Robinson, 2000). In Korea, heat waves are defined as two or more consecutive days with the daily maximum temperature

above 33 °C, which is the 95th percentile for temperature in the summer season. However, it is necessary to apply a bias correction because the NWP models produce systematic errors. Therefore, we consider heat waves as extreme events during which the daily maximum temperature exceeds the 95th percentile of the climatological probability density function (PDF). Here, the 95th percentile of the climatological PDFs was calculated from a 31-day moving window centred on each day. The PDFs were estimated for each forecast lead time using all ensemble members in each model for the summer seasons of 2011–2013, which were available as reference periods (Matsueda and Nakazawa, 2014).

2.2. Verification

In this study, the occurrence of heat waves over Korea was forecasted and verified for the period between July and August of 2013 using the TIGGE models under the three methods that are listed as follows:

1. TMAX (maximum temperature): forecast of heat waves when the daily maximum temperature exceeds the upper 95th percentile value of the observed climatological PDF
2. BCT (bias corrected temperature): forecast of heat waves when the daily maximum temperature

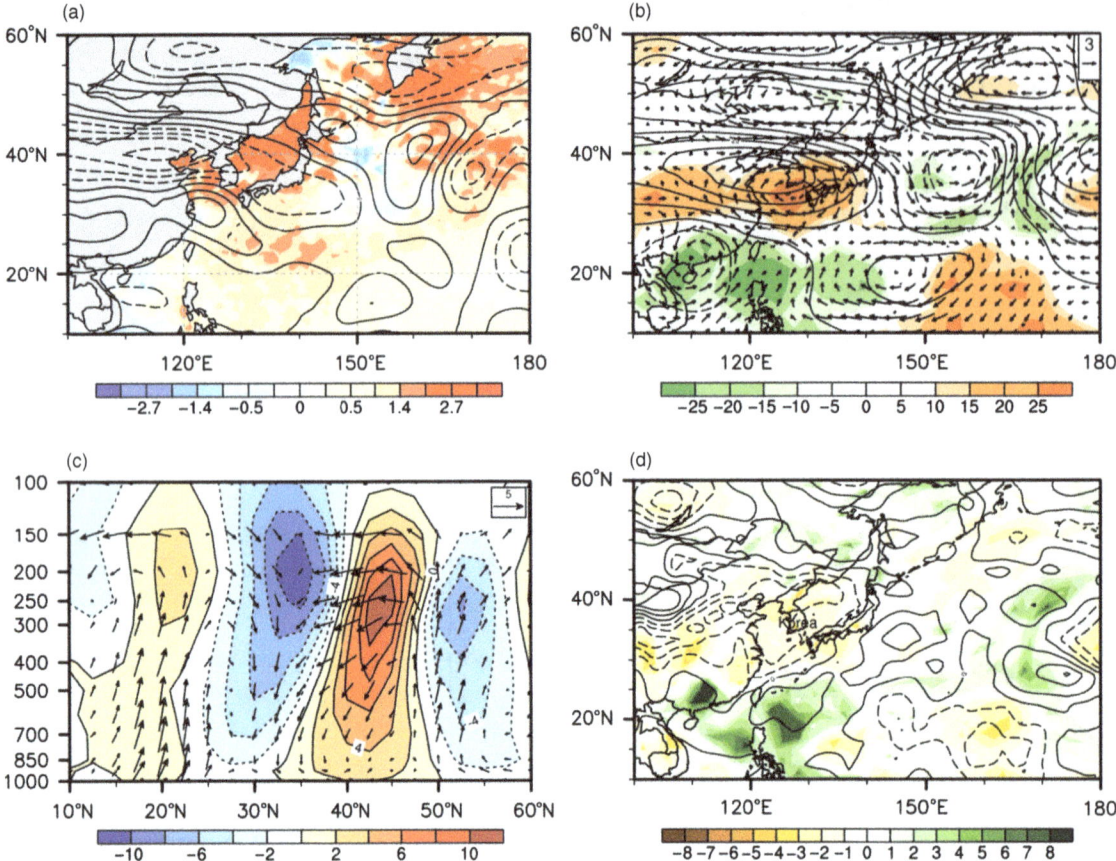

Figure 2. (a) SST (shading, °C) and 200 hPa (contours, 10^{-6} s^{-1} interval) vorticity anomaly from 6 to 23 August 2013 for relative to 1981–2010. (b) Same as (a) but for outgoing longwave radiation (shading, Wm^{-2}), 500 hPa geopotential height (contours, 10 m interval) and 850 hPa wind (ms^{-1}, arrows). (c) Meridional vertical circulation (vector) and zonal wind component (shading, ms^{-1}) over 110°–130°E. (d) Same as (a) but for precipitation (shading, mm day^{-1}) and the rate of change of total cloud cover (contours, 10% interval) compared to 1981–2000.

exceeds the value of the upper 95th percentile of each model's climatological PDF

3. HWI: forecast of heat waves when the HWI, which is based on large-scale circulation, is positive. Here, HWI is defined as difference in the 200 hPa vorticity between the average over 25°–30°N, 110°–130°E and the average over 35°–45°N, 120°–140°E.

To evaluate the performance of the heat-wave forecast, the percent correct (PC), threat score (TS) and equitable threat score (ETS) were employed from the two-by-two contingency table. These scores provide information on how well a forecasting method performs.

The PC calculates the accuracy based on the number of correctly forecasted and observed occurrences and the correctly forecasted and not observed occurrences divided by the total number of forecasts. The TS, known as the critical success index, is measured as the fraction of all events forecasted and/or observed that were correctly diagnosed. The ETS is equivalent to TS with a correction to remove the bias associated with a random chance (Mesinger and Black, 1992). This score is intended to offset the sensitivity of the TS to the underlying climatology of the event. The verification scores are respectively defined as follows:

$$PC = \frac{H + N}{H + F + M + N} \tag{1}$$

$$TS = \frac{H + N}{H + F + M} \tag{2}$$

$$ETS = \frac{H}{H + F + M} - H_{ref} = \frac{H - H_{ref}}{H + F + M - H_{ref}},$$
$$H_{ref} = \frac{(H + M)(H + F)}{H + F + N + N} \tag{3}$$

where, H is the number of cases where the events were observed and forecasted. M is defined by the number of cases where the events were observed but not forecasted and N is given by the count that the events were neither forecasted nor observed. F is the number of cases where the events were forecasted but never occurred.

3. Results

Figure 1(a) indicates the time series of the anomaly for the daily maximum temperature and the minimum temperature over Korea in summer 2013 relative to the 1981–2010 climatological normal. Korea recorded average maximum temperature approximately 30 °C during summer 2013, which was the second hottest summer since 1954. It was 1.6 °C above the normal of 1981–2010, i.e. of 28.3 °C. The minimum temperature of 21.7 °C and the mean temperature of 25.4 °C in the summer 2013 were the highest in Korea for almost the recent 30 years. The daily minimum temperature of summer 2013 was 1.9 °C above the 1981–2010 average of 23.6 °C. In July 2013, a heat wave began from southern Korea due to Changma front. Starting on 8 July and ending on 23 August 2013, a persistent heat wave caused by the high atmospheric pressure from the strengthening North Pacific High occurred across the country. A number of locations, Ulsan, Uljin and Namhea, had their hottest day record on 8 August, with highest recorded daily maximum temperature of 38.8 °C at Ulsan in the southern part of Korea. For the 2013 heat wave event (from 6 to 23 August), the daily maximum temperature anomaly was above 4.5 °C at several stations over the southern and eastern parts of Korea (Figure 1(b)).

Figure 2 shows the anomalies patterns for vorticity at 200 hPa, sea surface temperature (SST), geopotential height at 500, 850 hPa wind, total cloud cover and precipitation during July–August 2013 relative to 1981–2010. The intensification of the North Pacific High and convection activity over South China Sea (Figure 2(b)), which are important factors leading to heat wave, were enhanced during July–August 2013 (Tao and Zhu, 1964; Lau and Li, 1984; Yanai *et al.*, 1992; Park and Schubert, 1997). The convective activity closely associated with SST was enhanced and the anomalous convergence over the South China Sea was dominant, so that a downward motion was generated especially over the southern part of Korea (Figure 2(c)). Anomalous meridional circulation associated with the Korean heat wave links with the anomalous ascent of air at about 20°N. This leads to extreme (hot and dry) conditions over Korea. This warmer and drier air reduced precipitation and total cloud amount (Figure 2(d)). The reduced total cloud cover increased downward surface solar radiation over Korea.

As a result, the North Pacific High expanded westward and the SSTs around Korea were significantly higher than normal (Figure 2(b)). Therefore, an intense heat wave occurred. A noteworthy feature exhibited in the results is the north-south dipole mode that dominated the variability in the South China Sea, with opposite signs in the East Asia sectors (Figure 2(c)). The positive anomaly of vorticity at 200 hPa over the South China Sea induced more convection and adiabatic heating, which in turn became a source of Rossby wave-train along southerly wind patterns that generated positive geopotential height anomalies around Korea (Figure 2(b)).

On the basis of the dipole pattern, we proposed HWI, which is defined as the difference in the 200-hPa vorticity between the average over 25°–30°N, 110°–130°E and the average over 35°–40°N, 120°–140°E. If the HWI is greater than or equal to zero, it indicates high

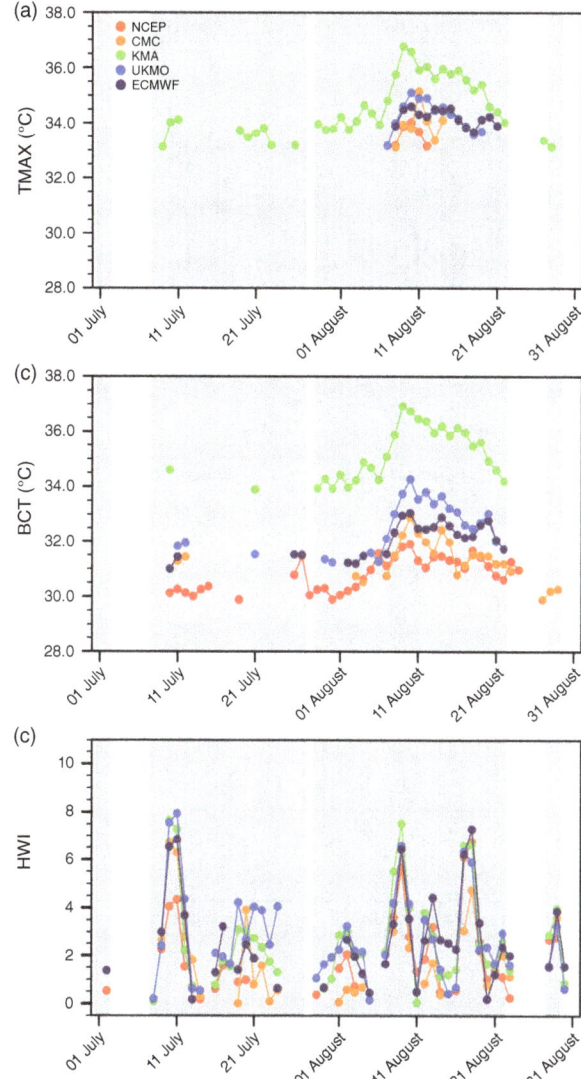

Figure 3. Occurrence probability of heat wave in summer 2013: (a) TMAX, (b) BCT and (c) HWI. The shading indicates the occurrence of an observed heat wave.

chance of heat wave over Korea. The HWI well represents the heat wave frequency in Korea not only at inter-annual scales (with the correlation coefficient of 0.81) but also at intra-seasonal scales (not shown).

Figure 3 shows the 5-day lead forecast of heat-wave occurrences from different models by using three methods (TMAX, BCT and HWI) from July to August 2013. The grey shading indicates occurrence of observed heat waves. In general, the early warning system for extreme events such as heat waves and floods rely on observation or short-range forecast of up to 2 days or less because the accuracy of forecast decreases a longer lead times (Thielen *et al.*, 2009; WHO, 2009). However, if the forecast lead time is too short, the user of forecast will not have enough time to confront the expected heat-wave.

When using TMAX, all models failed to forecast heatwave events, particularly in early summer. This is because most of the models, except of KMA, underestimated the daily maximum temperature during

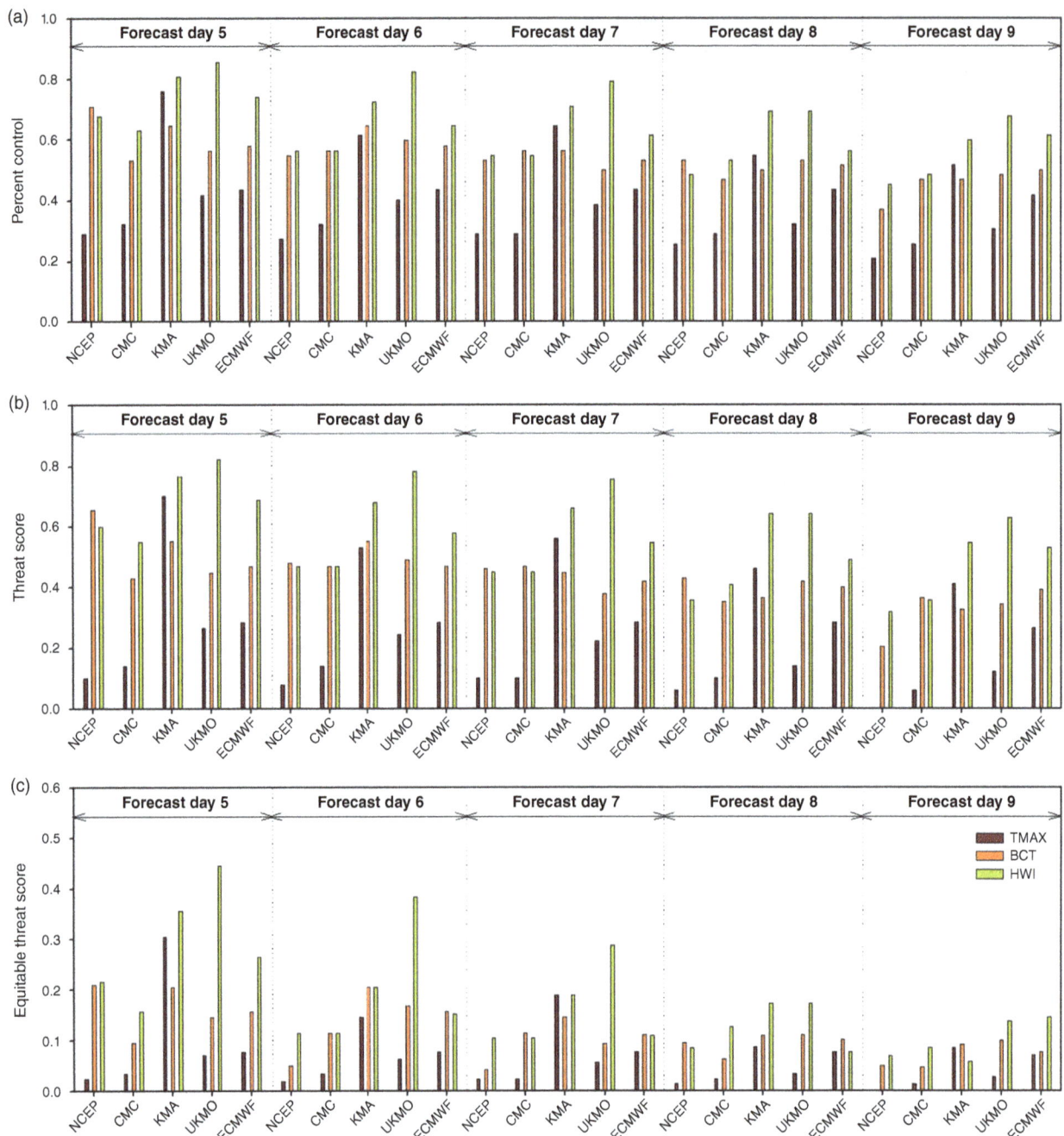

Figure 4. (a) Perfect correct, (b) threat score and (c) equitable threat score for heat wave forecast methods.

this period. Therefore, KMA seems to have a superior heat-wave forecasting. However, when the bias correction is applied, all the models show very similar results, but they do fail to predict the occurrence of heat waves in early summer. Compared to the other two methods determined by temperature, forecasting with HWI is able to forecast early summer heat waves well.

The PC, TS and ETS of the heat wave forecast from the three methods for the 5–9 days ahead are shown in Figure 4. The relatively higher skill of the KMA is well demonstrated by the results of verification scores. In Figure 4, it is seen that the KMA has a greater than 0.7 in PC value, whereas NCEP, CMC, UKMO and ECMWF indicate less than 0.4 in PC. This shows that

forecasting heat waves in July is difficult using daily maximum temperature. Also the KMA shows the best performance skill to forecast the occurrence probability of heat waves in TS and ETS values.

As shown in Figure 4(b), BCT provided better forecasting probability in NCEP, CMC, UKMO and ECMWF than the TMAX approach, while little difference was observed in KMA. In the case of NCEP, the results show that the method using BCT was more reliable than that using TMAX. In addition, the former can predict heat wave events up to 5 days ahead while the latter up to 3 days ahead. Not only the PC values significantly increased from 0.3 to 0.7 but also TS showed an improvement of the forecasting heat waves from 0.1 to 0.65 for BCT. Furthermore, the value of ETS for BCT was approximately 10 times as much as that of TMAX. The forecast of heat waves using BCT is more realistic than that using TMAX. In particular, NCEP predicts that there will be a probability for the heat wave from 9 July to 15 July when it is a difficult period to forecast heat wave using TMAX.

The highest skill in forecasting heat waves is given by HWI, even in the period from 8 July to 27 July, when it gives lower probability in comparison with the previous two methodologies (BCT, TMAX). In particular, the KMA shows better performance than other NWP centres in forecasting the starting time, ending time and persistence of heat waves. For 5-day ahead forecast, high resolution forecast models such as KMA, UKMO and ECMWF gave over 0.7 in TS value. The ETS, which is equivalent to TS with a correction to remove the bias associated with random chance, showed similar results.

The verification of the PC and the TS showed that the heat-wave forecast using HWI compared to others regardless of forecast time was successfully carried out. HWI is able to provide an early warning forecast for heat waves with 5 days up to 9 days forecast. It can be reliable information for decision makers to provide efficacious and timely actions to prepare for imminent heat waves (Rogers et al., 2010). The timing and duration of heat waves are important factors in terms of forecasting because they seem to have an impact on health and well-being. It is appropriate to determine the starting time and ending time of a heat-wave in Korea using HWI. Therefore, the HWI will aid in monitoring and making reasonable determinations about heat-wave occurrence in routine operations.

4. Conclusions

South Korea experienced an unusually strong heat wave in the summer of 2013, which was characterized by the enhancement of the Northwestern Pacific high resulting from an anomalous convergence over the South China Sea. As a result, we proposed the HWI to assist in the medium-range forecasting of heat waves over Korea. This study investigated the predictability of extreme heat waves in the summer of 2013 using five operational medium-range ensemble forecasts: CMC, ECMWF, KMA, NCEP and UKMO.

We compared the heat-wave forecast result with TMAX, BCT and HWI for the period of July to August 2013. By using TMAX, the KMA showed the best prediction capability for heat-wave occurrence. However, all models cannot predict the occurrence of a heat wave from 9 to 15 July with a 5-day lead time. Therefore, we should not expect the current models to always reproduce the daily maximum temperature.

The forecasting of heat waves using BCT is more realistic than using TMAX. In particular, NCEP predicts that there will be a heat wave from 9 to 15 July when it is a difficult period to forecast a heat wave using TMAX.

The value of ETS for BCT was approximately 10 times as much as that of TMAX

The highest skill in forecasting the heat waves was given by HWI even during the period from 8 to 27 July, when it had a lower probability, compared to the previous two methodologies (BCT, TMAX). For all lead times, all of the PC, TS and ETS of HWI were higher than those of BCT and TMAX. Results in predicting a heat-wave occurrence by HWI indicates significant skill in both short and medium-range.

The results might help us to get reliable information about heat waves in advance and this can provide more prevention time for the public in Korea. Therefore, advanced planning and preparedness are essential and can reduce the health impacts of exposure to extreme climate events like heat wave.

Acknowledgements

The authors are deeply grateful to two anonymous reviewers for their valuable comments and helpful advice. The authors acknowledge support from the APEC Climate Center.

References

Alexander L. 2011. Extreme heat rooted in dry soils. *Nature Geoscience* **4**: 12–13.

Boer GJ. 1984. A spectral analysis of predictability and error in an operational forecast system. *Monthly Weather Review* **112**: 1183–1197.

Grotjahn R. 2011. Identifying extreme hottest days from large scale upper air data: a pilot scheme to find California Central Valley summertime maximum surface temperatures. *Climate Dynamics* **37**: 587–604, doi: 10.1007/s00382-011-0999-z.

Kalnay E, Kanamitsu M, Kistler R, Collins W, Deaven D, Gandin L, Iredell M, Saha S, White G, Woollen J, Zhu Y, Leetmaa A, Reynolds R, Chelliah M, Ebisuzaki W, Higgins W, Janowiak J, Mo KC, Ropelewski C, Wang J, Jenne R, Joseph D. 1996. The NCEP/NCAR 40 year reanalysis project. *Bulletin of the American Meteorological Society* **77**: 437–471.

Lau K-M, Li MT. 1984. The monsoon of East Asia and its global associations. A survey. *Bulletin of the American Meteorological Society* **65**: 114–125.

Matsueda M. 2011. Predictability of Euro-Russian blocking in summer of 2010. *Geophysical Research Letters* **38**: L06801, doi: 10.1029/2011GL046557.

Matsueda M, Nakazawa T. 2014. Early warning productions for severe weather events derived from operational medium-range ensemble forecasts. *Meteorological Applications* **22**: 213–222.

Meehl GA, Tebaldi C. 2004. More intense, more frequent, and longer lasting heat waves in the 21st century. *Science* **303**: 1499–1503.

Mesinger F, Black TL. 1992. On the impact on forecast accuracy of the stemp-mouthtain(eta) vw. Singma coordinate. *Meteorology and Atmospheric Physics* **50**: 47–60.

Min SK, Kim YH, Kim MK, Park C. 2014. Assessing human contribution to the summer 2013 Korean heat wave [In "Explaining extreme events of 2013 from a climate perspective"]. *Bulletin of the American Meteorological Society* **95**(9): S48–S51.

Papanastasious DK, Melas D, Kambezidis HD. 2014. Heat waves characteristics and their relation to air quality in Athens. *Global NEST Journal* **16**(5): 919–928.

Park CK, Schubert SD. 1997. On the nature of the 1994 East Asian summer drought. *Journal of Climate* **10**: 1056–1070.

Pezza AB, van Rensch P, Cai W. 2012. Severe heat waves in Southern Australia: synoptic climatology and large scale connections. *Climate Dynamics* **38**: 209–224.

Robinson PJ. 2000. On the definition of a heat wave. *Journal of Applied Meteorology and Climatology* **40**: 762–775.

Rogers DP, Shapiro MA, Brunet G, Cohen JC, Connor SJ, Diallo AA, Elliott W, Haidong K, Hales S, Hemming D, Jeanne I, Lafaye M, Mumba Z, Raholijao N, Rakotomanana F, Teka H, Trtanj J, Whung PY. 2010. Health and climate-opportunities. *Procedia Environmental Sciences* **1**: 37–54.

Tao SY, Zhu FK. 1964. The 100-mb flow patterns in southern Asia in summer and their relation to the advance and retreat of the West-Pacific Subtropical Anticyclone over the Far East. *Acta Meteorologica Sinica* **34**: 385–395.

Thielen J, Bogner K, Pappenberger F, Kalas M, Medico MD, de Roo A. 2009. Monthly-, medium-, and short-range flood warning: testing the limits of predictability. *Meteorological Applications* **16**: 77–90.

WHO. 2009. Improving Public Health Responses to Extreme Weather/Heat-Waves-EuroHEAT-Technical Summary. Copenhagen, Denmark: WHO Regional Office for Europe.

Yanai M, Li C, Song Z. 1992. Seasonal heating of the Tibetan Plateau and its effects of the evolution of the Asian summer monsoon. *Journal of Meteorological Society of Japan* **70**: 319–351.

26

A geoengineering approach toward tackling tropical cyclones over the Bay of Bengal

S. Ghosh,[1,2]* A. Sharma,[1] S. Arora[1] and G. Desouza[1]

[1] School of Mechanical and Building Sciences, Vellore Institute of Technology, Vellore, Tamilnadu, India
[2] School of Earth and Environment, University of Leeds, UK

*Correspondence to:
S. Ghosh, School of Mechanical and Building Sciences, Vellore Institute of Technology, Tamil Nadu, 632014, India.
E-mail: satgleeds@gmail.com

Abstract

The concept of seeding giant-sized ocean salt water aerosols in the eye-wall of a cyclonic storm abruptly increasing cloud condensation nuclei (CCN) concentration is investigated. To bring this to effect, design of a novel injection mechanism – a modified naval artillery shell, tailor made for the Indian Navy fleet, containing sea-salt solution to disperse the CCN is proposed. The effect of the seeding is modeled using a robust optimized warm rain microphysical scheme – amenable for quick local forecasts within the Weather Research and Forecast framework. The combined protocol results in a significant decrease in precipitation tendencies upon landfall.

Keywords: cyclones; cloud seeding; aerosol injection; WRF; geoengineering

1. Introduction

The eastern coast of India is routinely battered by severe tropical cyclones (e.g. Cyclonic storm Thane, Nilam, Viyaru, Phailin and more recently Hudhud). This article focuses on Cyclone Thane (25–31 December 2011) which wreaked havoc along south eastern coastal India. The coast is home to highly populous cities – many of them vulnerable, with a risk of flooding, affecting 13 million people.

Recent geoengineering methods used in the modification of cloud albedo by controlled emission of sea-salt spray into the atmosphere to offset global warming (Latham, 2002; Boyd, 2008; Lenton and Vaughan, 2009; Latham et al., 2012a). Latham (1990) suggested an increase in the total droplet surface area of the cloud albedo by increasing the cloud droplet number concentration and reducing the droplet size by biasing the CCN to film mode sizes, a phenomenon known as marine cloud brightening (MCB). Latham et al. (2012b) explores the cooling effect of MCB over oceans as a method to mitigate cyclones. However, the most notable attempts at disrupting a cyclonic storm using geoengineering modifications are the STORMFURY experiments (Willoughby et al., 1985), which relied on the seeding of a large amount of supercooled water in the cyclone using silver iodide. STORMFURY relied on directly targeting the eye-wall of the cyclone to deplete it of its energy. This method was subsequently disproved with the discovery that cyclonic storms possessed only a small amount of supercooled water which froze below 0 °C (Andreae et al., 2004). Cotton et al. (2007) proposed the seeding of the peripheral rain bands of a cyclone with a large quantity of small-sized CCN to inhibit rain formation in these bands to inhibit cyclones. This method relies on intensification of the convection in the outer bands of the cyclone which in turn weakens the eye-wall by disrupting the convective forcing toward it. However, simulations of Typhoon Nuri (Krall and Cotton, 2012) show that the invasion of these small-sized CCN in the eye-wall has a synergistic effect on the cyclone intensification. There are other subsidiary articles that illustrate the concept of hygroscopic seeding of tropical cyclones (TCs) (Carrio and Cotton, 2010; Rosenfeld et al., 2012; Hebener et al., 2014). However, all these associated articles relate majorly to ice microphysics-mediated seeding hurricane Thane (as we discuss later) was predominantly governed by warm rain microphysics.

There is a rapid decrease in naturally produced sea-salt concentration above the mean sea level (MSL) (Blanchard et al., 1984). Further, de Leeuw et al., (2000) observed a logarithmic *decrease* in sea-salt concentrations with height. In this study, we explore the *effect of injecting artificially produced giant-sized sea-salt aerosol into the eye-wall of a cyclonic storm to enhance precipitation* prior to its landfall at a height of 200 m above MSL where contribution of sea-salt aerosols due to white capping is negligible. Like the previous attempts at cyclone modification, the central idea is the weakening of convection in the eye-wall of the cyclone. However, unlike STORMFURY, *this process seeks to induce greater precipitation using warm rain processes*. The cited tropical storms were mainly governed by warm rain microphysics. Enhancing warm rain within the eye-wall prevents ice-formation by averting the ascent of cloud droplets due to an early onset of precipitation. The only difference between a cloud seeding experiment and inducing precipitation in the eye-wall is with regard to the stronger up draughts and the rotational vorticity present in the latter. Therefore, shear generated turbulence inside the eye-wall

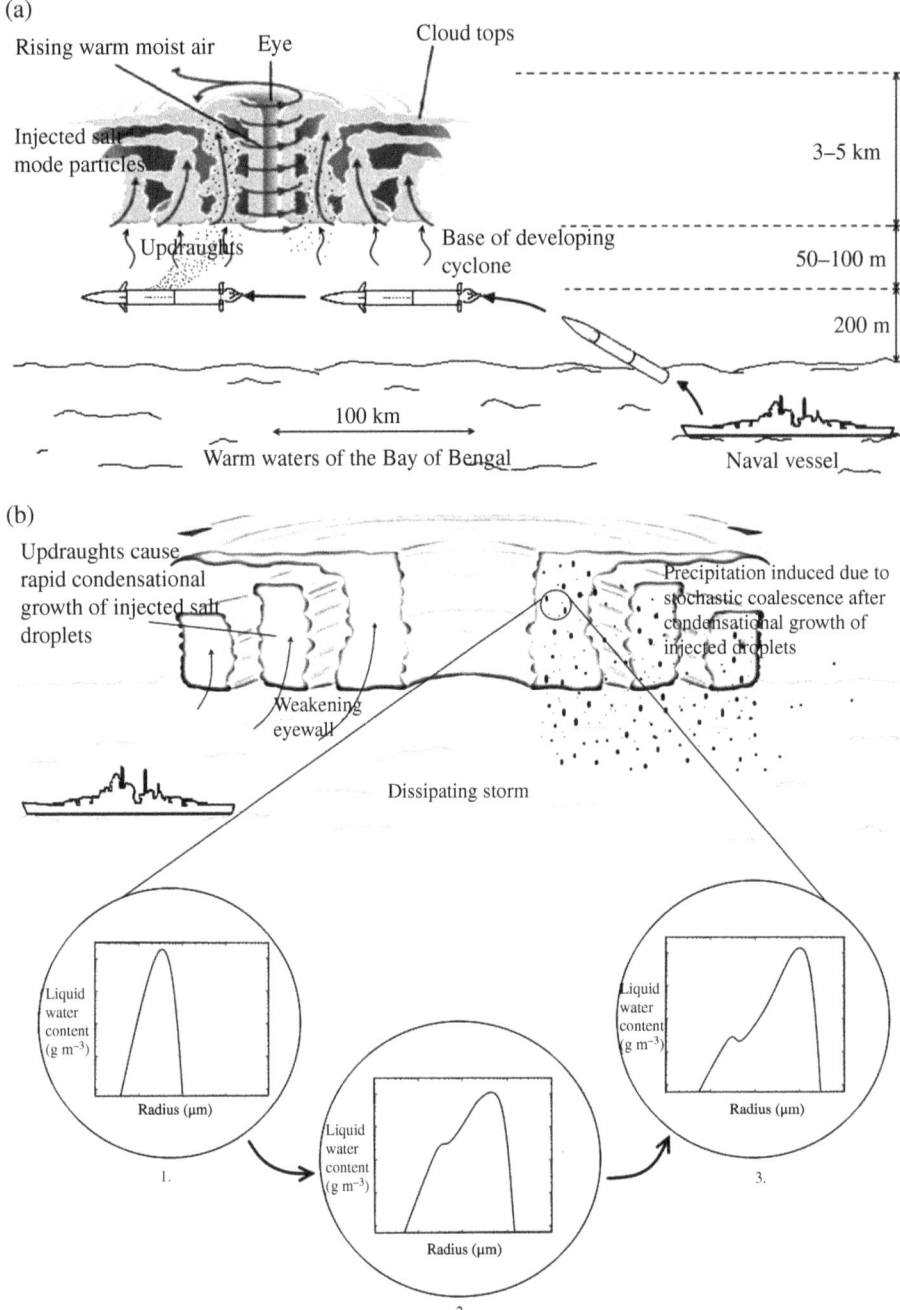

Figure 1. An outline of the proposed geoengineering mechanism scheme. (a) A light-weight artillery shell carrying the aerosol payload is launched from the vessel. Canards and fins are deployed: cruise altitude is reached as the propellant provides the necessary thrust. Sea-salt droplets in the jet mode are sprayed at the base of the developing cyclone which are then caught and carried along by the updraughts (figure not to scale) and (b) representative schematic of the time evolution of the droplet distribution leading to rain formation. 1) Initial single-humped distribution for liquid-water content. 2) Liquid-water distribution starts becoming double humped. 3) A fully developed bimodal spectrum causing rain (g m^{-3}).

would enhance the coalescence of all settling droplets (Ghosh *et al.*, 2005).

Another important difference between this study and the others is in the method of seeding. Traditional seeding experiments are aircraft dependent which restricts them to operate over safe areas. The use of unmanned, wind-driven Flettner-rotor vessels, which can be remotely steered beneath marine clouds to effect MCB, was proposed by Salter *et al.* (2008). For the present design, we propose a novel mechanism combining the flexibility afforded by aircrafts armed with autonomous capabilities.

The entire seeding process is summarized in Figure 1.

2. Microphysical characterization

A detailed microphysical chemical parcel model (henceforth called CPM) (O'Dowd *et al.*, 1999; Ghosh *et al.*, 2007) that employs the use of dynamical growth equations to predict the growth of aerosol solution droplets by the condensation of water vapor in an updraught, on a size resolved droplet spectrum is used in this study.

Keeping in mind that a strong updraught is present in the range $2.5-10\,ms^{-1}$ in a cyclone to ensure convective transport (Black *et al.*, 1994), the CPM is used to study the growth of natural and artificially injected sea-salt aerosols mimicking the conditions in the cyclonic rain bands. Even though Bay of Bengal (BOB) cyclones spend a majority of their lifespan over deep oceans, trace amounts of anthropogenic CCN such as sulfate aerosols are present in the local atmosphere of the cyclone (Jayaraman, 2001). These particles need to be added to the naturally present sea-salt aerosol spectrum (Figure 2(b)). When a cyclone approaches the coastline, aerosols get caught in the rising updraughts, initiating the process of condensational growth. The CPM incorporates the competing curvature (Kelvin effect) and solution effects on the equilibrium vapor pressure of the growing droplets based on the formation of a Kohler barrier, essentially the peak in the supersaturation versus radii curve (Seinfeld and Pandis, 1998). Our model results show that the Kohler barrier is exceeded for all the injected sea spray droplets with a median radius in the range of 15 μm. Eventually, these droplets attain the critical size of 20 microns required to initiate stochastic coalescence and cause rainfall (Ghosh *et al.*, 2005). The artificially injected spectrum is unique in the fact that the sizes and concentrations present work in tandem with the background spectrum to quickly achieve the requisite size for collision coalescence to set in.

The supersaturation curves (Figure 2(c)) obtained from the CPM reveal a clear cloud base 50 m above the point of injection.

2.1. Droplet trajectories and dispersion

The spray droplet trajectories were obtained using a commercial CFD code- Ansys Fluent– 15 on a $2\times2\times2\,m$ grid. A Discrete Phase Model (DPM) was used to simulate a secondary phase of discrete sea water droplets. Turbulence effects have been included by using the k-ε realizable turbulence model with default settings (ANSYS Fluent Theory Guide Release 12.1, 2009). The surface wave instability model is used, wherein the time of breakup and the resulting droplet size related to the fastest growing Kelvin–Helmholtz instability is accounted for (Reitz, 1987). The nozzle outlet is at the base of the domain. The trigger mass flow rate for spray ejection was set as $0.07\,kg\,s^{-1}$ after several numerical experiments with a cross-wind speed of $15\,ms^{-1}$ (India Meteorological Department, 2012; Vinod *et al.*, 2014). The droplets continue to rise with increasing height – as is expected for the small droplets as the drag caused by the updraught considerably exceeds the gravitational force. These trajectories are validated using the by Ghosh and Hunt (1998) model for droplet behavior of spray jets in cross-winds.

The droplet size distribution of a plain orifice atomizer for a particular liquid is dependent on the flow rate (pressure at inlet) and the orifice diameter (Ayres *et al.*, 2001; ANSYS Fluent Theory Guide, 2009). For the purposes of this study, an orifice of constant diameter (1 mm) was analyzed at varying flow rates to obtain the optimum CCN size distribution. These results were validated by comparing the Sauter mean diameter (SMD) from a similarly sized nozzle. The numerically predicted SMD was 113 μm, only 9 μm greater than that analytically calculated (Merrington and Richardson, 1947; Omer and Ashgriz, 2011) as shown in Figure S1 (Supporting Information).This size distribution is essential for the successful alteration of the required cloud microphysics as there are only specific size combinations that result in enhanced autoconversion.

The near field dispersion generated from ANSYS Fluent were used to configure the far field Gaussian dispersion using a Diffusion code (Picardo and Ghosh, 2011) in moderate crosswinds causing a reduction in the CCN concentration of three orders of magnitude at a height of 50 m.

2.2. Growth by stochastic collision coalescence and onset of precipitation

The injected sea-salt droplets are biased toward the larger size, with the reduced surface to volume ratio of these droplets – the condensational growth rates are slower when compared with smaller droplets. The production of large aerosol particles by collision coalescence for the enhancement in precipitation is of crucial importance for the proposed mechanism to be successful.

The progression of droplet size distribution due to collision coalescence of drops is given by the stochastic collection equation. Figure 3 shows the time evolution of liquid-water mass as a function of total particle radius calculated using the Hall collection kernel based on the flux method (Bott, 1997).

3. The spraying mechanism

The efficacy of the proposed geoengineering technique crucially rests upon the efficiency of the ballistic hygroscopic seeding operation. The spraying mechanism proposed here is a ship borne Extended Range Guided Munition (ERGM) (Graham, 2004) modified to carry sea water as its payload. This shell has a caliber of 5 inches and can be fired from the main gun of the Indian Navy's Kolkata Class Guided Missile Destroyers. The spray is ejected out from a number of orifices drilled on to the surface of the shell. The cylindrical shell (length 135 cm and diameter 12.5 cm) is designed to hold 5–7 l of liquid to be ejected from 80–100 orifices of 1 mm diameter (Figure 2(a)). A piston assembly (actuated by a small explosive charge) pushes the fluid out through the orifices regulated by a sleeve valve. The opening and closing of the valves is regulated by servo motors

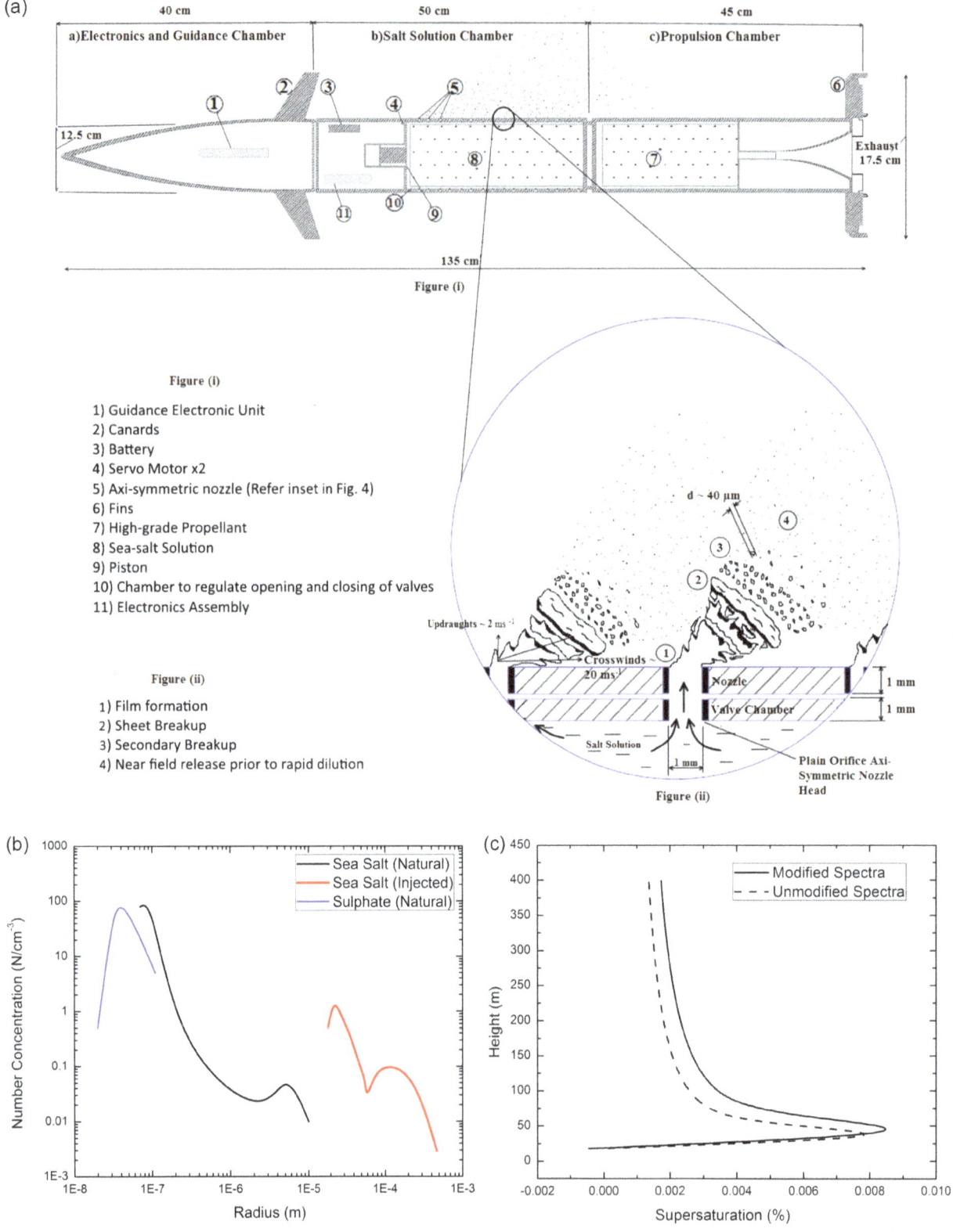

(a)

Figure (i)

1) Guidance Electronic Unit
2) Canards
3) Battery
4) Servo Motor x2
5) Axi-symmetric nozzle (Refer inset in Fig. 4)
6) Fins
7) High-grade Propellant
8) Sea-salt Solution
9) Piston
10) Chamber to regulate opening and closing of valves
11) Electronics Assembly

Figure (ii)

1) Film formation
2) Sheet Breakup
3) Secondary Breakup
4) Near field release prior to rapid dilution

Figure 2. (a) Spray ejection and droplet breakup through an artillery shell (i) longitudinal cross-section (ii) Magnified view of the orifice and droplet breakup. (b) Aerosol spectra over the Bay of Bengal. The first curve shows the natural sulphate spectrum and the second, the film and jet modes of the natural sea-salt spectrum (Jayaraman, 2001; Ghanti and Ghosh, 2010). The curve on the right shows the artificially injected spectra. The positioning of the injected spectra is crucial – if it were positioned on the left, the droplets would have competed with the natural spectra for moisture and eventually reduced in size. However, if it were positioned on the right, they would have quickly gathered the smaller droplets and fallen due to inertia without making a significant contribution to the rainfall rate. Panel (c) shows the supersaturation profile for the natural and injected spectra, as obtained from the chemical parcel model (O'Dowd et al., 1999). Note the rise in the height of the cloud base to 50 m for injected spectrum. This buffer height allows the injected droplets to activate and undergo condensational growth as they rise in the updraught.

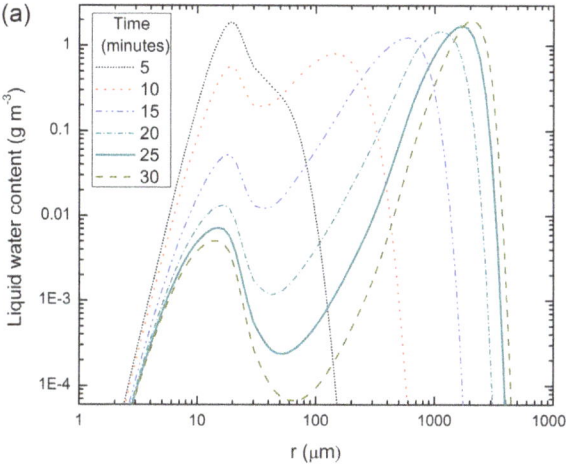

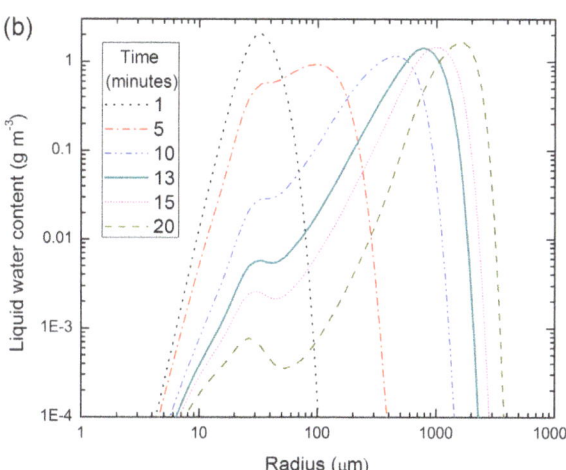

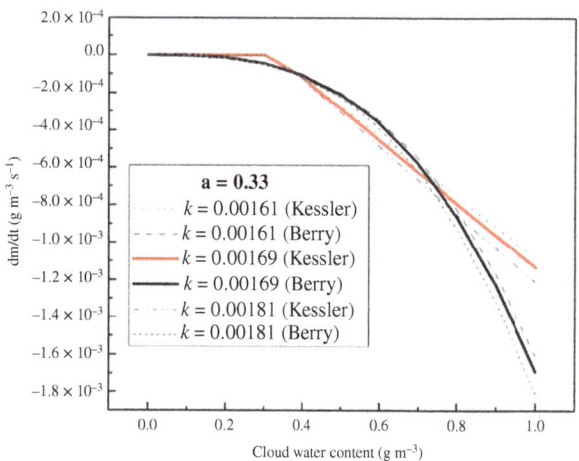

Figure 4. Overlapped mass-flux versus total cloud water content for Kessler schemes using Ghosh and Jonas (1998) (and references therein) – optimization of curves to include aerosol signatures into cyclone Thane microphysics. The bold lines show the closest correlation and have been used for the WRF simulations. The autoconversion rates and the threshold are represented by 'k' and 'a', respectively.

Figure 3. Time evolution of liquid-water mass. Panel (a) shows the unmodified spectra consisting only the naturally occurring sea-salt and sulphate aerosols (modal radius 18 μm). The time required to obtain a droplet radius of 1 mm is 25 min. Panel (b) shows modified spectra consisting of the injected and naturally occurring aerosols (modal radius 40 μm) for a total water content of 1.5 g m^{-3}. Note that the time to obtain a droplet radius of 1 mm through collision coalescence is reduced to 13 min for the modified spectrum after hygroscopic seeding, thus initiating hastened precipitation. The onset time of precipitation shows good agreement with analytical relations given by Ghosh and Jonas (1998).

situated at the extremities of the payload bay. Streams of uncharged salt water are sprayed through a series of plain orifice atomizers of 1 mm diameter and 2 mm length retrofitted in the orifices across the surface of the artillery shell. With the proposed design, it is estimated that the seeding of an annular ring of thickness 50 m around the eye of a cyclone (15 km in diameter) will require 3860 l of sea water. This equates to firing 640 shells (each shell can carry 6 l of payload). A deck gun of a conventional frigate has a fire rate of around 25 rounds per minute (Polmar, 2005). Thus, a constellation of 3–5 ships will theoretically complete the seeding operations in 5–10 min. The Indian Navy has about 24 frigates and destroyers in service as of today.

The shell will be guided to the precise location of the injection using a Guidance Electronic Unit (GEU), mounted in the nose cone of the projectile. Control

surfaces in the form of canards will allow the GEU to maneuver the shell. Sensors used to measure the altitude, velocity and temperature of the atmosphere around the shell will also be present in the nose cone. At the pre-fed coordinates, the sleeve valve will open and the piston charge will deploy, releasing the salt water droplets.

It may be noted here that we assume that the described system will function as simulated in this study.

4. Optimization of autoconversion parameters and WRF simulations

The Kessler warm rain microphysics scheme (Kessler, 1969) used by the WRF-3.4.1 does not explicitly take into account the droplet size distribution and their number concentrations while predicting rainfall. It is important to rectify this, where the rainfall rates have been synthetically modified. A combination of the Kessler and Berry schemes (Ghosh and Jonas, 1998; Fournier et al., 2005; Yin et al., 2015) is used calculate optimized values of autoconversion rate and threshold by taking into account the combined effects of number and size of the droplets present in the cloud base. The value of the autoconversion rate (k1) was calculated to be 1.7×10^{-4} s^{-1} which represents a 70% increase from the default value. An autoconversion threshold of 0.3 g kg^{-1} (as opposed to a default value of 1 g kg^{-1}) yielded the greatest agreement between the linear Kessler and cubic Berry schemes (Figure 4).

Based on the reflectivity data for cyclone Thane, the cloud base around the eye-wall was placed at about 200 m above the sea (India Meteorological Department, 2012) where the temperatures are non-freezing. NASA's TRMM satellites revealed that on the 27th and 29th December 2011 majority of

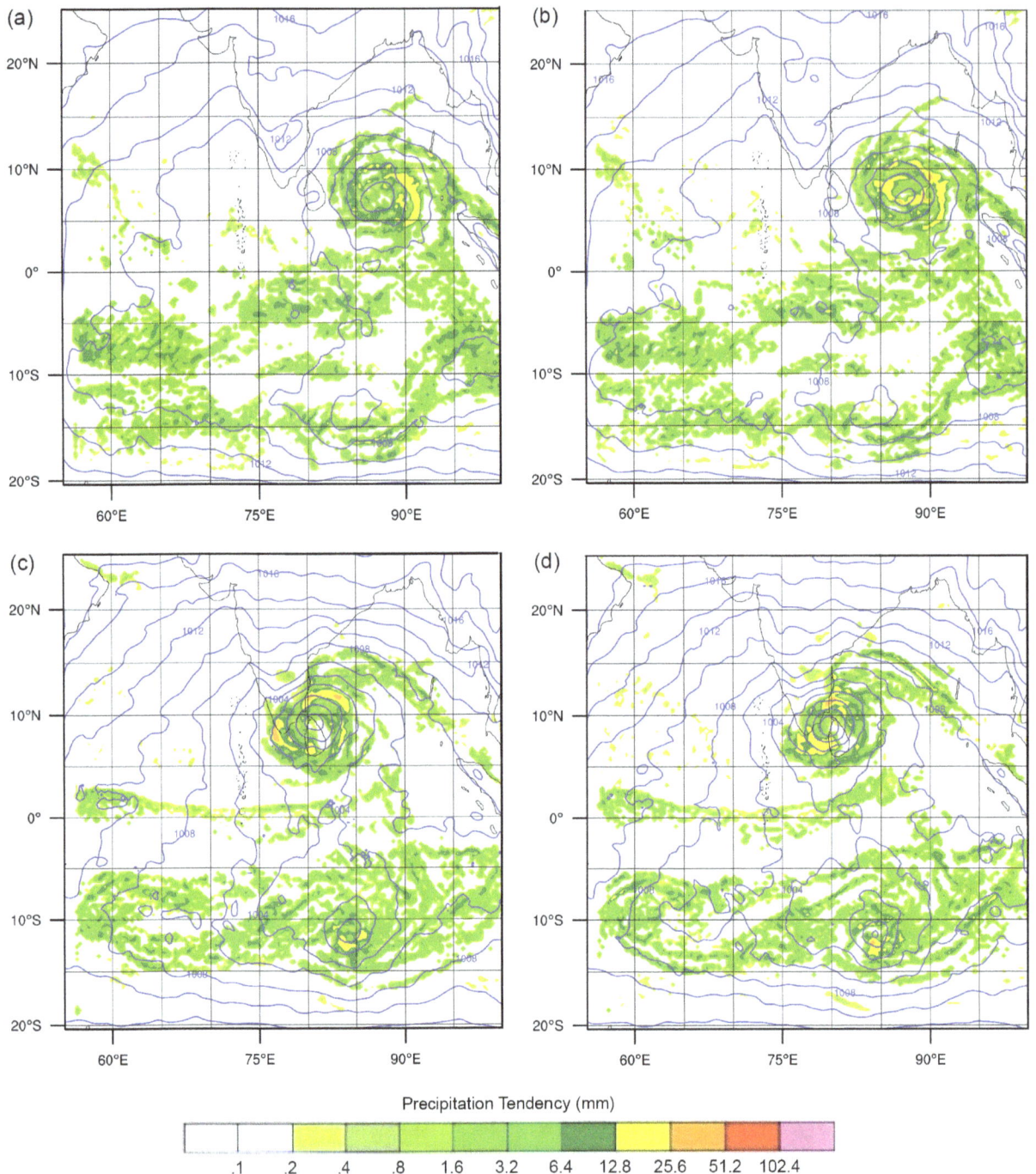

Figure 5. Precipitation tendencies (change of the intensity of precipitation during the last time period) for cyclone Thane over 6 h for (a) Run 1(Unseeded/control run): 25 December 2011 – 1500 UTC to 1800 UTC. (b) Run 2 (Seeded run): 2 December 2011 – 1500 UTC to 1800 UTC. (c) Run 1: 28 December 2011 – 2100 UTC to 29 December 2011 0000 UTC. (d) Run 2: 28 December 2011 2100 UTC – 29 December 2011 0000 UTC. Panel (b) shows an abrupt increase in precipitation in the final three hours of seeding. Just prior to landfall, a 50% decrease in precipitation tendency can be seen in the seeded run (d). As is expected, a marked increase in the rainfall rate is observed near the eye-wall of the cyclone seen as the bright yellow streak bordering the eye (a). The rainfall rate in Run-2 abates rapidly after the end of the seeding process and continues to reduce beyond the rates seen in Run 1 until the cyclone makes landfall. As can be seen in (d) the maximum precipitation tendency is restricted to 25 mm over the majority of the south eastern coast of India whereas in Run 1 (c) large swathes of land experience a precipitation tendency of 50 mm (marked orange). The spatial extent of the rain-bands is also visibly diminished as is seen when comparing (c) and (d).

the storm bands resided between 0 and 7 km above the mean sea level. The notion of using warm rain microphysics to simulate a cyclone was tested in a recent study by Wang (2001), who concluded that even though this led to early intensification of the cyclone, the final intensity was comparable with those produced by more complex models. Our own Thane simulations attests to this conclusion-runs with and without ice microphysics show only around 10% difference. The unification of the Kessler and Berry schemes by

Ghosh and Jonas (1998), along with the high degree of customizability offered by the WRF code allowed the modification of the autoconversion parameters. The nesting option provided within the WRF was used to pinpoint the seeded area, i.e. the eye-wall within the larger domain. Two-way feedback between the nested and the parent domains illustrated the effect of seeding the eye-wall exclusively. Validation runs using WRF Double Moment −5 microphysics with and without seeding were carried out (as shown in Figure S2), which show a similar trend as the warm rain runs, lending support to our view that computationally low-cost warm rain microphysics can indeed provide a fair estimate of cyclone progression and weakening.

The seeding is simulated for a period of 12 h, from 0600 UTC to 1800 UTC on the 25th of December in the test case (Cyclone Thane). The time step for the simulation was fixed at 180 s. The grid size used was 10 m for the larger domain and 2 m for the nested domain which corresponds to 19 and 4 km, respectively. The time step and geographical grid was scaled down for the nested domain by a factor of 7. Figure 5 compares the precipitation tendencies for the cyclone just after completion of the seeding.

5. Conclusions

In a seminal article on the geoengineering modification of marine cloud brightening, Latham *et al.* (2012a), showed how sea sprays can offset global warming. While in this process, precipitation was suppressed by the application of sea water, in the current research, precipitation in the cyclone is enhanced over preferred locations resulting in reduction of rainfall along the southeastern coast of India by up to 50%.

A carefully crafted design study is presented which accounts for the accurate delivery of sea-salt droplets into the eye-wall of a developing cyclonic storm. For the first time, the full spraying mechanism is articulated for a real-time ballistic discharge. Once the droplets permeate the cloud base, they are allowed to ascend to the cyclones womb through turbulent updraughts. This ascension and the subsequent activation process are rigorously modeled with the CPM yielding copious precipitation over the Bay of Bengal. Another important caveat explored is tied to the whole process of rapid dissemination of cyclone alerting mechanism (Ghosh *et al.*, 2014) using a well-tuned WRF run. This article marks out the conditions required for such an optimized WRF run as a function of the liquid discharge rate of the sprayed sea salt. Through a combination of these procedures, the article shows that Cyclone Thane when impregnated with the shown spectrum of sea water aerosols released 50 m below the eye-wall, when it is 900 km from the Indian coast, causes it to copiously precipitate over 24 h after injection. This causes a rapid depletion of precipitation tendencies at landfall. Finally, we wish to also add that observational studies with in-situ measurements of Hurricane Thane were very limited and future studies in this direction must focus on observational analyses as well.

Acknowledgements

The authors are grateful to Archit Verma for his help.

Supporting information

The following supporting information is available:

Figure S1. Correlation between numerically and analytically calculated Sauter mean diameters for varying flow rates.

Figure S2. Precipitation for the 25th of December for a full microphysics WDM 5 run including ice processes (left panel-a) contrasted with a Kessler only warm rain microphysics run (right panel-b). Note that almost identical precipitation tendencies are observed – the distribution of rain are very similar across peninsular India and the Indian Ocean for the two runs indicating that Cyclone Thane was operated upon by mainly warm rain microphysical processes.

References

Andreae MO, Rosenfeld D, Artaxo P, Costa AA, Frank GP, Longo KM, Silva-Dias MAF. 2004. Smoking Rain Clouds over the Amazon. *Science* **303**(5662): 1337–1342.

ANSYS® Academic Research. 2009. *Release 12.1, Help System, Fluent Theory Guide*. ANSYS Inc.

Ayres D, Caldas M, Semiaäo V, Da GracëaCarvalho M. 2001. Prediction of the droplet size and velocity joint distribution for sprays. *Fuel* **80**(2001): 383–394.

Black RA, Bluestein HB, Black ML. 1994. Unusually strong vertical motions in a Caribbean Hurricane. *Monthly Weather Review* **122**: 2722–2739.

Blanchard DC, Cipriano RJ, Woodcock AH. 1984. The vertical distribution of the concentration of sea salt in the marine atmosphere near Hawaii. *Tellus Series B Chemical and Physical Meteorology B* **36**: 118–125.

Bott A. 1997. A flux method for the numerical solution of the stochastic collection equation. *Journal of the Atmospheric Sciences* **57**(2): 284–294.

Boyd PW. 2008. Ranking geoengineering scheme. *Nature Geosciencei* **1**: 722–724.

Carrio GG, Cotton WR. 2010. Investigations of aerosol impacts on hurricanes: virtual seeding flights. *Atmospheric Chemistry and Physics* **11**: 2557–2567.

Cotton , Zhang H, McFarquhar GM, Saleeby SM. 2007. Should we consider polluting hurricanes to reduce their intensity? *Journal of Weather Modification* **39**: 70–73.

Fournier N, Weston KJ, Dore AJ, Sutton MA. 2005. Modeling of wet deposition of reduced nitrogen over the British Isles using a Lagrangian multi-layer atmospheric transport model. *Quarterly Journal of the Royal Meteorological Society* **131**: 703–722.

Ghanti R, Ghosh S. 2010. The Great Indian haze revisited: Aerosol distribution effects on microphysical and optical properties of warm clouds over peninsular India. *Advances in Geosciences* **25**: 51–54.

Ghosh S, Hunt JCR. 1998. Spray jets in a cross-flow. *Journal of Fluid Mechanics* **365**: 109–136.

Ghosh S, Jonas PR. 1998. On the application of the classic Kessler and Berry schemes in Large Eddy Simulation models with a particular emphasis on cloud autoconversion, the onset time of precipitation and droplet evaporation. *Annales Geophysicae* **16**: 628–637.

Ghosh S, Davila J, Hunt JCR, Srdic A, Fernando HJS, Jonas PR. 2005. How turbulence enhances coalescence of settling particles with applications to rain in clouds. *Proceedings of the Royal Society A: Mathematical Physical and Engineering Science* **461**(2062): 3059–3088.

Ghosh S, Smith MH, Rap A. 2007. Integrating biomass, sulphate and sea-salt aerosol responses into a microphysical chemical parcel model: implications for climate studies. *Philosophical Transactions of the Royal Society A* **365**: 2659–2674, doi: 10.1098/rsta.2007.2082.

Ghosh S, Vidyasagaran V, Sandeep S. 2014. Smart cyclone alerts over the Indian subcontinent. *Atmospheric Science Letters* **15**(2): 157–158.

Graham J. 2004. Extended Range Guided Munition (ERGM) Program",NDIA International Armaments Technology Symposium and Exhibition.

Herbener SR, van den Heever SC, Carrió GG, Saleeby SM, Cotton WR. 2014. Aerosol indirect effects on idealized tropical cyclone dynamics. *Journal of the Atmospheric Sciences* **71**: 2040–2055.

India Meteorological Department, 2012 - Very Severe Cyclonic Storm THANE over the Bay of Bengal (25–31 December, 2011); A Report.

Jayaraman A. 2001. Aerosol radiation cloud interactions over the Indian Ocean prior to the onset of the summer monsoon. *Current Science* **81**(11): 1437–1445.

Kessler E. 1969. On the distribution and continuity of water substance in atmospheric circulation. *Meteorological Monographs* **32**: Amer. Meteor. Soc.: 84.

Krall G, Cotton WR. 2012. Potential indirect effects of aerosol on tropical cyclone intensity: Convective fluxes and cold-pool. *Atmospheric Chemistry and Physics Discussions* **12**: 351–385.

Latham J. 1990. Control of global warming? *Nature* **347**: 339–340.

Latham J. 2002. Amelioration of global warming by controlled enhancement of the albedo and longevity of low-level maritime clouds. *Atmospheric Science Letters* **3**: 53–58.

Latham J, Bower K, Choularthon T, Coe H, Connolly P, Cooper G, Caft T, Foster J, Gadian A, Galbraith L, Iacovides H, Johnston D, Launder B, Leslie B, Meyer J, Neukemans A, Ormond B, Parkes B, Rasch P, Rush J, Salter S, Stevenson T, Wang H, Wang Q, Wood R. 2012a. Marine cloud brightening. *Philosophical Transactions of the Royal Society A* **370**: 4217–4262.

Latham J, Parkes B, Gadian A, Salter S. 2012b. Weakening of hurricanes via marine cloud brightening (MCB). *Atmospheric Science Letters* **13**(4): 231–237.

de Leeuw G, Neele FP, Hill M, Smith MH, Vignati E. 2000. Sea spray aerosol production by waves breaking in the surf zone. *Journal of Geophysical Research* **105**(29): 397–29, 409.

Lenton TM, Vaughan NE. 2009. The radiative forcing potential of different climate geoengineering options. *Atmospheric Chemistry and Physics* **9**: 5539–5561.

Merrington AC, Richardson EG. 1947. The break-up of liquid sheets. *Proceedings of the Physical Society of London* **59**(33): 1–13.

O'Dowd C, Lowe JA, Smith MH. 1999. Observations and modeling of aerosol growth in marine stratocumulus – case study. *Atmospheric Environment* **33**: 3053–3062.

Omer K, Ashgriz N. 2011. Chapter. 24: Spray nozzles. In *Handbook of Atomization and Sprays*. Springer: Boston, MA.

Picardo JR, Ghosh S. 2011. Removal mechanisms in a tropical boundary layer: quantification of air pollutant removal rates around a heavily afforested power plant. In *Air Pollution-New Developments*, Moldoveanu A (ed). InTech Publishers: Croatia, doi: 10.5772/18348.

Polmar N. 2005. *The Naval Institute Guide to the Ships and Aircraft of the U.S. Fleet*, 18 ed. Naval Institute Press: Annapolis, MD.

Reitz RD. 1987. Modeling atomization processes in high-pressure vaporizing sprays. *Atomization and Spray Technology* **3**: 309–337.

Rosenfeld D, Woodley WL, Khain A, Cotton WR, Carrió G, Ginis I, Golden JH. 2012. Aerosol effects on microstructure and intensity of tropical cyclone. *American Meteorological Society* **93**: 987–1001.

Salter S, Sortino G, Latham J. 2008. Sea-going hardware for the cloud albedo method of reversing global warming. *Philosophical Transactions of the Royal Society* **A366**: 3989–4006.

Seinfeld JH, Pandis SN. 1998. *Atmospheric Chemistry and Physics: From Air Pollution to Climate Change. Cloud Physics*. John Wiley & Sons, Inc.; 777–840.

Vinod KK, Soumya M, Tkalich P, Vethamony P. 2014. Ocean–atmosphere interaction during Thane cyclone: a numerical study using WRF. *Indian Journal of Marine Sciences* **43**(7): 1–5.

Wang Y. 2001. An explicit simulation of tropical cyclones with a triply nested movable mesh primitive equation model: TCM3 part II: model refinements and sensitivity to cloud microphysics parameterization. *Monthly Weather Review* **129**: 1370–1394.

Willoughby HE, Jorgensen DP, Black RA, Rosenthal SL. 1985. Project STORMFURY: a scientific chronicle. *Bulletin of the American Meteorological Society* **66**(5): 1962–1983.

Yin J, Wang D, Zhai G. 2015. An attempt to improve Kessler-type parameterization of warm cloud microphysical conversion process using CloudSat observations. *Journal of Meteorological Research (Elsevier)* **29**: 82–92.

27

Genesis of westerly wind bursts over the equatorial western Pacific during the onset of the strong 2015–2016 El Niño

Shangfeng Chen,[1]* Renguang Wu,[1] Wen Chen,[1] Bin Yu[2] and Xi Cao[1]

[1] Center for Monsoon System Research, Institute of Atmospheric Physics, Chinese Academy of Sciences, Beijing, China
[2] Climate Research Division, Environment and Climate Change Canada, Toronto, ON, Canada

*Correspondence to:
S. Chen, Center for Monsoon System Research, Institute of Atmospheric Physics, Chinese Academy of Sciences, No. 40, Huayanli, Beijing 100190, China.
E-mail:
chenshangfeng@mail.iap.ac.cn

Abstract

The strong 2015–2016 El Niño was initiated by several strong westerly wind bursts over the equatorial western Pacific in March and May 2015. These westerly wind bursts trigger eastward propagating warm Kelvin waves and lead to large sea surface temperature (SST) warming in the equatorial eastern Pacific. The first burst of westerly winds in early March was mainly induced by the Arctic Oscillation (AO) event. These westerly wind anomalies were enhanced subsequently due to the Madden-Julian Oscillation activity and northerly cold surges from East Asia-western Pacific in mid-March. Another westerly wind burst in May, induced by anomalous southerly winds from the Australian continent, further increased the SST anomaly in the equatorial eastern Pacific. This study provides an evidence of the AO influence on this strong El Niño-Southern Oscillation (ENSO) event and demonstrates the complexity in the genesis of westerly wind bursts during the El Niño outbreak, which may help improve the prediction of ENSO.

Keywords: 2015–2016 El Niño; westerly wind bursts; Arctic Oscillation; MJO

1. Introduction

The El Niño-Southern Oscillation (ENSO) is the strongest ocean-atmosphere coupled mode in the tropical Pacific on the interannual timescale. It exerts pronounced influences on weather and climate over many parts of the globe (e.g. Trenberth *et al.*, 1998; Wang *et al.*, 2000; Alexander *et al.*, 2002). Thus, a better understanding of the variability and mechanism of the formation of ENSO is of great importance for improving the ENSO prediction.

Observations show that a strong El Niño event occurred in the tropical central-eastern Pacific, with the maximum sea surface temperature (SST) anomalies reaching over 2.5 °C after May 2015 (http://www.cpc.ncep.noaa.gov/products/precip/CWlink/MJO/enso.shtml). The World Meteorological Organization reported that the current El Niño event is one of the strongest events since 1950 (http://www.wmo.int/pages/index_en.html). This ongoing strong El Niño has already been found to play an important role in a number of extreme weather and climate events. For instance, there have been a total of 21 storms in 2015 over the North Pacific, exceeding the record of 17 set in 1997, which was attributed to the effect of El Niño (Chu, 2004). The National Oceanic and Atmospheric Administration (NOAA) reported that, by the end of 2015, ocean conditions resulted largely from the ongoing El Niño event in the tropical central-eastern Pacific would cause almost 95% of the

US corals to bleach (http://coralreefwatch.noaa.gov/). The India Meteorological Department reported that the strong 2015–2016 El Niño contributes largely to the rainfall deficit in India during June–September 2015 (http://www.imdtvm.gov.in/). In view of the pronounced influences, it is an important issue to characterize and understand the outbreak of this strong El Niño event.

Previous studies demonstrated that westerly wind anomalies over the equatorial western-central Pacific play a key role in the outbreak of El Niño events (Yu and Rienecker, 1998; Huang *et al.*, 2001; Lengaigne *et al.*, 2004). The anomalous westerly winds trigger an eastward propagating and downwelling Kelvin wave, which results in SST warming in the equatorial central-eastern Pacific. The SST warming may further develop into an El Niño event via the Bjerknes positive feedback mechanism (Bjerknes, 1969). Hence, a key issue to understand the outbreak of the 2015–2016 El Niño is to identify the origin of equatorial Pacific wind anomalies.

Westerly wind anomalies over the equatorial western-central Pacific are influenced by several factors, such as the Madden-Julian Oscillation (MJO; Madden and Julian, 1972; Hendon *et al.*, 2007), northerly cold surges from East Asia (Li, 1990), southerly wind anomalies from the Australian continent (Chen and Wu, 2000; Xu and Chan, 2001) and the North Pacific Oscillation (NPO; Vimont *et al.*, 2003). For example, Li (1990) indicated that northerly cold surges originated from East Asian land may induce westerly wind anomalies over the tropical western

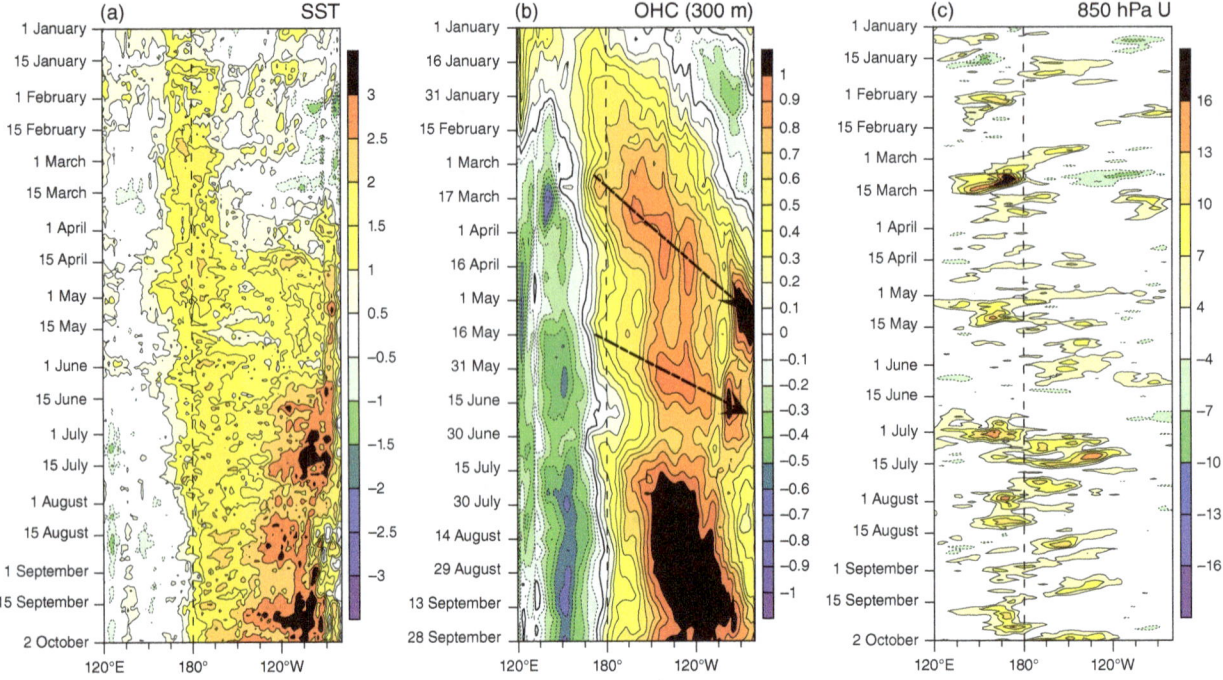

Figure 1. Time-longitude diagrams of the anomalous (a) SST (daily mean, °C) and (b) OHC in the upper 300 m (pentad mean, 10^8 J m^{-2}) averaged over 2°S–2°N in 2015. (c) Time-longitude diagram of the anomalous zonal wind at 850 hPa (daily mean, m s^{-1}) averaged over 5°S–5°N. Anomalies in (a) and (c) are calculated by removing the daily mean climatology over 1982–2011. Anomalies in (b) are calculated by removing the pentad mean climatology over 1982–2011. Arrows shown in (b) indicate propagations of warm Kelvin wave.

Pacific via modulating the local atmospheric convection. Vimont *et al.* (2003) demonstrated that the boreal winter NPO influences El Niño events in the subsequent winter via the seasonal footprinting mechanism (SFM). Nakamura *et al.* (2006) found that the spring Arctic Oscillation (AO) is an important trigger for the outbreak of an El Niño in the following winter via modulating westerly wind anomalies over the equatorial western Pacific. A recent study reported that the AO can induce westerly wind anomalies in spring over the equatorial western Pacific via the interaction between synoptic scale eddy and low frequency mean flow (Chen *et al.*, 2014). In addition, Chen *et al.* (2013) found that only when the spring AO is in its positive phase, the preceding winter positive NPO may lead to an El Niño event in the following winter via the SFM. In this study, we will examine which factor(s) contribute to the generation of westerly wind anomalies in the equatorial western Pacific during the onset of the strong 2015–2016 El Niño.

2. Data and methods

This study uses daily mean 850 hPa winds from the National Centers for Environmental Prediction-National Center for Atmospheric Research (NCEP-NCAR) reanalysis (Kalnay *et al.*, 1996) with a horizontal resolution of 2.5° × 2.5°. The daily mean SST data were derived from the NOAA Optimum Interpolation 1/4 Degree Daily Sea Surface Temperature analysis (Reynolds *et al.*, 2007), with a resolution of

0.25° × 0.25°. The pentad mean ocean temperature data were obtained from the NCEP Global Ocean Data Assimilation System reanalysis (Xue *et al.*, 2012). This oceanic reanalysis has a 1° × 0.33° horizontal resolution and 40 levels in the vertical direction. The ocean temperature is used to calculate the ocean heat content (OHC) in the upper 300 m.

The normalized daily mean AO index was derived from the website of NOAA Climate Prediction Center (http://www.cpc.ncep.noaa.gov/). We employ a real-time multivariate (RMM) MJO index developed by Wheeler and Hendon (2004) to examine the MJO activity (http://www.bom.gov.au/climate/mjo). This RMM index monitors well the real-time MJO activity (Wheeler and Hendon, 2004).

3. Results

3.1. Formation of the 2015 El Niño

Before late March 2015, negative and positive SST anomalies appeared in the equatorial eastern and western Pacific, respectively (Figure 1(a)). Correspondingly, negative and positive anomalies of OHC are observed in the equatorial eastern and western-central Pacific, respectively (Figure 1(b)). This zonal distribution of SST and OHC anomalies in the equatorial Pacific provides a favorable condition for the outbreak of an El Niño (Jin, 1997). In mid-April 2015, large positive SST anomalies, with the magnitude reaching 2.1 °C, occurred abruptly in the equatorial eastern Pacific around 90°W (Figure 1(a)). This was preceded by a strong warm Kelvin wave that was formed in the equatorial central Pacific in early March 2015 (Figure 1(b)). This eastward propagating warm Kelvin wave can be captured by the OHC anomalies (Figure 1(b)). As shown in Figure 1(b), positive OHC anomalies propagated eastward and reached the equatorial eastern Pacific in April 2015. Note that the above-mentioned warm Kelvin wave can also be detected by the sea surface height anomalies (not shown). The large positive SST anomalies in the equatorial eastern Pacific after the early April tend to be related to this eastward propagating warm Kelvin wave. The magnitude of SST anomalies in the equatorial eastern Pacific increases and reaches above 2.5 °C after the mid-June. As shown in Figure 1(b), another OHC anomaly with a smaller magnitude formed over the equatorial central Pacific in the mid-May and subsequently propagated eastward. This implies that the enhancement of SST anomalies after the mid-June is related to this latter eastward propagating Kelvin wave. The results hence suggest that the two eastward propagating warm Kelvin waves formed in March and May, respectively, play a key role in the outbreak of the 2015 El Niño.

Previous studies have demonstrated that an equatorial eastward propagating Kelvin wave can be resulted from tropical zonal wind stress anomalies or generated by the reflection of a Rossby wave at the western boundary (e.g. Matsuno, 1966; Huang et al., 2001). A comparison of Figure 1(b) and (c) indicates that the two above mentioned warm Kelvin waves were formed at a time of strong westerly wind anomalies over the equatorial western-central Pacific. Hence, the generation of these two eastward propagating Kelvin waves may be mainly attributed to the corresponding strong westerly wind anomalies in the early March and May. The westerly wind anomalies in the early March ($>16\,\mathrm{m\,s^{-1}}$) are larger than those in the early May ($>13\,\mathrm{m\,s^{-1}}$). Note that strong westerly wind anomalies are also observed in the equatorial western-central Pacific in the early July 2015, and propagate eastward to the tropical central-eastern Pacific in mid-July. This eastward propagating Kelvin wave may largely contribute to the strengthening of the current El Niño. As this study is focused on the onset phase of the 2015–2016 El Niño, we only investigate the factors that contributed to the formation of the westerly wind bursts in the early March and May.

3.2. Factors for the westerly wind burst in March 2015

Figure 2 displays the evolution of daily-averaged 850 hPa winds anomalies in March 2015. To help understand the formation of westerly wind anomalies over the equatorial western-central Pacific, we also show in Figure 3(a) the normalized time series of daily-averaged AO index from 15 February to 24 April, and the daily RMM index in Figure 3(b).

On 1 March, a strong anomalous anticyclone formed over the midlatitude of North Pacific, together with a weak anomalous cyclone over the subtropical central North Pacific (Figure 2(a)). Note that westerly wind anomalies are present over the equatorial central Pacific associated with an anomalous cyclone over the tropical South Pacific. This anomalous cyclone weakened in the following 2 days (Figure 2(b)). As such, the associated westerly wind anomalies over the equatorial central Pacific decayed. On 3 March, a clear dipole wind anomaly pattern is observed over the North Pacific, with a strong anomalous anticyclone in the midlatitude and a weak anomalous cyclone in the Tropics (Figure 2(b)). Correspondingly, weak westerly wind anomalies develop over the equatorial Pacific between 150° and 180°E. The dipole wind anomaly pattern maintains and enhances in the next 4 days (Figure 2(b)–(e)). On 7 March, the maximum wind speed over the equatorial western-central Pacific reaches $17\,\mathrm{m\,s^{-1}}$ (Figure 2(e)). The dipole pattern over the North Pacific bears a close resemblance to that associated with the spring AO (Gong et al., 2011; Chen et al., 2014). This indicates a possible contribution of AO to the formation of the anomalous cyclone over the tropical central North Pacific.

Chen et al. (2014) showed that the spring AO can exert a significant influence on the outbreak of an El Niño via modulating the westerly wind anomalies over the tropical western-central Pacific. Specifically, a significant anticyclonic circulation anomaly is observed over the midlatitude North Pacific and a pronounced cyclonic circulation anomaly is seen over the tropical central North Pacific in the positive phase of spring AO (please see their Figure 3). As such, westerly wind anomalies formed over the equatorial western-central Pacific to the south of the anomalous cyclone. The interaction between the synoptic scale eddy and low frequency mean flow plays a key role in the formation of the anomalous cyclone over the tropical central North Pacific in association with the spring AO. The normalized daily AO index shows values larger than 2 from 1 to 7 March (Figure 3(a)). In particular, the normalized AO index is larger than 5 on 7 March (Figure 3(a)). An enhancement of anomalous cyclone over the tropical central North Pacific and the associated westerly wind anomalies from 5 to 7 March (Figure 2(d)) is accompanied by an increase of the AO value (Figure 3(a)). Hence, AO plays a key role in the formation of the anomalous cyclone over the tropical central North Pacific and the anomalous westerly wind to its south over the equatorial western-central Pacific.

On 8 March, the anomalous cyclone over the tropical central North Pacific enhances, with the maximum wind speed reaching $20\,\mathrm{m\,s^{-1}}$ (Figure 2(f)). In addition, an anomalous equatorial twin cyclones pattern is clearly seen. The enhancement of this anomalous cyclone over tropical central Pacific is also related to a northerly cold surge penetrating into the tropical western-central Pacific from East Asia. Previous studies demonstrated that the northerly cold surge from higher latitudes could

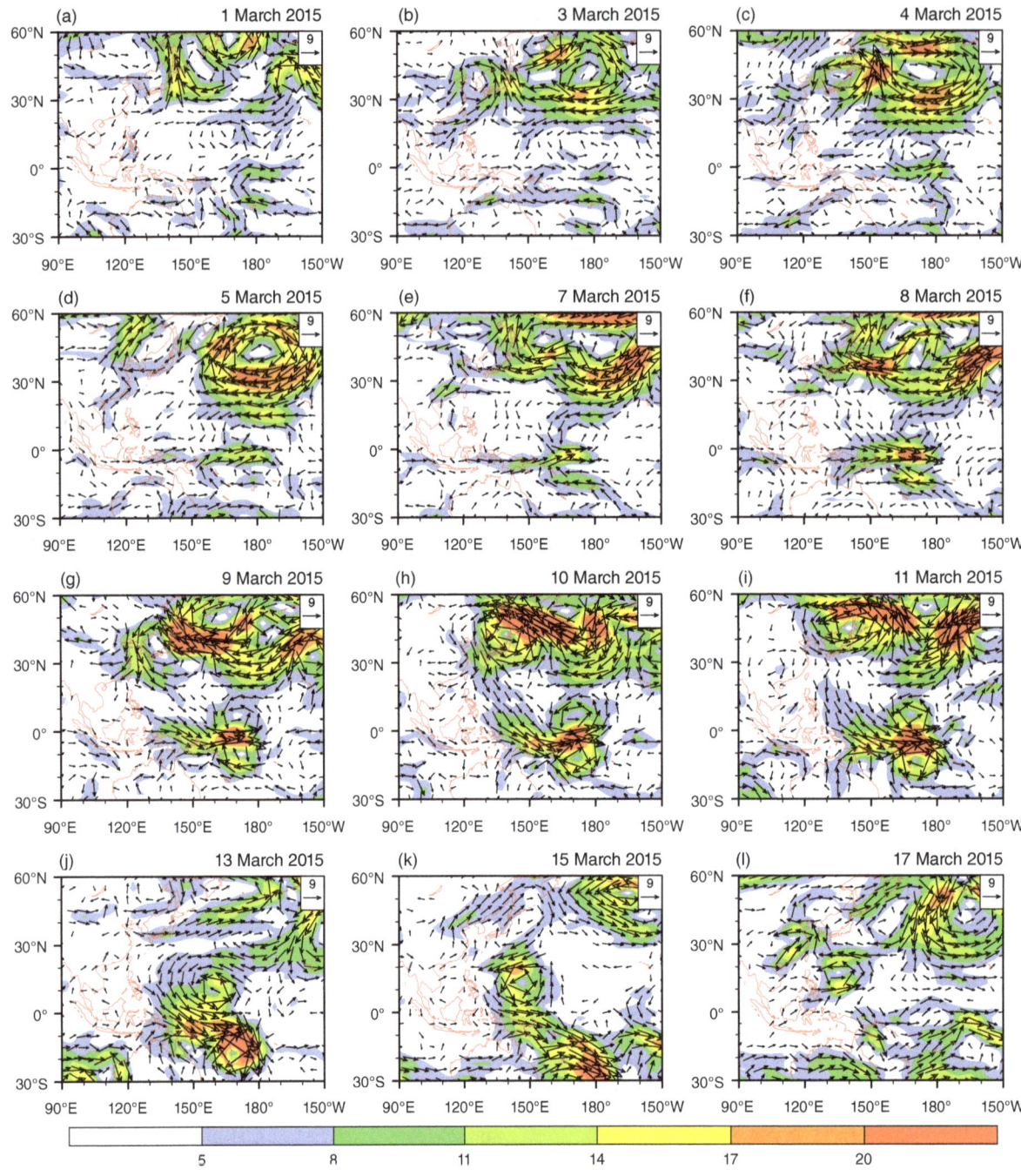

Figure 2. Anomalies of the daily mean 850 hPa wind (vector, m s⁻¹) and wind speed (shading, m s⁻¹) from 1 to 17 March in 2015. Anomalies are calculated by removing the daily mean climatology over 1982–2011.

result in deep convections in the Tropics and induce equatorial twin cyclonic anomalies (e.g. Li, 1990). Formation of the northerly cold surge over East Asia and western Pacific may be partly related to the spring AO (Nakamura *et al.*, 2007). Nakamura *et al.* (2007) indicated that spring AO could impact the westerly wind anomalies over the tropical western Pacific via modulating the Asian cold surge. Nevertheless, the physical process responsible for the influence of spring AO on the Asian cold surge is still unclear and remains to be explored. The northerly cold surge from East Asia is observed in the next 3 days (Figure 2(g)–(i)). This indicates a contribution of the cold surge to the

enhancement of westerly wind anomalies over the equatorial western-central Pacific.

The anomalous equatorial twin cyclones are stronger and their spatial scales are larger on 10 and 11 March compared to those on 8 and 9 March (Figure 2(g)–(i)). This enhancement may be related to the arrival of MJO into the tropical western Pacific. From 7 to 9 March, the RMM index enters phase 5 over the Maritime continent (Figure 3(b)). The anomalous cyclone to the northwest of Australia on 8 March may be related to the MJO activity (Figure 2(f)). After 10 March, the RMM index enters into phase 6 (Figure 2(h) and (i)). This indicates that the westerly wind anomalies associated with the

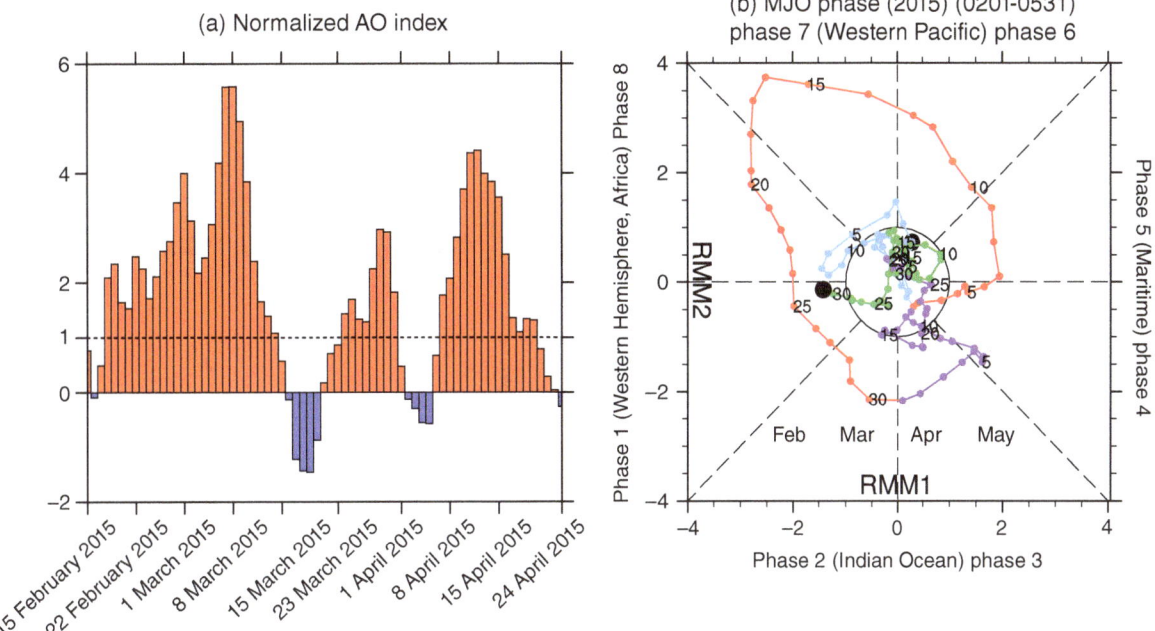

Figure 3. (a) Normalized time series of the daily mean AO index obtained from NOAA CPC. (b) (RMM1, RMM2) phase space diagram for the period from 1 February to 31 May in 2015.

MJO would enhance the westerly wind anomalous initiated by the AO and northerly cold surge. Thus, the further enhancement of westerly wind anomalies over the equatorial western-central Pacific on 10 and 11 March may be attributed to the MJO. After 13 March, the anomalous cyclone over the North Pacific propagates northwestward into subtropical western North Pacific, and the associated westerly wind anomalies are weakened and move away from the equator to around 10°N (Figure 2(j)–(l)). In comparison, the anomalous cyclone over the tropical South Pacific propagates southeastward into the subtropical southeastern Pacific (Figure 2(j)–(l)).

The earlier results indicate that the first appearance of westerly wind anomalies over the western Pacific is attributed to the AO. Subsequently, the enhancements of westerly wind anomalies are related to both the northerly cold surge originated from East Asia and an eastward propagating MJO. It should be noted that interactions may exist among AO, MJO and northerly cold surge. For example, Zhou and Miller (2005) indicated that a positive (negative) AO phase is more likely connected with a positive (negative) MJO phase in the boreal winter season. L'Heureux and Higgins (2008) found that the global circulation anomalies in association with AO bear a close resemble to those related to MJO during boreal winter. Murakami (1988) reported that the northerly surge from extratropics could excite pronounced convection related to MJO. In addition, the occurrence frequency of the East Asian cold surge during boreal winter is found to be below normal when AO is in its positive phase (e.g. Jeong and Ho, 2005). Further investigation is needed to clarify whether spring AO plays a role in the formation of the northerly cold surge and MJO over the tropical western Pacific in

March 2015 and to understand the related physical processes.

3.3. Factors for the westerly wind burst in May 2015

In the following, we further analyze the generation of westerly wind bursts over the tropical western-central Pacific in May 2015. Figure 4 displays the evolution of daily-averaged 850 hPa winds anomalies in May 2015. On 2 May, the wind anomalies over the tropical western-central Pacific are weak (Figure 4(a)). A strong anomalous cyclone (anticyclone) is observed over the midlatitudes of North (South) Pacific (Figure 4(a)). In addition, weak westerly wind anomalies are present over the equatorial western-central Pacific. An anomalous cyclone appears in the tropical South Pacific, but it is weakened and then disappears in the next 2 days (Figure 4(a) and (b)). On 4 May, the westerly wind anomalies are observed over the equatorial western Pacific extending eastward to the eastern Pacific, together with a weak anomalous cyclone seen in the tropical western Pacific around 140°E (Figure 4(b)). The anomalous cyclone intensifies from 4 to 5 May, which may be related to the arrival of northerly wind anomalies originating from the East China Sea (Figure 4(b) and (c)). Previous studies have demonstrated that northerly cold surges from East Asia would enhance the regional convection and subsequently lead to the formation of anomalous cyclones through a Matsuno-Gill type atmospheric response when it reaches the tropical western North Pacific (e.g. Li, 1990).

A southerly surge originating from East Australia along 150°E starts to penetrate into the equatorial

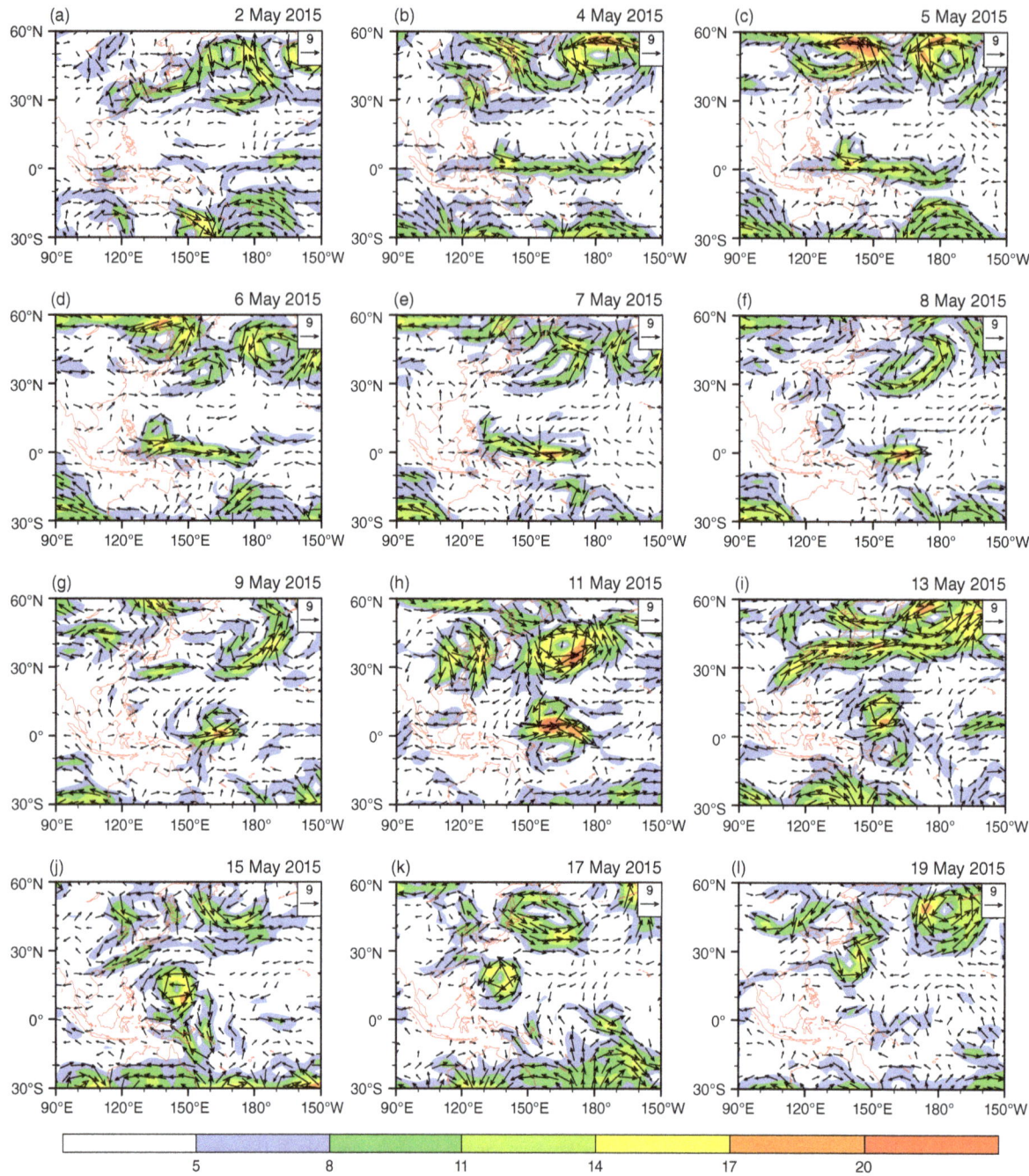

Figure 4. Anomalies of the daily mean 850 hPa wind (vector, m s^{-1}) and wind speed (shading, m s^{-1}) anomalies from 2 to 19 May in 2015. Anomalies are calculated by removing the daily mean climatology over 1982–2011.

region on 5 May (Figure 4(c)). As this southerly surge associated with the Australian monsoon continues for the next 4 days, a clear and strong anomalous cyclone forms over the tropical western Pacific between 150° and 170°E (Figure 4(d)–(g)). Xu and Chan (2001) have reported that anomalous southerly winds associated with the Australian monsoon could produce strong convergence over the equatorial western-central Pacific, which would subsequently lead to the intensification of tropical westerly wind anomalies. Correspondingly, the westerly wind anomalies around 160°E significantly increase (Figure 4(d)–(g)). On 11 May, a northerly surge originating from the midlatitude North Pacific

penetrates deep into the equator and converges with the southerly surge from the Australian continent. This convergence leads to the generation of the anomalous equatorial twin cyclones and enhancement of the westerly wind anomalies between 150° and 170°E (Figure 4(h)). The southerly surge from Australia and the anomalous northerly wind weaken on 13 May and disappear on 15 May (Figure 4(i) and (j)). Subsequently, the anomalous cyclone over the tropical western-central Pacific weakens and propagates poleward, and the westerly wind anomalies over the tropical western-central Pacific decay (Figure 4(k) and (l)). An examination of the RMM index in May shows that the

MJO is inactive in May (Figure 3(b)). This implies that MJO may not play a significant role in the generation of westerly wind bursts over the tropical western-central Pacific in May 2015.

The results suggest that the southerly surge from the Australian continent plays a key role in the generation of the westerly wind anomalies over the equatorial western-central Pacific in May. This confirms the finding of previous studies that the southerly surge from East Australia can contribute to the development of the ENSO (Chen and Wu, 2000; Xu and Chan, 2001).

4. Summary and discussion

A strong El Niño event occurred in the tropical central-eastern Pacific in 2015. Our study indicates that the onset of this strong El Niño is closely related to two eastward propagating warm Kelvin waves induced by strong westerly wind anomalies in early March and May. The first appearance of westerly wind anomalies over the tropical western-central Pacific in March is found to be induced by the AO. The westerly wind anomalies are further enhanced by the MJO activity and the northerly cold surge from East Asia. Thus, the strong westerly wind anomalies generated jointly by the AO, MJO and northerly cold surge trigger eastward propagating warm Kelvin wave to initiate the SST warming in the equatorial eastern Pacific. In May, the generation of the westerly wind anomalies is mainly related to the southerly surge associated with the Australian monsoon activity. This study provides another evidence confirming that the AO can exert significant influences on the outbreak of an El Niño via modulating westerly wind anomalies over the equatorial western-central Pacific. Yet, it also indicates a combined role of other factors that provides a more favorable condition for the El Niño occurrence.

AO, MJO and cold surges from extratropics are important for the outbreak of the strong 2015–2016 El Niño. Nevertheless, it should be noted that other factors may also play a role in influencing this El Niño event. For example, previous studies indicated that the preceding NPO (Vimont *et al.*, 2003), North Atlantic Oscillation (Oshika *et al.*, 2015), North Atlantic SST (Wang *et al.*, 2011; Ham *et al.*, 2013) and warm water in the tropical Pacific (Anderson *et al.*, 2013) may also influence the subsequent winter ENSO variability. In addition, preliminary results indicate that the preceding AO also plays an important role in the formation of westerly wind bursts over the equatorial western-central Pacific in the outbreak of the 1997–1998 El Niño (not shown). A comparison of the AO–ENSO connection between these two strong El Niño cases is ongoing.

Acknowledgements

We thank two anonymous reviewers for their constructive suggestions and comments, which helped to improve the paper. We declare that we have no conflict of interest. This study is supported by the China Postdoctoral Science Foundation (2015M581151) and the National Natural Science Foundation of China grant (41230527 and 41530425).

References

Alexander MA, Bladé I, Newman M, Lanzante JR, Lau NC, Scott JD. 2002. The atmospheric bridge: the influence of ENSO teleconnections on air–sea interaction over the global oceans. *Journal of Climate* **15**: 2205–2231.

Anderson BT, Perez RC, Karspeck A. 2013. Triggering of El Niño onset through trade wind-induced charging of the equatorial Pacific. *Geophysical Research Letters* **40**: 1212–1216, doi: 10.1002/grl.50200.

Bjerknes J. 1969. Atmospheric teleconnections from the equatorial Pacific. *Monthly Weather Review* **97**: 661–686.

Chen L, Wu R. 2000. The role of the Asian/Australian monsoons and the southern/northern oscillation in the ENSO cycle. *Theoretical and Applied Climatology* **65**: 37–47.

Chen SF, Chen W, Yu B, Graf HF. 2013. Modulation of the seasonal footprinting mechanism by the boreal spring Arctic Oscillation. *Geophysical Research Letters* **40**: 6384–6389, doi: 10.1002/2013GL058628.

Chen SF, Yu B, Chen W. 2014. An analysis on the physical process of the influence of AO on ENSO. *Climate Dynamics* **42**: 973–989, doi: 10.1007/s00382-012-1654-z.

Chu PS. 2004. *ENSO and Tropical Cyclone Activity*, edited. Columbia University Press: New York, NY; 297–332.

Gong DY, Yang J, Kim S, Gao Y, Guo D, Zhou T, Hu M. 2011. Spring Arctic Oscillation–East Asian summer monsoon connection through circulation changes over the western North Pacific. *Climate Dynamics* **37**: 2199–2216, doi: 10.1007/s00382-011-1041-1.

Ham YG, Kug JS, Park JY, Jin FF. 2013. Sea surface temperature in the north tropical Atlantic as a trigger for El Niño/Southern Oscillation events. *Nature Geoscience* **6**: 112–116.

Hendon HH, Wheeler MC, Zhang C. 2007. Seasonal dependence of the MJO–ENSO relationship. *Journal of Climate* **20**: 531–543.

Huang RH, Zhang RH, Yan B. 2001. Dynamical effect of the zonal wind anomalies over the tropical western Pacific on ENSO cycles. *Science in China Series D: Earth Sciences* **44**: 1089–1098.

Jeong JH, Ho CH. 2005. Changes in occurrence of cold surges over East Asia in association with Arctic Oscillation. *Geophysical Research Letters* **32**: L14704.

Jin FF. 1997. An equatorial ocean recharge paradigm for ENSO. Part I: conceptual model. *Journal of the Atmospheric Sciences* **54**: 811–829.

Kalnay E, Kanamitsu M, Kistler R, Collins W, Deaven D, Gandin L, Iredell M, Saha S, White G, Woollen J. 1996. The NCEP/NCAR 40-year reanalysis project. *Bulletin of the American Meteorological Society* **77**: 437–471.

L'Heureux ML, Higgins RW. 2008. Boreal winter links between the Madden-Julian Oscillation and the Arctic Oscillation. *Journal of Climate* **21**: 3040–3050.

Lengaigne M, Guilyardi E, Boulanger JP, Menkes C, Delecluse P, Inness P, Cole J, Slingo J. 2004. Triggering of El Nino by westerly wind events in a coupled general circulation model. *Climate Dynamics* **23**: 601–620.

Li C. 1990. Interaction between anomalous winter monsoon in East Asia and El Niño events. *Advances in Atmospheric Sciences* **7**: 36–46.

Madden RA, Julian PR. 1972. Description of global-scale circulation cells in the Tropics with a 40–50 day period. *Journal of the Atmospheric Sciences* **29**: 1109–1123.

Matsuno T. 1966. Quasi-geostrophic motions in the equatorial areas. *Journal of the Meteorological Society of Japan* **44**: 25–42.

Murakami T. 1988. Intraseasonal atmospheric teleconnection patterns during the Northern Hemisphere winter. *Journal of Climate* **1**: 117–131.

Nakamura T, Tachibana Y, Honda M, Yamane S. 2006. Influence of the Northern Hemisphere annular mode on ENSO by modulating westerly wind bursts. *Geophysical Research Letters* **33**: L07709, doi: 10.1029/2005GL025432.

Nakamura T, Tachibana Y, Shimoda H. 2007. Importance of cold and dry surges in substantiating the NAM and ENSO relationship. *Geophysical Research Letters* **34**: L22703, doi: 10.1029/2007GL031220.

Oshika M, Tachibana Y, Nakamura T. 2015. Impact of the winter North Atlantic Oscillation (NAO) on the Western Pacific (WP) pattern in the following winter through Arctic sea ice and ENSO: Part I – observational evidence. *Climate Dynamics* **45**: 1355–1366.

Reynolds RW, Smith TM, Liu C, Chelton DB, Casey KS, Schlax MG. 2007. Daily high-resolution-blended analyses for sea surface temperature. *Journal of Climate* **20**: 5473–5496.

Trenberth KE, Branstator G, Karoly D, Kumar A, Lau NC, Ropelewski C. 1998. Progress during TOGA in understanding and modeling global teleconnections associated with tropical sea surface temperatures. *Journal of Geophysical Research* **103**: 14291–14324.

Vimont DJ, Wallace JM, Battisti DS. 2003. The seasonal footprinting mechanism in the Pacific: implications for ENSO. *Journal of Climate* **16**: 2668–2675.

Wang B, Wu R, Fu X. 2000. Pacific-East Asian teleconnection: how does ENSO affect East Asian climate? *Journal of Climate* **13**: 1517–1536.

Wang X, Wang C, Zhou W, Wang D, Song J. 2011. Teleconnected influence of North Atlantic sea surface temperature on the El Niño onset. *Climate Dynamics* **37**: 663–676.

Wheeler MC, Hendon HH. 2004. An all-season real-time multivariate MJO index: development of an index for monitoring and prediction. *Monthly Weather Review* **132**: 1917–1932.

Xu J, Chan JC. 2001. The role of the Asian-Australian monsoon system in the onset time of El Niño events. *Journal of Climate* **14**: 418–433.

Xue Y, Balmaseda MA, Boyer T, Ferry N, Good S, Ishikawa I, Kumar A, Rienecker M, Rosati AJ, Yin Y. 2012. A comparative analysis of upper-ocean heat content variability from an ensemble of operational ocean reanalyses. *Journal of Climate* **25**: 6905–6929.

Yu LA, Rienecker MM. 1998. Evidence of an extratropical atmospheric influence during the onset of the 1997–98 El Niño. *Geophysical Research Letters* **25**: 3537–3540.

Zhou S, Miller AJ. 2005. The interaction of the Madden-Julian Oscillation and the Arctic Oscillation. *Journal of Climate* **18**: 143–159.

28

The relationships between temperature gradient and wind during cold frontal passages in the Eastern United States: A numerical modeling study

Robert Conrick,[1]* Nathan L. Curtis,[2] Paul W. Staten[2] and Cody Kirkpatrick[2]

[1] Department of Atmospheric Sciences, University of Washington, Seattle, WA, USA
[2] Department of Geological Sciences, Indiana University, Bloomington, IN, USA

*Correspondence to:
R. Conrick, Department of
Atmospheric Sciences, University
of Washington, Box 351640,
Seattle, WA 98195-1640, USA.
E-mail: rconrick@uw.edu

Abstract

Cold frontal passages are a common occurrence throughout the eastern United States. Previous observational research showed the surprising result that there is only a very weak, statistically nonsignificant relationship between a cold front's maximum 2-min sustained wind and the across-front temperature gradient. By using the WRF-ARW model to simulate eight cold fronts, we re-examine the relationship between temperature gradient and wind near the surface, and provide additional analysis. Results confirm previous observational research that found no relationship between wind speed and cross-frontal temperature gradient. The agreement between studies suggests that the lack of relationship may be physical and not a result of undersampling.

Keywords: cold front; front; wind; temperature gradient; correlation

1. Introduction

Midlatitude cyclones and their associated cold fronts are among the most common meteorological events in the eastern United States. With many events occurring each year, these systems were among the first atmospheric phenomena studied. Early cyclone models outlined characteristics of the midlatitude cyclone and associated cold front (Bjerknes, 1919; Bjerknes and Solberg, 1922; Henry, 1922; Godske et al., 1957), and are frequently cited in modern discussions of cyclogenesis. With such a long history of study, cold fronts have received wide attention with many theories describing their movement and behavior (see Smith and Reeder (1988) for an in-depth discussion of several theories of frontal motion).

Early cyclone models often describe cold fronts as discontinuities between air masses. Sharp gradients in temperature, moisture, precipitation, and wind direction are well-documented (e.g. Brundidge, 1965; Shapiro, 1984; Moore, 1985; Crook, 1987; Smith and Reeder, 1988; Mass and Schultz, 1993; Sanders, 1999; Schultz, 2004; Zhang et al., 2009; Payer et al., 2011; Sinclair et al., 2012; Sinclair, 2013; Clark and Parker, 2014). However, relatively little research is available on the relationships between these quantities – in particular between temperature gradient and wind – along the cold-frontal interface.

Frontal theory offers two important hypotheses regarding the relationship between temperature gradients, wind, and wind gradients. First, the speed of cold frontal propagation – in the absence of convection – should depend on wind speeds immediately following and normal to the cold front (Bluestein, 1993). Assuming cold fronts have characteristics similar to density currents, then the speed of frontal propagation will partially depend upon the virtual potential temperature gradient across the front (Simpson, 1987) due to the temperature/density gradient at the frontal interface promoting an across-front pressure gradient force (PGF). Sinclair and Keyser (2015) showed that cold fronts can exhibit dynamical regimes similar to density currents if simulated at high resolution with a modern Planetary Boundary Layer (PBL) scheme, and further concluded that in such simulations, PGF dominates force balances at the frontal interface. This framework provides a possible link between temperature gradient and wind speed at the frontal interface.

The second theory presented is based on the two-dimensional frontogenetical function. For an idealized cold front oriented in the north–south direction, this function simplifies to:

$$F = \frac{1}{|\nabla \theta|} \frac{\partial \theta}{\partial x} \left\{ \frac{1}{C_p} \left(\frac{p_0}{p} \right)^K \left[\frac{\partial}{\partial x} \left(\frac{\partial Q}{\partial t} \right) \right] - \left(\frac{\partial u}{\partial x} \frac{\partial \theta}{\partial x} \right) \right\}$$

(Bluestein, 1993), where terms follow convention. Frontogenesis thus depends on the strength of the temperature gradient and the front-normal wind gradient. Therefore, one can expect coincident gradients in wind and temperature at the cold frontal interface. Moreover, the temperature gradient at any point along a cold front should, by definition, point toward warm air and be normal to isotherms along the front. Considering these viewpoints, it is expected that the temperature gradient at any point along a cold front should point in the direction of frontal motion, with strong wind and wind gradients associated with strong cold front temperature gradients.

Strong wind and wind gradients associated with cold fronts are directly and indirectly documented on numerous occasions (Shapiro, 1984; Smith and Reeder, 1988; Friedrich *et al.*, 2008; Ma *et al.*, 2010; Sinclair *et al.*, 2012). Shapiro (1984) observed a cold front as it passed over an instrument tower in Colorado – finding temperature gradients overlapped front-normal wind gradients along the frontal interface (see his Figure 3). Sanders (1999) observed a temperature gradient of 18 °C per 100 km that coincided with damaging wind gusts ranging from 22 to 31 m s^{-1}. Pryor *et al.* (2014), however, showed only weak relationships exist between a cold front's temperature gradient and maximum 2-min observed wind. Pryor *et al.* (2014) analyzes the 'scale and intensity' of extreme wind in the eastern United States. Among many results presented, they discuss the relationship between temperature gradient and wind observed along 35 cold fronts between January 2012 and September 2013. Their analysis reveals a weak positive trend, but no statistically significant relationship. We choose to verify and expand upon their results in part to address the following: (1) data are chosen from surface observing stations, which are often separated by more than 100 km – thus a gradient through several stations may not be representative of what occurs along the frontal interface; (2) the calculation of temperature gradients at only 1200 UTC; (3) only wind speed – not direction or gradient – is considered. We address these points by using high-resolution simulations to spatially and temporally increase the amount of data along the frontal interface. Furthermore, we consider wind gradients due to their connection to frontogenesis, and include the direction of wind relative to temperature gradient in order to provide a complete analysis.

Our aim is to investigate the relationships between horizontal temperature gradients (dT), wind (Wind), and wind gradients (dWind) at the frontal interface. Beyond better understanding cold fronts, our work provides insight into sensible weather quantities, which may prove useful for short-term forecasting and understanding surface weather.

This article is structured as follows. Section 2 describes model configuration, methodology, and data filtering. Section 3 presents results/relationships found in the model output. Section 4 offers concluding remarks.

2. Methods

2.1. Cold front cases and model configuration

We identified eight midlatitude cyclones based on the location and orientation of their associated cold fronts (Table 1). We selected cyclones that approximately followed the structure common to the eastern United States and outlined in Bjerknes (1919) in which the cold front extends from the center of low pressure in a southwestward direction. The National Oceanic and

Table 1. A listing of cold front cases, including model initialization time, length of simulation, and maximum analyzed dT value.

Case	Start time, UTC	Duration, h	Maximum dT, °C km^{-1}
14 April 2012	1200	84	3.4564 (LMH)
10 March 2013	0000	84	2.5088 (LMH)
10 April 2013	0000	72	5.0622 (SFC)
17 April 2013	1200	57	5.0240 (LMH)
22 April 2013	1200	84	2.9360 (LMH)
12 June 2013	0000	84	2.5932 (LMH)
04 October 2013	1200	84	1.6788 (LMH)
04 December 2013	0000	72	1.8742 (LMH)

Atmospheric Administration Weather Prediction Center (WPC) Daily Weather Maps were used to identify cold fronts which occurred between April 2012 and December 2013 – chosen to overlap with many of the fronts analyzed in Pryor *et al.* (2014). We chose a particular cyclone if the cold front: (1) traveled, in part, from west-to-east across the model domain; (2) impacted the majority of the eastern United States; (3) remained mostly unoccluded while over the domain; and (4) exhibited a sufficiently large temperature gradient. This fourth criterion was achieved by choosing the eight fronts with the largest temperature gradients that met the other criteria.

The Advanced Weather Research and Forecasting Model version 3.5 (WRF-ARW; Skamarock *et al.*, 2005) is used to simulate all cases. We integrate the model over a domain extending from approximately 105° to 66°W (Figure 1; similar to Pryor *et al.* (2014)) with a horizontal grid spacing of 2 km and 31 vertical levels. The Noah land-surface model (Ek *et al.*, 2003) is used with high-resolution 30-s topography data. Initial and boundary conditions are from the North American Mesoscale (NAM) Model (Janjić *et al.*, 2005) 40-km analysis. Boundary conditions are updated every 6 h from the corresponding NAM forecast. Other model parameters include the Goddard Microphysics Scheme (Tao *et al.*, 2003), the Dudhia longwave radiation scheme (Dudhia, 1989), and the Mellor-Yamada-Janjić (Janjić, 2002) planetary boundary layer/surface layer schemes. No cumulus scheme is used. Simulation durations are outlined in Table 1.

2.2. Data filtering and processing

We process data at 1 h increments starting at forecast hour 02 and ending at the last hour of simulation. Near-surface temperature gradients (dT), wind (Wind), and wind gradients (dWind) are the fields of interest for this study. 'Near-surface' refers to either the surface (SFC) or the lowest model height (LMH). We define 'Wind' as the magnitude of the vector consisting of the *u*- and *v*-component wind at a particular level and location. The horizontal gradient is defined as

$$\nabla F = \frac{\partial F}{\partial x}\hat{\mathrm{I}} + \frac{\partial F}{\partial y}\hat{\mathrm{J}} \qquad (1)$$

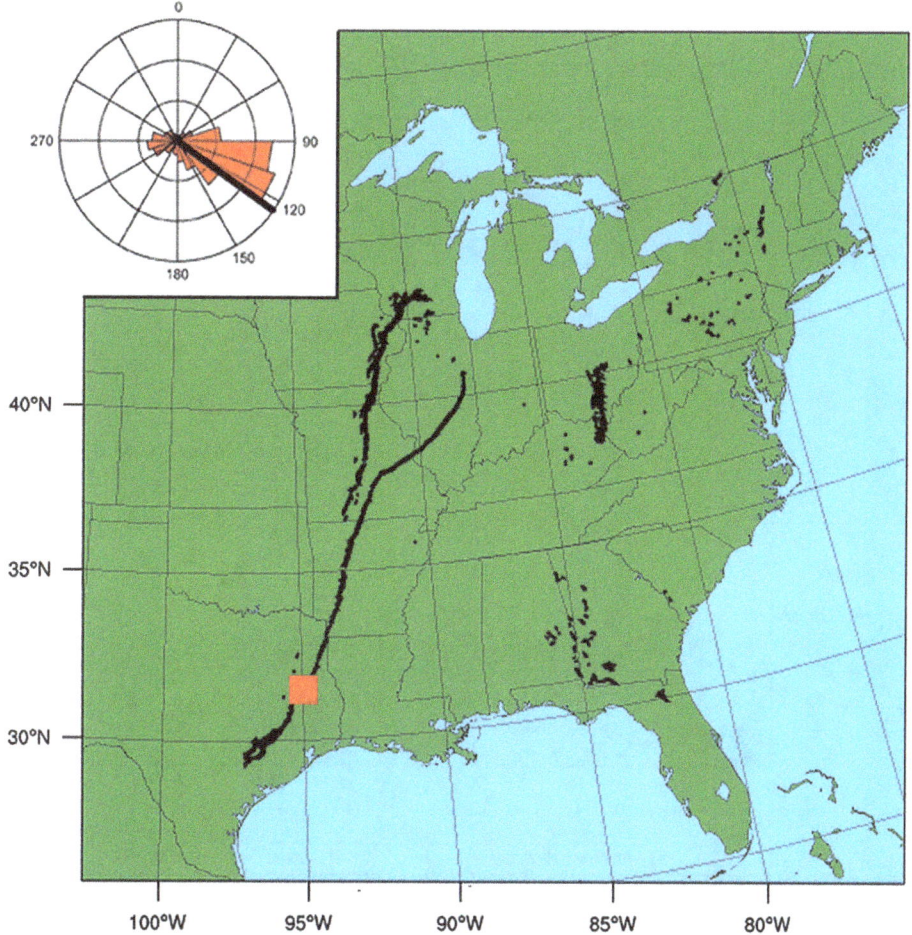

Figure 1. The model domain for all simulations, including the filtered dT field (black dots) of hour 26, 17 April 2013. The inset shows directions of all dT points - the bold line corresponds to the direction of the point of maximum dT (red box on map).

Table 2. Correlation coefficients (r^2), normalized regression coefficients (b), and p values for the results outlined in Section 3, including 100-km regional averaging. Relationships are organized by model level (surface [SFC] or lowest model height [LMH]) and by the dependent variable being compared to temperature gradient (dT).

Level	Independent variable	Dependent variable	r^2/b	p	$r^2_{regional}/b_{regional}$	$p_{regional}$
SFC	dT	Wind	0.2084 / 0.44 ± 0.35	<0.001	0.3487 / 12.6 ± 4.39	<0.001
		dWind	0.1415 / 0.36 ± 0.07	<0.001	0.0317 / 3.80 ± 0.46	0.0094
		dT-parallel Wind	0.2061 / 0.43 ± 0.34	<0.001	0.1528 / 8.34 ± 4.79	<0.001
		dT-normal Wind	0.0608 / 0.24 ± 0.26	0.003	0.1801 / 7.84 ± 5.32	<0.001
		dT-parallel dWind	0.1314 / 0.35 ± 0.07	<0.001	0.0231 / 3.24 ± 0.40	0.0270
		dT-normal dWind	0.0424 / 0.20 ± 0.04	0.003	0.0137 / 2.50 ± 0.37	0.0894
LMH	dT	Wind	0.1951 / 0.41 ± 0.33	<0.001	0.3170 / 11.42 ± 4.49	<0.001
		dWind	0.1062 / 0.30 ± 0.06	<0.001	0.0430 / 4.21 ± 0.48	0.0019
		dT-parallel Wind	0.0971 / 0.29 ± 0.29	<0.001	0.1801 / 8.61 ± 4.64	<0.001
		dT-normal Wind	0.1303 / 0.34 ± 0.29	<0.001	0.1077 / 6.65 ± 5.36	<0.001
		dT-parallel dWind	0.0450 / 0.20 ± 0.05	0.002	0.0285 / 3.43 ± 0.41	0.0117
		dT-normal dWind	0.0817 / 0.27 ± 0.05	<0.001	0.0225 / 3.04 ± 0.41	0.0253

where F is a scalar field in the x-y plane. The gradient is calculated using a central difference for interior points ($dx = 4$ km). To isolate cold fronts from the model output and remove data affected by other processes (e.g. moist convection), we apply data filters, including gradient thresholds, edge detection, and a terrain filter, to the dT fields. An example of a filtered dT field can be found in Figure 1. For a full description of data filters

applied, see Appendix S2, Supporting Information. Filtered data were inspected manually, and only hours with a clearly detected cold front were chosen (SFC $n = 212$; LMH $n = 222$). Because surface temperature (2-m) and wind (10-m) are computed in the surface- and boundary-layer parameterization schemes, we consider dT, Wind, and dWind at the surface and LMH – thus giving attention to parameterized surface data and

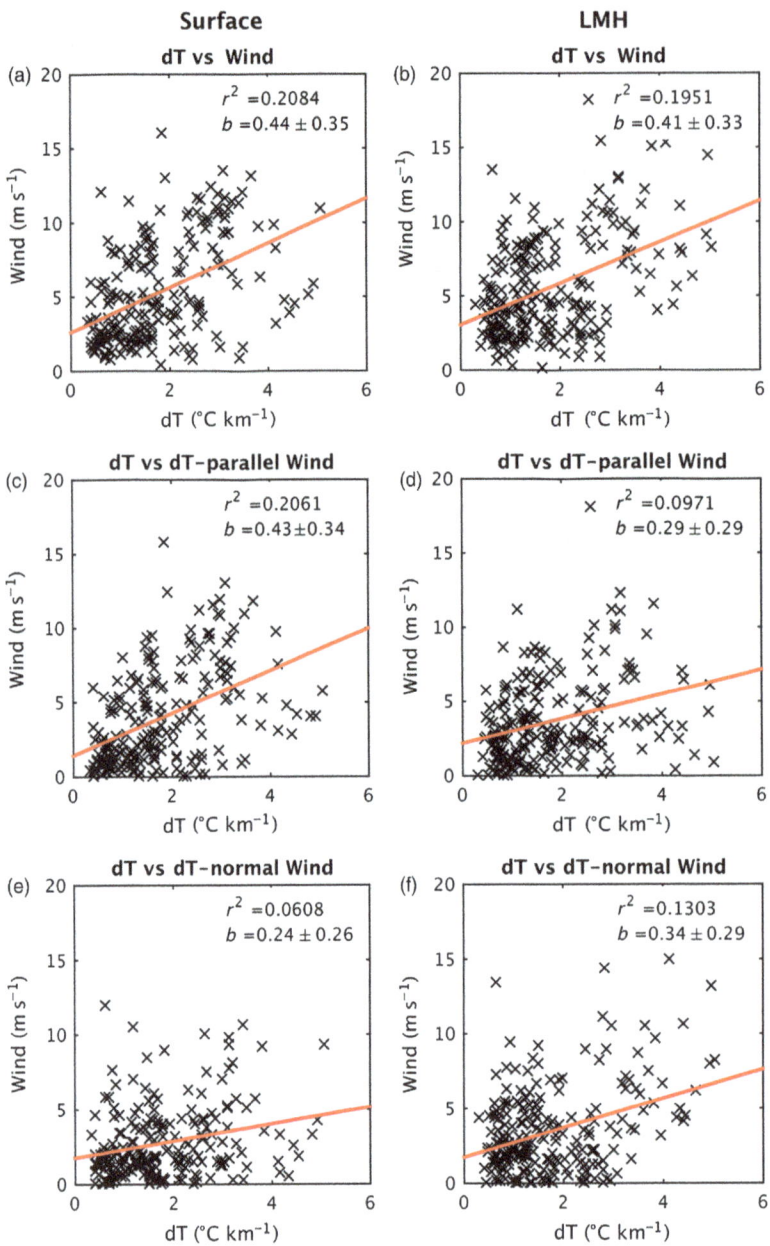

Figure 2. Scatterplots of dT vs. Wind. The left column shows surface relationships and the right column shows LMH relationships. Regression lines, regression coefficients (b), and correlation coefficients (r^2) are displayed.

less-parameterized LMH data. Our chief interest is identifying correlations between these quantities, therefore we perform linear regression to determine whether relationships exist. Correlation coefficients (r^2) and normalized regression coefficients (with uncertainty bounds derived from a t-statistic) are computed as metrics of relationship strength.

3. Results and discussion

We calculate two sets of correlations (dT vs Wind; dT vs dWind) at the surface and LMH. In these analyses, the point of maximum dT in the domain is chosen with collocated Wind or dWind selected for each 1-h increment of each model run (i.e. dT is the independent variable; henceforth referred to as the dT-Wind analysis).

Assuming that the direction of the dT vector is primarily oriented across the front (Appendix S2 presents a qualitative look at dT direction), then components of Wind and dWind are examined to assess whether their across-front (dT-parallel) or along-front (dT-normal) components exhibit a stronger relationship. Directions are computed by projecting the Wind vector onto the dT vector: the projection itself is dT-parallel and the normal component is dT-normal. While it is hypothesized that dT-parallel wind will be better correlated, we include dT-normal wind to present a complete argument. Correlations, regression coefficients, and p values are listed in Table 2.

Correlations between dT and Wind are low at both the surface ($r^2 = 0.2084$; Figure 2(a)) and LMH ($r^2 = 0.1951$; Figure 2(b)). When dT is correlated

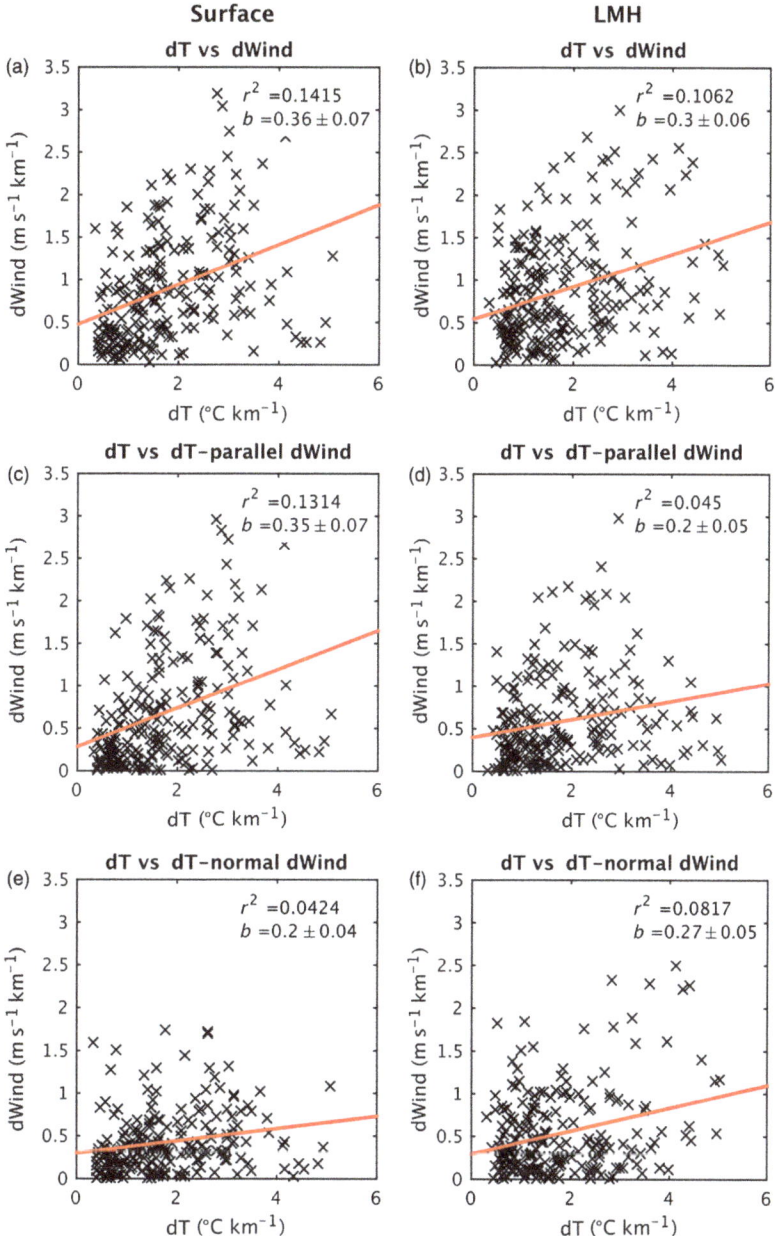

Figure 3. Scatterplots of dT vs. dWind. The left column shows surface relationships and the right column shows LMH relationships. Regression lines, regression coefficients (b), and correlation coefficients (r^2) are displayed.

with dWind, results are similar (SFC $r^2 = 0.1415$; LMH $r^2 = 0.1062$; Figure 3(a) and (b), respectively). Associated p values suggest significance ($p < 0.05$), however low correlation coefficients imply that no strong linear relationships exist between dT and Wind or dT and dWind. An examination of Figures 2 and 3 confirm no obvious relationship among the data. It is possible, however, that either the along-front or across-front component of the wind is correlated with the temperature gradient, even if the wind speed is not. To investigate this possibility, we consider the parallel and normal components of Wind and dWind relative to dT in order to determine whether correlations are stronger in the across-front (dT-parallel) or along-front (dT-normal) directions. At the surface and LMH,

the correlations between dT and dT-parallel Wind is also weak (SFC $r^2 = 0.2061$ and LMH $r^2 = 0.0971$; $p < 0.05$ for both; Figure 2(c) and (d), respectively). For relationships between dT and dT-normal Wind, correlations are the weakest observed (Figure 2(e) and (f)). Correlations between dT and components of dWind are similarly weak with $r^2 > 0.0424$. For all regressions, regression coefficients are also computed. However, the substantial overlap in uncertainty bounds across all regressions implies that the regression slopes do not significantly differ. All correlations and regression coefficients are summarized in Table 2 and displayed on Figures 2 and 3.

A natural question to ask is whether inspecting a region along the front, rather than one grid point, will lead to better sampling. Thus, we inspect an average of dT, Wind, and dWind over a 100 km region centered on the points of greatest dT for each hour. While results are similar to those presented earlier, correlations between dT and Wind magnitude increase – possibly due to including many points with dT similar in magnitude to the point of maximum dT, but with stronger Wind. We summarize the results in Table 2.

The results of these analyses show that while trends are positive for all comparisons, no strong correlations exist between dT, Wind, and dWind. The conclusion applies for both the surface and the LMH, and indicates that the magnitude of dT is not a reliable indicator of the magnitude of near-surface Wind or dWind observed along a cold front. These results are consistent with previous studies.

4. Conclusion

This study builds on previous observational work on relationships that exist between temperature gradient and wind observed during cold frontal passages. Eight carefully selected cold front cases that impacted the eastern United States are simulated. We use linear regression to examine potential correlations between near-surface temperature gradients, wind, and wind gradients. Scatterplots of these variables exhibit no obvious linear or nonlinear relationship. Low correlation coefficients, coupled with statistically indistinct regression coefficients, indicate a lack of strong relationships.

The results of the dT-Wind analysis imply that the influence of near-surface temperature gradients on wind or wind gradients observed during frontal passage is likely negligible (all $r^2 \leq 0.2084$). These findings are contrary to theory presented in Section 1, but agree with results of Pryor et al. (2014). These weak relationships may be physical – rather than the result of analysis or undersampling – and it is likely the case that other mechanisms exist which accelerate air along/across cold fronts. Future work could focus on these mechanisms. For instance, near-surface wind may be more closely related to synoptic than near-surface conditions. Furthermore, boundary layer friction may play a role by lessening the impact of frontogenesis. We reiterate that the agreement between observations and simulations is of great importance to understanding these results, and conclude that cold fronts with strong temperature gradients will not necessarily yield strong near-surface wind.

Acknowledgements

We thank the two anonymous reviewers for providing constructive feedback, which helped us greatly improve the manuscript. This research was supported in part by Lilly Endowment, Inc., through its support for the Indiana University Pervasive Technology Institute and the Indiana METACyt Initiative (also supported by Lilly Endowment, Inc). This material is based upon work supported by National Science Foundation under grant no. CNS-0521433. Any opinions, findings and conclusions, or recommendations expressed in this material are those of the author(s), and do not necessarily reflect the views of the National Science Foundation (NSF).

Supporting information

The following supporting information is available:

Appendix S1. Data filtering

Appendix S2. Inspection of the direction of temperature gradients

References

Bjerknes J. 1919. On the structure of moving cold fronts. *Mon. Weather Rev.* **47**: 95–99, doi: 10.1175/1520-0493(1919)47<95:OTSOMC>2.0.CO;2.

Bjerknes J, Solberg H. 1922. Life cycle of cyclones and the polar front theory of atmospheric circulation. *Geofysiske Publikasjoner* **3**: 1–18.

Bluestein HR. 1993. *Synoptic-Dynamic Meteorology in Midlatitudes.* Oxford University Press: Oxford, UK.

Brundidge KC. 1965. The wind and temperature structure of nocturnal cold fronts in the first 1,420 feet. *Mon. Weather Rev.* **93**: 587–603, doi: 10.1175/1520-0493(1965)093<0587:TWATSO>2.3.CO;2.

Clark MR, Parker DJ. 2014. On the mesoscale structure of surface wind and pressure fields near tornadic and nontornadic cold fronts. *Mon. Weather Rev.* **142**: 3560–3585, doi: 10.1175/MWR-D-13-00395.1.

Crook NA. 1987. Moist convection at a surface cold front. *J. Atmos. Sci.* **44**: 3469–3494, doi: 10.1175/1520-0469(1987)044<3469:MCAASC>2.0.CO;2.

Dudhia J. 1989. Numerical study of convection observed during the winter monsoon experiment using a mesoscale two-dimensional model. *J. Atmos. Sci.* **46**: 3077–3107, doi: 10.1175/1520-0469(1989)046<3077:NSOCOD>2.0.CO;2.

Ek M, Mitchell KE, Lin Y, Rogers E, Grunmann P, Koren V, Gayno G, Tarpley JD. 2003. Implementation of the Noah land surface model advances in the National Centers for Environmental Prediction operational mesoscale Eta Model. *J. Geophys. Res.* **108**: 8851, doi: 10.1029/2002JD003296.

Friedrich K, Kingsmill DE, Flamant C, Murphey HV, Wakimoto RM. 2008. Kinematic and moisture characteristics of a nonprecipitating cold front observed during IHOP. Part I: across-front structures. *Mon. Weather Rev.* **136**: 147–172, doi: 10.1175/2007MWR1908.1.

Godske CL, Bergeron T, Bjerknes J, Bundgaard RC. 1957. *Dynamic Meteorology and Weather Forecasting.* American Meteorological Society Press: Boston, MA, USA.

Henry AJ. 1922. J. Berjnes and H. Solberg on the life cycle of cyclones and the polar front theory of atmospheric circulation. *Mon. Weather Rev.* **50**: 468–473, doi: 10.1175/1520-0493(1922)50<468:JBAHSO>2.0.CO;2.

Janjić ZI. 2002. Nonsingular implementation of the Mellor-Yamada level 2.5 scheme in the NCEP Meso Model. *NCEP Office Note 437.*

Janjić ZI, Black YL, Pyle ME, Chuang HY, Rogers E, DiMego GJ. 2005. The NCEP WRF-NMM core. *2005 WRF/MM5 User's Workshop*, Boulder, CO, University Corporation for Atmospheric Research. http://www2.mmm.ucar.edu/wrf/users/workshops/WS2005/abstracts/Session2/9-Janjic.pdf (accessed 30 July 2015).

Ma Y, Huang X, Mills GA, Parkyn K. 2010. Verification of mesoscale NWP forecasts of abrupt cold frontal wind changes. *Weather Forecast.* **25**: 93–112, doi: 10.1175/2009WAF2222259.1.

Mass CF, Schultz DM. 1993. The structure and evolution of a simulated midlatitude cyclone over land. *Mon. Weather Rev.* **121**: 889–917, doi: 10.1175/1520-0493(1993)121<0889:TSAEOA>2.0.CO;2.

Moore GWK. 1985. The organization of convection in narrow cold-frontal rainbands. *J. Atmos. Sci.* **42**: 1777–1791, doi: 10.1175/1520-0469(1985)042<1777:TOOCIN>2.0.CO;2.

Payer M, Laird NF, Maliawco RJ Jr, Hoffman EG. 2011. Surface fronts, troughs, and baroclinic zones in the Great Lakes region. *Weather Forecast.* **26**: 555–563.

Pryor SC, Conrick R, Miller C, Tytell J, Barthelmie RJ. 2014. Intense and extreme wind speeds observed by anemometer and seismic networks: an Eastern U.S. case study. *J. Appl. Meteorol. Climatol.* **53**: 2417–2429, doi: 10.1175/JAMC-D-14-0091.1.

Sanders F. 1999. A short-lived cold front in the Southwestern United States. *Mon. Weather Rev.* **127**: 2395–2403, doi: 10.1175/1520-0493 (1999)127<2395:ASLCFI>2.0.CO;2.

Schultz DM. 2004. Cold fronts with and without prefrontal wind shifts in the Central United States. *Mon. Weather Rev.* **132**: 2040–2053, doi: 10.1175/1520-0493(2004)132<2040:CFWAWP>2.0.CO;2.

Shapiro MA. 1984. Meteorological tower measurements of a surface cold front. *Mon. Weather Rev.* **112**: 1634–1639, doi: 10.1175/1520-0493(1984)112<1634:MTMOAS>2.0.CO;2.

Simpson JE. 1987. *Gravity Currents: in the Environment and the Laboratory.* Ellis Horwood: Cheichester, UK.

Sinclair VA. 2013. A 6-yr climatology of fronts affecting Helsinki, Finland, and their boundary layer structure. *J. Appl. Meteorol. Climatol.* **52**: 2106–2124, doi: 10.1175/JAMC-D-12-0318.1.

Sinclair VA, Keyser D. 2015. Force balances and dynamical regimes of numerically simulated cold fronts within the boundary layer. *Q. J. R. Meteorol. Soc.* **141**: 2148–2164.

Sinclair VA, Niemelä S, Leskinen M. 2012. Structure of a narrow cold front in the boundary layer: observations versus model simulation. *Mon. Weather Rev.* **140**: 2497–2519, doi: 10.1175/MWR-D-11-00328.1.

Skamarock WC, Klemp JB, Dudhia J, Gill DO, Baker DM, Wang W, Powers JG. 2005. A description of the Advanced Research WRF version 2. *NCAR Technical Note NCAR/TN-468+STR*. http://www.mmm.ucar.edu/wrf/users/docs/arw_v2.pdf (accessed 30 July 2015).

Smith RK, Reeder MJ. 1988. On the movement and low-level structure of cold fronts. *Mon. Weather Rev.* **116**: 1927–1944, doi: 10.1175/1520-0493(1988)116<1927:OTMALL>2.0.CO;2.

Tao WK, Simpson J, Baker D, Braun S, Chou MD, Ferrier B, Johnson D, Khain A, Lang S, Lynn B, Shie CL, Starr D, Sui CH, Wang Y, Wetzel P. 2003. Microphysics, radiation and surface processes in the Goddard Cumulus Ensemble (GCE) model. *Meteorog. Atmos. Phys.* **82**: 97–137, doi: 10.1007/s00703-001-0594-7.

Zhang Y, Smith JA, Ntelekos AA, Baeck ML, Krajewski WF, Moshary F. 2009. Structure and evolution of precipitation along a cold front in the Northeastern United States. *J. Hydrometeorol.* **10**: 1243–1256, doi: 10.1175/2009JHM1046.1.

PERMISSIONS

All chapters in this book were first published in ASL, by John Wiley & Sons Ltd.; hereby published with permission under the Creative Commons Attribution License or equivalent. Every chapter published in this book has been scrutinized by our experts. Their significance has been extensively debated. The topics covered herein carry significant findings which will fuel the growth of the discipline. They may even be implemented as practical applications or may be referred to as a beginning point for another development.

The contributors of this book come from diverse backgrounds, making this book a truly international effort. This book will bring forth new frontiers with its revolutionizing research information and detailed analysis of the nascent developments around the world.

We would like to thank all the contributing authors for lending their expertise to make the book truly unique. They have played a crucial role in the development of this book. Without their invaluable contributions this book wouldn't have been possible. They have made vital efforts to compile up to date information on the varied aspects of this subject to make this book a valuable addition to the collection of many professionals and students.

This book was conceptualized with the vision of imparting up-to-date information and advanced data in this field. To ensure the same, a matchless editorial board was set up. Every individual on the board went through rigorous rounds of assessment to prove their worth. After which they invested a large part of their time researching and compiling the most relevant data for our readers.

The editorial board has been involved in producing this book since its inception. They have spent rigorous hours researching and exploring the diverse topics which have resulted in the successful publishing of this book. They have passed on their knowledge of decades through this book. To expedite this challenging task, the publisher supported the team at every step. A small team of assistant editors was also appointed to further simplify the editing procedure and attain best results for the readers.

Apart from the editorial board, the designing team has also invested a significant amount of their time in understanding the subject and creating the most relevant covers. They scrutinized every image to scout for the most suitable representation of the subject and create an appropriate cover for the book.

The publishing team has been an ardent support to the editorial, designing and production team. Their endless efforts to recruit the best for this project, has resulted in the accomplishment of this book. They are a veteran in the field of academics and their pool of knowledge is as vast as their experience in printing. Their expertise and guidance has proved useful at every step. Their uncompromising quality standards have made this book an exceptional effort. Their encouragement from time to time has been an inspiration for everyone.

The publisher and the editorial board hope that this book will prove to be a valuable piece of knowledge for researchers, students, practitioners and scholars across the globe.

LIST OF CONTRIBUTORS

Theodore L. Allen and Brian E. Mapes
Department of Meteorology and Physical Oceanography, Rosenstiel School of Marine and Atmospheric Science, University of Miami, FL, USA

Nicholas Cavanaugh
Climate and Ecosystem Sciences, Lawrence Berkeley National Laboratory, Berkeley, CA, USA

Jae-Won Choi and Yumi Cha
Research Planning and Management Division, National Institute of Meteorological Sciences, JeJu, Korea

Elissavet Galanaki
National Observatory of Athens, Institute for Environmental Research and Sustainable Development, Greece
Department of Physics, Laboratory of Atmospheric Physics, University of Patras, Greece

Emmanouil Flaounas, Vassiliki Kotroni and Konstantinos Lagouvardos
National Observatory of Athens, Institute for Environmental Research and Sustainable Development, Greece

Athanassios Argiriou
Department of Physics, Laboratory of Atmospheric Physics, University of Patras, Greece

Juan Feng
College of Global Change and Earth System Science (GCESS), Beijing Normal University, China
Joint Center for Global Change Studies, Beijing, China

Jianlei Zhu
State Key Laboratory of Numerical Modeling for Atmospheric Sciences and Geophysical Fluid Dynamics, Institute of Atmospheric Physics, Chinese Academy of Sciences, Beijing, China

Yan Li
Key Laboratory of Semi-Arid Climate Change of Ministry of Education, College of Atmospheric Sciences, Lanzhou University, China

Kirsty E. Hanley and Humphrey W. Lean
University of Reading, Reading, UK

Andrew I. Barrett
Department of Meteorology, University of Reading, Reading, UK

Subin Jose, Biswadip Gharai, P. V. N. Rao and C. B. S. Dutt
Atmospheric and Climate Sciences Group (ACSG)-ECSA, National Remote Sensing Centre (NRSC), ISRO, Hyderabad, India

Melanie K. Karremann
Institute for Geophysics and Meteorology, University of Cologne, Cologne, Germany
Now at: Institute of Meteorology and Climate Research, Karlsruhe Institute of Technology, Karlsruhe, Germany

Margarida L. R. Liberato
Escola de Ciências e Tecnologia, Universidade de Trás-os-Montes e Alto Douro, Vila-Real, Portugal
Instituto Dom Luiz (IDL), Universidade de Lisboa, Lisboa, Portugal

Paulina Ordóñez
Escola de Ciências e Tecnologia, Universidade de Trás-os-Montes e Alto Douro, Vila-Real, Portugal
Centro de Ciencias de la Atmósfera, Universidad Nacional Autónoma de México, Mexico City, Mexico

Joaquim G. Pinto
Institute for Geophysics and Meteorology, University of Cologne, Cologne, Germany
Department of Meteorology, University of Reading, Reading, UK

Chiung-Wen June Chang
Department of Atmospheric Sciences, Chinese Cultural University, Taipei, Taiwan

S.-Y. Simon Wang
Department of Plants, Soils and Climate and Utah Climate Center, Utah State University, Logan, UT, USA

Huang-Hsiung Hsu
Research Center for Environmental Changes, Academia Sinica, Taipei, Taiwan

P. A. Mooney, D. O. Gill and C. L. Bruyère
Mesoscale and Microscale Meteorology Laboratory, National Center for Atmospheric Research (NCAR), Boulder, CO, USA

F. J. Mulligan
Department of Experimental Physics, Maynooth University, Kildare, Ireland

Giovanni Dolif Neto
Centre for Weather Forecasting and Climate Studies, National Space Research Institute (CPTEC INPE), Cachoeira Paulista, SP, Brazil
Centre for Natural Disaster Monitoring and Early Warning – CEMADEN, São José dos Campos, SP, Brazil

Patrick S. Market
Department of Soil and Atmospheric Sciences, University of Missouri, Columbia, MO, USA

Alexandre Bernardes Pezza
School of Earth Sciences, The University of Melbourne, Australia
Environmental Science Department, Greater Wellington Regional Council (GWRC), Wellington, New Zealand

Carlos Augusto Morales Rodriguez
Atmospheric Sciences Department, Institute of Astronomy, Geophysics and Atmospheric Sciences, University of São Paulo (IAG-USP), São Paulo, SP, Brazil

Leonardo Calvetti
Meteorological Service of Parana State (SIMEPAR), Curitiba, PR, Brazil

Pedro Leite da Silva Dias
National Laboratory for Scientific Computing (LNCC), Institute of Astronomy, Geophysics and Atmospheric Sciences, University of São Paulo (IAG-USP), Petrópolis, RJ, Brazil

Gustavo Carlos Juan Escobar
Centre for Weather Forecasting and Climate Studies, National Space Research Institute (CPTEC INPE), Cachoeira Paulista, SP, Brazil

Shira Raveh-Rubin
Institute for Atmospheric and Climate Science, ETH Zurich, Switzerland

Emmanouil Flaounas
National Observatory of Athens, Greece

Hong-Chang Ren
Collaborative Innovation Center on Forecast and Evaluation of Meteorological Disasters, Nanjing University of Information Science and Technology, China

Weijing Li and Jinqing Zuo
Collaborative Innovation Center on Forecast and Evaluation of Meteorological Disasters, Nanjing University of Information Science and Technology, China
Laboratory for Climate Studies, National Climate Center, China Meteorological Administration, Beijing, China

Hong-Li Ren
Laboratory for Climate Studies, National Climate Center, China Meteorological Administration, Beijing, China
Joint Center for Global Change Studies (JCGCS), Beijing, China

Xiang Wang
Center of Data Assimilation for Research and Application, Nanjing University of Information Science and Technology, China
Key Laboratory of South China Sea Meteorological Disaster Prevention and Mitigation of Hainan Province, Haikou, China

Yifang Ren
Jiangsu Meteorological Service, China Meteorological Administration, Nanjing, China

Xun Li
Hainan Meteorological Service, China Meteorological Administration, Haikou, China

Alexander J. Roberts, John H. Marsham, Douglas J. Parker, Luis Garcia-Carreras, Matthew Hobby, James B. McQuaid and Philip D. Rosenberg
Institute for Climate and Atmospheric Science, School of Earth and Environment, University of Leeds, UK

Peter Knippertz
Institute of Meteorology and Climate Research, Karlsruhe Institute of Technology, Germany

Mark Bart
Air Quality Ltd., Auckland, New Zealand

Daniel Walker
National Centre for Atmospheric Science, University of Leeds, Leeds, UK

Djordje Romanić, Miroljub Zarić, Miloš Lompar and Ilija Jovičić
Department of Meteorology, Republic Hydrometerological Service of Serbia, Belgrade, Serbia

Mladjen urić
Institute of Meteorology, Faculty of Physics, University of Belgrade, Belgrade, Serbia

Yousuke Sato
Riken Advanced Institute for Computational Science, Kobe, Japan
Graduated School of Simulation Study, University of Hyogo, Kobe, Japan

Shin-ichiro Shima
Graduated School of Simulation Study, University of Hyogo, Kobe, Japan
Riken Advanced Institute for Computational Science, Kobe, Japan

Hirofumi Tomita
Riken Advanced Institute for Computational Science, Kobe, Japan

Dirk Schindler, Christopher Jung and Alexander Buchholz
Environmental Meteorology, Albert-Ludwigs-University of Freiburg, Freiburg, Germany

Yuhei Takaya, Yutaro Kubo and Shoji Hirahara
Meteorological Research Institute, Japan Meteorological Agency, Ibaraki, Japan
Climate Prediction Division, Japan Meteorological Agency, Tokyo, Japan

Shuhei Maeda
Climate Prediction Division, Japan Meteorological Agency, Tokyo, Japan

Michael A. Walz
School of Geography, Earth and Environmental Science, University of Birmingham, UK
Institut fuer Meteorologie, Freie Universität Berlin, Germany

Tim Kruschke
Institut fuer Meteorologie, Freie Universität Berlin, Germany
Department of Ocean Circulation and Climate Dynamics, Research Unit Marine Meteorology,

Henning W. Rust and Uwe Ulbrich
Institut fuer Meteorologie, Freie Universität Berlin, Germany

Gregor C. Leckebusch
School of Geography, Earth and Environmental Science, University of Birmingham, UK

Yaping Wang and Yongjie Huang
Key Laboratory of Cloud-Precipitation Physics and Severe Storms (LACS), Institute of Atmospheric Physics, Chinese Academy of Sciences, Beijing, China
University of Chinese Academy of Sciences, Beijing, China

Xiaopeng Cui
Key Laboratory of Cloud-Precipitation Physics and Severe Storms (LACS), Institute of Atmospheric Physics, Chinese Academy of Sciences, Beijing, China
Collaborative Innovation Center on Forecast and Evaluation of Meteorological Disasters, Nanjing University of Information Science and Technology, Nanjing, China

Benjamin Wong and Ralf Toumi
Blackett Laboratory, Space and Atmospheric Physics Group, Imperial College London, UK

Akio Yamagami and Hiroshi L. Tanaka
Center for Computational Sciences, University of Tsukuba, Japan

Mio Matsueda
Center for Computational Sciences, University of Tsukuba, Japan
Department of Physics, University of Oxford, UK

Adrián E. Yuchechen and Pablo O. Canziani
Universidad Tecnológica Nacional, Facultad Regional Buenos Aires, Consejo Nacional de Investigaciones Científicas y Técnicas (CONICET), Unidad de Investigación y Desarrollo de las Ingenierías, Buenos Aires, Argentina

Jinlin Zha and Jian Wu
Department of Atmospheric Science, Yunnan University, Kunming, China

Deming Zhao
Key Laboratory of Regional Climate-Environment for Temperate East Asia, Institute of Atmospheric Physics, Chinese Academy of Sciences, Beijing, P. R. China

Hyun-Ju Lee, Woo-Seop Lee and Jin Ho Yoo
Climate Research Department, APEC Climate Center, Busan, Republic of Korea

S. Ghosh
School of Mechanical and Building Sciences, Vellore Institute of Technology, Vellore, Tamilnadu, India
School of Earth and Environment, University of Leeds, UK

A. Sharma, S. Arora and G. Desouza
School of Mechanical and Building Sciences, Vellore Institute of Technology, Vellore, Tamilnadu, India

Shangfeng Chen, Renguang Wu, Wen Chen and Xi Cao
Center for Monsoon System Research, Institute of Atmospheric Physics, Chinese Academy of Sciences, Beijing, China

Bin Yu
Climate Research Division, Environment and Climate Change Canada, Toronto, ON, Canada

Robert Conrick
Department of Atmospheric Sciences, University of Washington, Seattle, WA, USA

Nathan L. Curtis, Paul W. Staten and Cody Kirkpatrick
Department of Geological Sciences, Indiana University, Bloomington, IN, USA

Index